Undersea Lightwave
Communications

FRONTIERS IN COMMUNICATIONS

Undersea Lightwave Communications

Edited by
Peter K. Runge
Patrick R. Trischitta
AT&T Bell Laboratories

Published for the IEEE Communications Society by the IEEE PRESS

The Institute of Electrical and Electronics Engineers, Inc., New York

IEEE Order Number: PC01933

Library of Congress Cataloging-in-Publication Data

Undersea lightwave communications.

(Frontiers in communications)
Includes index.
1. Cables, Submarine. 2. Telephone cables.
3. Optical communications. I. Runge, Peter K.
II. Trischitta, Patrick R., 1958–   . III. IEEE
Communications Society. IV. Series.
TK6377.U63 1986   621.38′0414   86-10670
ISBN 0-87942-201-7

The Contributors

Ali Adl
AT&T Bell Laboratories

Mamoru Aiki
Nippon Telegraph and Telephone Corp.

Cleo D. Anderson
AT&T Bell Laboratories

Peter J. Anslow
Standard Telecommunication Laboratories, Ltd.

Kenichi Asakawa
Kokusai Denshin Denwa Co., Ltd.

Fridolin Bosch
AT&T Bell Laboratories

Christodoulos Chamzas
AT&T Bell Laboratories

Ta-Mu Chien
AT&T Bell Laboratories

Martin Chown
Standard Telephones and Cables PLC

Tek-Che Chu
AT&T Bell Laboratories

Paul A. Dawson
British Telecom Research Laboratories

S. Wallace Dawson, Jr.
AT&T Communications

Robert D. Ehrbar
Formerly with Bell Laboratories–Retired

David G. Ehrenberg
AT&T Bell Laboratories

Yoshihiro Ejiri
Kokusai Denshin Denwa Co., Ltd.

Jeffrey G. Farrington
Standard Telecommunication Laboratories, Ltd.

Daniel A. Fishman
AT&T Bell Laboratories

Kenneth D. Fitchew
British Telecom Research Laboratories

Jean Yves Fourrier
CIT-Alcatel

Pierre Franco
CIT-Alcatel

J. Lance Fromme
AT&T Bell Laboratories

Hiroshi Fukinuki
Nippon Telegraph and Telephone Corp.

Kahei Furusawa
Kokusai Denshin Denwa Co., Ltd.

Robert F. Gleason
AT&T Bell Laboratories

Isobel J. Goddard
Standard Telecommunication Laboratories, Ltd.

Yoshihiro Hayashi
Nippon Telegraph and Telephone Corp.

George A. Heath
Standard Telephones and Cables PLC

Glen M. Homsey
AT&T Bell Laboratories

Toshihito Hosaka
Nippon Telegraph and Telephone Corp.

Paul T. Hutchison
AT&T Bell Laboratories

Yasutaka Ichihashi
Nippon Telegraph and Telephone Corp.

Tetsuhiko Ikegami
Nippon Telegraph and Telephone Corp.

Koushi Ishihara
Nippon Telegraph and Telephone Corp.

Takeshi Ito
Nippon Telegraph and Telephone Corp.

Yoshinao Iwamato
Kokusai Denshin Denwa Co., Ltd.

Genzo Iwane
Nippon Telegraph and Telephone Corp.

Tsutomo Kamoto
Nippon Telegraph and Telephone Corp.

Shuichi Kanamori
Nippon Telegraph and Telephone Corp.

Haruo Kano
Nippon Telegraph and Telephone Corp.

Stanley Kaufman
AT&T Bell Laboratories

David L. Keller
AT&T Bell Laboratories

Iwao Kitazawa
Nippon Telegraph and Telephone Corp.

Richard Klinman
AT&T Bell Laboratories

Junichi Kojima
Kokusai Denshin Denwa Co., Ltd.

Masakuni Kuwazuru
Kokusai Denshin Denwa Co., Ltd.

Jean-Claude Lacroix
CIT-Alcatel

George C. Loeffler
AT&T Bell Laboratories

Tadashi Matsumoto
Nippon Telegraph and Telephone Corp.

John J. McNulty
AT&T Bell Laboratories

Lewis E. Miller
AT&T Bell Laboratories

Kiyofumi Mochizuki
Kokusai Denshin Denwa Co., Ltd.

Yasuji Murakami
Nippon Telegraph and Telephone Corp.

Suzanne R. Nagel
AT&T Bell Laboratories

Shinji Nakamura
Nippon Telegraph and Telephone Corp.

Yoshinori Nakano
Nippon Telegraph and Telephone Corp.

Yoshinori Namihira
Kokusai Denshin Denwa Co., Ltd.

Yukiyasu Negishi
Nippon Telegraph and Telephone Corp.

Yasuhiko Niiro
Kokusai Denshin Denwa Co., Ltd.

Shigendo Nishi
Nippon Telegraph and Telephone Corp.

Juichi Noda
Nippon Telegraph and Telephone Corp.

Makoto Nunokawa
Kokusai Denshin Denwa Co., Ltd.

Mamoru Ohara
Nippon Telegraph and Telephone Corp.

Masanori Ohkubo
Furukawa Electric Co.

Haruo Okamura
Nippon Telegraph and Telephone Corp.

Pierre Ollion
Les Cables de Lyon

G. M. Palmer
AT&T Bell Laboratories

Robert M. Paski
AT&T Bell Laboratories

Michael W. Perry
AT&T Bell Laboratories

Jean Pierre Pestie
CIT-Alcatel

Gordon A. Reinold
AT&T Laboratories

Robert L. Reynolds
AT&T Laboratories

Jose Riera Riera
CTNE

S. Paul Rogerson
British Telecom Research Laboratories

Robert L. Rosenberg
AT&T Bell Laboratories

David G. Ross
AT&T Bell Laboratories

Peter K. Runge
AT&T Bell Laboratories

Shigeki Sakaguchi
Nippon Telegraph and Telephone Corp.

Charles D. Sallada
AT&T Bell Laboratories

Yutaka Sasaki
Nippon Telegraph and Telephone Corp.

Yuichi Shirasaki
Kokusai Denshin Denwa Co., Ltd.

Michael L. Snodgrass
AT&T Bell Laboratories

Elaine K. Stafford
AT&T Bell Laboratories

Hiromi Sudo
Nippon Telegraph and Telephone Corp.

Masatoyo Sumida
Nippon Telegraph and Telephone Corp.

C. Burke Swan
AT&T Bell Laboratories

Nobuharu Takahara
Nippon Telegraph and Telephone Corp.

Ko-Hichi Tatekura
Kokusai Denshin Denwa Co., Ltd.

Jean Thiennot
CNET

W. R. Throssell
Standard Telecommunication Laboratories, Ltd.

M. D. Tremblay
AT&T Bell Laboratories

Jean-Pierre Trezeguet
Les Cables de Lyon

Patrick R. Trischitta
AT&T Bell Laboratories

Shunsuke Tsutsumi
Nippon Telegraph and Telephone Corp.

Richard E. Wagner
AT&T Bell Laboratories

Hiroharu Wakabayashi
Kokusai Denshin Denwa Co., Ltd.

R. L. Williamson
Standard Telephones and Cables PLC

Peter Worthington
Standard Telephones and Cables PLC

Yoshihiko Yamazaki
Kokusai Denshin Denwa Co., Ltd.

Hiroaki Yano
Kokusai Denshin Denwa Co., Ltd.

Paul A. Yeisley
AT&T Bell Laboratories

Nobuyuki Yoshizawa
Nippon Telegraph and Telephone Corp.

Contents

Part I: Background

Part II: Undersea Lightwave Systems

Part III: Undersea Fiber

Part IV: Undersea Fiber Cables

Part V: Undersea Lightwave Repeater Design

Part VI: Undersea Electrical Components

Part VII: Undersea Optical Devices

Part VIII: Assuring the Reliability of Undersea Electrical and Optical Devices

Part IX: Supervisory, Control, and Terminal Equipment

Part X: Future Undersea Lightwave Systems

Series Foreword
Frontiers in Communications

Frontiers in Communications brings to engineers in the communications profession technical coverage of rapidly changing areas.

Although each volume consists of a number of chapters written by different authors, it is more than a collection of reprints. Each volume is loosely based on one or more outstanding issues of the IEEE JOURNAL ON SELECTED AREAS IN COMMUNICATIONS. Every contribution is carefully selected for its technical quality and the quality of the presentation: material from the open technical literature is also considered. New material, where necessary, has been written to provide a comprehensive treatment of the topic. The editors are chosen for their outstanding expertise as world leaders in their specialty in communications.

Perhaps the greatest challenge to the editors is to cover the topic, even though it is a narrow one, in just a single volume. Emphasis has been placed on clarity of presentation. In many cases original papers have been re-edited to update the material and to delete material that is extraneous or redundant within the context of the volume.

This first volume in the series is based on the November 1984 issue of the IEEE JOURNAL ON SELECTED AREAS IN COMMUNICATIONS, published jointly as the December 1984 issue of the IEEE JOURNAL OF LIGHTWAVE TECHNOLOGY. The second volume, on "Local Area Networks", will draw heavily from the November 1983 and November 1984 issues of the IEEE JOURNAL ON SELECTED AREAS IN COMMUNICATIONS and the June 1985 issue of the IEEE JOURNAL OF LIGHTWAVE TECHNOLOGY. Other volumes on "Packet Switched Communications" and "Military Communications" are in planning stages. It is anticipated that up to four topics will be chosen from the IEEE JOURNAL ON SELECTED AREAS IN COMMUNICATIONS for development as volumes in Frontiers in Communications.

JOHN O. LIMB
Series Editor

Dedication

This book is dedicated to our wonderfully supportive wives,
Elaine Trischitta and Ilse-Lore Runge.

Preface
Undersea Lightwave Communications

I. Introduction

LIGHTWAVE systems are being installed on land at a phenomenal pace. The selection of lightwave technology for communication systems to be installed in Europe, Japan, and the United States is exemplified in the title of Mort Schwartz's article "Optical Fiber Transmission-From Conception to Prominence in 20 Years," in the Centennial Issue of the IEEE COMMUNICATIONS MAGAZINE. With this prominence on land, the superficial lightwave observer could assume that it would be an easy task to apply terrestrial lightwave technology to crisscross the world's oceans and seas. This is far from being true. The technology necessary for an undersea lightwave communication system is very different from the terrestrial lightwave system technology. The purpose of this book is to describe the technology necessary for the realization of undersea lightwave communication systems.

II. Reliability, Reliability, Reliability

The basic difference between a terrestrial lightwave system and an undersea lightwave system is the stringent reliability requirement on the undersea system. Since a deep-water failure results in a high repair cost and a huge loss of revenue, an undersea system must be made orders of magnitude more reliable than a terrestrial system. Terrestrial systems are more easily repaired and traffic can be rerouted via the national networks. The reliability target for the first transatlantic lightwave system is fewer than three system repairs in the 25 years of system life. This translates into reliability requirements for the components of typically a few FIT's, failures in 10^9 component hours.

The stringent reliability requirement is inherent in every element of an undersea lightwave system. Parts count must be minimized, dictating a single-mode fiber long-wavelength laser diode system to minimize the number of repeaters. A high level of circuit integration is required to minimize the number of active components within the repeaters. Each component in turn must undergo a rigorous qualification program during its development phase to assure that the components are capable of meeting their reliability objectives. Then, in manufacture, all components are inspected, burned in, and tested according to a carefully developed certification program which will assure that the very components to be committed to the undersea environment will meet the reliability requirement. In addition, redundancy of critical components is employed to further enhance the system reliability. The redundancy can include, the sparing of all system elements for which no previous undersea experience exists, i.e., the laser transmitter, the fiber, and the receiver, through span by span standby sparing or sparing only the laser transmitters with correspondingly higher risk for system failures.

Undersea fiber must be strong enough to survive cable laying and deep-sea recovery operations. Unlike terrestrial fiber, undersea fibers must be proof tested at levels which only a few yers ago were thought impossible to obtain. Today, these fibers are routinely proof tested at 2-percent elongation and yield unbroken lengths determined only by the fiber preform capacity. Thus the development of undersea lightwave systems has advanced the state of the art of high strength fiber design. Cable design and repeater physical design must protect the fiber and the optical component from the rigors of laying and recovery operations as well as the deep-sea environment.

III. Why Undersea Lightwave Systems?

During the last 20 years the international message traffic over the North Atlantic has been growing at a rate of 20–30 percent annually and this growth is expected to continue well into the 1990's. In the last decade, the telephone traffic was shared equally by coaxial undersea cables and satellites. This split was largely a function of service diversification and political considerations.

It was recognized in the late 1970's that single-mode long-wavelength lightwave technology would provide an opportunity to develop undersea systems which would be cheaper and provide more usable channel capacity than previous coaxial undersea systems or new shuttle-launched communication satellites. The world's oceans and seas have now become a marketplace for undersea lightwave systems, and vigorous development activities around the world are targeted for that market. But more than that, the early development efforts for single-mode long-wavelength systems for the undersea market have stimulated the development of terrestrial applications as well, and have led to a general rapid growth of this new technology.

IV. About the Book

This book is a comprehensive study of the progress made worldwide during the last five years on the realization of undersea lightwave technology for use in transoceanic undersea communication systems. It also offers an outlook on technology options available for future developments in undersea lightwave systems.

We thank the authors from England, France, Japan, Spain, and the United States for their contribution to this book on a specialized but important area in lightwave communications.

The book contains 43 chapters divided into ten parts. Part I contains background information about previous undersea systems and the merging of lightwave technology with undersea communication technology. Part II contains eight chapters describing undersea lightwave communication systems being developed, some nearing their targeted installation dates. Efforts in this country center around the development of the SL system. Chapters from France, England, and a joint Spain/U.S. paper discuss European development and installation efforts and plans. Two chapters from Japan describe the domestic and international systems developed there. Finally a military application of a repeaterless undersea lightwave communication system concludes the systems part.

In the next six parts, some elements of undersea lightwave communication systems are examined in detail. Part III describes design issues of fibers for undersea use. Special focus is given to the issues of fiber strength as well as issues concerning the effects of hydrogen on fiber loss.

Part IV contains chapters describing the undersea fiber cables being developed around the world including a chapter describing a fiber tether cable for an unmanned submersible vehicle. In Part V the physical design requirements of the undersea lightwave repeater are discussed. A chapter on an undersea branching repeater introduces the feature of multiple landing points and undersea networking of undersea communications systems.

Part VI addresses undersea electrical components and includes chapters on high-speed integrated circuits and their reliability issues, surface acoustic wave (SAW) filters and retiming issues in long haul systems, and regenerators using these components.

Part VII focuses on optical devices for undersea systems. It includes a chapter on laser and detector reliability, and chapters on transmitters, system penalties stemming from spectral anomalies of transmitters, receivers, and redundancy switches developed for undersea lightwave systems, and is separated into Part VIII on assuring the reliability of these optical devices.

Part IX deals with supervision and control of undersea lightwave systems. Finally Part X offers a look at the technology options for future, next generation undersea lightwave systems.

PETER K. RUNGE
AND PATRICK R. TRISCHITTA
Editors

Part I
Background

Undersea cables have played a central role in the evolution of worldwide telecommunications long before the existence of lightwave technology. We begin this book by looking back at the experience gained in the evolution from the first undersea telegraph cables and establish the technology requirements necessary to introduce newly developed lightwave technology to the harsh ocean environment. We conclude this part on background material with the first bonafide system proposal written on applying lightwave technology to undersea communications.

1
Undersea Cables for Telephony

R. D. EHRBAR

THE BEGINNING

In 1858, the desire for rapid communication between North America and Great Britain was satisfied by an undersea (submarine) telegraph cable. This was only 26 years after Morse invented the telegraph. It was almost a hundred years later, in 1956, that voice transmission by cable was achieved between the same points, though Alexander Graham Bell had invented the telephone in 1876. The reason for this long delay was that a satisfactory voice signal was much more difficult to transmit than a telegraph signal. In fact, a long period of development and several inventions (such as the electron tube, negative feedback, and the carrier principle) were required before economical long distance telephone transmission was achieved, even on land.

Overseas radio telephony was introduced in 1927. A number of subsequent technical developments—such as high-frequency transmission, single sideband with suppressed carrier, and steerable antennas—improved the quality and reliability of this service. However, as service increased, frequency space rapidly filled up and ionospheric disturbances were a source of trouble. These factors kept usage low and the cost relatively high.

In the early 1930's, the number of people who continued to use the transatlantic telephone despite its cost and shortcomings indicated that good transatlantic telephone service could be profitable. At that time, the Bell System made a proposal to the British Post Office for a continuously loaded cable across the Atlantic to carry a single voice channel without amplification. Such a submarine installation had been used in 1921 between Florida and Cuba for telegraph and voice circuits. The Great Depression and the ability of the shortwave radio to meet the demand prevented the transatlantic project from being completed.

Bell Laboratories then launched a long-range program of research and development in electron tube and component technology to assure an operating life of at least 20 years in the undersea environment. They also designed repeaters (amplifiers) to transmit multi-channel signals on a coaxial cable. The designs used the techniques being applied on land, but there were many new restrictions to be considered.

By 1942, a plan was in existence for a 12-channel system, with repeaters at 50-mile intervals, that could cross the Atlantic. The plans for the installation of long, deep-water systems were always based on continuous laying without stopping payout or man-

SOME IMPORTANT DATES IN AMERICAN UNDERSEA CABLE HISTORY

1950	Florida–Cuba. First SA system.
1952	Negotiations start on TAT-1. Exploratory development of next generation.
1956	Scotland–Newfoundland (TAT-1) in service. First SB system.
	Final development of next system started.
1957–1960	France–Newfoundland (TAT-2), Washington–Alaska, Florida–Puerto Rico, California–Hawaii (HAW-1) (SB).
1959	3-kHz-spaced channels used. TASI-A service. Cable protection studies.
	First man-made cable break.
1961	United Kingdom–Canada (CANTAT-1). First transatlantic British system.
1962	Florida–Jamaica–Panama. First SD system.
1963	United States–United Kingdom (TAT-3)—(SD).
1964	California–Hawaii (HAW-2), Hawaii-Japan (TP-1), Florida–St. Thomas (St. T-1) (SD).
1965	United States–France (TAT-4) (SD).
1967	First major Sea Plow system operated (Sea Plow-I).
1968	Florida–St. Thomas (St. T-2). First transistor system (SF).
1970	United States–Spain (TAT-5)(SF).
1974	California–Hawaii (HAW-3)(SF).
1975	Hawaii–Japan (TP-2) (SF). Sea Plow IV.
1975	TASI-B operational. Vancouver–Sydney.
1976	United States–France (TAT-6). First SG system.
1980	St. Thomas–Venezuela, St. Thomas–Brazil (SF).
1981	Guam–Taiwan (SF). TASI-E operational.
1982	Florida–St. Thomas (St. T-3) (SG). Greece–Egypt (SF).
	Lightwave demonstration—Lightwave final development.
1983	United States–United Kingdom (TAT-7) (SG).

handling the repeaters. These features seemed essential to the safety of personnel and equipment and to the achievement of predictable performance. Therefore, the repeater was designed to be only slightly larger than the cable and was made of articulated sections so that it could traverse the drum of the cable engine just as the cable did. In the first design, the active length was about eight feet and the overall length, including the cable tails, was approximately 70 feet. The extremely limited space in the repeater and the long electrical lengths dictated rather narrow bandwidths and eliminated any fancy circuitry such as that required for bidirectional transmission. This further dictated the need for a separate cable for each direction of transmission. While this simplified the repeater design, it added the expense of a second cable to achieve two-way transmission.

THE FIRST BELL SYSTEM REPEATERED CABLE PROJECT (FLORIDA–CUBA)

World War II interfered with any further work for some time. Work resumed at the end of the war and two cables were laid between Florida and Cuba to carry 24 telephone channels. Both cables were very similar to telegraph cables, but polyethylene insulation was used instead of gutta percha. In addition, an outer conductor was added to form a complete coaxial structure. Each cable was 120 nautical miles (nmi) in length, with three repeaters at depths of about 950 fathoms (1737 meters). This was called the SA System. It went into service in May 1950 and continues to be used today.

Power for the repeaters was transmitted as dc along the center conductor of the cable. It was very accurately regulated to a constant current. The electron tube heaters were in series and the anode potential was derived from the total voltage drop across the repeater. While many new technologies and devices had been developed during the war years, they were not considered ready to meet the reliability essential for long-life underwater use.

The First Transatlantic System (TAT-1)

In 1952, negotiations were begun with the British Post Office for a telephone cable between the United States and the United Kingdom. The British had pioneered shallow water repeaters and short systems serving other countries on the North Sea or across the English Channel. They used a rigid repeater that required the cable drum to be bypassed when laying the repeater. This required stopping the ship, and was not considered by Bell System representatives to be a satisfactory method of operation for long, deep-water links.

A 60-channel system had been designed and built. In this system, both directions of transmission were carried in the same cable. This became the final goal of all designers.

The basic route for the system was decided easily—the shortest practical deep-water crossing, in order to minimize the hazards of high voltage. Power to operate the repeaters was fed from the terminal stations. The voltage required was a direct function of distance. There was to be a link between Oban, Scotland and Clarenville, Newfoundland (1950 nmi or 3611 km), and one between Clarenville and Sydney Mines, Nova Scotia in Cabot Strait (270 nmi or 500 km).

Many conferences and much correspondence followed, which included disagreements along with a high degree of cooperation. It was realized that compromise was essential. Innovation was too risky. The phrase "proven integrity" was coined early in the project, and became the guiding philosophy. An agreement was signed on November 27, 1953 which specified the Bell System repeater design (patterned after the United States/Cuba

Fig. 1. The SB one-way repeater in its shipping crate. Its active length (electron tubes and components) was 8 ft. The total length, including coiled cable tails, was 70 ft. The shipping container was 34 ft in length and weighed 1000 lbs.

Fig. 2. British repeater used on TAT-1. The unit is about 9 ft long and 10-1/2 inches in diameter. Its weight in air is 1150 lbs.

design) for the deep-water transatlantic link (see Fig. 1). This was to transmit 36 4-kHz-spaced channels and form a part of the transatlantic system. The 60-channel British system (a repeater is shown in Fig. 2) would be installed between Newfoundland and Nova Scotia where the shallow water presented fewer hazards. The same cable was used for both systems (see Fig. 3). This cable had a diameter over the dielectric of 0.620 inches. The system was carried to the United States from Nova Scotia by a dedicated microwave link. The entire facility, called TAT-1, was installed in 1955 and 1956 and went into service on September 25, 1956. It was taken out of service in 1979 after exceeding its 20-year design life. There were no system outages due to electronic or mechanical failures, despite numerous breaks by fishing trawlers.

The details of this accomplishment are extremely interesting. Many new techniques were introduced and many hazards were dealt with successfully most of the time. New manufacturing methods, particularly in the area of quality control, were required. The ship encountered a hurricane during the laying of the first transatlantic link, adding to the excitement. Details have been covered in the *Bell System Technical Journal* [1]; this reference will provide the complete story.

It was known that the flexible repeater and the systems built around it would have very little use when systems of wider bandwidth for larger numbers of circuits were needed. No one was quite sure what was ahead, but they were sure that additional circuit capacity and longer systems would be needed. Work on these new systems was proceeding, as you will see in a later section. The French and Germans wanted direct circuits to the continent and there were other demands in the Pacific and Caribbean. Before the flexible repeater era was over, systems were installed to France (TAT-2), Alaska, Hawaii, and Puerto Rico.

One catastrophe occurred during the installation of TAT-2: the cable ship *Ocean Layer* caught fire in open water. Although the ship was destroyed, no one was seriously injured. This experience was a powerful reminder that things were not always simple. About 8000

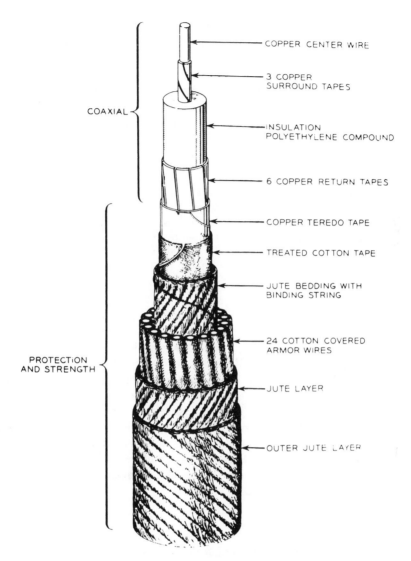

COPPER CENTER WIRE

3 COPPER SURROUND TAPES

COAXIAL

INSULATION POLYETHYLENE COMPOUND

6 COPPER RETURN TAPES

COPPER TEREDO TAPE

TREATED COTTON TAPE

JUTE BEDDING WITH BINDING STRING

24 COTTON COVERED ARMOR WIRES

PROTECTION AND STRENGTH

JUTE LAYER

OUTER JUTE LAYER

Fig. 3.   Armored cable for deep sea.

system nmi of SB were installed before the era came to a close. As of January 1983, almost 3500 system nmi were still in service.

DEVELOPMENTS IN OTHER COUNTRIES

Although the British Post Office and other foreign development agencies (in France, Germany, and Japan, for instance) were following parallel paths, it would be very complicated to try to cover all these events. Their importance, however, should not be underestimated. For continuity, the next big milestone in transatlantic communications should be noted, as it had considerable impact on the future.

This milestone was a system designated CANTAT-I. It ran between Oban, Scotland, and Corner Brook, Newfoundland, and was turned up for service in 1961. It was a single-cable system developed by the British to carry 60 4-kHz-spaced channels over about 2100 nmi (3889 km). The system used the lightweight cable developed by the British Post Office, which was similar in design to the cable we were planning for our next system (we called it armorless cable). The arrangement on shipboard also included the start of automation of the cable handling process, although laying had to be stopped to get the repeater around the cable engine. The link was financed by the British Commonwealth and was expected to care for the buildup of traffic through 1963.

HOW TO SAVE BANDWIDTH

It is important to mention two developments at this point, as a short digression from the orderly evolution of undersea cable systems. It was only natural that people start thinking of ways to save money as the number of undersea cables grew. Transmission circuits were very costly and this made it attractive to use terminal innovations to provide more efficient use.

3-kHz-Spaced Voice Channels

The first approach was to change the channel spacing. Four-kHz spacing had always been used in the and plant, permitting the use of inexpensive filtering and modulation techniques which, in the case of short channels, was very important. In the expensive undersea plant, this was not the case. Money could be well spent in the terminal to get more voice channels on a given system. For this reason, a channel multiplex was designed with more complicated circuitry; it would provide 16 channels spaced at 3 kHz in the spectrum formerly occupied by 12 channels. This equipment was more expensive but very cost effective. Furthermore, the impairment in transmission quality was trivial since the actual voice band of the 3-kHz-spaced channels was almost as great as that of the earlier 4-kHz-spaced channels. The 16-channel equipment was never manufactured in the United States, but was available from the French, British, and Japanese. Most undersea systems starting in 1959 were to use 3-kHz-spaced channels.

TASI

TASI (Time Assignment Speech Interpolation) is another idea that had been looking for an application. TASI is an electronic circuit multiplier used to expand the transmission capacity of four-wire analog facilities. It depends on the fact that, in normal conversation, the average talker speaks less than 40% of the time. By using fast switches and good speech detectors, the system permits voice circuits (trunks) to time share a smaller number of channels. TASI equipment is expensive to develop and manufacture, but it is extremely cost-effective for undersea use. The first TASI system was put into service in 1959. Many improvements have been made as circuit techniques and components have improved. (These improvements have produced a cheaper system with many new and useful features.) The latest version, TASI-E, went into service in 1981. It and other equipment using similar techniques will probably see service over the entire undersea network. It has been described extensively in the literature [2]. One impact of both of these band-saving schemes was to increase the total broadband signal power. This in turn increased the required load handling capacity of the transmission system. As usual, you cannot get something for nothing.

THE NEXT MAJOR CHANGE

The remarkable success of the first Transatlantic Telephone Cable was not to be established for many years, but everyone sensed the impending communications explosion, as long distance calling was booming. Limitations of the planned SB system were recognized, so it was time to take the next step. Exploratory work was started in 1952, but a new system was not completed until late 1962. This is indicative of the long lead time in development and manufacture required to produce the necessary integrity and reliability in an undersea system.

As this new phase of exploratory work began, there was general agreement in the Bell System on the following points:

1) A rigid (or lumped) repeater had to replace the flexible repeater to provide a form factor that would permit wide bandwidth and bidirectional transmission on a single cable. These features were needed to meet the growing channel demand as economically as possible.
2) A new cable design was required that would put the strength member inside the inner conductor, and thus allow a larger coaxial cable in a given outside diameter. A larger coaxial could, in part, offset the higher losses associated with higher frequencies.
3) The functions of cable and repeater stowage, cable payout control, and overboarding should be reexamined and approached as a major system problem.

There were also many fundamental arguments, including schedules. These involved:

- Electron tubes or transistors.
- Was there a need for TV transmission?
- Redundancy to improve reliability.
- The desirability of a new cable ship.

While these points were being studied, work began on development of the cable laying system that we thought was needed to accommodate the proposed physical design.

The repeaters would be stored on the cable working deck and the associated connecting leads carried into the cable tank in an orderly, preplanned fashion. The cable restraining device (crinoline) had to be designed to allow these leads to leave the tank without assistance wherever they occurred in the normal coiling pattern of the cable. They would pick up the repeater positioned on the cable working deck. A cable engine was needed at the stern to handle the cable and repeaters without interruption in a straight line (called a linear cable engine) and avoid multiple turns around a drum. Lastly, a chute was proposed for the stern of the cable ship to avoid a large-diameter sheave.

The development process started with the construction of a $\frac{1}{4}$-scale working model of the basic ship arrangements. A part of this model is shown in Fig. 4.

Two development processes were now proceeding simultaneously: a new system was taking shape and we were designing a new ship to place it. It would be difficult to further discuss the ship without discussing the general physical characteristics of the system it was to handle. Figures 5 and 6 illustrate the cable and repeater. The next step in the ship development was construction of a full-scale mock-up on a hillside in Chester, NJ. This very valuable facility was called the *C.S. Fantastic*; it taught us many lessons and established guidelines for the design of the cable-handling features of a new cable-laying ship. Two views of the facility are shown in Figs. 7 and 8.

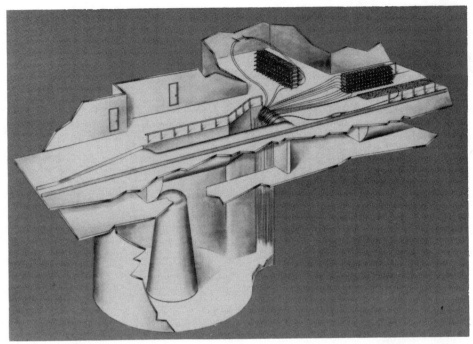

Fig. 4.   Quarter-scale model of a typical cable tank used to test design concepts.

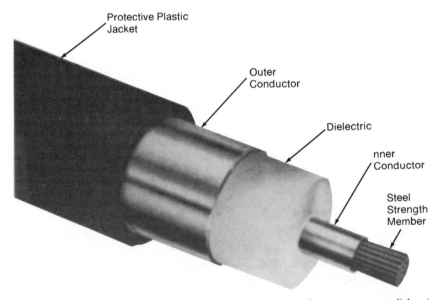

Fig. 5.   Armorless cable for 1 MHz system. Outer and inner conductors are copper, dielectric is polyethylene, 1 inch diameter.

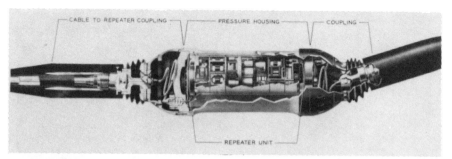

Fig. 6.   The basic repeater design for SD. Its maximum diameter is about 13 in. The distance between gimbal pivot points is 42 in. Its weight is about 650 lbs.

Fig. 7.   A view of the full size cable tank mock-up looking toward the cable slot in the main cable working deck. The top of the cone can just be seen. This varied from 10 ft in diameter at the top to 15 ft at the bottom of the cable tank. The various parts can be identified from Fig. 4.

Fig. 8.   Another view of *C.S. Fantastic* showing the repeater headed for a dry ocean.

Original plans contemplated modification of the British cable ship *Monarch*, but these were never carried out. Instead, the Bell System decided to build its won cable ship, the *C.S. Long Lines*.

The design and construction of the *Long Lines* makes for an interesting tale; one which is particularly exciting since the Bell System had never been associated with anything quite like it before. The *Bell System Technical Journal* contains a number of papers on the subject [3]. To provide some notion of this, three views are given in Figs. 9, 10, and 11. The ship contains sophisticated and complicated electrical and mechanical equipment, and much has been added since she sailed to her first cable laying job in July 1963. (The launching occurred on September 24, 1961). Much work went into reliability studies of materials and methods. One can argue that the ship and its equipment are two com-

Fig. 9.   The *C.S. Long Lines*. Overall length, 511 ft; beam, 69 ft; cruising speed, 15 kn; draft (fully loaded, 26 ft; cable capacity (1-1/4 in diameter), 2000 nmi; bow sheaves—center, 11 ft; others, 10 ft; forward cable engines, drum diameter, 12 ft; first cable operation, 1963.

Fig. 10.   Stern chute to replace a large diameter sheave.

plicated, but now that *C.S. Long Lines* has laid over 45 000 nmi of cable, it is hard not to be impressed.

THE SECOND GENERATION, OR SD SYSTEM

The previous paragraphs indicate that an argument mentioned earlier had been resolved, that is, a new cable ship would be built. Some of the other arguments were not so easily resolved, but the need for a new system forced some compromises and agreements. The goal was determined to be a single cable system with a bandwidth of 1 MHz carrying 128 3-kHz-spaced channels. The system was to be capable of direct connection between the United States mainland and the United Kingdom. Repeater spacing was to be 20 nmi. System operation was set for 1963, with a shorter pilot system set for late 1962.

Electron tubes were to be used as the active devices, and the same amplifier would be used for both directions of transmission. A second parallel amplifier was to be provided for redundancy, since high transconductance tubes with their close grid-to-cathode spacings could not yet be trusted. Transistors were also considered, but their reliability was not yet established. The main problem was to establish testing procedures to determine the transistors' life expectancy. The general configuration of the repeater is shown in Fig. 12; the complete repeater and cable are shown in Figs. 5 and 6. This system was designated SD.

The repeater was contained in a rigid housing that could be laid at high speed from the *C.S. Long Lines*. The housing was made from a special heat-treated copper beryllium

Fig. 11. The linear cable engine, the heart of the cable laying system. It is made of two grips similar to the threads of a Caterpillar tractor. These are forced together to tightly grip the cable, but will widen to accept or "swallow" the large diameter repeater when necessary. Major technical questions to be considered were cable gripping, and a method to permit the track to effectively lengthen to match the normal cable elongation under tension. A major reliability testing program was required because of the large number of parts (bearings, rollers, and others) where one failure could be disastrous. The basic engine is 42 ft long, 8 ft 6 in wide, and 15 ft high. It weighs 180 000 lbs.

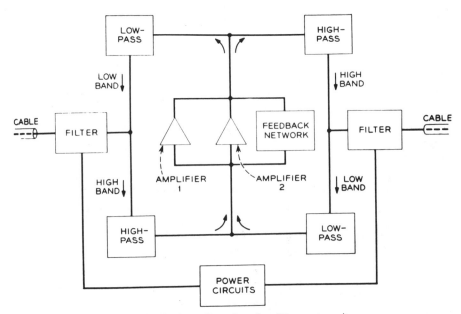

Fig. 12. Basic configuration of an SD repeater unit.

alloy, chosen for its resistance to corrosion. A pilot system was planned for Florida–Jamaica, with a transatlantic system planned for the next year. Everything went according to plan except that the completion of the *Long Lines* was delayed, and it was necessary to use the British cable ship *Alert* to lay the Florida–Jamaica link. This meant that the first operation of the *Long Lines* was a big one, TAT-3.

TAT-3 was a 3500 nmi (6480 km) system between Tuckerton, NJ, and Cornwall, England. Everything turned out satisfactorily and, after final equalization, 138 channels were acceptable for service. There are now more than 19 000 nmi (35 000 km) of SD in service. The design and manufacture of the SD system emphasize reliability and reproducibility, and make an interesting story, told more completely in [3].

As in all long analog systems, the equalization plan is of particular interest. The desire was to obtain a signal-to-noise ratio (SNR) consistent with the requirements of the network. For a number of reasons the goals are set rather high. The job in undersea cable systems is simplified by the fact that the environment is very stable for most of the transmission components. However, there is extremely limited access to any part of the system except the ends after installation.

In SD, the total transmission loss in 3500 nmi is 9000 dB at the highest transmitted frequency. This must be matched by the gain of 180 or more repeaters. Obviously, very small deviations must be corrected at intervals along the system to prevent accumulation of very large misalignments.

Basic equalization of the cable loss is accomplished by shaping the repeater gain to match the cable loss at standardized temperature and pressure. The individual section lengths of manufactured cable are trimmed in terms of expected temperature and pressure of the specific site. The objective is to obtain the desired loss at the top transmitted frequency. All residual deviations are then equalized at 200 nmi intervals by an adjustable ocean-block equalizer. The proper adjustment is determined from careful analysis of data taken as the system is installed. In this system, the equalizer is adjusted just before it goes overboard. After the system is installed, the signals at the terminals are shaped in level so that the optimum signal-to-noise performance is obtained.

The SF System

With the high quality circuits provided by TAT-3 and subsequent SD systems, overseas traffic continued to expand. There was ample evidence that new technology should be exploited as rapidly as possible within the boundaries dictated by undersea use. Exploratory efforts had started in the late 1950's on transistor characterization and the development of amplifier circuits for a new wideband undersea system. By 1963, a detailed development of a system called SF was scheduled. The first system was installed between Florida and the Virgin Islands in 1968. In 1970, the TAT-5 SF system established a link between the United States and Spain.

About this time (1967), communication satellites began to make available a substantial number of circuits for overseas traffic. In fact, various mixes of cable and satellite circuits dictated by United States Government regulation had begun to have a substantial effect on the installation and development of new cable systems. A comparison of the two transmission systems is not intended here; technically and economically, there is a need for both systems.

The SF system [4] closely follows the SD design, both physically and electrically (see Fig. 13). New transistor active devices and many problems are associated with the wider

Fig. 13.   View of the SF repeater unit.

bandwidth. The parallel amplifier used in the SD repeater was omitted. The top frequency is 6 MHz and repeaters are spaced at 10 nmi intervals. The cable has a diameter over the dielectric of 1.5 inches, but is otherwise very similar to the SD design. The overall loss of a 3500 nmi system at the top frequency is about 15 000 dB. Misalignment is controlled by ocean block equalizers at about 200 nmi intervals. Initially, the capacity objective was 720 3-kHz-spaced channels, but about 845 have been achieved in practice. The system continues to be used and installed. By the end of 1983 there will be over 17 000 nmi (31 500 km) in operation. A very recent version of SF, designated SF-2, uses silicon transistors and has a repeater spacing of 12 nmi.

The SG System

The demand for cable circuits continued to increase, though many thought satellite communications was the ultimate answer. With satellites, there was the long inherent transmission delay to contend with, and vulnerability to interception to be considered. In addition, the economics were not conclusive or easy to understand. Cables were not destined to "go away."

Exploratory work started in about 1966 to determine what the latest technology would suggest as the next broadband system. Further challenges were introduced early in 1970 when it was decided that the next Bell System design, called SG, would be a "shared development" with the British Post Office (now British Telecom). The division of work was changed during the early meetings, and the French Ministry of Postes and Telecommunications was given the responsibility for terminal development and brought into the joint development team. A formal agreement for the shared development was signed on May 2, 1973. The first system was to be installed between Green Hill, RI, and St. Hilaire, France, to be completed in 1976. The details of this rather complex arrangement and the final system design are covered in [5].

The SG system is an equivalent 4-wire design, as was SF. The initial capacity was set at 4000 3-kHz-spaced channels. The top frequency was 29.5 MHz. The noise objective was 1 picowatt per kilometer (pW/km) applied to the average noise in all channels, with no channel greater than 2 pW/km. The repeaters are spaced at 5.1 nmi (9.5 km) intervals.

The outside is very similar to the SF repeater, but newer techniques have been used inside to meet the needs of a higher top frequency and greatly increased bandwidth (see Fig. 14). These factors complicated the design of many passive components and required a high degree of sophistication in the development and measurement of the cable dielectric. The cable diameter over the dielectric was increased to 1.7 inches to reduce the cable loss. The equalization plan for SG followed the same pattern as SF, but a new

Fig. 14.   View of the SG repeater unit.

equalizer was added late in the development to handle unknown changes in cable loss with time. This equalizer was controllable from the shore terminal station.

The first SG system worked very well, but did not quite meet the noise objectives. This should not be a problem in future systems since the cause is understood and has been eliminated. Attempts to improve the noise performance with new circuitry at the terminals have been partially successful. Actually, the circuits are very high quality and acceptable to most users.

By the end of 1983, there will be about 8000 nmi (14 800 km) of SG installations in service. Figure 15 shows a summary of all the American systems living and dead.

THE CLIMAX OF THE ANALOG ERA?

The SG system is probably the last major analog design to be realized. However, before attention was turned to digital systems, some exploratory work was done on an analog successor to SG, called SH. It had a top frequency of 125 MHz and would have been capable of transmitting 16 000 3-kHz-spaced channels. The repeater spacing was short (about 2-1/2 miles), and transmission problems were tough. As noted later in this chapter, the SH design was dropped in favor of a lightwave program.

Up to this point, emphasis has been on the evolution of new undersea analog cable systems to meet an increasing demand for communication facilities over the world. Now,

System	First Service	VF Channels		Top Freq.	Cable			Repeater		Active Device
		3kHz	4kHz	MHz	No.	Type[1]	DOD[2]	Type	Spacing	
									NM	
Analog Systems										
SA	1950		24	.108	2	(A)	.46 in.	Flex	40	E-Tube
SB	1956	48	36	.164	2	(A)	.62 in.	Flex	36	E-Tube
SD	1962	138	100	1	1	A	1.0 in.	Rigid	20	E-Tube
SF	1968	845	640	6	1	A	1.5 in.	Rigid	10	Transistor
SG	1976	4200	3150	30	1	A	1.7 in.	Rigid	5.1	Transistor
Digital Lightwave System										
SL	1988				(UNDER DEVELOPMENT)					

[1](A) = Armored; A = Armorless
[2]Diameter Over Dielectric

Fig. 15.   American undersea systems.

it may be beneficial to look at some of the other major technical problems which had to be faced in building up this worldwide undersea network.

PROTECTING THE CABLE SYSTEM

After completion of TAT-1 in 1956, the cables that were installed worked very well and everyone was happy with their performance. However, in February 1959, a cable break off Newfoundland caused the first interruption of service. Since a Russian trawler (fishing ship) was presumed to have done the damage, the story received considerable coverage in the press. This was only the beginning!

The tremendous fishing activity in shallow water and the impact of the rapid post-war development of large, powerful trawlers had not been fully appreciated. It was an expensive matter in at least three aspects:

- the cost and hazards associated with cable repair,
- the amount of time required for repair that resulted in lost revenue or the cost
- of alternate circuits, and
- the potential electrical damage to the system caused by rapid discharge of the cable.

Repair time in particular became a great factor as the number of circuits increased. It did not take long to realize that a major offensive was required fast.

One promising plan was to bury the cable in the hazardous areas, and so an extensive program was undertaken to study methods for burying cables by water jetting or by

Fig. 16. One shot from an underwater TV survey of cable in place on the ocean bottom.

Fig. 17. One of the latest Sea Plows developed to bury cable and repeaters as they are laid. It will bury to a depth of about 2 ft in a trench 16 in wide at a speed of 1 kn. The vehicle is 32 ft long, 10 ft wide, and 10 ft high. It weighs 35000 lbs in water.

plowing. Cable laying procedures were reviewed to specify what had to be done to insure coverage of the bottom with no suspensions or tangles, if possible. An underwater TV survey of the cable in the shelf area (see Fig. 16) was undertaken to see exactly how things stood. A plan to use redundant shore connections was given extensive study. This involved an underwater switching device about 200 miles off shore. A large effort was also undertaken to educate fishermen by giving them cable locations and other information. Payment was offered if they would leave their trawling gear on the bottom. This idea was not very popular.

The use of jetting was not regarded as a fully satisfactory solution to the burying problem. A number of soils which were encountered were not easily moved. Development work concentrated on a plow and several versions were built. One plow in current use, shown in Fig. 17, will bury cable and repeaters during the laying operation. So far, this seems to have reduced manmade cable breaks to practically zero. This plow, the Sea Plow IV, is described in the *Bell Laboratories Record* [6].

The plow is limited in operational depth to 500 fathoms (914 m). At times, it is also necessary to recover a buried cable or bury a cable already on the bottom; for these and many other situations, the tethered, unmanned, submersible SCARAB was developed (see Fig. 18). The SCARAB, which has been described extensively in the literature [7], also serves as an excellent survey vehicle to examine the proposed routes for new cables within its operational depth of 3000 ft (5486 m).

A FEW NASTY SURPRISES

In an undertaking as large and complex as undersea cable systems, there are likely to be surprises. By careful design, quite a long development interval, and enormous amounts of testing before and during manufacture, we have managed to side step many, but not

Fig. 18. SCARAB (Submersible Craft Assisting Recovery and Repair), an unmanned submersible developed by a consortium of cable owners to perform almost any operation that might be required on the ocean bottom.

all, of them. It is not unexpected that most of these surprises show up when a complete system is assembled, or after several years of system operation. This article would not be complete without mentioning a few.

Trawler Damage

This is an instance of not understanding the environment or realizing the seriousness of the threat to service.

Nonlinear Sing

Nonlinear sing is a phenomenon found in equivalent fourwire systems using a single amplifier for both directions of transmission. Under certain conditions these systems can lock up in a so-called noise sing. This is generally the result of gross overload in the presence of large misalignment. It was encountered in testing the first SD system with much of the cable on shipboard. The problem was attacked methodically. Computer programs now exist to study repeater behavior under various conditions [8], programs which permit prediction of the nonlinear sing margin in a system for different misalignments.

Excess Noise

Higher than expected noise was encountered in the first (TAT-6)SG system under noise loading tests after completion [5]. This noise occurred in the top third of the high band,

and was due to third-order modulation behavior that had not shown up in single repeater tests or systems tests of ten connected repeaters. Since the noise originated in repeater intermodulation, it was possible to cancel much of the excess noise by means of a specially designed distortion canceller in the high-band receiving terminal of TAT-6. Such excess noise will not be a problem in future SG systems because of transistor improvement that permits removal of output protector diodes that were a major source of the modulation problem.

Corrosion

Materials were carefully picked to avoid corrosion in sea water. However, one situation escaped our attention, namely, the electric current generated by ocean currents traveling in the earth's magnetic field. In many areas of the ocean, particularly around the mid-Atlantic ridge and similar formations, the rate of the ocean current is significant and varies from place to place. The current generates a potential between points and causes an electrical current to flow in the cable's outer conductor. This causes corrosion and loss of the outer conductor of the repeater pigtail after many years of operation. In systems such as the SF and SG, loss of the outer conductor changes the cable characteristic at higher frequencies. The effect becomes serious after a number of repeaters have been attacked. This effect was first discovered on TAT-5 and was corrected by picking up and relaying a very long section of the system. Once the problem is known there are ways to avoid it.

What Next?

The worries over analog or digital systems are settled by the emergence of new system tools such as a very wide-band low-loss medium (glass fiber) and a wide-band transmission technique (light). These are perfect for digital transmission. Other factors favoring digital transmission are simplified connectivity and channel sharing. Since each system of many thousand channels can be handled on a very small fiber pair, it should be possible to route systems directly to their destination even if they travel most of the way in the same cable with other systems.

There are many enthusiastic people working on this problem everywhere in the world. The nature of the first long, deep water system to be installed is beyond the scope of this paper. One can guess that the late 1980's will be the earliest date for a transatlantic system. It will be called TAT-8, and the details will appear in the literature as agreement is reached among the owners. There have been trials and there will be more. The Bell System has demonstrated a repeater-cable combination that can be laid in deep water. It will work and can be recovered, and this is very encouraging. However, some anticipated problems are worth mentioning.

Lightwave System Problems and Challenges

  1) System architecture—what is needed; what is cost effective?
  2) Strength and quality control of glass fibers.
  3) Repeater-cable connection.
  4) Component reliability-proof.
  5) Digital multiplex compatibility.
  6) System repairs.
  7) New cable ship tools and techniques.

These are formidable problems and their solution will require a great deal of effort in design and manufacture. The end result has so much promise in reduced cost and increased circuit capacity, improved quality and flexibility of operation, that the technology must be harnessed for the undersea network. We did not solve the problems of our present network in a short time, and the next step will take time and dedicated effort, but we will do it and it will be worth it.

REFERENCES

[1] *Bell Syst. Tech. J.*, vol. 36, no. 1, Jan. 1957.
[2] "TASI-E Communication System," *IEEE Trans. Commun.*, vol. COM-30, no. 4, Apr. 1982.
[3] *Bell Syst. Tech. J.*, vol. 43, no. 4, part 1, July 1964.
[4] *Bell Syst. Tech. J.*, vol. 49, no. 5, May–June 1970.
[5] *Bel Syst. Tech. J.*, vol. 57, no. 7, part 1, Sept. 1978.
[6] "Sea Plow IV—Digging in the newest transatlantic cable," *Bell Lab. Rec.*, Sept. 1976.
[7] *Bell Lab. Rec.*, Sept. 1981; *OTC 3995*, 1981 Offshore Technology Conference, Houston, TX.
[8] *Bell Syst. Tech. J.*, vol. 48, no. 6, pp. 1853–1864, July–Aug. 1969.

2
Technology Requirements for Optical Fiber Submarine Systems

KENNETH D. FITCHEW

Introduction

Repeatered analog FDM submarine coaxial cable systems were first employed in commercial services in 1950, and although there were several notable milestones in achievement, such as the first transatlantic system in 1956, development progressed in a gradual manner from the early 36 circuit repeaters to the present day systems of 5520 two-way circuits. Each new system was strongly based on previous development; the only two really fundamental changes in technology were the introduction, in 1960, of lightweight cable with a central strength member and the change in the late 1960's from vacuum tubes to transistors.

The potential of optical fiber transmission to provide high capacity, reduced circuit cost, and compatibility with digital networks makes it very attractive for submarine systems. However, the development of optical fiber submarine systems requires the simultaneous introduction of a whole range of technologies which are not only new in the submarine field, but in many cases have not yet become established in terrestrial systems.

This paper seeks to outline the technical developments being undertaken in this field. The scope of this study has been primarily confined to what are considered to be the first generation of "main-line" systems [1], i.e., repeatered systems operating at 1.3 μm on monomode fiber, carrying data or telephony at rates in excess of 100 MBd, using binary intensity modulation.

Specific Requirements for Submarine Systems

The practical operational requirements of submarine systems are rather different from those of land systems, and, in consequence, the design constraints on specific parameters also differ.

Reliability

Network planners have come to expect a high reliability over a 25-year system life from FDM submarine systems and wish to see optical systems aim at similar targets. The

23

requirement for low fault rates (a typical target being two or three failures during system life) is due to the considerable expense involved in a repair operation; these costs arise from: provision of alternative traffic paths, the repair operation itself (use of cable ships, etc.), and, in the case of deep-water repairs, the additional cable (and repeater) which must be added (usually equivalent to about twice the depth of water). The requirement for longevity arises since the capital cost of planning, manufacturing, and laying a submarine cable system is so much greater than the annual running costs that there are considerable economic advantages in extending the life of a system for as long as is practical.

The target reliability requirements are generally taken to apply to the submerged plant alone since special measures can be taken in shore-based terminal equipment. Furthermore, cable has generally been excluded from reliability calculations since it has been assumed that any acceptable design of cable will have an expected "intrinsic" fault rate which is far too low to be quantified meaningfullly. It should be noted that faults due to external causes, e.g., trawlers or anchors, are excluded from these reliability targets although they dominate decisions such as choice of route, degree of armoring to be provided, and desirability of burial. It is possible that it may be necessary to consider a small but quantified probability of failure for a jointed optical cable, but this would not materially alter the submerged repeater reliability requirements which are considered in detail in the section on repeaters found on page 29.

Physical Environment

The most obvious requirement of a deep-sea plant is that it must be capable of withstanding hydrostatic pressures [~ 70 MPa (10 000 psi) at a depth of 4000 fathoms]. In addition, it must have tensile strength characteristics such that it can be layed and recovered, the latter operation resulting in a standing force with a cyclic variation (period several seconds) superimposed. In other ways, the deep-sea environment is quite benign, the temperature being constant (2°C North Atlantic, 13°C Mediterranean) and there being little danger of mechanical damage from outside forces (provided that care is taken to avoid areas prone to deep-water currents, or movement of the earth's crust).

Shallow-water plants may have to cope with considerably greater tensile forces, both during the recovery of a buried or silted cable and also in the event of snagging by a trawl. Danger of trawler damage also results in a requirement either for abrasion resistance [2] or for ability to withstand the operation of burial below the sea bed. Shallow-water cables are also prone to seasonal temperature variations, a typical value being ±7°C around a mean of 10°C for the southern North Sea; mean temperatures well in excess of 20°C may occur, for example, in the South China Sea.

Repeater Subsystem Design

Owing to the absence of accessible test points along a submarine cable, it is essential that the supervisory system can locate a faulty unit accurately and with a high level of confidence. The system should be designed such that a level of accuracy as high as possible is maintained even if one or more supervisory units should fail.

Repeater circuits must also be designed to be compatible with requirements for remote power feeding.

On most routes, repeater spacings can be arranged to be close to nominal, and so variability of section lengths is not a major factor in determining the dynamic range requirements of repeaters. Margin must, however, be allowed for system repairs, though for deep-water repairs the amount of excess cable added may well demand the insertion of an additional repeater.

FIBER

Choice of Parameters

Parameters of monomode fibers are selected with a view to meeting the following requirements at the wavelength of interest:

- low loss of primary coated fiber,
- low incremental loss due to cabling,
- low splice loss, and
- low dispersion.

For submarine applications, the operating wavelength is likely to remain fixed during system life, at least for repeatered systems, since it would be uneconomic to recover a system to upgrade it later.

During the last two or three years, monomode fiber development has been closely linked to the achievement of the lowest primary coated fiber loss near 1.3 and 1.55 μm [3], [4]. Within that context, various schemes have been suggested to minimize the chromatic pulse dispersion for operation within the 1.55 μm loss window [5], [6]. It has become clear, however, that in addition to primary loss and bandwidth, it is necessary to design the fiber to ensure low to zero incremental cabling loss [7] and minimum fusion splice loss [8]; various trade-offs thus become necessary. A high degree of geometric control is required at the preform fabrication stage, in particular to minimize the concentricity error contribution to splice loss [4]; active alignment during splicing to overcome significant concentricity offsets is not acceptable for shipboard splices.

As long lengths of sufficiently high strength monomode fiber, which can be incorporated into cable with high yield, become available, the splice frequency can be reduced thus easing the design requirements on the fiber. Concern has been expressed that quite modest levels of stress birefringence in the fiber would yield significant monochromatic dispersion of the LP_{01} modes [9], thus restricting the bandwidth of long links. In fact, with current monomode fiber technologies, any remnant birefringence effects appear to have a negligible effect on link bandwidths [10].

To identify the fiber design window for optimum submerged cabled and spliced monomode link performance, a measurement/theoretical-based feedback loop needs to be established to evaluate the dependence and sensitivity of link properties on fiber design (refractive-index profile and geometry). The criteria for a standard monomode 1.3 and 1.55 μm operation fiber design have now been established [11], and cabled and multi-spliced monomode links more than 30 km long have been constructed with an overall loss of 0.55 dB/km at 1.3 μm and < 0.5 dB/km near 1.55 μm (despite a mean splice spacing ~ 1.75 km, deliberately more exacting than the submarine requirement) [12], [13]. In this context, the ESI (equivalent step index) approach has become the most appropriate method of forward prediction to the performance of the final cabled and spliced link from individual primary fiber measurements [14], [15]. Monomode link technology is now fairly well placed to meet the optical performance requirements of submerged optical systems planned for the next few years.

Fiber Reliability

The strength of fused silica fibers, with *undamaged* surfaces, has been found to be about 6000 MPa [16], corresponding to a breaking strain of about 8.5 percent. This strength has been found to be insensitive to fiber diameter, but very sensitive to even the slightest surface abrasion or other surface damage. Consequently, it is imperative that optical fibers be given a protective coating immediately after drawing. For this primary coating, soft silicone [17], ethylene vinyl acetate [17], $U-V$ curable polymers [18], metal coatings [19], silicon-nitride [20], and other materials have various attractions and some disadvantages.

Most of the silicia fiber currently made has strengths which are considerably less than 6000 MPa and also very variable. It is generally accepted that this is due to the presence of flaws at the surface of the fibers [16], [21]–[23]. Possible causes are

- fiber drawing temperature insufficiently high,
- air-borne particles impinging on the hot fiber during drawing, and
- inadequacies in the fiber coating.

The first of these may be difficult to overcome since an increase in drawing temperature can result in increased loss [24]. The second is more amenable to improvement since fiber strength can be improved considerably by keeping airborne particles away from the drawing area [25]. Improvements are being sought in the fiber coating material and its adhesion to the fiber; $U-V$ curable polymers have shown considerable promise, and other coatings, such as metals, may become attractive if the problems of applying them can be overcome economically.

A Weibull plot of fiber strength allows different failure modes to be isolated [26], and also facilitates the estimate of failure probability of long lengths of fiber from measurements on short lengths. This usually brings to light the disastrous effect of the low-strength tail of the statistical strength distribution. In order to improve the failure probability, it is necessary to eliminate all the weaker parts of fiber, by proof testing [27].

A further complication arises from strength deterioration. It is well known [16], [28], [29] that silica fibers suffer a time dependent deterioration in strength under conditions of sustained tensile stress in a moist environment. This effect, commonly known as "static fatigue," is now attributed mainly to stress corrosion at surface flaws [21], [23]. Since it is not convenient to measure time to failure under constant strain directly, it is customary to calculate this from dynamic fatigue measurements [30]–[32] of time to failure under ramp loading. Some long-term static fatigue data are also necessary because there are uncertainties in predictions from relatively short-term dynamic fatigue data.

The rate of deterioration of strength with time is described by the "stress corrosion susceptibility index" n in the empirical crack growth equation

$$v = AK_I^n$$

where v is the crack growth velocity, A is a material property constant, and K_I is the mode I crack stress intensity factor ($K_I < 1$) [27]. When the silica fibers are exposed to 100 percent relative humidity (which is quite likely in many designs of submarine cable), n can be as low as 15 [26] for the low-strength tail of the strength distribution. Certain fiber coatings, particularly metal coatings [19] and silicon oxynitride [20], can increase n considerably, and, thereby, greatly reduce the rate of strength deterioration, but it is not yet certain to what extent these could be relied on for a service life of 25 years. n is also increased if the humidity can be kept low, but until cable designs with moisture control have been proven, it would be prudent to design with low n, say 15.

The level of strain to be used in proof testing is of great significance. If a relatively high proof strain, e.g., more than 1.5 percent, can be used without disastrously reducing the yield of fiber, then a relatively cheap cable may be adequate, even for the extreme loading case of recovery from deep water. Use of a lower figure demands the use of a stronger cable, leading to a trade-off between fiber cost and cabling cost.

Fiber Splicing

Fiber splicing is required both at the cable manufacturer's plant and at sea during installation or repair. In each case, there is a requirement for equipment capable of producing consistent, reliable, low-loss splices by operators with average skills. The splicing of monomode as opposed to multimode fiber calls for tighter tolerances both in the fiber concentricity and in the mechanics of the splicing operation. With current technology, fusion splices of 0.25 dB or less are readily obtainable [33], [34]. However, a considerable degree of care and skill is required in the various operations of stripping the coatings, cleaving the ends such that they are flat and normal to the axis, cleaning the ends, inspecting the end surfaces (measurements of end angle to 0.25° are possible [35]), and fusion. The loss of a splice can be measured at the factory using standard techniques. At sea, the problem of assessing splice loss is more difficult since the remote ends of the fiber are not available. However, a skilled operator can often give a good assessment of his splice, and in the future it may be possible to confirm this by detailed visual inspection [36].

Splices must be strong enough to withstand all the stresses to which they may be subjected during the system life. They are usually protected by a moulding over the splice area, and, where possible, by a rigid sleeve. It may be possible to increase the inherent strength of fusion splices by taking steps to minimize the surface-OH content during the splicing operation [37].

CABLE

Experimental optical submarine cable trials have been reported from the UK [38] and Japan [39], [40]. Details of proposed cable designs have also been published by the U.S.A. [41] and France [42], and both these countries have now also conducted sea trials. While some of these designs are early examples, they do serve to indicate some of the requirements of cable technology, as follows.

Reduction of Cabling Loss

While the choice of fiber parameters allows some scope for minimizing the additional loss introduced by microbending, it is nonetheless important that such microbending is itself minimized. This is often achieved by incorporating a soft cushion layer into the fiber coating, but other techniques include embedding primary coated fibers in a cushion layer [40] or allowing the fibers freedom of movement in a loose fitting tube or groove [42].

Residual Strain

To achieve minimum strength deterioration due to static fatigue, the residual strain left in the optical fibers after cabling should be as small as possible. Typically, values of residual strain of less than 0.05 percent have been achieved [38].

Pressure-Resisting Housing

The pressure-resisting housing ensures that microbending of the optical fibers, due to external pressure, is minimized. In most designs, the pressure-resisting housing serves also as a water barrier, and as a substantial part of the power feeding conductor, and is in fact a composite structure in which more than one component contributes to the pressure resistance. In the design described in [38], for example, the main pressure resistance is provided by the thick-walled aluminum alloy tube, but the steel wires also contribute significantly by virtue of the arch effect; at deep-water pressure, the reduction in bore diameter is only about 0.1 percent.

Water Blocking

If the cable becomes damaged by a ship's anchor or by trawl fishing gear, it is desirable that ingress of sea water should be prevented, or at least restricted to a short length of cable (a few tens of meters). One approach is the use of intermittent water blocks to restrict water ingress to a given length of cable. A more attractive solution is continuous water blocking, and in many cable designs this requires a material which can be injected into all interstices during manufacture with no detrimental effect on the properties of the cable. The evaluation of such materials is currently receiving attention.

Steel Wire Strength Member

For high-tensile steel, the usual material for the strength member wires, the safe working stress is such that the strain in the cable under the maximum tension load condition would be around 0.7–0.8 percent. With a relatively simple cable design, with no strain relief for the optical fibers, the strain in the fibers would be of a similar magnitude and the duration of this high strain could be some 30–40 hours during a deep water repair. The level of proof strain needed to ensure that the optical fibers would survive a tensile strain of 0.8 percent for 40 h would be just under 2 percent.

There are two approaches which allow the proof strain of the fiber to be reduced. One is to increase the strength of the cable such that the strain at the maximum load is compatible with the proof strain of the available fiber. The other is to design the cable for strain relief so that when the cable is tensioned the optical fibers experience a strain smaller than that in the strength member [42].

Jointing

The aims in cable joint design are that the joint should have the following:

- a strength nearly as good as (say 90 percent of) the strength of the parent cable,
- the smallest splice loss which the current technology of fiber-splicing permits,

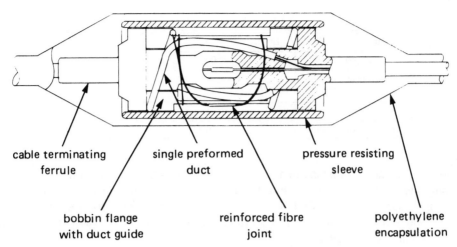

cable terminating   single preformed   pressure resisting
ferrule   duct   sleeve

bobbin flange   reinforced fibre   polyethylene
with duct guide   joint   encapsulation

Fig. 1.   Joint housing for optical fiber submarine cable.

- a sufficiently large bending radius of the fibers that the residual strain in the fibers will not lead to failure within the design life,
- a low-resistance through path for the power feed conductor, and
- minimum increase in overall cable diameter.

No details have yet been published of a satisfactory joint for factory use, meeting all these requirements. The main difficulty arises from the disposition of the fibers within the strength member. Since it may not be possible to ensure that all the fiber splices are completed successfully at the first attempt, it becomes necessary to stow a certain amount of "spare" fiber. In order to do this without violating the minimum safe bending radius, the diameter of the strength member has to be locally increased.

For shipboard joints, the requirement for minimum overall diameter is somewhat less important. Figure 1 shows a possible design for a shipboard joint which has considerable capacity to accommodate excess fiber. The two swaged ferrules are connected by a central strength member equipped with flanges that form the ends of a pressure resistant housing. The minimum bend radius of the fibers is controlled with a bobbin and duct system. After splicing, the fibers are wound and secured on the bobin and a steel tube is positioned over the assembly to complete the joint housing. A polyethylene moulding is then used to provide high-voltage insulation.

REPEATERS

Repeater design involves drawing on a number of different technologies, some of which are outlined below. However, reliability is a theme that is common to all of them, and it is therefore appropriate at this stage to look at the magnitude of the reliability requirement:

Reliability

Component Reliability Targets: It is useful to divide the system reliability targets among the major component families in order to get reliability targets [43]. Table I shows arbitrary but reasonable division.

TABLE 1

Component Family	Expected Failures in 25 Years
Optical Source	1.00
Receiver Devices (e.g., PIN, FET)	0.25
Silicon IC	0.25
Passive Component	0.50

These targets may then be divided by the numbers of each component in the system to give component reliability targets. For example, consider a prospective one-way repeater containing 1 laser, 3 p-i-n or FET devices, and 5 IC's. Then, for a six-fiber transatlantic system (6000 km) consisting of 175 sections, each 35 km long, Table II shows the necessary component targets.

Component Selection Philosophy: At an early stage in the circuit design, it is necessary to limit the number of component families, if the cost of the reliability exercise is not to be prohibitively high. This is essential even if it involves some compromises in the circuit design. For use in submarine systems, components are usually derated from the conditions acceptable for other applications. As for all other high-reliability applications, the selection of manufacturing source for each component family necessitates a dedicated high-reliability production line, and favorable performance of product in both realistic and overstress life tests. All significant device parameters should be monitored at intervals throughout the tests, and large test batches are necessary in order to observe any examples of "rogue" behavior. It is essential to prove that the chosen overstress (e.g., temperature or voltage) accelerates the normal wear-out failure mechanism rather than producing a new one.

Reliability Improvement: It may readily be shown that redundancy is not the panacea for all reliability problems; however, a simple reliability model can quantify the value of any proposed redundancy scheme. Two forms of implementation are possible active redundancy, in which two or more devices are connected so that the combination will operate until all fail, and standby redundancy, in which one or more identical devices are held inoperative until the working device fails [43].

For n devices in active redundancy, the unreliability at time t for the combination $q_a(t)$ is related to the unreliability of one device $q(t)$ by the relation $q_a(t) = q^n(t)$.

The relation between these quantities for standby redundancy is more complex, but for most devices which exhibit log normal failure characteristics, a simple "rule-of-thumb" exists

$$q_s(nt) < q(t) \qquad (\text{for } q(t) < 0.5).$$

TABLE II
COMPONENT RELIABILITY TARGETS

Component	Number	Expected Failures in 25 Years (per device)	Equivalent FIT's (per device)
LASER	1,050	.0001	4.0
PIN or FET	3,150	.00008	0.4
IC	5,250	.00005	0.2

It has been suggested [52] that standby redundancy may be useful when there is insufficient time to prove the reliability of a novel device, e.g., a laser with an advanced structure.

Component Screening: Submarine-quality components are subjected to a screening comprising mechanical overstress tests, measurement of the major device parameters, and a burn in. For the new components, screening tests must be designed on the basis of the results of the large-scale life tests performed prior to the approval of the device family. In addition, the measurement of extra noncritical parameters may reveal "rouges" which might otherwise appear acceptable (for example, supply current analysis as a screen for integrated circuits [44]).

Transmitters

For monomode submarine systems, the following characteristics are desirable in a laser:

- low threshold current (preferably < 100 mA at 20°C),
- high available power in zero-order transverse mode (say > 10 mW/facet),
- narrow line width at modulation rates of several hundred MBd (particularly important for 1.55 μm systems),
- operation up to heat-sink temperatures of about 40°C (requiring as high as possible a value of T_0), and
- high reliability (associated with as small as possible an increase in threshold over 25 years).

GaInAsP/InP lasers have been available for approximately three and a half years. Early devices were double-heterostructure stripe contact lasers with threshold currents of 140–200 mA and maximum zero-order transverse mode powers of less than 3 mW. More advanced structures have now been produced, having better lateral confinement of light and current, as well as strong vertical confinement provided by the double heterostructure. The current density required for efficient lasing operation can thus be obtained with lower threshold currents of 10–80 mA. Zero-order mode power has also been increased to over 10 mW because of real guiding in the plane of the junction. The transient characteristics of these lasers can be adequate for NRZ binary digital modulation at over 1 Gbit/s. Structures typical of those reported include buried crescent [45], inverted rib [46], and planar buried heterostructure [47].

For high bit rate 1.55 μm systems, single-longitudinal mode operation is likely to be necessary to reduce the dispersion over long lengths. Possible approaches include the use of more complex laser structures, such as distributed feedback [48] or Bragg reflector lasers, or the use of external cavities or injection-locking to constrain the lasing mode.

Figure 2 shows the light/current characteristics of a typical advanced structure 1.3 μm laser. As threshold currents are temperature dependant and may increase with time, a control system is required to adjust the bias current such that the laser is always modulated from near threshold. Modulation from too far away from the threshold point results in either a seriously distorted signal or a poor extinction ratio and possible laser damage. Several control schemes have been proposed [49], [50], some of which also compensate for changes in laser slope efficiency. For submarine systems, the simplest practicable control system is to be preferred.

Since the first 0.85 μm lasers were reported, several degradation mechanisms have been identified, time to failure has been increased by two orders of magnitude and the rate of

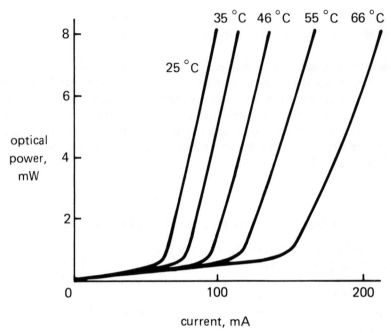

Fig. 2.   Typical characteristic of advanced structure 1.3 μm laser.

increase of threshold current reduced to below 1 percent/1000 h [51]. Although there is far less reliability information available for long wavelength lasers, several laboratories have reported that these show much better performance. However, the lack of quantitative evidence has provoked an interest in the possibility of introducing laser standby redundancy into the transmitter. The optical coupling between several lasers and one monomode fiber introduces penalties of its own, but may be achieved by means of an optica switch [52] or a passive coupler [53].

A major component in the design of a transmitter package is the technique for launching light into a monomode fiber. Three aspects need to be addressed: the launch optics to optimize the power coupled into the fiber, the alignment and fixing technology, and the effect on the laser of optical power coupled back into the cavity.

The launch optics must attempt to match the planar waveguide of the laser to the cylindrical fiber, the critical parameters being the dimensions of the laser guide or the beam waists in orthogonal directions, and the V-value of the fiber. Saruwatari and Nawata [54] have given a detailed theoretical analysis which compares well to available experimental results. With a butt launch, 10–15 percent efficiency may be achieved. A cylindrical lens as a separate element, or a hemicylindrical lens formed on the end of the fiber, improves this to 30–50 percent. Further improvement requires more complex optics providing beam expansion in one plane without astigmatism. Alignment tolerances for a 1 dB reduction in launched power are typically ±2 μm with a butt launch, and ±1 μm or less with lenses. Such alignment accuracy is best achieved with active positioning, i.e., with the laser emitting power into the fiber. Fixing technology using solders is available and other techniques are being developed.

Optical power coupled back into the laser cavity from any perturbations of the fiber, such as joints, can alter the threshold current and produce wavelength instabilities and wideband noise. The magnitude of these problems depends on the laser cavity Q and on the coupling efficiency. The resulting penalty has been shown to be small even under worst conditions at 140 Mbits/s [55], but at higher data rates it may become more significant. Reducing the laser-to-fiber coupling, or using optical isolators, are possible solutions [56].

Receivers

For submarine systems, reliability and, hence, simplicity become important factors in receiver design, along with the requirements of sensitivity, dynamic range, and bandwidth. Receivers for operation at 1.3 μm or 1.55 μm may use either p-i-n photodiodes or avalanche photodiodes (APD's).

In an APD receiver, the photodiode has internal current gain so the signal level at the input of the first amplifier is higher, for a comparable optical input, than with the p-i-n photodiode. Hence, amplifier noise is less significant, allowing a simpler amplifier design to be used. Little increase in sensitivity can be achieved by reducing amplifier noise below a certain limit because noise sources in the APD dominate. Of these, the two major ones are shot noise from the multiplied leakage current and multiplication noise caused by the stochastic nature of the avalanche gain. At 1.3 μm and 1.55 μm the only established APD's are based on germanium and, while exhibiting good bandwidth, the leakage current (typically 250 μA at the operating bias voltage) and multiplication noise of these devices are high compared to silicon APD's which were used below 1 μm. In the search for improved performance, much effort has been expended on research into APD's made of III-V compounds, principally InGaAsP. To date, no high bit rate system results have been reported, although there is much data available on the devices themselves. Near the breakdown voltage, the reverse bias leakage current increases rapidly because of band-to-band tunnelling, which becomes the dominant noise contribution. In addition, the excess noise properties of III-V APD's do not compare favorably with silicon. It is possible that GaAlSb may be a better material [57]. Generally, low noise APD's are expected to be extremely critical devices both to fabricate and to operate near optimum sensitivity.

p-i-n diodes responding to 1.3 μm or 1.55 μm can have a relatively simple structure, and several laboratories have reported good results using GaInAs p-i-n diodes made either on an InP or on a graded GaAs substrate [58]. Most devices reported to date have a mesa etched structure which, because of its sensitivity to surface contamination, might show an increase in surface leakage with time. With the planar structures now under development the possibility of this effect is eliminated.

For p-i-n photodiodes fabricated on an InP substrate, the quantum efficiency can be increased by introducing the light through the substrate thus avoiding absorption in undepleted GaInAs.

The sensitivity of p-i-n receivers depends largely upon the noise of the first-stage amplifier and this can be minimized by using a high-impedance design [59]. Best sensitivity results from such a receiver have been obtained using a GaAs MESFET, as this design leaves the FET channel noise as the dominant noise source [58]. If a bipolar transistor is used as the first-stage device, the noise will be higher, and the overall sensitivity will not be as good as that of a p-i-n FET receiver. However, this penalty may be outweighed by advantages in staying with silicon technology rather than having to establish the reliability of GaAs FET's.

The simple bias requirement of p-i-n diodes (5–15 V) is an advantage in submarine systems where simple power supply arrangements are desirable. APD's require a carefully controlled bias voltage of between 25–35 V in order to compensate for the effect on responsivity and noise of changes in temperature. Both APD and p-i-n receivers are capable of providing adequate bandwidth for operation up to 640 Mbits/s and APD receivers have been used at over 1 Gbit/s [60].

Electrical Regenerator

The electrical regenerator for use in a first-generation optical system is similar, in principle, to those used for many years for PCM transmission on coaxial cable or metallic pairs. However, the practical implementation for high bit rate submarine systems differs markedly. The most significant difference stems from the reliability problems mentioned above. Since digital regenerators use far more active devices than analog repeaters, the number of discrete transistors needed for a digital optical system might be up to five or six times as high as that required for an analog coaxial system of similar capacity. This problem can be approached in the three following ways:

- use of discrete devices of improved reliability (but the cost of providing large numbers of devices screened to this higher level makes this unattractive),
- reduction of the circuit complexity (but this is not possible beyond a certain limit, with current technology, and so must wait for the development of "all optical" regenerators), and
- use of integrated circuits to condense the discrete transistor circuitry [43].

The third approach is being actively pursued [61], [62] and promises to reduce the number of active devices needed by nearly an order of magnitude. There are, however, several ways of implementing this step. One area of choice is the level of integration. A very high level of integration will result in the most marked reduction in the number of devices needed, but the benefits of this will partially offset by an increase in the testing time required for each chip, and possibly a lower yield. A second variable is the approach to the chip layout and circuit design. The optimum performance can be achieved by a full custom design for each circuit function, but several fairly lengthy iterations may be required before the design can be finalized. A partially committed mask-programmable or p-i-n programmable approach will show advantages in flexibility (and possibly cost) though at some penalty in performance. A third area of choice lies in the selection of a suitable high-speed process. A considerable program of work is required for the validation of a new process [63]. For first-generation systems, it may be possible to build on one of the silicon bipolar processes already established for coaxial submarine systems (see, for example, [64]), although the different requirements of digital circuits must be taken into account.

Although some work has been done on untimed, or hybrid systems [65], it is currently assumed that long-haul systems will employ retiming in each regenerator. The timing extraction may be performed either with an active circuit (e.g., a phase-locked loop or injection-locked oscillator) or with a "passive" resonant circuit. In the case of the latter, various types of resonators are available; at high bit rates these include surface-acoustic wave (SAW) devices, the technology of which is becoming well established and offers a fair degree of freedom in the characteristics which can be provided.

The interconnection of the various components could be achieved by means of printed-circuit boards, but hybrid technology is now well established and appears to be more appropriate for at least parts of the regenerator circuits. The introduction of laser trimming has allowed the required tolerances to be achieved on thick film resistors, and either thick- or thin-film technology is suitable in principle. However, to ensure reliability, specifications of minimum line widths and track separations, resistor aspect ratios, film resistivities, etc., must be established.

Housings

The designs of pressure housing used by the different manufacturers of analog coaxial submarine systems differ considerably in the choice of materials, technique of cable-anchorage, and high-voltage insulation configuration. For optical systems, the following new factors must be brought into the design requirements:

- mechanical protection of the fibers at the point of cable-termination,
- provision of a pressure-resistant bulkhead gland for fiber entry, and
- provision of a path of low thermal resistance from the regenerator circuits to the sea.

The last of these requirements assumes significance since the total power dissipated in an optical repeater housing may be more than four times greater than that dissipated in current designs of coaxial repeaters. If simple direct current series power feeding of the regenerator circuits is to be employed, then an assembly is required which exhibits the somewhat difficult combination of low thermal resistance but high electrical insulation resistance.

Future Developments

Two possible steps beyond the introduction of first-generation systems may be foreseen. In the short-to-medium term, the basic system configuration is likely to be maintained, but optimization may include: the introduction of 1.55 μm systems, the use of higher bit rates using more advanced Si or possibly GaAs technology, and the possible use of wavelength multiplexing where this offers cost benefits.

In the longer term, it may be possible to exploit the potential benefits of coherent (homodyne or heterodyne) detection, but this involves a significantly different optical configuration, and several important problems must first be solved. The linewidth of the semiconductor laser must be reduced to 50 kHz–1 MHz, depending on the modulation format used; the local oscillator laser must be made to track the signal in frequency; the polarization of the signal must be controlled; and a suitable optical mixer with phase-front matching devised. Laser linewidth down to 40 kHz has been reported [66] using a filter in an external cavity. Frequency tuning may be accomplished by temperature and drive current control. Polarization-maintaining fiber has been made [67], but so far the increased attenuation far outweighs the benefits sought by using coherent detection. Various optical mixers have been proposed, although at present all cause a few dB of attenuation. The advantages of coherent detection could be considerable: a 10–15 dB increase in receiver sensitivity and the possibility of using FSK or PSK modulation and, hence, the possibility of more efficient multiplexing.

Another type of coherent optical system employs injection-locked laser repeaters [68]. A narrow line width laser source (such as an injection-locked laser) is frequency modulated and its output is coupled into another laser after tens of km of fiber so as to produce FSK modulation within the locking bandwidth. Such a system may offer very much simplified repeaters (although considerable feedback control of each laser is necessary) in exchange for a reduction in repeater separation.

ACKNOWLEDGMENT

The author of this chapter wishes to thank all those who provided contributions for the text; in particular, D. Borley, J. Chick, I. Garrett, P. Jenkins, D. King, G. Lee, A. Mitchell, D. Monro, S. Taylor, and C. Todd, of British Telecom Research Department, Submarine Systems and Optical Communications Technology Divisions.

Thanks are also expressed to the Director of Research of British Telecom for permission to publish this Chapter.

REFERENCES

[1] I. Yamashita, Y. Negishi, M. Nunokawa, and H. Wakabayashi, "The application of optical fibres in submarine cable systems," *ITU J.*, Feb. 1980.

[2] G. E. Morse, "The development of a trawler resistant submarine cable," in *Proc. IEE Conf. Submarine Telecommun. Syst.*, London, England, pp. 82–85, Feb. 1982.

[3] T. Miya, Y. Teranuma, T. Hosaka, and T. Miyashita, "Ultimate low-loss single-mode fibre at 1.55 μm," *Electron. Lett.*, vol. 15, pp. 106–108, 1979.

[4] B. J. Ainslie, K. J. Beales, C. R. Day, and J. D. Rush, "The design and fabrication of monomode optical fiber," *IEEE J. Quantum Electron.*, vol. QE-18, Apr. 1982.

[5] K. I. White and B. P. Nelson, "Zero total dispersion in step-index monomode fibres at 1.30 and 1.55 μm," *Electron. Lett.*, vol. 15, pp. 396–397, 1979.

[6] T. Miya, K. Okamoto, Y. Ohmori, and Y. Sasaki, "Fabrication of low-dispersion single-mode fibers over a wide spectral range," *IEEE J. Quantum Electron.*, vol. QE-17, pp. 858–861, 1981.

[7] S. Hornung and M. H. Reeve, "Single-mode optical fibre microbending loss in a loose tube coating," *Electron. Lett.*, vol. 17, pp. 774–775, 1981.

[8] D. B. Payne and D. J. McCartney, *Proc. Int. Conf. Commun.*, vol. 2, Denver, CO, June 14–18, 1981.

[9] M. J. Adams, D. N. Payne, and C. M. Ragdale, "Birefringence in optical fibers with elliptical cross-section," *Electron. Lett.*, vol. 15, pp. 298–299, 1979.

[10] K. I. White, S. Hornung, J. V. Wright, B. P. Nelson, and M. C. Brierley, "Characterization of single-mode optical fibres," *Radio Electron. Eng.*, vol. 51, pp. 385–391, 1981.

[11] I. Garrett and C. J. Todd, "Components and systems for long wavelength monomode fibre transmission," *Opt. Quantum Electron.*, vol. 14, 1982.

[12] D. J. McCartney, D. B. Payne, and P. Healey, *Electron. Lett.*, vol. 18, pp. 82–84, 1982.

[13] S. Hornung and B. P. Nelson, to be published.

[14] C. A. Millar, "Direct method of determining equivalent-step-index profiles for monomode fibres," *Electron. Lett.*, vol. 17, pp. 458–460, 1981.

[15] C. A. Millar, to be published.

[16] B. A. Proctor, I. Whitney, and J. W. Johnson, "The strength of fused silica," in *Proc. Roya Soc.*, A297, 1967, pp. 534–557.

[17] T. J. Miller, A. C. Hart, W. J. Vroom, and M. J. Bowden, "Silicone and ethylene vinyl acetate-coated laser drawn silica fibres with tensile strengths > 3.5 GN/m^2 (500 kpsi) in > 3 km lengths," *Electron. Lett.*, vol. 14, pp. 603–605, 1978.

[18] H. Schonhorn *et al.*, "Epoxy acrylate-coated fused silica fibres with tensile strengths > 500 kpsi (3.5 GN/m^2) in 1 km gauge lengths," *Appl. Phys. Lett.*, vol. 29, pp. 712–714, 1976.

[19] D. A. Pinnow, J. A. Wysocki, and G. D. Robertson, "Hermetically sealed high strength, fibre optical waveguides," in *Proc. Int. Conf. Int. Opt. and Opt. Fiber Commun.*, Tokyo, Japan, 1977, p. 335.

[20] W. J. Duncan, P. W. France, and K. J. Beales, "Effect of service environment on proof-testing of optical fibres," in *Proc. 7th Euro. Conf. Opt. Commun.*, Copenhagen, Denmark, Sept. 1981, pp. 4.5-1–4.5-4.

[21] S. M. Wiederhorn, *Fracture Mechanics of Ceramics*, vol. 4. New York: Plenum, 1974, p. 549.

[22] R. Adams and P. W. McMillan, "Review: Static fatigue in glass," *J. Mater. Sci.*, vol. 12, pp. 643–657, 1977.

[23] J. E. Ritter, J. M. Sullivan, and K. Jakus, "Application of fracture mechanics theory to fatigue failure of optical glass fibres," *J. Appl. Phys.*, vol. 49, pp. 4779–4782, Sept. 1978.

[24] B. J. Ainslie, K. J. Beales, C. R. Day, and J. D. Rush, "Interplay of design parameters and fabrication conditions on the performance of monomode fibers made by MCVD," *IEEE J. Quantum Electron.*, vol. QE-17, pp. 854–857, June 1981.

[25] B. K. Tariyal, P. L. Narasimkam, and D. L. Myers, "Effect of clean environment on fiber strength and proof-test performance," in *Proc. 7th Euro. Conf. Opt. Commun.*, Copenhagen, Denmark, Sept. 1981, pp. 4.6-1–4.6-4.

[26] J. D. Helfinstine and R. D. Maurer, "Effect on flaw distribution on fatigue characterization," in *Proc. 6th Euro. Conf. Opt. Commun.*, York, England, Sept. 1980, pp. 117–120.

[27] A. G. Evans and S. M. Wiederhorn, "Proof-testing of ceramic materials—An analytical basis for failure prediction," *Int. J. Fracture*, vol. 10, pp. 379–392, 1974.

[28] R. D. Maurer, "Strength of optical waveguides," in *Dig. Topical Meet Opt. Fibre Trans. II*, Williamsburg, VA, 1977.

[29] T. T. Wang and H. M. Zupko, "Long-term mechanical behavior of optical fibers coated with a UV-curable epoxy acrylate," *J. Mater. Sci.*, vol. 13, pp. 2241–2248, 1978.

[30] A. G. Evans, "Slow crack growth in brittle materials under dynamic loading conditions," *Int. J. Fracture*, vol. 10, pp. 251–259, 1974.

[31] D. Kalish and B. K. Tariyal, "Static and dynamic fatigue of a polymer-coated fused silica optical fibers," *J. Amer. Ceram. Soc.*, vol. 16, pp. 518–523, 1978.

[32] J. E. Ritter, "Probability of fatigue failure in glass fibers," *Fiber Int. Opt.*, vol. 1, pp. 387–399, 1978.

[33] D. B. Payne and D. J. McCartney, in *Proc. Int. Conf. Commun.*, Denver, CO, June 1981, vol. 2, pp. 27.6.1–27.6.5.

[34] J. S. Leach *et al.*, "Low-loss splicing of a 62.4 km single-mode-fibre link," *Electron. Lett.*, vol. 18, pp. 697–698, 1982.

[35] C. A. Millar, "A measurement technique for optical fibre break angles," *Opt. Quantum Electron.*, vol. 13, 1981.

[36] D. Marr, private communication, 1981.

[37] J. T. Krause, C. R. Kurkjian, and U. C. Pack, "Tensile strengths > 4 GPa for lightguide fusion splices," *Electron. Lett.*, vol. 17, pp. 812–813, 1981.

[38] P. Worthington, "Design and manufacture of an optical fibre cable for submarine telecommunications systems," in *Proc. 6th Euro. Conf. Opt. Commun.*, York, England, 1980, pp. 347–350.

[39] N. Kojima *et al.*, "Sea trial of submarine optical fiber cable," *6th Euro. Conf. Opt. Commun.*, York, Sept. 1980, see under Post-Deadline Papers, no page nos.

[40] M. Washio *et al.*, "400 Mb/s submarine optical fibre cable transmission system field trial," in *Proc. 8th Euro. Conf. Opt. Commun.*, Cannes, France, Sept. 1982, pp. 472–477.

[41] R. F. Gleason, R. C. Mondello, B. W. Fellows, and A. Hadfield, "Design and manufacture of an experimental lightguide cable for undersea transmission systems," in *Proc. 27th Int. Wire Cable Symp.*, Nov. 1978, pp. 1299–1303.

[42] G. Le Noane and M. Lenior, "Submarine optical fibre cable development in France," presented at the *Int. Conf. Commun.*, Philadelphia, PA, June 1982.

[43] D. E. N. King, "Reliability engineering for submarine cables—A simple approach," in *Proc. IEE Conf. Sub. Tele. Syst.*, London, England, Feb. 1980, pp. 131–134.

[44] A. J. Melia, "Supply current analysis (SCAN) as a screen for bipolar integrated circuits," *Electron. Lett.*, vol. 14, pp. 434–436, 1978.

[45] W. J. Devlin *et al.*, "Low threshold channel substrate buried crescent InGaAsP lasers emitting at 1.54 μm," *Electron. Lett.*, vol. 17, pp. 651–653, 1981.

[46] S. E. H. Turley *et al.*, "Properties of inverted rib-waveguide lasers operating at 1.3 μm wavelength," *Electron. Lett.*, vol. 17, pp. 868–870, 1981.

[47] I. Mito *et al.*, "InGaAsP planar buried heterostructure laser diode (PBH-LD) with very low threshold current," *Electron. Lett.*, vol. 18, pp. 2–3, 1982.

[48] T. Matsuoka *et al.*, "CW operation at DFB-DH GaInAsP/InP lasers in 1.5 μm wavelength region," *Electron. Lett.*, vol. 18, pp. 27–28, 1982.

[49] D. W. Smith and T. G. Hodgkinson, "Laser level control for high bit rate optical fibre systems," presented at the 13th Int. Symp. Circuits Syst., Houston, TX, 1980.

[50] R. E. Epworth, "Sub-systems for high-speed optical links," in *Proc. 2nd Euro. Conf. Opt. Commun.*, Paris, France, 1976, pp. 377–382.

[51] A. R. Goodwin *et al.*, "Narrow stripe semiconductor laser for improved performance of optical communication systems," in *Proc. 5th Euro. Conf. Opt. Commun.*, Amsterdam, The Netherlands, Sept. 1979, pp. 4.3-1–4.3-4.

[52] C. D. Anderson, R. F. Gleason, P. T. Hutchison, and P. K. Runge, "An undersea communication system using fiberguide cables," *Proc. IEEE*, vol. 68, pp. 1299–1303, Oct. 1980.

[53] R. Kishimoto, "Optical coupler for laser redundancy system," *Electron. Lett.*, vol. 18, pp. 140–141, 1982.
[54] M. Saruwatari and K. Nawata, "Semiconductor laser to single-mode fiber coupler," *Appl. Opt.*, vol. 18, pp. 1847–1856, 1979.
[55] K. H. Cameron, P. J. Chidgey, K. R. Preston, D. W. Smith, and M. R. Matthews, "A laser transmitter module for monomode fibre transmission systems," presented at the IEE Colloq. Opt. Fibre Syst. Present and Future, London, England, May 29, 1981.
[56] T. Kanada and K. Nawata, "Injection laser characteristics due to reflected optical power," *IEEE J. Quantum Electron.*, vol. QE-15, pp. 559–565, 1979.
[57] O. Hildebrand, W. Knebrat, K. W. Benz, and M. H. Pilkuhn, "GaAlSb avalanche photodiodes: Resonant impact ionization with very high ratio of ionization coefficients," *IEEE J. Quantum Electron.*, vol. QE-17, pp. 284–288, 1981.
[58] R. C. Hooper *et al.*, "PIN-FET hybrid optical receivers for longer wavelength optical communications systems," in *Proc. 6th Euro. Conf. Opt. Commun.*, York, England, Sept. 1980, pp. 222–225.
[59] S. D. Personick, "Receiver design for digital fibre optic communications systems," *Bell Syst. Tech. J.*, vol. 52, pp. 843–886, 1973.
[60] J. J. Yamada, S. Machida, and T. Kimuro, "2 Gbit/s optical transmission experiments at 1.3 μm with 44 km single-mode fibre," *Electron. Lett.*, vol. 17, pp. 479–480, 1981.
[61] D. W. Faulkner and R. J. Hawkins, "A single-chip integrated circuit regenerator capable of operation in the range 2–320 Mbit/s," in *Proc. Euro. Solid States Circuits Conf.*, 1981, pp. 211–213.
[62] D. Ross, "Integrated circuits for high speed digital transmission on optical fiber," presented at the High Speed Dig. Technol. Conf., San Diego, CA, Jan. 1981.
[63] M. F. Holmes, "Active element in submerged repeaters: First quarter century," *Proc. IEE*, vol. 123, no. 10R, pp. 1081–1112, 1976.
[64] D. Baker, "High reliability transistors for submarine systems," in *Proc. Inst. Phys. Conf.*, Serial no. 40, 1978, pp. 87–105.
[65] J. J. O'Reilly and P. Cochrane, "Potential role of untimed repeaters in optical submarine systems," in *Proc. IEE Conf. Sub. Tele. Syst.*, Feb. 1980, pp. 165–169.
[66] S. Saito and Y. Yamamoto, "Direct observation of Lorentzian line-shape of semiconductor laser and linewidth reduction with external grating feedback," *Electron. Lett.*, vol. 17, pp. 325–327, 1981.
[67] T. Okoshi, "Single polarisation single mode optical fibers," *IEEE J. Quantum Electron.*, vol. QE-17, pp. 879–884, 1981.
[68] D. W. Smith and D. J. Malyon, "Experimental 1.51 μm monomode fibre link containing an injection-locked oscillator," *Electron. Lett.*, vol. 18, pp. 43–45, 1982.

3
An Undersea Communication System Using Fiberguide Cables

CLEO D. ANDERSON, ROBERT F. GLEASON,
PAUL T. HUTCHISON, AND PETER K. RUNGE

INTRODUCTION

This chapter describes a digital undersea lightwave cable system which is now in an exploratory development phase at Bell Laboratories. Since the first voice-quality transatlantic cable system was installed in 1956, the yearly growth in overseas traffic has averaged more than 22 percent, significantly more than the growth rate within the continental U.S. The first system, called SB, had a capacity of only 48 3-kHz channels,[1] but as technology improved, succeeding systems; SD in 1963,[2] SF in 1970, and SG in 1976 had capacities of about 140, 840, and 4200 3-kHz channels,[3] respectively. The increases in capacity were realized by using larger bandwidths which require larger coaxial cable and shorter repeater spacings. Many people feel that for analog frequency-division-multiplexed systems, the SG system[4] has about reached the limit of practicality. Others feel that a larger system with 16 000 3-kHz channels would be practical, even though the repeater spacing would be only about 5 km. An optical 2-level PCM system (called SL) described in this chapter will be the first of a Bell System series of new digital undersea systems. Future systems, because of improvements in technology, will eventually operate at much higher baud rates than SL and will also have increased capacity by wavelength division multiplexing and possibly multilevel transmission. Digital transmission is desirable because its signal-to-noise ratio requirements are increasing rapidly and intermixing voice, data and TV is accomplished with essentially no interaction.

For an optical system to be attractive to the Bell System, it must be cost competitive, must use devices that will be available with proven reliability in time to be used in a transatlantic system in the late 1980's, and must have the capacity at least three times that of the existing SG system. Studies have been made on a number of potential systems that would use laser transmitters, light-emitting-diode (LED) transmitters, multi-mode fiber, single-mode fiber, high-bit-rate modulation, low-bit-rate modulation, 0.89-μm wavelength,

---

[1] In this Chapter, this always means two-way channels.
[2] Overseas satellite service started in 1965.
[3] All values here are without circuit multipliers such as Time Assignment Speech Interpolation (TASI).
[4] The British NG2 system can provide 5500 3-kHz transatlantic channels.

1.3-μm wavelength, 1.6-μm wavelength, etc. The results of these studies showed the following points.

1) The system should operate with a minimum number of fibers, each carrying the highest practical bit rate. We think that something around 300 Mbit/s with 2-level PCM is suitable. In our analysis, we use the Bell System standard of 274 Mbit/s[5] which can provide 4032 64-kbit/s channels with no speech processing such as digital TASI.
2) With the high bit rate, the injection laser is the only suitable optical source.
3) A single-mode fiber operating at 1.3 μm provides long repeater spacing and the low dispersion reduces requirements on the laser's spectral purity. It does, however, require exacting methods of fiber splicing. We consider 1.6 μm as too risky because reliable lasers at this wavelength are not yet available.
4) From the standpoints of cost and reliability, nearly all circuits must be integrated. Selection and qualification will be the only way to realize the required fractional FIT[6] rates and this is very expensive when many components are required.
5) If present-day technology is at all indicative of things to come, the reliability of lasers used in systems in the late 1980's will require the use of cold spares. Even with spare lasers, we will require a mean time to failure of over 1-million h.
6) Speech processing (digital TASI) should be used. In our cost comparisons, we assumed a very conservative circuit multiplication of three for the digital TASI.

Thus the proposed SL system will be made up of subsystems each of which has 274-Mbit/s bit streams transmitted from 1.3-μm lasers over single-mode fibers. An SL system can be made up of one, two or three subsystems depending on load requirements. Our cost estimates show that an SL transatlantic system with two subsystems (8000 64-kbit/s channels or 24000 TASIed voice circuits) would cost only 40 percent as much as an analog system with the same number of TASIed voice circuits. Since so much of the system cost is associated with the nonfiber part of cable, the fixed terminal equipment, and the system's installation, an SL system with three subsystems will cost only about 12 percent more than one with two subsystems.

Since the subsystems are electrically independent, except for a shared power supply, an SL system can be made with one or both ends split so that different subsystems go to different countries. This is very important in Europe because it avoids transfer charges for traffic from say France routed through the U.K. The end links will be made with a full complement of fibers and regenerators to provide flexibility, restorability, and additional communication between the end-link countries.

CABLE

The undersea lightguide cable design is based in part on technology developed for analog coaxial undersea cables. A cross-sectional view of the experimental cable design is shown in Fig. 1. In essence, it consists of a power cable with a fiber-containing core placed at the center.

The cable has been designed to provide the necessary strength and weight for ocean installation and recovery operations. If it is necessary to provide additional abrasion

---

[5] 274.176 Mbit/s is exact value.
[6] FIT means failures in 10^9 h.

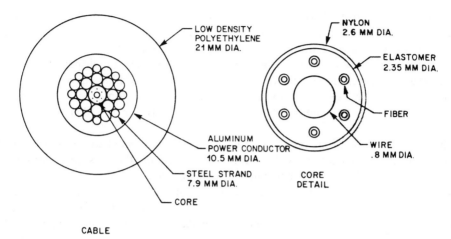

Fig. 1.   Undersea lightguide cable.

protection for the cable, a 2-mm-thick jacket of high density polyethylene, not shown in Fig. 1, can be included. For shallow water use, the cable will be armored with one or two layers of jute- or polymer-bedded steel armor wires.

The thickness of the polyethylene insulation was chosen to give an electric field strength of no more than 3.0 kV/mm in the insulation at the expected maximum terminal power supply voltages. The power conductor dc resistance is approximately 0.65 Ω/km. This value was chosen on the basis of both economic and system performance considerations. The ocean provides the dc return path for the power.

The power conductor is formed from a continuous aluminum tube which also acts as a barrier to water diffusing through the insulation. This prevents buildup of water in the vicinity of the fibers which, if permitted, could lead to static fatigue failure of the fibers.

The steel-strand wires are carefully arranged and tightly toleranced to provide a compact structure which isolates the fiber core from the sea pressure. In one experiment, a one-meter length of cable in which the fiber core could be slid axially within the steel strand by hand was subjected to 75 MPa pressure.[7] After two months at this pressure, it was still possible to slide the fiber core by hand. In addition, experiments conducted in the new Ocean Simulating Facility at Bell Labs in Holmdel, NJ, showed no change in fiber attenuation at 63 MPa pressure for short times and at 48 MPa (average deep water pressure for a transatlantic system) for three months.

The fiber core is placed at the center of the cable to minimize fiber strain due to cable bending and twisting as well as to provide pressure protection. The details of the fiber core construction are shown in Fig. 1. The king wire provides a tension member for handling the fiber core through the manufacturing line. In addition, in the completed cable, the king wire and the power conductor from a coaxial line which will be used in the supervisory system discussed in the section "Supervisory System" on page 45.

The fibers are encased in an elastomer which serves as a buffer to limit microbending as a result of cable manufacturing operations and subsequent handling. The fibers are stranded with the same lay length as that of the steel strand, 230 mm, to minimize bending of the fibers due to the slight deformation of the fiber core by the strand wires which occurs during manufacture.

---

[7] MPa is megapascals; 1 MPa = 145.04 lbs/in^2.

The elastomer is covered with a thin jacket of nylon to maintain structural integrity when the elastomer softens due to the heat of subsequent cable manufacturing operations and of molding operations required to make a splice in the cable.

Single-mode fibers were chosen to take advantage of the very low dispersion attainable at operating wavelengths near 1.3 μm. In addition, single-mode fibers have been found to exhibit significantly less sensitivity to microbending than do mutlimode fibers. Our goal is to have at 1.3 μm an attenuation of no more than 1.0 dB/km for fiber in an installed cable.

Although repair operations on an installed system are expected to be infrequent, it is necessary to design the cable to withstand recovery tensions. Under such conditions, the cable can be strained up to 1.0 percent. In this cable design, the strain in the fibers is essentially the same as that in the cable. As a consequence, we must develop fibers with more than 1 percent strain capability (approximately 750-MPa tensile strength) in continuous lengths of 35 km which meet the attenuation and dispersion requirements of the system and can be produced in large quantity at reasonable cost. This is one of the most challenging aspects of this project.

Samples of the cable as long as 1 km have been manufactured by Simplex Wire and Cable Company in Newington, NH. Fibers designed for and tested at 0.86-μm wavelength[8] that have an average uncabled loss of about 4 dB/km show an average loss of 4.2 dB in a cable. We believe that correspondingly low cabling loss will be observed at 1.3 μm, but data are not yet available.

Results of temperature tests between 3°C and 20°C (typical deep sea temperature) and of tension tests up to 1-percent strain are encouraging for tests at 0.86 μm. There is enough variability in the results and uncertainty in extrapolations, that confidence in the design cannot be established until extensive testing at 1.3 μm is completed.

REPEATERS

A repeater is made up of the housing that protects the electronic and fibers from sea pressure, optical high-speed regenerators, supervisory circuits, and power circuits. Power-dissipation and space limitations in the high-pressure housing (similar to the one used in the SG system) limit to six the maximum number of one-way optical regenerators in a repeater. Thus the SL system can be equipped to handle one, two or three 274-Mbit/s subsystems.

Optical High-Speed Regenerators

Figure 2 shows a block diagram of the experimental undersea optical regenerator. The incoming single-mode fiber couples to a p-i-n detector which is in intimate contact with a silicon integrated circuit preamplifier. Although an avalanche photodiode (APD) might have a sensitivity advantage of between 3 and 6 dB over a p-i-n detector, the current planning still prefers the p-i-n. Currently available avalanche photodiodes for the 1.3- to 1.6-μm wavelength range do not exhibit sufficient reliability to be committed for undersea cable use. In addition, these diodes require high-voltage power supplies and temperature control circuitry which also have to meet very stringent reliability requirements. From a systems point of view, an optimized p-i-n detector with high reliability, therefore, is preferred over an APD.

[8]1.3-μm test equipment was not available at the time the experiments were performed.

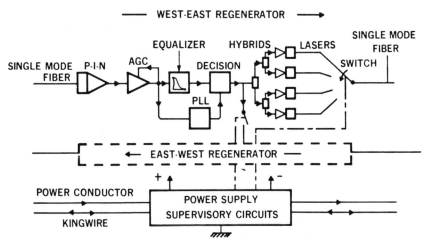

Fig. 2. Exploratory SL regenerator.

The output signal of the receiver is amplified in an AGC amplifier, then filtered and fed into a decision circuit. The timing information is recovered by a phase-locked-loop circuit. The output of the decision circuit is split four ways through hybrids to drive four lasers, one hot and three cold standbys. The laser output is coupled to the outgoing single-mode fiber through a single-mode fiber-optic switch. The switch in turn is controlled through a supervisory circuit from either shore terminal. The supervisory system is described in detail in the next section.

The transatlantic SL system should have an expected life of 24 years and a reliability of 8 years mean time before failure. The undersea portion of the system should during its entire system life require no more than three repairs involving the use of a cable ship. This is a formidable reliability requirement which calls for careful reliability engineering of this system. As a consequence the level of integration of the regenerator should be as high as possible. It is expected that all high-speed functions can be combined on five integrated circuits total, corresponding to the building blocks indicated in Fig. 2. The reliability of these integrated circuits should be in the order of 0.5 FIT's each, a degree of reliability that can only be achieved through extremely careful manufacturing procedures and controls.

Given the current level of confidence in the reliability of the integrated circuits and the uncertainty in the reliability of injection lasers, two of the three system failures mentioned above were allocated to laser failures leaving one failure for the remainder of the system. Even with that assumption, up to four laser diodes per regenerator might still be required because of potential reliability problems with injection lasers.

One possible implementation of the required sparing scheme for laser diodes is a single-mode fiber-optic switch as indicated in Fig. 3. The basic switch concept is the same as that suggested by Kummer et al. [1], for multimode fibers. One way to operate this switch is to apply force near the center and perpendicular to the longitudinal axis. Since the ends of the switch are fixed, the bending causes the movable output fiber to move to one of the corners of the square glass tube. An initial experiment conducted with such a switch built for single-mode fibers indicates that insertion losses of less than 2 dB are possible. We expect to be able to develop switches based on this principle which will have sufficient reliability for the undersea cable environment and provide the needed re-

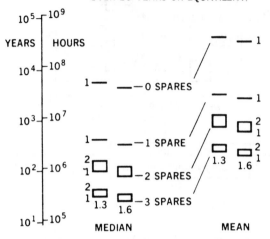

1. LOG-NORMAL DISTRIBUTION ($\sigma = 2$)
2. PLUS INFANT MORTALITY (1%/YEAR DECREASING LINEARLY OVER 20 YEARS OR EQUIVALENT)

TIME TO FAILURE

Fig. 3.   Laser sparing.

dundancy for injection lasers. In addition to the mechanical fiber-on-fiber switch, other fiber switches are being studied. One possibility is a laser sparing scheme based on electro-optic switches having no movable parts.

Figure 4 shows the required laser reliability for injection lasers in a transatlantic system with 180 repeaters (35-km spacing). Assumed is a log normal failure distribution for lasers with a spread factor of 2. In addition, an infant mortality is superimposed starting at a rate of 1 percent per year and decreasing linearly over a period of 20 years. Shown are the required median time to failure (time at which 50 percent of the devices have failed) and the required mean time to failure (MTTF). With three spares, the minimum required reliability, therefore, is in the order of 2×10^6 hours MTTF for 1.3-μm lasers and slightly

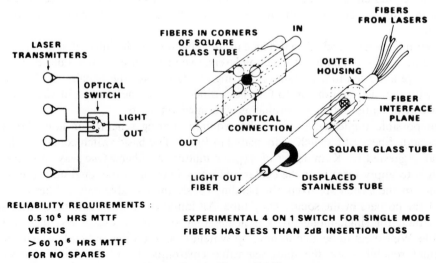

Fig. 4.   Required laser reliability.

less for 1.6-μm lasers because of the potentially longer repeater spacing at that wavelength. For fewer spares the required laser reliability increases rapidly.

Today's best lasers operating at 1.3 μm with a single transverse mode have expected reliabilities that are inferior to the minimum required reliability by about a factor of 2 to 5. However, the technology is progressing at a very rapid pace, and it is expected that the actual laser reliability will surpass the minimum required level within a very short time.

Supervisory Circuits

The supervisory circuits in a repeater are described in the section "Supervisory System."

Power Circuits

The present estimate of power per optical regenerator is about 4 to 4.5 W (about 300 mA at 14 V) and the supervisory circuitry will require about 2 W. Since each repeater contains up to 6 optical regenerators, there is considerable flexibility in the design of the biasing arrangement. Optical regenerators may be arranged in parallel, series, or power. Remember the dc resistance of the power conductor is about $0.65 \times 35 \approx 23 \ \Omega$ for a 35-km repeater spacing, so at 300 mA per regenerator, the minimum cable loss is 2 W/repeater span. A decision on regenerator arrangement has not been made, but it will probably be series parallel.

SUPERVISORY SYSTEM

The proposed supervisory system will be capable of 1) localizing faults to within one repeater section, 2) switching spare lasers into operation upon failure or degradation of an in-service laser, 3) transposing signal paths, within one direction of transmission, to minimize lost service due to multiple faults, and 4) monitoring laser bias and receiver AGC voltages.

Supervisory information will be conveyed between computer-controlled terminal equipment in the shore stations and each repeater by low-baud-rate electrical pulses transmitted over the coaxial cable described in the section "Cable." The relatively high resistance of the center conductor (king wire) and the high capacitance of the fiber-supporting insultant limits the baud rate to about 100 bit/s for the expected repeater spacing of 35 km. The proposed signal format is binary, asynchronous, unipolar, and pulsewidth modulated.

Each repeater contains a supervisory signal regenerator and associated circuits as indicated in Fig. 5. Supervisory signals (commands), transmitted from either terminal as a group of 24 pulses, are regenerated and retransmitted by each repeater. A pulse group is decoded as a string of binary numbers and stored in a shift register. Nine of the 24 bits identify the repeater being addressed and the remaining 15 specify the commands. If the address portion of the received pulse group corresponds to that permanently stored in the decoder, the "operation" portion of the command is executed and a confirming response in the form of another group of pulses is retransmitted towards the originating terminal.

Supervisory commands determine the configuration of the repeater, that is, which of the four lasers in each regenerator is in operation, which, if any, loopback connections for fault localization exist and which, if any, signal paths are transposed. The memory which stores the configuration state of a repeater should not be altered by loss of system power. Miniature magnetic latching relays, up to four per regenerator, are likely candidates for this function.

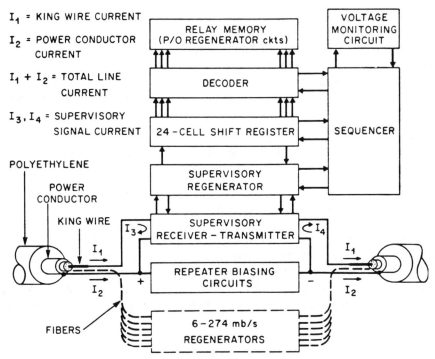

I_1 = KING WIRE CURRENT

I_2 = POWER CONDUCTOR CURRENT

$I_1 + I_2$ = TOTAL LINE CURRENT

I_3, I_4 = SUPERVISORY SIGNAL CURRENT

Fig. 5.   Block diagram—Repeater supervisory circuit.

The response to a command to monitor either the AGC or laser bias current in a particular regenerator will be in the form of a string of "zero" pulses appended to the normal 24-pulse response. The number of appended pulses will be the analog of the monitored voltage or current. By periodically monitoring and recording these measurements, trends may be established which could, for example, indicate rate of laser degradation.

In the absence of supervisory signaling currents, there is no voltage between the king wire and the main power conductor. The longitudinal component of current supplies the biasing power to the repeaters. The return path is via the sea. This line current divides between the king wire and the power conductor inversely as their respective resistances. Transmitted supervisory signals, typically 3-V pulses, increase the center-conductor current and decrease the outer conductor current. The supervisory receiver senses only this transverse component of the total current.

Except for the receiver-transmitter circuit, the supervisory circuits will be completely integrated in CMOS or I^2L. Because this circuit must compensate for loss of the cable span, which will probably be one of the last system parameters to be fixed, it will probably contain several discrete components in the equalizing networks.

TERMINAL EQUIPMENT

The repeaters of undersea cable system are powered in series over the power conductor by dc power plants at the terminals. On long systems, both terminals apply voltage, one positive and the other negative, with respect to the common sea return. The supplies furnish a constant current to the system; terminal voltages may be as high as 7 or 8 kV.

The undersea system will be connected to the land-line extensions by standard digital multiplexing equipment. The line bit rate has tentatively been selected to match the standard Bell System T4 rate,[9] 274 Mbit/s. Because the digital standards of many countries differ from those in the U.S., the interconnection will probably be made at the channel level where there is a common 64-kbit/s rate. Broad-band interconnections, such as TV, will require special code translating equipment.

System Assumptions

The following parameters have been assumed for the optical system:

1) Average optical power coupled into a single-mode fiber, between 0 and −3 dBm. Single-transverse-mode lasers will have built-in waveguide structures for mode confinement and shaping, and high coupling efficiencies to single-mode fibers are expected.
2) Spliced and cabled fiber loss, between 0.7 and 1 dB/km.
3) Insertion loss of the optical switch or order device, between 1 and 2 dB.
4) System margin, between 3 and 5 dB.
5) Receiver sensitivity, between −37 and −42 dBm for nonavalanching photodiodes.

Taking the extreme values of all assumed parameters, a possible range of repeater spacing will be between 27 and 54 km. For initial system evaluations, we have assumed a nominal 35-km repeater spacing.

Status of Development

Bell Laboratories is actively engaged in the exploratory development of an optical undersea cable system. Major areas of activities are high-reliability, single-mode 1.3-μm lasers and corresponding optical receivers; high-speed and high-reliability medium-scale silicon integrated circuits; low-loss and high-strength single-mode fibers, exploratory undersea cable and exploratory undersea repeaters.

In each of these areas, difficult technological problems need to be solved before a specific system should be committed to final development. However, the fiber-optic technology has been progressing with an ever increasing speed over the last decade and spectacular results have been achieved. The potentially very large economic payoffs are expected to accelerate the development progress even more.

We are convinced that the technological problems in the above-mentioned areas are severe, but surmountable. We therefore, seriously expect to be able to commission an optical transatlantic undersea cable system in the present decade.

Reference

[1] R. B. Krummer, S. C. Mettler, and C. M. Miller, "A mechanically operated four-way optical fiber switch," *Opt. Commun. Conf.*, (Amsterdam, The Netherlands), Sept. 17–19, 1979.

[9] A bit rate of twice the European standard of 140 Mbit/s could also be used.

Part II
Undersea Lightwave Systems

With development efforts nearing completion around the world, undersea lightwave communication systems are nearly ready for transoceanic system deployment. This part, made up of eight chapters, describes in detail the various systems designed and manufactured in the United States, Great Britian, France, and Japan. Also deployment of short distance undersea lightwave systems in Belgium, Spain, France, and Japan are announced.

4
The SL Undersea Lightwave System

PETER K. RUNGE AND PATRICK R. TRISCHITTA

INTRODUCTION

The SL Undersea Lightwave System is a large capacity digital fiber-optic transmission system capable of spanning the world's largest oceans and seas [1]–[2]. The system is now being manufactured for use in telecommunications systems planned for the Atlantic and Pacific Oceans. In the Atlantic, the SL system's undersea lightwave technology will be proven-in on what will be the world's first deep-water repeatered undersea lightwave system to be installed. This SL short system to be installed in 1985, will link two islands of the Canary Island chain and after initially serving as an SL system experiment will provide commercial service for Compañia Telefonica Nacional de España [3]. This short system installation will be the test vehicle for the eighth transatlantic telephone cable (TAT-8) to be placed in service in 1988.

The SL system was selected by the TAT-8 co-owners to provide 87 percent of the TAT-8 system. In Fig. 1 the route of the TAT-8 system is shown with the SL system, spanning from Tuckerton, NJ, to an SL branching repeater [4] off the coast of Europe. STC and Submarcon will provide their NL2 [5][1] and S280 [6] systems, respectively, to complete TAT-8 into Widemouth, England, and to Penmarch, France.

Each system is warranted differently. The SL system is warranted for 10 years while the British and French systems are each warranted for 2 years. The dollar amounts depicted in Fig. 1 represent the value of the contract awarded to each supplier.

In the Pacific, SL is proposed for a U.S. domestic application linking Hawaii with the Western United States (HAW-4). It is proposed that SL along with the KDD OS-280M system [7] be used for the third transpacific cable system (TPC-3) linking Hawaii with Japan and Guam.

In this chapter, we will discuss how the risks associated with using relatively new lightwave technology were minimized to design a digital undersea communication system which excels in reliability, availability, and performance.

---

[1] The NL2 system differs from NL1 in order to adopt a common line rate and to ensure that all supervisory signaling systems are compatible with all three suppliers of TAT-8.

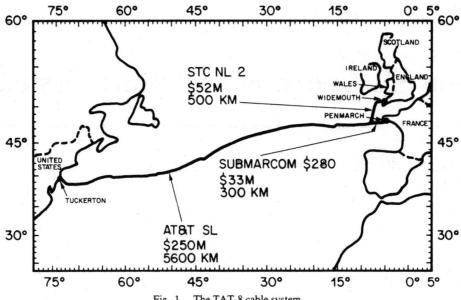

Fig. 1.   The TAT-8 cable system.

System Design and Capabilities

The SL Lightwave System uses proven fiber-optic technology to the maximum extent possible and includes the following features:

- transmission at 1.3 μm over depressed cladding single-mode fibers [8];
- a cable design with high strength fibers tightly coupled to the cable structure to provide high reliability in service and during cable recovery and repair [9];
- transmitters with buried double heterstructure lasers [10];
- ultrareliable silicon microwave integrated circuits [11]; individual section redundancy for both regenerators and fibers [12], [13];
- in-service monitoring of system performance [13], [14]; and
- proven mechanical design [15].

Each element of the SL system is designed to meet specific performance and reliability requirements. These combine to produce a system with substantial margins to allow for the uncertainties associated with applying fiber technology to the rigors of the undersea environment and cable ship operations for the first time. Fig. 2 shows how the system reliability requirement for a complete TAT system of less than three repairs by a cable ship in 25 years is divided between uncertainties and projected failure rates. The large margin prudently reserved for uncertainties is obtained by two key redundancy features: section-to-section span redundancy and additional laser transmitter redundancy in each regenerator. The high degree of reliability makes the SL system a very available system with a projected outage of much less than 115.6 min/year for a full length TAT system. This is shown in Fig. 3 which again indicates the large margin reserved for uncertainties.

The transmission capacity of SL is 557 Mbit/s. This is handled over two fiber pairs, each operating at the line rate of 295.6 Mbit/s using a 24B1P 24 information bits, 1 purity bit code. One fiber pair is standby for section redundancy. Although the SL system

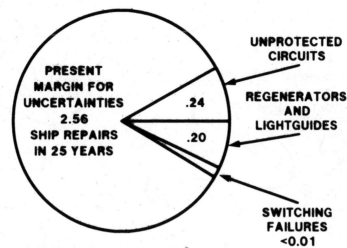

- **LASER MEDIAN LIFETIME 10^6 HOURS**
- **SECTION REDUNDANCY AND ADDITIONAL LASER TRANSMITTER REDUNDANCY THROUGHOUT**

Fig. 2. Reliability of SL system ⩽ repairs by a cable ship in 25 years.

is specified between CEPT4 (139.264 Mbit/s) interfaces, a total system capacity of 40 000 voice channels can be realized using ADPCM codecs and digital TASI equipment designed for use with SL [16].

Multiple use of branching repeaters [4] allows the SL system to be configured with multiple landing points on both sides of the ocean. For example, the TAT-8 cable system will use an SL branching repeater to have landing points both in England and France.

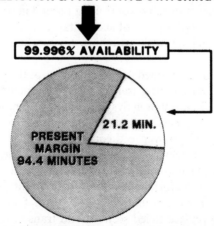

- **AUTOMATIC FAULT DETECTION, LOCATION, CORRECTION**
- **FAILURE PREDICTION & PREVENTIVE SWITCHING**

TAT-8 REQUIREMENT: 115.6 MINUTES/YR

Fig. 3. SL in-service performance monitor and control keeps low system voltage.

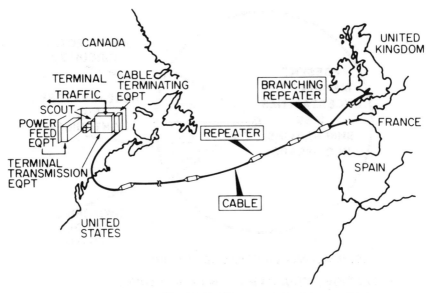

Fig. 4.   The SL system.

System Description

The SL system diagramed in Fig. 4, consists of the submersible equipment and the terminal equipment. The submersible equipment consists of the undersea lightwave cable, the undersea repeaters, and the undersea branching repeaters. The lightwave cable provides the transmission and supervisory paths for traffic and supervisory signaling via two working fiber pairs. One additional fiber pair provides the capability for standby redundancy in all repeater sections. The cable also conducts 1.6-A constant current for repeater powering. The undersea repeaters contain six optoelectronic regenerators to regenerate the optical signals, including two standby regenerators to provide regenerator span redundancy. Supervisory circuitry is contained in each repeater for monitoring the performance of each regenerator and controlling redundancy switching. The repeaters are protected against power surges in case of accidental cable cuts. The SL branching repeater provides lightguide branching with switching to permit routing of all traffic to any branch.

The terminal equipment diagramed in Fig. 5 consists of the cable terminating equipment, power feed equipment, terminating transmission equipment [14], and Surveillance and Control of the Undersea Transmission System (SCOUT). The cable terminating equipment terminates and couples the fibers and power conductor into the undersea cable structure and provides test access. The power feed equipment provides a constant current of 1.6 A to power a transoceanic system. This requires a high voltage supply of up to ± 7500 V with redundancy, overvoltage, overcurrent, and surge protection features.

SCOUT monitors error performance and critical circuit parameters in each undersea regenerator. SCOUT responds to undersea faults with fault recognition, fault location, and allows switching to replace failed sections and transmitters.

The terminating transmission equipment, diagramed in Fig. 6, multiplexes and demultiplexes between the 140 Mbit/s CEPT4 interface and the 295.6 Mbit/s line rates, adds

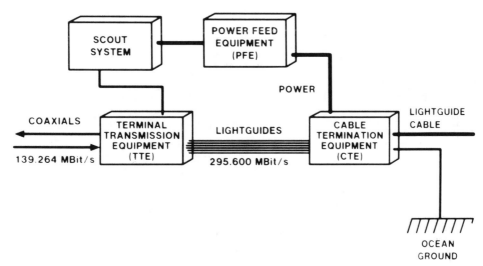

Fig. 5. The SL terminal.

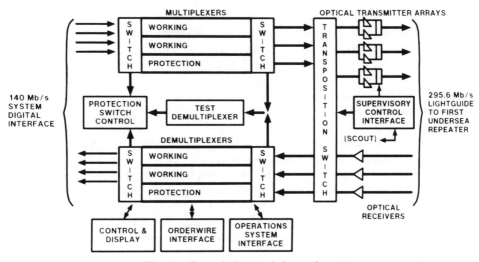

Fig. 6. SL terminal transmission equipment.

parity bits for supervisory signaling and repeater performance monitoring [13] and performs the electrical/optical conversions. In addition, the terminal monitors the end-to-end transmission performance of the system, includes both redundancy and automatic protection switching, and provides orderwire and interface to local and remote operations systems.

THE SL FIBER

The SL fiber is a depressed-cladding single-mode fiber with a refractive-index profile shown in Fig. 7(b) [8]. The depressed-cladding index profile allows more versatility and flexibility in designing the fiber's light propagation parameters than the common single-

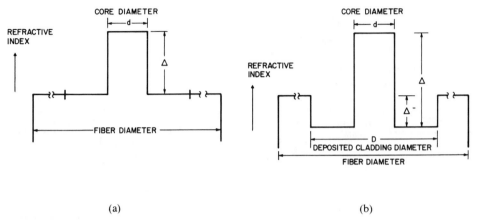

(a)   (b)

Fig. 7. Comparison of (a) common single-mode fibers with (b) depressed cladding single-mode fibers.

mode fiber profile shown in Fig. 7(a). In the common single-mode fiber, the fiber design parameters of mode field radius ω_0, zero dispersion wavelength λ_0, and high-order-mode cutoff wavelength λ_c, are all coupled and dependent on the core/cladding index difference Δ and the core diameter d. This results in a single-mode fiber design where these parameters must be traded off against one another. In the depressed-cladding single-mode fiber design each parameter can be optimized individually, producing a fiber which is optimum for transmission system applications. For the depressed cladding design ω_0 is still dependent on Δ and d but λ_0 is dependent on Δ, d, and Δ^-. This additional degree of freedom allows for system optimization of λ_0. λ_c is dependent on four parameters: Δ, d, Δ^-, and D/d which allows the design freedom for optimization of λ_c. A comparison of the depressed cladding and common cladding designs is given in Table I, which illustrates the design constraints of the common single-mode fiber profile.

The transmission requirements on the SL fiber for a transoceanic application are shown in the first column of Table II. In the second column are the measured results of the 217 km of fiber used for the SL deep-water sea trial [18]. The loss change due to hydrogen evolution and radiation were measured 2.5 years after manufacture. The sea trial results demonstrate that the SL fiber can now be produced which meets all undersea transmission performance requirements.

TABLE I
COMPARISON OF DEPRESSED CLADDING AND COMMON CLADDING SINGLE-MODE FIBER DESIGNS

	Zero Dispersion Wavelength (nm)	Mode Field Radius (μm)	Cutoff Wavelength (μm)
Depressed	1310	4.5	1250
Case 1 Common	1310	4.5	1470
Depressed	1310	4.5	1250
Case 2 Common	1310	5.5	1250
Depressed	1310	4.5	1250
Case 3 Common	1330	4.5	1250

TABLE II
SL FIBER REQUIREMENTS FOR THE 1290–1330-nm SPECTRAL WINDOW

	Present technology	Sea trial results
Average loss before cabling (Including splices) at 1300 nm	$\leqslant$ 0.48 dB/km	0.40 dB/km
Average increase in loss due to cabling at 1300 nm	$\leqslant$ 0.02 dB/km	0.008 dB/km
Average increase in loss due to environmental effects (Temperature, pressure, tension) at 1300 nm	$\leqslant$ 0.05 dB/km	0.005 dB/km
Average loss variation in spectral window	$\leqslant$ 5%	5%
Minimum dispersion wavelength	1310 + 10 nm	1311.5 + 2 nm
Dispersion over 1290–1330 nm	$\leqslant$ 2.8 ps/(km–nm)	< 2.0 ps/(km–nm)
Average loss change due to hydrogen evolution	$\leqslant$ .01 dB/km in 25 years	.002 dB/km in 25 years
Average loss change due to undersea radiation	$\leqslant$ .008 dB/km in 25 years	$\leqslant$ .007 dB/km in 25 years

TABLE III
SL FIBER EXPECTED SERVICE CONDITIONS

		Strain (%)	Time	Required Proof Test Strain, %
Installation		0.44	6 Hrs	0.89
Cable as installed		0.04	25 Yrs	0.14
Recovery 4 to 6m wave height 5.5 km depth	Lift	0.87	5 Hrs	1.94
	Hold	0.54	48 Hrs	1.35
	Relay	0.44	6 Hrs	0.89
Cable after recovery and relaying		0.24	25 Yrs	1.14
Repeater and splice box as installed radius of curvature 30 mm		0.21	25 Yrs	0.97

To commit optical fibers to the undersea environment requires fibers of high strength to allow recovery and laying of the fiber-optic cable under the worst sea conditions in 5.5-km depth and 4–6-m wave heights. This fiber reliability requirement is assured by a combination of static fatigue data, understanding of fracture mechanics theory, and knowledge of expected cable strains under service conditions. In Table III the expected service condition for various cable ship operations are shown along with the proof-test strain required to assure fiber survivability in that condition. Note that a deep-sea recovery operation puts the most strain on the fibers and this is responsible for a 2-percent strain proof-test requirement on all SL fibers.

The span lengths of SL are such that splicing of the single-mode fibers are necessary. The fiber splice must also meet the fiber strength and reliability requirements. This is accomplished by flame fusion splices [18] which are proof tested at 3-percent strain.

THE UNDERSEA CABLE

The SL undersea cable is designed to protect the single-mode fiber from excessive strain during laying and recovery, from the pressure of the deep ocean, and from external aggression [9]. A cross-sectional view of the embedded fiber core cable is shown in Fig. 8.

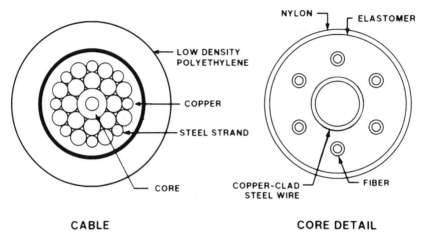

CABLE   CORE DETAIL

Fig. 8.   Embedded fiber core SL cables.

This cable structure consists of a central core 2.6 mm in diameter containing six fibers embedded in an elastomer and helically wound around a central steel wire. The elastomer is used as a cushion for the fibers and to reduce cabling induced microbending losses. The elastomer core is surrounded by a thin covering of nylon and then a series of steel strands which provide cable strength. Surrounding these strength members is a continuously welded copper cylinder which is a water and hydrogen diffusion barrier [19] and a power conductor for the undersea repeaters. Surrounding the conductor is a layer of low-density polyethylene providing cable insulation and abrasion resistance. The outer diameter for the completed deep-water cable as 21 mm. The small size of the fiber-optic cable results in material cost savings, and in easier deployment and storage aboard the cable ship.

The embedded fiber core cable design features low fiber curvature (1506-mm radius) and steel strands which act as a pressure vessel. Also the fibers and strands move together easing deep-water repairs. This cable design is such that the increased fiber loss due to cabling is less than 0.02 dB/km and the increased loss due to environment effects (temperature, pressure, tension, and hydrogen diffusion) are less than 0.05 dB/km. Fig. 9

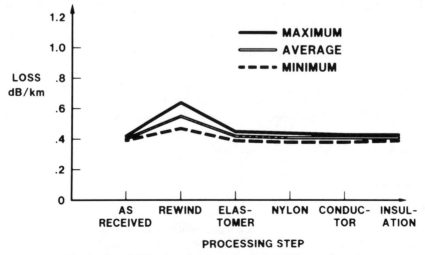

Fig. 9.   Sea trial fiber loss characteristics during cable manufacturing.

Fig. 10. An artist's rendering of an SL repeater being laid into the sea.

shows the fiber loss for a 19-km cable manufacturing run at each manufacturing step. The loss of the fibers increase when they are rewound at high tension on cable machine bobbins to start the cabling process but return to their original value after the elastomer is added. The results at $\lambda = 1.3$ μm show that the cable manufacturing process increased the loss of the fibers by only 0.009 dB/km. This 19-km cable was then used in the SL deep-water sea trial [20]. This cable was layed to an ocean depth of 5.5 km and despite an 18-h hold and two recovery operations, there was no damage to the cable and fibers and no increase in transmission loss. In fact, the combined environment effects of temperature, pressure and tension changed the fiber loss less than 0.005 dB/km during the trial. The SL deep-water sea trial successfully demonstrated the cable's features.

THE UNDERSEA REPEATER

The SL repeater shown in Fig. 10 is designed to protect the regenerator circuits from the ocean environment and securely mount them. It also provides bend restraint and connection to the cable. Mechanically the repeater features a low 5°C internal temperature rise to enhance component reliability and makes the maximum use of previous undersea technology. The repeater is evacuated and then back-filled with dry nitrogen. To lay the SL repeater requires no cable ship modification and provides a rugged cable to repeater coupling.

Electrically, the repeater is designed with high reliability by using highly integrated circuits [11] and redundancy in the form of a standby line and transmitter sparing (see Fig. 11) [21]. The repeater contains two regenerators operating at 295.6 Mbit/s in each direction with a standby 295.6/Mbit/s regenerator in each direction. Common supervisory circuitry and powering and surge protection circuits complete the repeater.

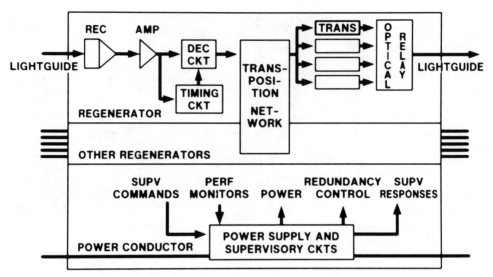

Fig. 11.   The SL repeater block diagram.

THE SL REGENERATOR

The regenerator design uses a high level of integration using proven silicon bipolar technology [11]. In Fig. 11 we show a block diagram of the SL regenerator [12]. The receiver contains an integrated silicon transimpedance amplifier for high reliability, high sensitivity, better than -34 dBm at 10^{-9} BER and a greater than 18-dB dynamic range [22]. The output of the receiver is coupled into the equalizer-AGC amplifier. This silicon IC with two added components equalizes the signal pulse shape to a raise cosine and maintains a constant output level over a 40-dB input voltage range.

One output of the dual output AGC amplifier goes into the retiming circuit. The retiming circuits contain a passive prefilter for spectral shaping at the half baud rate and an integrated squarer to produce the baud frequency. A highly reliable SAW filter [23] extracts the retiming component and an integrated amplifier with stable temperature characteristics produces a low jittered recovered clock.

The other output of the AGC amplifier goes to a single silicon decision circuit IC. The decision circuit produces regenerated data based on decisions made on the data at the optimum sampling time. The decision circuit also provides error monitor data to the supervisory system [13]. Also, as a system reliability feature, the decision circuit will output reshaped but not retimed data in the event of a clock recovery failure. This data will then be retimed at the next regenerator.

The regenerated output signal of the decision circuit is led through the transposition network into a transmitter array which contains up to four laser transmitters [10] connected to the output fiber by a 1×4 optical relay [21].

The transposition network permits the signal to be switched to the standby line, thereby bypassing a failed repeater section. This provides the SL system's major redundancy feature. The network itself consists of sealed contact electromagnetic latching relays with proven reliability through flawless service in previous undersea systems. These relays also permit signal loop-back for potential fault location purposes.

Supervisory System

In order to minimize the number of ship repairs the SL system must have an effective means of handling the rare faults which may occur and maximize service availability. The SL supervisory system provides the means of remotely controlling and monitoring the performance of the SL system [13] from the shore terminals.

The SL supervisory system remotely controls:

the switching of the optical relay to standby lasers; and the switching of a complete standby regenerator into a working line.

The SL supervisory system remotely monitors the performance of the SL system by

- measuring the laser bias current of each transmitter (this is an indication of change in the lasers);
- measuring the receiver AGC voltage of each regenerator provides a measure of the usable optical data (this power at each regenerator);
- counting the number of parity block errors over given time intervals giving a good indication of BER while in service; and
- indicating the states of the protective switching function.

Supervisory information to and from the repeaters is carried digitally by a group of baseband pulses which modulate the data streams in the following ways. To send a command to a repeater the baseband pulses modulate (ASK) a 13.2-kHz subcarrier, which in turn modulates a parity bit in every fourth multiplex frame [13]. To receive a response from a repeater the baseband pulses modulate (ASK) a 26.4-kHz subcarrier which in turn phase modulates the high-speed data. Both the sending and receiving of commands and responses are done in service with no degradation to the high-speed data signals.

The SL Transmission Engineering Plan—Bit Error Rate

The SL transmission engineering plan provides the method that translates end-to-end system transmission performance requirements on bit-error ratio (BER) and jitter accumulation to design objectives for individual components.

The TAT-8 repeater line operating at 295.6 Mbit/s must meet the following objectives for the 25-year life of the system:

- mean BER averaged over any 24-h period $< 4.4 \times 10^{-8}$;
- minutes with BER $> 10^{-6}$ averaged over any 30-day period $< 8.2/$day;
- seconds with BER $> 10^{-3}$ averaged over any 30-day period $< 2.56/$day;
- outage < 0.0022 percent of time averaged over 1 year.

In order to meet the system BER requirement of $< 4.4 \times 10^{-8}$ the individual spans are designed to have an end of life BER performance of better than 1×10^{-9}. This span requirement translates directly to a loss and dynamic range budget on a nominal deep-water span.

The output of the optical relay which is connected to up to four laser transmitters will be greater than -2.0 dBm. Presently the measured receiver sensitivity for a 10^{-9} BER on the first six production models of SL regenerators is better than -34.2 dBm with a receiver overload power of greater than -16.2 dBm. The regenerator will be operated at a nominal receiver power greater than -31.0 dBm. The resulting 29.0 dB power budget is allotted as follows: 1.0 dB for penalties including dispersion, optical feedback, timing error, and mode partition noise; 5.0 dB is budgeted for aging of system components. This leaves 3.0 dB for fiber loss. At 0.45-dB/km fiber loss, this results in nominal deep-sea span lengths in excess of 50.0 km.

THE SL TRANSMISSION ENGINEERING PLAN—JITTER ACCUMULATION

In a chain of regenerators the only quantity that accumulates along the chain is jitter. The SL jitter accumulation will be controlled by

1) making sure the RMS jitter will accumulate as the square root of the number of regenerators, not exponentially;
2) lowering the amount of jitter produced in each regenerator; and
3) making sure that every SL regenerator can tolerate with a < 0.1-dB BER penalty the maximum expected $P - P$ jitter.

Each issue is attacked separately.

Jitter will be accumulated roughly as the square root of the number of regenerators if there is no jitter peaking in the jitter transfer function of the retiming circuit. Jitter

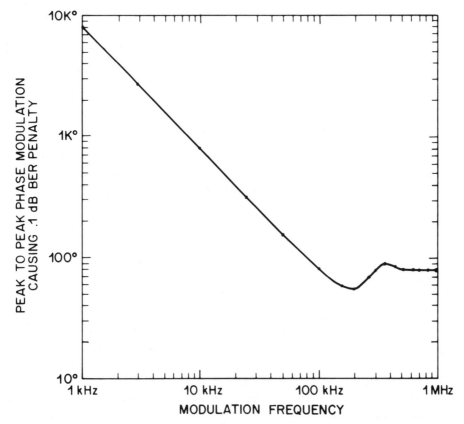

Fig. 12.   Measured jitter tolerance of typical SL regenerator.

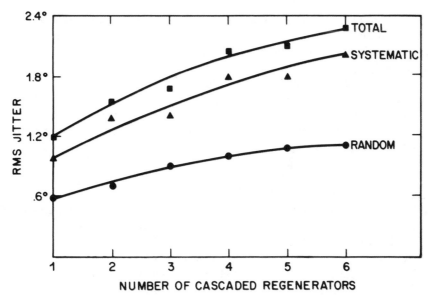

Fig. 13.   Jitter accumulation measurement made on a chin for SL regenerators.

peaking arises from passband ripple and mistuning of the SAW filter [23]. The jitter transfer function of the regenerator will be measured at 295.6 Mbit/s and ± 10 kHz from the line rate. By accurately measuring and controlling the amount of jitter peaking we are confident that the jitter will accumulate as $\sqrt{N}$.

Controlling the rate of accumulation does not limit the amount of jitter accumulation. This is done by careful design of the equalizer, prefilter, and retiming circuit of each regenerator [12]. Currently we are producing regenerators with jitter power spectral densities of less than 10 $\deg^2$/MHz. Even with this small amount of jitter generated at each regenerator and a $\sqrt{N}$ accumulation law, we must be sure that each regenerator is

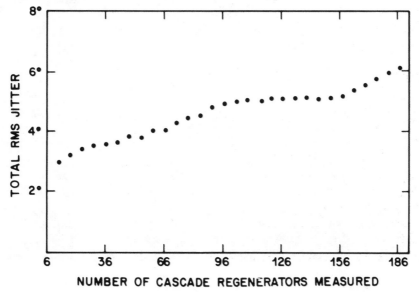

Fig. 14.   Circulating loop measurement using six SL regenerators and 217 km of SL fibers.

optimized to tolerate the maximum amount of jitter without an appreciable BER penalty. This is accomplished by proper phase adjustment between the clock and data and proper equalization to minimize ISI. Fig. 12 shows a typical jitter tolerance curve on an SL regenerator.

So far we have measured jitter accumulation on a chain of six production model SL regenerators. The RMS jitter accumulation is shown in Fig. 13. The total RMS jitter is increasing as $\sqrt{N}$ verifying the 0-dB peaking measurements. But a chain of six is a poor estimate for a transoceanic chain of 200 regenerators, therefore, a circulation loop measurement [2], [25] was performed on this chain of six regenerators to simulate a chain of 186 regenerators taken six regenerators at a time. Fig. 14 shows the $\sqrt{N}$ accumulation showing jitter is well controlled and understood on the SL system.

The jitter accumulation issue is complicated by the political solutions to the problems of providing an international undersea lightwave system when the terminal countries have all developed their own undersea lightwave systems. Politically the connections of two or three different systems at some undersea location provides a way to install a system without excluding a particular country's own system. But the technical problems are multiplied. In particular, the jitter accumulation requirements of the three systems which join together at the "wet connection" to form TAT-8 must be more closely defined and monitored than when a complete system is provided by only one supplier.

Fig. 15. Location of deep-water SL sea trial.

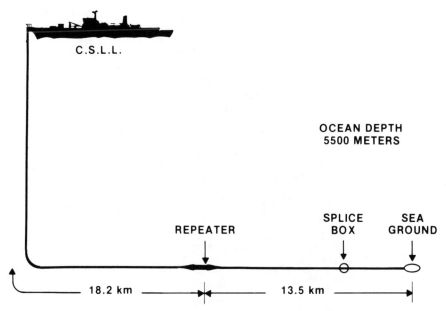

Fig. 16. Diagram of the SL deep-water sea trial.

THE SL SEA TRIAL

In order to demonstrate the ruggedness of the SL system elements and their readiness for underseas applications, a deep-sea trial of the SL Undersea Lightwave System was completed in September 1982. This deep-water sea trial was the first ever of any lightwave system. An 18-km length of the lightwave cable plus a repeater were laid in over 5.5-km deep water in the Atlantic Ocean 500-nm ENE from Bermuda (see Figs. 15 and 16).

The sequence of operations at the test site were

- survey of sea trial site;
- lay tail cable with splice box, repeater, and 10.4 km of active cable;
- hold in this configuration for 18 h;
- recover 10.4 km of cable;
- lay 17.5 km of active cable;
- hold in this configuration for 2 h; and
- recover active cable, repeater, and tail cable.

A photograph of the repeater breaking the surface after successfully laying and recovery is shown in Fig. 17.

The results of the Deep Water Sea Trial were the following:

18-km cable and repeater performed as expected during laying, holding, and recovery operations in 5.5-km deep water.

Error-free transmission of 274 Mbit/s and 420 Mbit/s was demonstrated during 1-h test periods in all phases of operations.

Maximum loss change due to temperature, tension, and pressure was less than 0.1 dB under all conditions.

The cable, repeater, and splice box were recovered in original working condition.

Fig. 17. The SL repeater breaking the surface after successful laying and recovery to a 5.5-km depth in the North Atlantic.

CONCLUSIONS

We have described the SL Undersea Lightwave System now in production for use in spanning the Atlantic. This system, when installed, will set a new milestone in the history of fiber-optic system design, introducing lightwave technology to the undersea.

REFERENCES

[1] C. D. Anderson, R. F. Gleason, P. T. Hutchison, and P. K. Runge, "An undersea communications system using fiberguide cable," this book, see ch. 3, p. 39.

[2] P. K. Runge and P. R. Trischitta, "The SL undersea lightguide system," *IEEE J. Selected Areas Commun.*, vol. SAC-1, no. 3, Apr. 1983.

[3] S. W. Dawson, J. Riera, and E. K. Stafford, "CTNE undersea lightwave inter-island system," this book, see ch. 8, p. 109.

[4] M. W. Perry, G. A. Reinold, and P. A. Yeisley, "Physical design of the SL branching repeater," this book, see ch. 25, p. 365.

[5] R. L. Williamson and M. Chown, "The NL1 submarine system," this book, see ch. 9, p. 119.

[6] P. Franco, J. P. Trezeguet, and J. Thiennot, "S280—A new submarine optical system," this book, see ch. 6, p. 83.

[7] Y. Niiro, "The OS-280M optical fiber submarine cable system," this book, see ch. 7, p. 95.

[8] S. R. Nagel, "Review of the depressed cladding fiber design and performance for the SL undersea system," this book.

[9] A. Adl, T. M. Chien, and T. C. Chu, "Design and testing of the SL cable," this book, see ch. 16, p. 233.

[10] F. Bosch, G. M. Palmer, C. D. Sallada, and C. B. Swan, "Compact 1.3 μm laser transmitter for the SL undersea lightwave system," this book, see ch. 31, p. 445.

[11] L. E. Miller, "Ultra high reliability, ultra high-speed silicon integrated circuits for undersea optical communications," this book, see ch. 30, p. 433.

[12] D. G. Ross, R. M. Paski, D. G. Ehrenburg, and G. M. Homsey, "A highly integrated regenerator for 295.6 Mbit/s undersea optical transmission," this book, see ch. 26, p. 377.

[13] C. D. Anderson and D. L. Keller, "The SL supervisory system," this book, see ch. 40, p. 567.

[14] J. L. Fromme and M. D. Tremblay, "The SL transmission terminal equipment," this book, see ch. 42, p. 589.

[15] M. W. Perry, G. A. Reinold, and P. A. Yeisley, "Physical design of the SL repeater," this book, see ch. 25, p. 365.

[16] W. R. Daumer and J. L. Sullivan, "Subjective quality of several 9.6–32 Kb/s speech coders," in *ICASSP'82* (Paris, France).

[17] P. K. Runge, "Deep-sea trial of an undersea lightwave system," in *Tech. Dig. Opt. Fiber Commun. Conf.* (New Orleans, LA), 1983, paper MD2, 8.

[18] J. T. Krause and C. R. Kurkjian, "Improved high strength flame fusion single mode splices," in *Tech. Dig. Fourth IOOC*, (Tokyo, Japan), 1983, paper 29A 4–6, pp. 96–97.

[19] E. W. Mies, D. L. Phelen, W. D. Reents, and D. A. Meade, "Hydrogen susceptibility studies pertaining to optical fiber cables," in *Tech. Dig. Opt. Fiber Commun. Conf.*, Postdeadine WI3.1–4, (New Orleans, LA), 1984.

[20] H. J. Schulte, "Transmission tests during the SL lightwave submarine cable system sea trial," in *Proc. SPIE*, vol. 425 (San Diego, CA), Aug. 1983.

[21] S. Kaufman, R. L. Reynolds, and C. Loeffler "Optical switch for the SL undersea lightwave system," this book.

[22] M. L. Snodgrass and R. Klinman, "A high reliability, high sensitivity lightwave receiver for undersea lightwave systems," this book, see ch. 34, p. 487.

[23] R. L. Rosenberg, C. Chamzas, and D. A. Fishman, "Timing recovery with SAW transversal filters in the regenerators of undersea long-haul fiber transmission systems," pp. 957–965, this issue. C. J. Byrne, B. J. Karafin, and D. B. Robinson, "Systematic jitter in a chain of digital regenerators," *Bell Syst. Tech. J.*, vol. 42, Nov. 1963.
E. Roza, "Analysis of phased-locked loop timing extraction circuits for pulse code transmission," *IEEE Trans. Commun.*, vol. COM 22, Sept. 1974.
J. M. Manley, "The generation and accumulation of timing noise in PCM systems," *Bell Syst. Tech. J.*, vol. 48, Mar. 1969.

[24] D. A. Fishman, R. L. Rosenberg, and C. Chamzas, "Analysis of jitter peaking effects in digital long-haul transmission systems using SAW-filter retiming," submitted to *IEEE Trans. Commun.*

[25] C. C. Chamzas and P. R. Trischitta, "Simulation of a chain of digital optoelectronic regenerators," *Modelling and Simulations Conf.* (Athens, Greece), June 27–29, 1984.

5
The FS-400M Submarine System

HIROSHI FUKINUKI, TAKESHI ITO, MEMBER, IEEE, MAMORU AIKI, AND YOSHIHIRO HAYASHI

INTRODUCTION

With the growing need for more advanced global telecommunication services, worldwide development of a repeatered submarine optical-fiber system is underway [1]–[4]. NTT is also developing a repeatered submarine optical-fiber system named the FS-400M system.

The FS-400M system is going to be applied to domestic use in Japan. As is well known, Japan is composed of four main islands and is surrounded by many small adjacent islands, which are densely populated. This requires the FS-400M system to have a high capacity. It does not necessarily have a transmission length as long as that of a transoceanic system. The FS-400M system, however, is based on technology applicable to a transoceanic system.

Development of the F-400M terrestrial long-haul trunk transmission system began in 1979. Two years later, development began on the FS-400M system. Therefore, the FS-400M system is being developed as a daughter system of the F-400M system. In the first stage of development, the repeater was composed mainly of discrete transistors and the submarine cable was developed only for shallow sea use. Using such equipment, a field trial including a long-term stability test was carried out from January 1982 to February 1984 in Sagami Bay off the coast of Yahatano on the Izu Peninsula [5]. It demonstrated the capability of a repeatered submarine optical system. Encouraged by the field trial's results, development of a final model commenced in 1982. Its repeater is composed of integrated circuits, and submarine cable that is applicable for deep-sea use. To confirm transmission characteristics in a deep-sea environment, a sea trial was carried out in the fall of 1984. A commercial test for the FS-400M system is also planned for 1986. Major events in the system's development thus far are summarized in Table I.

This chapter describes the major system parameters, and outlines the system design and repeater circuit configurations. In addition, transmission characteristics of the system are discussed including the results of the long-term stability test obtained in the field trial described above. Aspects of the submarine cable, repeater housing, and key devices for the FS-400M system are discussed in other chapters of this book.

TABLE I
NOTEWORTHY EVENTS IN DEVELOPMENT OF FS-400M SYSTEM

Cable Evaluation	June 1981 (copper pipe)
Dry System Evaluation	December 1981 (discrete Tr)
Field Trial	Jan. 1982 ~ Feb. 1984 (45 km, 2 repeaters)
Cable Evaluation	Oct. 1982 (aluminium pipe)
Dry System Evaluation	July 1983 (integrated repeater with redundant optical source)
Sea Trial	Fall 1984 (8000 m water depth)
Commercial Test (scheduled)	1986 ~

MAIN FEATURES

The FS-400M system uses the low-loss and low-dispersion characteristics of single-mode fibers in the 1.3-μm band to economically provide a high-quality highly reliable submarine digital transmission line. Its main parameters are shown in Table II.

Taking into account the traffic demands and its linkage with the F-400M system, the system capacity is 17 280 telephone channels. More specifically, the transmission capacity per fiber is 5760 channels which corresponds to an information bit rate of 397.200 Mb/s. By using the 10B1C code as a line code [6], the line bit rate increases to 445.837 Mb/s. The FS-400M system is easily connected to the F-400M system at a shore terminal, as

TABLE II
FS-400M SYSTEM PARAMETERS

Maximum Transmission Capacity		17 280 telephone channels (5760 × 3)
Information Bit Rate Line Bit Rate Line Format		397 200 Mb/s 445 837 Mb/s 10B1C code, RZ pulse
Transmission Length Design		1000 km
Repeater Spacing Design		40 km
Water Depth Design		8000 m
Allowable Error Rate Allowable Timing Jitter		10^{-9}/1000 km, 10^{-11}/repeater 15° rms/1000 km, 1° rms/repeater
Transmission Medium		Single-mode fiber
Wavelength		1.3 μm
Power Feed	Format	Parallel for two-way repeater and series for three systems
	Voltage	±1600 V
Line Supervision		10B1C code rule check (in service)
Fault Location		Error rate check and operating condition monitor by electrical loop back (out of service)
System Life Design System Reliability Design		25 years MTBF 10 years

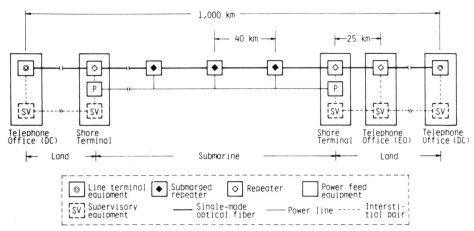

Fig. 1. Transmission-line configuration for the FS-400M system. The F-400M system is used on land.

shown in Fig. 1, and is terminated at a telephone office in rank of a district center. Power is supplied to each repeated by power feed equipment installed at both shore terminals. The power conductor is made of aluminum pipe containing steel strands and is included in a submarine cable.

The transmission length and applicable water depth are designed to be 1000 km and 8000 m, respectively. System reliability is designed to have an MTBF of ten years, taking into account system unavailability. From the viewpoint of economics, the use of highly reliable devices for a transoceanic system seems to be preferred to the exclusive use of the moderately reliable devices that are most suitable for a domestic system which is less demanding. Accordingly, the MTBF of the FS-400M system is actually expected to be longer than that of system life.

Because it gives the lowest loss value, the use of a 1.55-μm wavelength is preferable for the purpose of maximizing repeater spacing. But at 1.55 μm there is a repeater spacing limitation due to mode partition noise [7]. To avoid this, a single-frequency laser diode and/or a zero dispersion shifted fiber have to be used. Technology is not yet established for the use of the 1.55-μm wavelength. On the other hand, conventional components such a 1.3-μm laser diode or a 1.3-μm zero dispersion fiber have already proven their feasibility in the field on terrestrial transmission systems. Consequently, a 1.3-μm wavelength can be used.

It is forecast that the present worth of annual charges can be reduced by nearly 25 percent compared with those of the existing coaxial submarine system. This is a result of the reduction in repeater number due to longer repeater spacing.

SYSTEM DESIGN

Repeater Spacing

For a high-bit-rate transmission system, repeater spacing is limited by mode partition noise. This noise is caused by a conventional laser diode emitting several different wavelengths of light in a constant power series fluctuation. Repeater spacing is shown in Fig. 2, calculated on the assumption that there is an allocated line loss of 21.7 dB, a

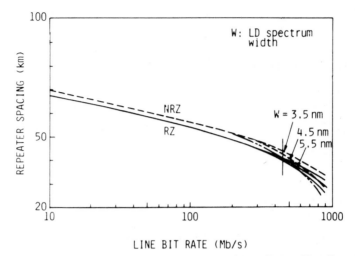

LINE BIT RATE (Mb/s)

Fig. 2. Calculated repeater spacing, assuming fiber loss of 0.45 dB/km, fiber dispersion of 3 ps/km·nm, repeater output of −8.4 dBm, and repeater receiving power of −33.1 dBm. *W* shows the spectrum width. Solid lines correspond to an RZ pulse system, and broken lines to an NRZ pulse system.

fiber loss of 0.45 dB/km, and a fiber chromatic dispersion of 3 ps/km·nm. As the peak power of a laser diode is limited to a given value in order to guarantee a long life, the average output power of a laser diode in an NRZ system is larger than that of the RZ system by 3 dB. At a lower bit rate, the NRZ pulse is preferable to the RZ pulse because of its average power benefit and that there is no power penalty due to mode partition noise. On the other hand, RZ pulse is preferable to NRZ pulse at a high bit rate, because an NRZ system is sensitive to mode partition noise. Therefore, the FS-400M system uses the RZ pulse. Repeater spacing is 40 km.

Level Diagram

Average output power and average received power of a repeater is −8.4 and −33.1 dBm, respectively, as shown in Fig. 3. The guaranteed peak power of a laser diode is 7 dBm. So, modulated light power becomes 1 dBm because of the duty ratio of 0.5 and the mark ratio of 0.5. It decreases to −8.4 dBm due to loss of the laser-diode-single-mode-fiber coupler, the polarization-sensitive optical coupler, and the feedthrough. The difference of 24.7 dB between −8.4 and −33.1 dBm is allocated to a line loss of 21.7 dB and system margin of 3 dB. Line loss is mainly from the fiber loss of 0.45 dB/km.

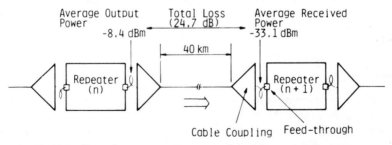

Fig. 3. Allocated average output power and receiving power of a repeater.

TABLE III
SENSITIVITY DESIGN

Ideal Sensitivity (dBm)		−42.1	
Sensitivity Degradation (dB)	Initial	6.1	
	Mode partition noise	1.5	9.0
	Aging	1.4	
Average Received Optical Power (dBm)		−33.1	

Sensitivity

Assuming a quantum efficiency of 100 percent, no dark current and an excess noise factor of 0.9 for the APD, sensitivity is calculated to be −42.1 dBm at 446 Mb/s. It is initially degraded by a quantum efficiency of less than 100 percent and by the existence of dark current for the APD, amplifier noise, and imperfections such as uncertainty of the decision level. Furthermore, it is degraded by the mode partition noise that accompanies a 40-km-long repeater spacing, and aging of the system which has a life expectancy of 25 years. Total of degradation is estimated to be 9 dB, as shown in Table III.

Reliability

Reliability of a fiber break for the system life of 25 years is ignored, because a fiber strength is guaranteed by a tensile proof test of 2.2-percent strain [8].

On the other hand, considering the quality of the present technology, laser diode reliability prospects are poor compared with those for components such as IC's, resistors, and so on. So, it is very effective to use a backup laser diode system [9]. A laser diode backup system brings about the moderation of requirement to a laser diode reliability. Also it reduces system cost compared with a full duplication system, which has backup circuits and fibers as well as a backup laser diode. For this reason, the optical sources of the FS-400M system are composed of an operating and a backup laser diode. In order to expect an MTBF of longer than ten years for the FS-400M system, a failure rate of less than 80 FIT's is required for a one-way repeater. It allows only 30 FIT's for a laser diode for a nonredundant system, but 300 FIT's for a laser diode backup system.

Supervision

A supervision system is very important to realize a highly reliable long-haul submarine repeatered system. Supervision is generally necessary for both line conditions and repeater hardware conditions. In the FS-400M system, line conditions are supervised while in use by a error-rate monitoring of a code rule at both end terminals [10]. When failure occurs, a line of the FS-400M system is automatically switched to an extra line via the other route. On the other hand, the FS-400M system has a standby laser diode, but not a standby repeater circuit nor a pair of spare fibers to fall back on, as previously described. When the drive current of an operating laser diode increases abnormally, it is automatically switched to the backup laser diode. Furthermore, it is clarified that transmission characteristics of a single-mode fiber transmission system are stable and an occasional failure has never been found in a long-term stability test for the F-400M and

TABLE IV
SUPERVISORY AND FAULT LOCATING SYSTEM

Loop Back Format		Main signal loop back at electrical stage
Configuration		
Power Feed		Parallel for two-way repeaters
Components		Monolithic Crystal Filter Monolithic ICs (SV_1, SV_2)
Monitoring and Control Function	Monitoring:	LD biasing current APD biasing voltage
	Control:	Repeater identification Loop back gate switching Monitor point selection Switching to standby LD (automatic, remote switching)
Repeater Identification Code		5 bit binary code

FS-400M systems. These fact render in use repeater hardware condition monitoring unnecessary. But a precise fault locating system is important.

Fault location is carried out by an error-rate counting system. If a fault occurs in a particular section, the signal is looped back at the preceding repeater circuit thus locating the fault in a particular section. An electrical loop-back procedure is provided, but not an optical loop-back procedure, because of hardware complications and the absence of the required mechanical switching. To match with an electrical loop back, power is fed in parallel for two-way repeaters. Its fault locating system is also capable of monitoring laser diode biasing currents and APD biasing voltages in order to distinguish a repeater circuit failure from a fiber break. This system also has other necessary control functions for achieving the loop-back fault locating and monitoring. Exclusive transmission media, receivers, and transmitters are unnecessary for the supervisory and fault locating system, because supervisory signals are transmitted over a pair of fibers on which information signals are also transmitted. These features are summarized in Table IV.

REPEATER CIRCUIT

Configuration

Figure 4 shows the repeater block diagram for the FS-400M system using developed monolithic IC's. Ge-APD and InGaAsP/InP laser diodes are used as an optical detector and an optical source, respectively. The LD wavelength is tuned to 1.31 μm at 10°C, because that is the deep-sea operating temperature for a repeater LD in a repeater. This is because chromatic dispersion of a conventional single-mode fiber is minimized at that wavelength. In addition, the LD spectrum width is limited to 3.5 nm. Considering the state of the art, main parameters shown in Table V are used.

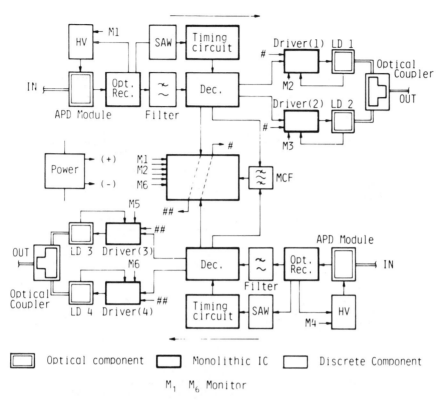

Fig. 4. Repeater circuit configuration. HV: High voltage circuit for APD biasing. SAW: Surface acoustic wave filter for timing extraction. Filter: 5th order Thomson low-pass filter for reshaping. MCF: Monolithic crystal filter for supervisory signal extraction. Monolithic IC's are referred to Table VI.

It is desirable that a repeater circuit be composed of one monolithic integrated circuit. At the present state of the art, however, this is especially difficult for a high-speed repeater. So, a high-speed repeater circuit is divided into the smallest possible number of MIC's based on technical requirements such as the power dissipation limit of 1 W, the avoidance of interferences, and so on. Considering the above requirements, six monolithic integrated circuits have been recently developed [11]. These are an optical receiver, a decider, a timing amplifier, a laser diode driver, and two types of supervisory circuits (SV_1, SV_2). Table VI shows the main characteristics, number of transistors, and their power dissipations. All IC's are fabricated by a shallow junction silicon bipolar process.

This repeater has a redundant optical source consisting of two LD modules with polarization holding fibers and a polarized optical coupler. The polarized optical coupler insertion loss is low, because there is no branching loss; it is 3 dB less than that of a conventional optical coupler. Automatic switching to the backup LD is performed electronically, because an electronic switch seems to be more reliable than a mechanical switch.

Supervisory Circuit

Figure 5 shows a block diagram of a fault locator at a shore terminal and a supervisory circuit in a submarine repeater [12]. The supervisory circuit consists of a monolithic crystal filter (MCF) and two monolithic IC's (SV_1, SV_2).

TABLE V
REPEATER PARAMETERS

Optical Transmitter	Average Output Optical Power	− 8.4 dBm
	Wavelength	1.30 ~ 1.32 μm
	Extinction Ratio	> 15 dB
	Optical Loss of Coupler	< 2 dB
	Spectrum Half Width	< 3.5 nm
Optical Receiver	Sensitivity	− 33.1 dBm
	Quantum Efficiency of APD	> 0.75
	Dark Current of APD	< 0.3 μA
	Pre-Amp Noise	< 10 pA/$\sqrt{\text{Hz}}$
System Margin		3.0 dB
LD Redundancy	ILD + Driving circuit (LD: cold standby)	
Power Consumption	6V, 0.9A for one-way repeater	
Failure Rate	80 Fit/one-way repeater	

TABLE VI
FAILURE RATE ALLOCATION

Component	Number (One Way Repeater)	Failure Rate (Fit)
Optical Components	LD: 2	300 (each LD)
(LD, APD, Optical Coupler etc.)	APD: 1 PD: 2 Coupler: 1	13
Electrical Active Components (Monolithic IC's, Diodes Transistors etc.)	IC: 6 Diode: 2 Tr: 1	14
Electrical Passive Components (Coils, Resistors Capacitors, MCF SAW Filter, etc.)	R: ~160 C: ~150 L: ~13 SAW: 1 MCF: 1/2	36
Soldered Joints	~ 700	7

Monitored information is transmitted to a shore terminal through the loop-back transmission line. Monitored analog voltage changes are transformed to frequency changes at the $V - F$ convertor and, then, the 446-Mb/s loop-back pulse stream is mark density modulated by the frequency of $V - F$ convertor output.

Control signals configuration is explained in Fig. 6. Marks of ⓐ, ⓑ,..., and ⓘ corresponds to the marks of ⓐ, ⓑ,..., and ⓘ shown in Fig. 5. Control signals are pulsewidth modulated 11-bit codewords. Each codeword is composed of a mark density modulated 446-Mb/s pulse stream at a supervisory clock frequency of 12.384 MHz. In

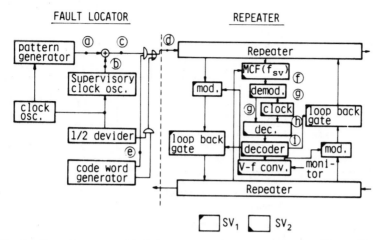

Fig. 5. Supervisory system block diagram. Waveforms at points of (a), (b),..., (i) are shown in Fig. 6.

11-bit codewords, 5 bits are applied for repeater identification and the following 4 bits are used for controls. The first and final bits are guard bits to identify the control signals. The pulsewidth of the codewords are designed to be 5.5 and 0.5 ms corresponding to codes "1" and "0", respectively, for the time slot of 10 ms. The time slot and pulsewidths have a margin of approximately 100 percent against undesirable changes of the clock signal for the codeword decision.

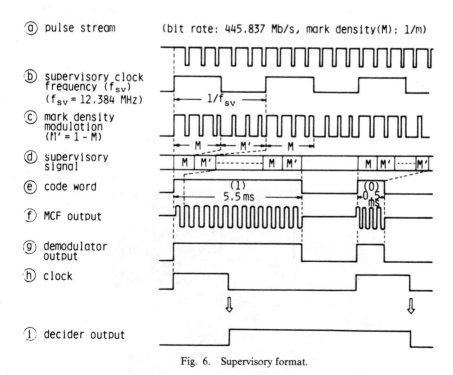

Fig. 6. Supervisory format.

TABLE VII
DEVELOPED MONOLITHIC INTEGRATED CIRCUITS

MIC		Characteristics	Number of Trs.	Power
Optical Receiver (OR)		Noise: $7pA/\sqrt{Hz}$ Max. Trans-impedance: 35.5 kΩ AGC range: 25 dB	180	600 mW
Decider (DEC)		Level decision sensitivity: 6 mV$_{0-p}$ Rise/fall time: 500 ps Output amp.: 800 mV$_{0-p}$	91	900 mW
Timing Amp. (TIM)		Gain: 55 dB Phase deviation: 13 degrees Dynamic range: 30 dB	96	900 mW
LD Driver (DRV)		Driving pulse: 30 mA$_{0-p}$ D.C. bias current: 0–70 mA	85	600 mW
Supervisory Circuit	SV_1	Loop back gate switching Mark density modulation (12,834 MHz)	120	640 mW
	SV_2	Repeater identification Loop back control LD switching Monitoring functions	1,350	290 mW

Allowable Failure Rate

In order to achieve a failure rate of less than 80 FIT's for a one-way repeater, the allocated component failure rate is shown in Table VII. The target of the laser diode failure rate is 300 FIT's.

TRANSMISSION CHARACTERISTICS

Bit Error Rate

A 446-Mb/s optical repeater was realized by using the above described components. Figure 7 shows the integrated optical repeater. The integrated optical repeater bit error

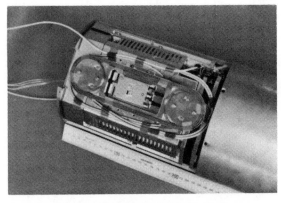

Fig. 7. Photograph of a repeater unit.

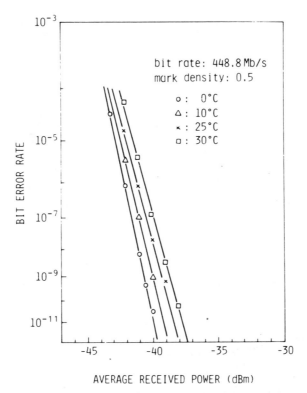

Fig. 8. Observed error-rate performance.

rate at 445.837 Mb/s was measured in the ambient temperature range of 0–30°C. The results are shown in Fig. 8. The repeater achieved a sensitivity of -37.5 dBm at 30°C. However, the minimum received optical power for a 10^{-11} bit error rate was degraded 2 dB for the temperature increment of 30°C. It is estimated that the main factor causing the 2-dB degradation is a dark current increase of Ge-APD.

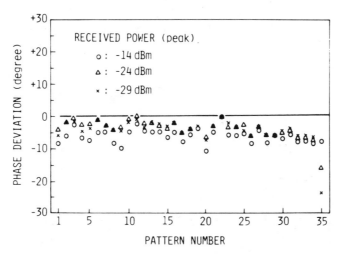

Fig. 9. Observed jitter performance.

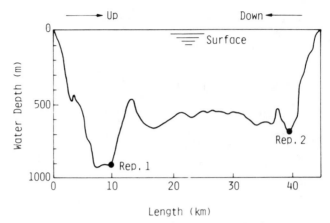

Fig. 10.   Field trial laying route water depth.

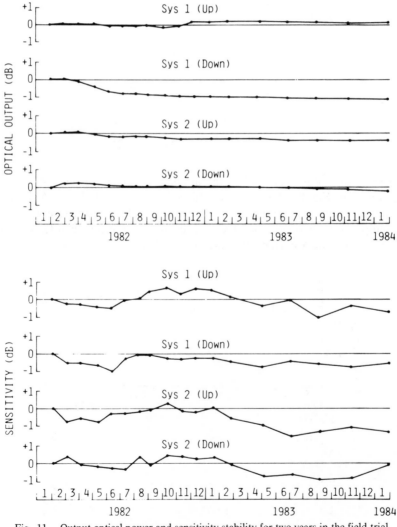

Fig. 11.   Output optical power and sensitivity stability for two years in the field trial.

Fig. 12. The FS-400M repeater laying into a 8000 m depth in the sea from the "Kuroshiomaru" cable ship.

Jitter

Jitter performance is an important factor on a repeatered line. In order to evaluate overall jitter performance, including that of the optical receiver, timing circuit, decider, and LD driver, static phase shifts were measured for optical input power of −14 and −29 dBm, corresponding to electrical gain-AGC and multiplication factor-AGC conditions, respectively. A dynamic jitter of 0.3° rms was estimated from the experimental results shown in Fig. 9, which satisfied the system requirement for less than 1° rms.

The performance described above fully satisfies system requirements, and is almost equal to that of a repeater composed of discrete transistors.

Deep Sea Trial

The 8000 m deep sea trial was carried out in the Ogasawara trench in November 1984 [13]. Two repeaters and submarine cable of 28 km long were laid, holded, and recovered. The results obtained in this trial were almost the same results described above. A photograph of the repeater laying into a 8000 m depth in the sea is shown in Fig 12.

Long-Term Stability

The field trial had been conducted for two years from Jan. 1982 to Feb. 1984, in order to confirm the stability of the performance of submarine repeaters and cables. Fig. 10 shows the water depth of the cable route. The field trial transmission line had a total cable length of 45 km. Its two repeaters were laid on the sea floor at 900 and 700 m.

Transmission tests had been conducted over two years, and periodical testing of the repeatered line including the output power, the wavelength, the sensitivity of submarine repeaters, etc., occurred every month or two [14]. Fig. 11 shows aging characteristics of submarine repeater output optical power and sensitivity. The output optical power changes were less than 0.4 dB except for the repeater Sys-1(Down). The output power of the Sys-1(Down) repeater deteriorated during the first six months, but, later stabilized. The optical sensitivity changes shown in Fig. 11 were observed. Sensitivity changed approximately 2 dB over two years, and had the same tendencies according to season. It was estimated that the changes were caused by the ambient temperature dependence of the optical repeater sensitivity described above. It has thus been confirmed that installed repeaters can operate reliably on the sea floor.

CONCLUSION

System design and transmission equipments for the FS-400M system have been described. It has been confirmed by the preliminary field trial that transmission characteristics were stable for a period of two years. Due to recent technological advances, in addition to the field trial results, the FS-400M system is now well established. It will be introduced for domestic commercial use by the end of 1986. This will produce great cost savings for submarine transmission systems.

ACKNOWLEDGMENT

The authors express their thanks to Dr. E. Iwahashi, for his guidance, and to Dr. S. Shimada for his encouragement. They would also like to thank their colleagues for fruitful discussions and, in particular, to Mr. S. Tsutsumi for the field trial data.

REFERENCES

[1]   K. D. Fitchew, "Technology requirements for optical fiber submarine systems," this book, see ch. 2, p. 23.
[2]   R. L. Williamson and M. Chown, "The NL1 submarine system," this book, see ch. 9, p. 119.
[3]   P. K. Runge and P. R. Trischitta, "The SL undersea lightguide system," *IEEE J. Selected Areas Commun.*, vol. SAC-1, pp. 459–466, 1983.
[4]   Y. Niiro, "Optical fiber submarine cable system development at KDD," *IEEE J. Selected Areas Commun.*, vol. SAC-1, pp. 467–478, 1983.
[5]   M. Washio, I. Kitazawa, H. Tsuji, and K. Takemoto, "400 Mb/s submarine optical fiber cable transmission system field trial," in *Proc. 8th ECOC* (Cannes, France), Sept. 1982, pp. 472–477.
[6]   T. Ito, K. Nakagawa, and Y. Hakamada, "Design and performances of the F-400M trunk transmission system using a single mode fiber cable," in *Conf. Rec. ICC '82* (Philadelphia, PA) June 1982, 6D.1.1–6D.1.5.
[7]   Y. Okano, K. Nakagawa, and T. Ito, "Laser mode partion noise evaluation for optical fiber transmission," *IEEE Trans. Commun.*, vol. COM-28, pp. 238–243, Feb. 1980.
[8]   Y. Miyajima, "Studies on high-tensile proof tests of optical fibers," *IEEE J. Lightwave Technol.*, vol. LT-1, pp. 340–346, June 1983.
[9]   K. Aida and M. Amemiya, "Submarine transmission system reliability with laser diode stand-by redundancy optical repeaters," *IEEE Trans. Reliability*, to be published.
[10] N. Yoshikai, K. Katagiri, and T. Ito, "mBlC code and its performance in an optical communication system," *IEEE Trans. Commun.*, vol. COM-32, pp. 163–168, Feb. 1984.
[11] M. Aiki, T. Tsuchiya, and M. Amemiya, "446 Mbit/s integrated optical repeater," *IEEE J. Lightwave Technol.*, to be published.
[12] Y. Hayashi, Y. Miyawaki, and T. Asari, "Repeater supervision for a submarine optical transmission system," in *Proc. Tech. Group Commun. Syst. IECE Japan*, Sept. 1983, paper CS83-112.
[13] S. Tsutsumi, I. Kitazawa, H. Kimoto, and O. Kawata "Operational trial of 400 Mb/s submarine optical transmission system," *Electron. Lett.*, vol. 21, no. 5, p. 182, 1985.
[14] S. Tsutsumi, "Transmission characteristics of the submarine optical repeatered system in field trial," in *Proc. Nat. Conv. IECE Japan* (Tokyo, Japan), March 1984, p. 2627 (in Japanese).

6
S 280—A New Submarine Optical System

PIERRE FRANCO, JEAN-PIERRE TREZEGUET,
AND JEAN THIENNOT

INTRODUCTION

As of the end of 1970's the results obtained on optical fibers and components operating at a wavelength of 1.3 m showed very clearly that an optical submarine system could present in a very near future a definite technical and economical advantage with regard to coaxial analog systems as far as capacities greater than 1000 to 1500 channels would be required (see Fig. 1).

Consequently French Industry in coordination with Post and Telecommunication Administration decided in early 1980 to initiate a five-year development program for an optical-fiber submarine system called S 280 to substitute for its analog systems range S 5, S 12, and S 25.

OVERALL SYSTEM DESCRIPTION

S 280 is a digital undersea transmission system operating at a bit rate of 280 Mbits/s ($2 \times 139\ 264$ kbits/s, that is, 3840 bidirectional channels) through each pair of single-mode fibers.

Cable and repeaters may include up to three pairs of fibers and regenerators giving a possible capacity of 11 520 channels, without speech concentration such as CELTIC.

Following the TAT 8 co-owners' agreement for designing a heterogeneous system between the three suppliers under the responsibility of AT&T, it was decided to adopt the line frame structure of the United States' "SL" system, leading to a line bit rate of 295.6 MBd.

Cable structure is provided to protect fibers against pressure and elongation strains, and against contact with seawater. The cable may be so laid, in the same conditions as coaxial cables, at sea depths reaching 6500 m.

The repeaters are spaced at least 45 km apart, and comprise a regenerator module associated with each fiber, each having four laser transmitters, one active and three on cold standby, operating at the wavelength of 1.3 μm.

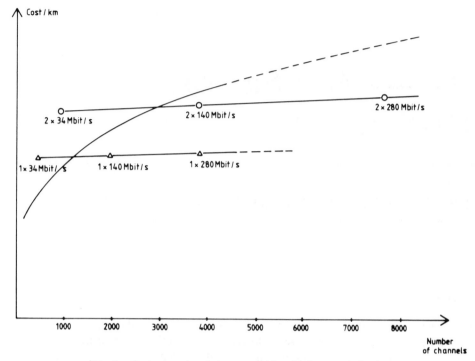

Fig. 1.   Cost comparison between coaxial and 1.3-μm optical systems.

Associated with this redundancy, all component technologies are designed and will be manufactured and selected in order to ensure the reliability objectives for a maximum length of 7500 km. The reliability objective is a 25-year life with no more than two ship repairs due to component failure during this period.

A supervisory system is designed in order to meet the following objectives: supervision of transmission performances on each repeater section, location of "soft" or "hard" permanent or intermittent faults on the link, and switching control of laser diode redundancy. The basic principle of this system is the transmission of brief specific "messages" by substitution to the traffic, transmitted towards or from repeaters.

OPTICAL FIBERS AND CABLE STRUCTURE

Single-mode fibers will be used, having a loss of 0.38 dB/km at 1.31 μm and 0.27 dB/km at 1.60-μm wavelength. The chromatic dispersion target, between 1.29 and 1.33 μm will be less than 3.5 ps/nm·km. These performances will be obtained using a special index profile (matched cladding) in order to maintain a total index difference compatible with a good resistance to bending and microbending effects.

Cable Structure

The cable is designed to protect the fibers against pressure, excessive elongation, and water ingress, without any optical loss impairment, so that the overall system performance requirements can be met throughout the life of the system.

Strength central member   Wires steel strand

Sheath   Copper tube

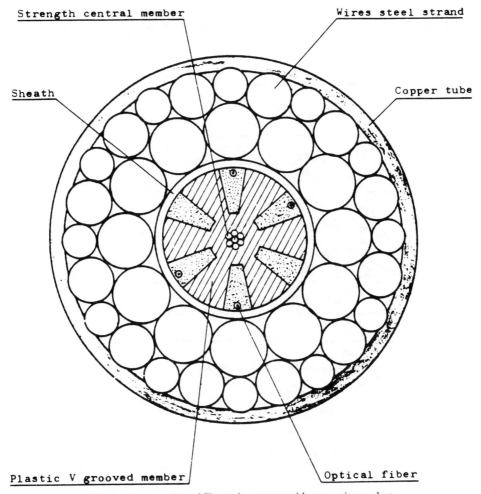

Plastic V grooved member   Optical fiber

Fig. 2.   Cross section of fiber unit structure with composite conductor.

The cross section of the fiber unit structure is shown in Fig. 2, and that of the deep-sea cable is shown in Fig. 3.

The design uses strong fibers (proof test 0.9 percent) which are assembled with a 1.1-percent slack, in helical grooves of a plastic core filled with a nonhygroscopic compound, around which a plastic sheath is then extruded. This optical core is surrounded by a vault of steel wires which protects the fibers from pressure and gives the cable tensile strength. This vault is covered by a copper tube, which conducts the power feeding current, and a polyethylene insulator. Elastomer blocks are periodically placed in the vault to provide longitudinal watertightness.

The fiber slack is designed in order to be, added to its screen test stress, greater than the steel wire strand elongation at break. Thus, no stress is transmitted to the fibers during the cable life except due to an external aggression. Particularly, after any elongation due to any external cause without cable break, no residual fiber strain is remaining, and fiber lifetime will not be reduced.

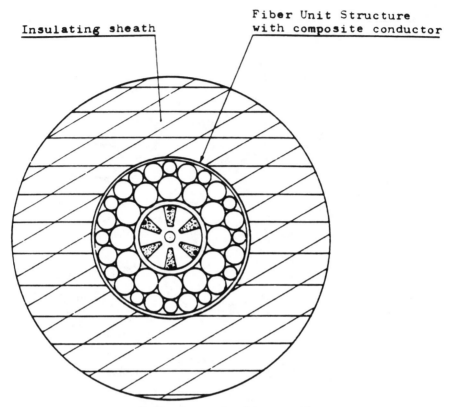

Fig. 3.   Cross section of deep-sea cable.

Under these conditions, the expected average of fiber attenuation, including cabling, fiber splices, and half attenuation difference between any two fibers, within the operational wavelength range (1.29–1.33 μm), is less than 0.46 dB/km.

SUPERVISORY SYSTEM

The messages carrying the supervisory information are of the three following types.

Alarm Messages: Sent by the repeater, by substitution to the frame, of short enough duration (~ 650 ns), they inform, through the return path, the corresponding terminal supervisory receiver of the number of the repeater and of the kind of observed fault (see Fig. 4).

These kinds of alarms are a "laser alarm" indicating that a fixed threshold current value is exceeded and a "margin alarm" indicating a margin canceling.

These alarms are in-service information, their duration being short enough to avoid any frame desynchronization.

Remote Control Messages: They are sent from the terminal supervisory transmitter in order to operate, in an assigned repeater, either loopback of the transmitting path in the return path or laser redundancy switching.

They always cause some traffic interruption. These messages are about 3 ms long and generate specific low frequency spectral lines recognizable by the addressed repeater and related to the kind of operation.

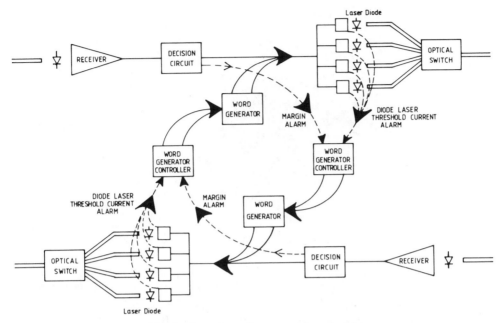

Fig. 4. Supervisory alarm generation principle.

In-Service Common Messages: Each terminal supervisory unit can send either one "full-mark" word of about 430 ns long to reset the alarm of both paths of all repeaters, or two such full marks, 70-μs spaced, for in-service margin testing.

This double emission of full marks causes in all repeaters, and simultaneously in the two paths, a temporary impairment (2 ms long) of 1 dB of the regenerator margins. Thus, each repeater having margins less than 1 dB sends the specific margin alarm.

Another kind of margin measurement is provided out of service, by operating, from the transmit terminal, progressive and calibrated shifts of clock frequency in order to reduce artificially the margins of all repeaters until each of them sends its own "margin alarms" on the return path. This procedure is available in service for margin impairment limited to 1 dB, and thus may be combined with the above-mentioned "double full-mark" operation for detecting every margin below 2 dB.

The necessary information exchanges between the two terminal supervisory units are transmitted through a special ancillary channel.

This supervisory system allows very simply the location of all kinds of happening or imminent hard or intermittent failures without any interrogation cycle delay.

THE REPEATERS

The mechanical design is based on the same proven principles already used for 20 years for analog repeaters (see Fig. 5). The steel casing protecting the optical unit against 650 bar pressure with a safety factor higher than 3 is at the local power-feeding potential. The voltage insulation against the sea potential is obtained by a polyethylene coating, ensuring at the same time watertightness and corrosion protection.

This casing is maintained inside an epoxy impregnated fiberglass housing supporting a tensile stress higher than 200 kN.

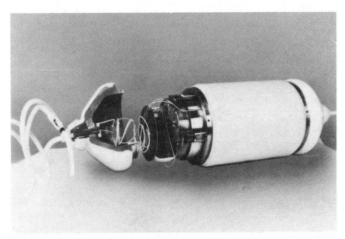

Fig. 5. Regenerator optical unit in its steel casing.

A special watertight optical-fiber throughput has been developed and tested during several thousand hours under 3000 bar pressure without any impairment.

The housing may contain, with a modular disposition, up to three pairs of regenerators such as are described in the diagram of Fig. 6.

The buried heterostructure 1.3-μm laser diodes equipping each transmitter (see Fig. 8) have a typical behavior shown in Fig. 7.

Since the temperatures allowed during the aging tests of this kind of laser diode cannot be higher than about 70°C, sufficient information of lifetime will not be available before its use in first commercial links. Reliability calculations show that redundancy will be provisionally necessary, with three cold standby transmitters.

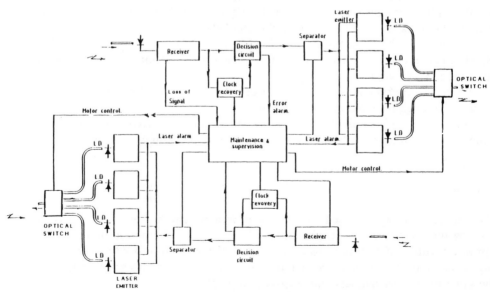

Fig. 6. Block diagram of a regenerator.

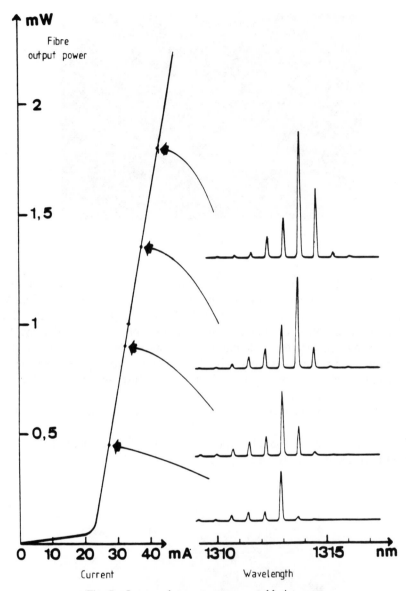

Fig. 7. Laser package output power and lasing spectra.

Three optical switches are provided (see Fig. 9), specially designed and developed, and controlled through magnetic action by a dc motor identical to those of fully proven technology used over the last 10 years on analog systems in remote-controlled equalizers.

The receiver function uses the p-i-n-silicon bipolar amplifier arrangement. It is considered to be the most reliable solution compatible with high performance.

The InGaAsP p-i-n photodiode is a back-illumination planar device providing very low capacitance and high sensitivity. It is combined with a fully integrated trans-impedance preamplifier having a usable bandwidth of 300 MHz. The receiver sensitivity will be potentially -35.5 dBm (BER $=10^{-9}$) and nominally about -34.5 dBm.

Fig. 8.   A view of an opened laser package.

All the functions of the regenerators are integrated with a level of integration sufficient for meeting the overall system reliability requirements, but allowing an efficient selection method for rogue rejection. The very fast ECL bipolar IC technology "DIFOX", developed as early as 1976 for digital terrestrial transmissions, was derived from the process used to manufacture the wide-band transistors for 60-MHz analog systems and the fully proven submarine transistors. This technology, currently used for 560-Mbit/s terrestrial systems is able to work up to 1 GHz; thus, its use for 300-MHz devices leaves good safety margins to the limits.

Figure 10 shows an example of a subunit realization using high reliability thick film hybrid technology.

Table I recapitulates the reasonably expected nominal performances of the system and the potential performances. (Also see Fig. 11.)

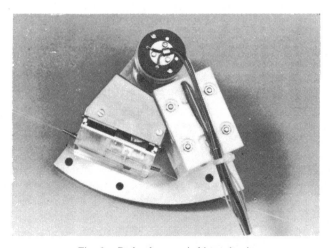

Fig. 9.   Redundancy switching subunit.

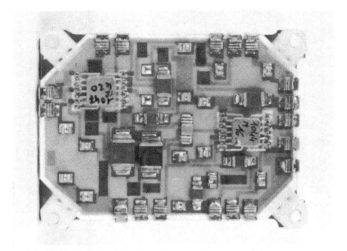

Fig. 10. Example of a hybrid, integrated subunit.

SYSTEM EXPERIMENTS

September 1982: A 20-km experimental link in depths of 1100 m was laid between Juan-les-Pins and Cagnes/Mer on the South Coast of France to test cable structure and laying. Without repeaters, and using two pairs of multimode fibers connected to a 1.3-μm laser diode, 34-Mbit/s terminal equipment for further traffic transmission, and one pair of single-mode experimental fibers, TV color transmission at 34 Mbits/s has been successfully achieved through a loop of 40 km (about 35 dB).

After 18-month experiments, no change in fiber attenuation has been observed.

November 1983: A 1.3-μm 340-Mbit/s experiment was conducted in the laboratory with buried heterostructure lasers and 0.39-dB/km loss single-mode fibers.

A BER better than 10^{-9} was achieved with 81 km length. The receiver sensitivity for 1×10^{-9} error rate was -33.4 dBm [2].

April 1984: On April 9, 1984, the first repeated experimental optical link was powered up between Antibes and Port-Grimaud on the French Riviera. This 80-km-long

TABLE I
SYSTEM CHARACTERISTICS

	Nominal	Potential
Average optical power into cable	-4.5 dBm	-2.5 dBm
Receive sensitivity (BER 10^{-9})	-34.5 dBm	-35.4 dBm
System margins	6 dB	6 dB
Average cabled fiber attenuation	.5 dB/km	.46 dB/km
Repeater span	48 km	58.7 km

Fig. 11.   Photograph of terminal equipment.

cable was laid by the cable ship "Raymond Croze" at a maximum depth of 1800 m, and was equipped with two pairs of single-mode fibers and two line repeaters, each submerged at a depth of about 1000 m. This system will provide a transmission of twice 280 Mbits/s at 1.3 μm with a 344-Mbd line bit rate, a p-i-n Ge-bipolar silicon preamplifier, and a redundancy of three laser diode cold standby transmitters.

Several months' transmission testing over a 320-km loop arrangement will be achieved before delivery for domestic telecommunications.

CONCLUSION

The S 280 system is in final development. The first S 280 lightwave commercial link is planned between the mainland (Marseille) and Corsica (Ajaccio) by the end of 1985, with a total length of about 400 km and a repeater spacing of 45 km.

We therefore have great confidence that we will be ready for the new generation of lightwave submarine systems which are on the verge of supplanting the analog coaxial ones. In particular, our participation in the TAT 8 project will be achieved in 1988.

REFERENCES

[1] P. Franco and M. Laurette, "Optical fibre submarine links," *Commutation Transmiss.*, no. 2/3, pp. 105–118, 1982.
[2] G. Bourret, A. Doll, G. Bassier, G. Raffin, and M. Jurczyszyn, "81 km submarine optical repeater trial," *Electron. Lett.*, vol. 19, no. 24, pp. 1053–1055, 1983.

7

The OS-280M Optical-Fiber Submarine Cable System

YASUHIKO NIIRO

INTRODUCTION

In 1976 the basic research of the OS-280M optical-fiber submarine cable system capable of transoceanic service started [1] and designed prototypes of optical-fiber submarine cables and repeaters were available in 1979. After each cable and repeater was evaluated in the laboratory, the experimental optical-fiber submarine cable system, a 50-km cable and two repeaters operating at 1.3 μm and 300 Mbits/s, was laid in June 1982. All objectives for the sea trial were accomplished, and the experimental system was recovered in October 1983. The conclusions and main results of this trial are as follows:

1) Durability of the optical cable and repeaters was confirmed during laying and recovering operations in a sea depth of up to 1300 m.
2) The capability of jointing the cable to the repeater on board the cable ship was demonstrated.
3) Fiber loss changes of less than ± 0.01 dB/km were observed after 1.5 years in the laid cable.
4) Optical repeaters operated properly and error free during the 1.5-year period.

In 1983, an optical cable with water blocking and high strength optical fibers and a repeater using monolithic IC's for the improvement of reliability were manufactured and evaluated.

The accelerated life tests of the optical and electrical components of the repeater circuits started. The total number of the components for the reliability tests is 2440, and all results will be gathered in the middle of 1984.

A deep-sea trial in up to 7000-m sea depth was successfully carried out in February 1984. The results of these experiments demonstrate that the cables and repeaters fit with the conventional laying method and are usable in deep water.

A major event of the development program is a short-haul commercial test system laid in early 1986 and finally, a long-haul commercial system to be laid in 1988.

TABLE I
OS-280M SYSTEM DESIGN PARAMETERS

	Item	Parameters
1	Transmission bit rate	280 Mb/S/1 subsystem
2	Transmission capacity	3780 ch (64 kb/S)/subsystem
3	Transmission signal Line code Bit error rate Jitter	Scrambled binary NRZ $\leqslant 10^{-11}$/Repeater $\leqslant 1°$ rms/Repeater
4	Number of subsystems	1–2
5	Maximum system length	8000 km
6	Maximum sea depth	8000 m
7	Repeater section length	50 km nominal
8	Optical fiber type	Single mode optical fiber
9	Optical wavelength	1.31 μm
10	System design life	25 years
11	Reliability	Less than 3 ship repairs in service life
12	Repeater supervisory Fault location Monitoring	Remote controlled Optical loop back APD BIAS, operating laser status Bit error rate
13	Redundancy	1 LD spare cold standby

SYSTEM DESIGN

The tentative design parameters of the OS-280M system are shown in Table I. A capacity of 280 Mbits/s per subsystem was chosen to provide a ×2 interface with the 140-Mbits/s CCITT CEPT hierarchy [2]. Bit error rate (BER) and jitter requirements for a long-haul submarine cable system are being studied to adopt the CCITT revised recommendation that will be available in the near future.

The repeater section length depends on the fiber loss, repeater gain, and system margin. Due to insufficient data on the long-term stability of cables and repeaters, the decision of the system margin is a subject for future study. At this stage, we use 8 dB for system margin, 33 dB for repeater gain, and 0.5 dB/km for fiber loss at 1.3-μm wavelength.

In case of a working laser diode failure, a redundant cold standby laser diode will be switched automatically to the operation condition. In-service error monitoring at each repeater enables rapid and accurate fault location for intermittent failures. An out-of-service optical loop-back system is capable of fault location to the accuracy of one repeater section. The input and output power of each repeater can be measured using both the optical loop back and the APD bias monitoring circuits in an out-of-service condition. This function is used to confirm the loss budget of the cable system after laying.

The mechanical design of the repeater housing with a gimbal joint was based on the required characteristics for proper winding around a 3-m-diameter sheave.

Fig. 1.   Optical-fiber submarine cable.

OPTICAL-FIBER SUBMARINE CABLE

The following points were intensively considered in the cable design:

1) metal shell structure to protect optical fibers from outer forces such as high water pressure, high tension, etc.;
2) cable structure with a large strength/weight ratio to prevent elongation of optical fibers as much as possible;
3) metal pipe as a barrier to water diffusing through the insulation and compound filled in air space in the cable as water blocking [3], and;
4) stable structure for more than 25 years of use.

The OS-280M cable for deep-sea use is shown in Fig. 1. Figure 2 shows the fiber strain during deep-sea laying and recovering. High strength fibers are required to survive cable laying and recovering in the deep sea. To survive, the fibers must be able to withstand more than a 50-h tensile load during recovery operations after 25 years, which requires the fibers to be proof tested at a stress of 2 percent/1 s [4].

The cross section of the compound filling the air space in the cable is shown in Fig. 3. Figure 4 shows an example of water propagation characteristics. The experimental results can be explained well by the formula $L = k\sqrt{t}$ where the length of the water run is L, time elapsed is t, and k is constant. Fig. 5 shows the water pressure characteristics of the cable. Loss changes at the 800 kg/cm^2 were less than 0.005 dB/km. Loss changes of water blocking cable under the high water pressure were very small and the water run length will meet the requirement.

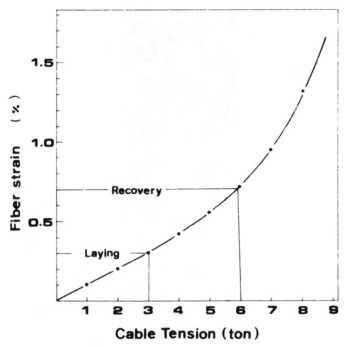

Fig. 2.   Fiber strain versus cable tension.

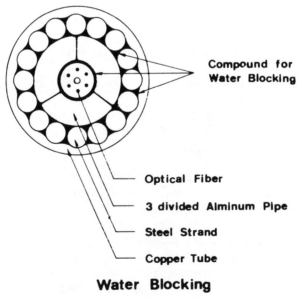

Fig. 3.   Cross section of a cable.

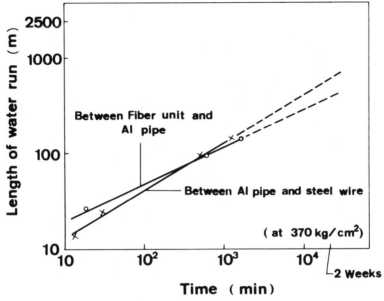

Fig. 4.   Length of water run.

IC REPEATER

The six types of integrated circuits will be used for the optical repeater shown in Fig. 6. Four IC's, the EQL-amplifier, timing circuit, decision circuit, and LD driver, have been fabricated. Experimental regenerators using those IC's have been developed for the evaluation of the transmission characteristics. An experimental IC regenerator circuit is shown in Fig. 7.

Figures 8 and 9 show measured average input optical power versus BER and the static pattern jitter characteristics of the experimental integrated regenerator, respectively. The minimum receiving level was less than -37 dBm at a BER of 10^{-11}, and the jitter was less than $1°$ rms.

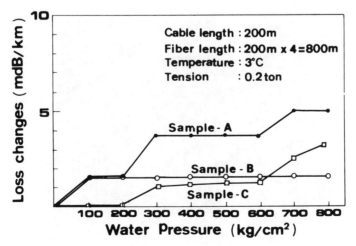

Fig. 5.   Loss changes versus water pressure.

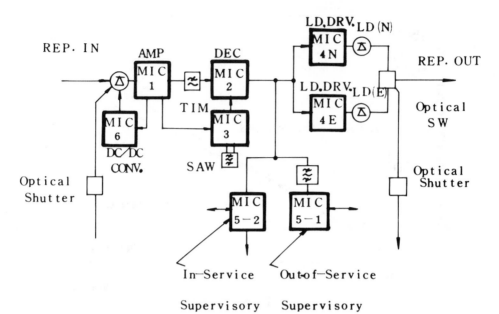

Fig. 6.   Block diagram of IC regenerator.

Several repeater supervisory methods for the submarine cable system were considered with respect to the transmission media of supervisory signal and operation modes (in service or out of service). The basic philosophy of repeater supervision of the OS-280M system is as follows:

1) It will not be economical to have supervisory signal paths of exclusive use because the supervisory transmission line requires almost as much reliability as the main transmission line.
2) Switchover to the spare laser diode should be performed in service (the laser diode of each repeater is switched automatically in service in OS-280M).

Fig. 7.   IC regenerator.

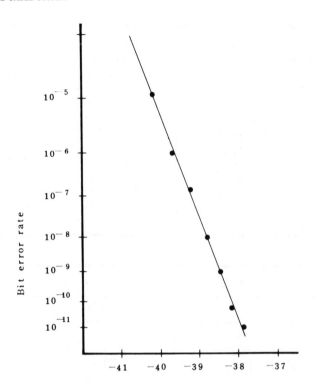

Average input optical power(dB)

Fig. 8.   BER versus input power.

3) It will be necessary to adopt an in service error monitoring method at each repeater which can locate intermittent faults.

4) The optical loop-back method enables us to locate the faulty section.

From the above point of view, OS-280M adopts an in-service and out-of-service supervision method which uses main signal paths as supervisory signal lines. Figure 10 shows how supervision and control are performed in OS-280M. An assembled optical regenerator is shown in Fig. 11.

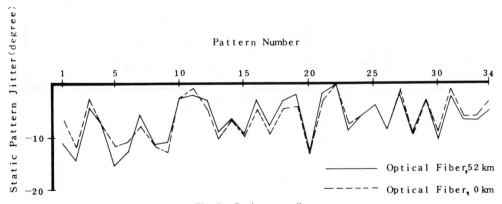

Fig. 9.   Static pattern jitter.

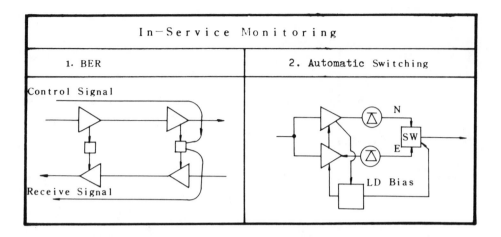

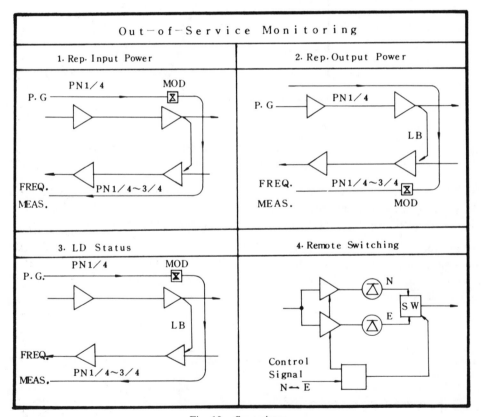

Fig. 10. Supervisory.

Fig. 11.   An optical regenerator.

RELIABILITY TEST

Repeater components such as the laser diode, APD, and IC's are the key devices needed to realize long-haul optical fiber submarine cable systems. Very high reliability such as 20 fits/reg is required, supposing a system length of 8000 km and a long life of 25 years. Because the reported reliability data are still insufficient to realize a long-haul system, KDD has been conducting the aging test of components to evaluate their reliability.

The accelerated aging test being conducted by KDD is classified into two steps. 200 fits/reg corresponding to a 700-km system length will be achieved in the first step. In the second step the final goal of 20 fits/reg will be achieved. The main purposes of the first step are as follows: 1) estimation of activation energy; 2) clarification of degradation modes; and 3) establishment of screening techniques.

The number of test samples and the reliability objectives are summarized in Table II. The number of the test sample was determined so as to ensure its reliability objectives after aging of 300 h.

All samples are screened before the aging test, and the differences in their characteristics after aging are examined. The typical characteristics of each device are measured continuously and/or periodically during the accelerated aging test.

Finally, all samples will be examined in detail, and the degree of degradation will be measured. The test conditions are the combination of the high temperature test, low temperature test, and the heat cycle test.

The test result of the first step will be obtained in the middle of 1984 and an evaluation of long-haul systems is scheduled to be completed by early 1985.

DEEP-SEA TRIAL

The experimental optical-fiber submarine cable system consisting of a 24-km optical cable and two submarine repeaters operating at 1.3 μm and 280 Mbits/s was laid in February 1984. After being held more than 10 h the system was recovered on board

TABLE II
RELIABILITY TEST IN THE FIRST STEP

DEVICES			RELIABILITY OBJECTIVES (Fits)	SAMPLE NUMBERS
1.3 µm LD			300	400
LD module				60
Ge-APD			10	200
APD module				60
Opt. switch			10	60
Opt. shutter			10	60
EQL	AMP	MIC	10	600
DEC		MIC	10	200
TIM		MIC	10	200
LD	DRI	MIC	10	600
	TOTAL			2,440

successfully from the 7000-m seabed as shown in Fig. 12. Three 8-km optical-fiber cable pieces and two optical repeaters were connected on board the cable ship in January 1984. Figure 13 shows the splicing between repeater and cable coupling on board. Each optical cable accommodated six single-mode optical fibers which were screened by the 2-percent proof test, and splices in the cable were tested by 2.5-percent proof. Three optical fiber loops, including four monolithic IC regenerators, were used for measurements of optical loss, elongation, and BER.

The laying, holding, and recovering test was carried out in the Ryukyu trench about 200 km off the shore of the Okinawa islands. The maximum sea depth was 7010 m and the average sea depth was 6900 m. Laying speed for the 16-km working cable and two repeaters averaged about 3 knots except when the repeater was paid out. After the

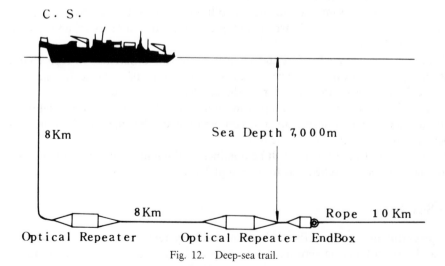

Fig. 12.   Deep-sea trail.

Fig. 13.   Splicing between repeater and cable coupling.

holding of 11 h, the experimental cable was recovered at the speed of 0.9–3 knots (average 2 knots). The tension applied to the cable was 4 tons, plus or minus 1 ton according to the recovering speeds. No fiber breaks occurred in this experiment. The optical repeater recovered from 7000-m sea depth is shown in Fig. 14. Loss changes of the cable are shown in Fig. 15. The change of cable loss through the 16-km fiber loop was less than 0.2 dB, which is mainly attributed to temperature changes. The measuring accuracy was less than ± 0.02 dB.

Fig. 14.   Optical repeater recovered from 7000-m sea depth.

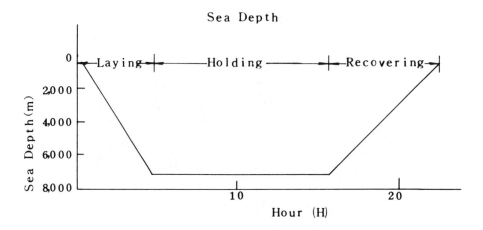

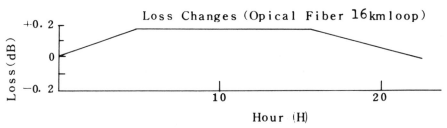

Fig. 15. A test result of the deep-sea trail.

Monolithic IC regenerators operated free from error during this experiment, although some mechanical shocks were applied by laying mechanisms. The result of the experiment demonstrates that the mechanical design of cable and repeater fits the conventional laying methods and is satisfactory for use in the deep sea. Transmission characteristics of the cable and monolithic IC repeater were also proven to be good for the deep-sea environment.

CONCLUSION

The OS-280M system is going to be deployed as a transoceanic long-haul commercial system in 1988. To realize a highly reliable system, a water blocking cable using high strength optical fibers and IC regenerators has been manufactured for experimental use. The deep-sea trial using that cable and those repeaters was carried out successfully in February 1984.

A large number of optical devices for repeater circuits is being tested at the factory and the improvement of device reliability is continuing.

In order to place the OS-280M system into commercial service, the following item must be solved to realize a highly reliable system: 1) high strength and long length optical fibers; 2) high strength fiber splices; 3) highly reliable 1.3-μm LD and optical devices; and 4) laying and repairing methods for optical-fiber systems.

ACKNOWLEDGMENT

The author would like to express appreciation to Director Ishikawa for encouragement and to members of the Optical Cable Development Group for their helpful assistance.

The author would also like to thank the Nippon Telegraph and Telephone Public Corporation for their cooperation, the Nippon Electric Company, Ltd., and Fujitsu Ltd. for manufacturing repeaters, and the Furukawa Electric Company, Ltd., Sumitomo Electric Industries, Ltd., Fujikura Cable Works, Ltd., and Ocean Cable Co., Ltd. for manufacturing cable.

REFERENCES

[1] Y. Niiro, "Optical fiber submarine cable system development at KDD," this book, see ch. 7, p. 95.
[2] R. L. Williamson and M. Chown, "The NL1 submarine system," this book, see ch. 9, p. 119.
[3] K. D. Fitchew, "Technology requirements for optical fiber submarine systems," this book, see ch. 2, p. 23.
[4] P. K. Runge and P. R. Trischitta, "The SL undersea lightguide system," *IEEE J. Select. Areas Commun.*, vol. SAC-1, no. 3, Apr. 1983.

8
CTNE Undersea Lightwave Inter-Island System

S. WALLACE DAWSON, JR., MEMBER, IEEE, JOSE RIERA RIERA, AND ELAINE K. STAFFORD

INTRODUCTION

The world's first deep-water undersea repeatered lightwave system will be installed in the Canary Islands in 1985. The system will connect the two major islands in the Canaries, Tenerife, and Gran Canaria, and will be used initially as a testbed to prove-in the SL Undersea Lightwave System [1]. At a later date, the system will provide approximately 557 Mb/s of transmission capacity for voice, data, and video services to Compania Telefonica Nacional de España (CTNE) over two fiber pairs operating at 295.6 Mb/s. This chapter describes the system route and architecture; highlights its performance, reliability and maintenance features; outlines the test program; and describes the commercial application of the system.

CABLE ROUTE

The general geographic orientation of the cable route is shown in Fig. 1. A potential route was initially selected based on a study of available bathymetric data, the location of existing active undersea cables, available information on seafloor characteristics, and similarity to the proposed TAT-8 system route. Subsequent deep-water and shore-end surveys were carried out for the purpose of finalizing the route. The depth profile for the route is shown in Fig. 2.

The land route is a duct route between each beach joint and cable station. The cable station on Tenerife is Candelaria, which is located 1.8 km from the beach joint. The terminal station on Gran Canaria is Altavista, which is 3.8 km from the beach joint.

The system will be initially laid with two repeaters. A system block diagram showing cable types and lengths is given in Fig. 3. The cable types designated are SL lightweight (LW), SL light wire armored (LWA), and SL double armored (DA) [2]. Shortly after the initial installation, a trial repair operation will be conducted. At this point in time, the cable will be cut and a third repeater will be added to the system. A system block diagram showing cable types and lengths after the planned repair trial is given in Fig. 4. Further details on the planned repair are given in subsequent sections.

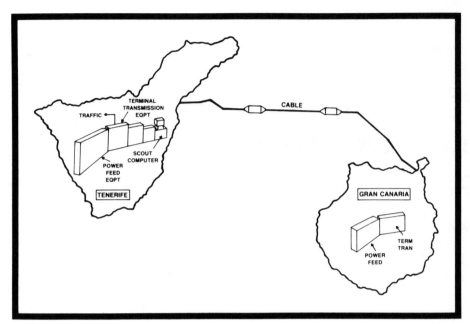

Fig. 1.   Cable route.

The power conductor and lightguides will be separated at the beach joints. Connection from each beach joint to its associated cable station will then be by two cables, one for carrying dc current to the repeaters and one for transmission (lightguides). The transmission cable will be a stranded fiber unit cable similar to single-mode cable planned for AT&T's domestic network [3], but with SL fiber [4]. The submarine feeder cable is a commercially available cable with a stranded copper center conductor and a helically wrapped copper tape outer conductor.

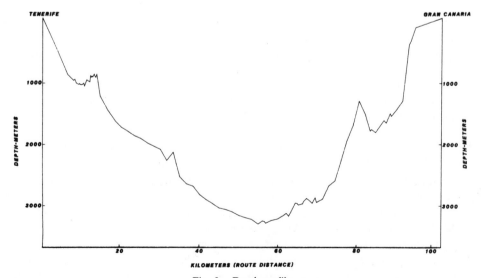

Fig. 2.   Depth profile.

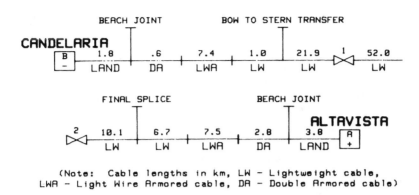

(Note: Cable lengths in km, LW - Lightweight cable,
LWA - Light Wire Armored cable, DA - Double Armored cable)

Fig. 3. System block diagram for initial installation.

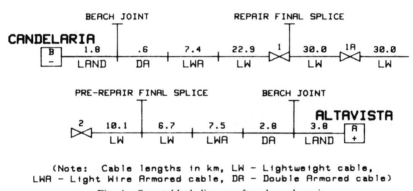

(Note: Cable lengths in km, LW - Lightweight cable,
LWA - Light Wire Armored cable, DA - Double Armored cable)

Fig. 4. System block diagram after planned repair.

Two types of cable protection are planned. First, armored cable is used from the beach joints out to depths of approximately 920 m to provide abrasion protection. Second, articulated pipe protectors will be applied to the cable near the landing at Gran Canaria.

Undersea Lightwave Cable

The SL undersea cable used in this system contains six high-strength fibers which are tightly coupled to the cable structure. This ensures reliability during high-tension cable operations. The fibers, proof-tested to 2-percent elongation, have losses ranging from 0.46 to 0.49 dB/km at the operating wavelength of 1.3 μm. The central core of the cable concentrically mounts these fibers in a hard elastomer compound. This core is surrounded by helically wound steel wires which provide strength to the cable. A welded copper sheath surrounds the steel, forming a power conductor for carrying current to the repeaters. The copper is surrounded by a polyethylene high-voltage insulator. In shallow water, the cable is surrounded by steel armor wires to protect it from abrasion.

Undersea Repeaters

Each of the three SL repeaters includes six 295.6-Mb/s digital regenerators, with their associated supervisory and power circuits, all enclosed in a high-pressure beryllium copper vessel. The repeater interfaces with the cable in a sealed splice chamber on each end of the high-pressure vessel via a gimbal-type joint [5].

Each regenerator [6] consists of a receiver, an automatic gain control (AGC) amplifier, a timing recovery circuit with a surface acoustic wave filter, a decision circuit that regenerates the digital signal, and redundant laser transmitters—one of whose outputs are switched to the output fiber by a mechanical optical relay. Two of the six regenerators, together with two of the six fibers in the cable, serve as redundant transmission paths for the other four regenerators and fibers. These redundancy features are further described in the discussion on reliability.

The supervisory circuitry in each repeater measures the inservice error ratio of the data signal, the gain of the AGC amplifier (which is inversely proportional to the received optical power), and the bias current of the laser transmitter at each regenerator. The repeaters can be polled for this information from the shore terminal through a channel which utilizes parity bits on the in-service optical line. Responses from the repeaters are returned through controlled phase modulation of the in-service optical line rate signal. This same supervisory system is used to control redundancy switching in the regenerators [7].

Terminal Equipment

The terminal stations house the SL terminal equipment and the digital transmission and multiplex equipment that interface with the terrestrial facilities.

The SL terminal equipment consists of three basic equipments: the Terminal Transmission Equipment (TTE) [8], the Surveillance and Control of Undersea Transmission system (SCOUT), and the Power Feed Equipment (PFE). In the Canary Islands system, identical TTE's and PFE's will be located in both the Tenerife station and the Gran Canaria station. The SCOUT system is located only at the Tenerife station.

The TTE equipment interfaces with the terrestrial equipment at 139.264-Mb/s CEPT-4 standards. Four CEPT-4 signals are multiplexed by the TTE to two 295.6-Mb/s optical signals. The multiplexed line-rate signal includes six 64-kb/s order-wire channels, and parity bits for end-to-end performance monitoring, in-service error-ratio monitoring, regenerator, and supervisory signaling. A redundant multiplexer and demultiplexer protect the working muldexes through automatic protection switching. A central microprocessor controls all operations and maintenance features of the terminal.

The TTE can be configured as a sophisticated test set by simple card and firmware replacement. The CEPT-4 interface cards of the TTE can be replaced with pseudo-random word generators and detectors and maintenance firmware can be replaced with test equipment firmware. In this configuration, known as System for Assembly and Laying Testing (SALT), the equipment can be used for system testing during cable-repeater integration, ship loading, and laying.

The SCOUT system is a mini-computer based system responsible for maintenance of the optical system including the terminal regenerators at both stations. If a fault occurs on the system, SCOUT locates and corrects it, when possible, through redundancy switching. In addition, SCOUT can be used to monitor the performance of the system on a regular basis through the supervisory system. The data collected by these routine measurements allow an operator to examine the performance history of the system and potentially predict a failure before it occurs. User-friendly screens allow an operator to make supervisory measurements and control redundancy switches. Thus degrading components can be manually switched out of service before failure—minimizing the effect on service.

The computer system can be loaded with an alternate set of software that is used to test the system in conjunction with the SALT equipment. These programs enable operators to make specialized tests during cable-repeater integration, loading, and laying. When the computer is used for these tests, it is termed the SALT computer.

The Power Feed Equipment in each station contains a single 48-V converter, which outputs several hundred volts at 1.6 A ($+/-2$ percent). Either converter can power the entire undersea system, which only requires approximately 175 V. Normally, the two stations will share the load; one will automatically take over in the event that the other fails. (Each repeater drops approximately 21 V and the cable resistance is approximately 0.7 Ω/km.) The PFE's have an automatic shutdown feature in case of over-voltage and are designed with a key interlock system to protect operators from high voltage. Primary lightning protection is done at the ocean ground distribution panel.

Digital multiplex equipment and transmultiplex equipment will be used to connect the system to domestic facilities. The digital multiplex will operate at CEPT standard transmission rates. Transmultiplexers will be used to interconnect the digital multiplex operating at 2.048 Mb/s to standard analog 60-channel supergroups. These multiplexes will be located at the Altavista terminal station, which is also a main switching center, and in Santa Cruz, the main city of Tenerife. The Santa Cruz office and Candelaria terminal stations will be connected via an existing coaxial cable, which will be upgraded to carry 140-Mb/s traffic.

System Transmission Performance

BER Requirements

The transmission performance requirements of the system are based on CCITT draft recommendations (G821). Specifically, the requirements are derived from the recommendation for the national portion of the international hypothetical reference circuit, prorated with distance. The Canary Islands system requirements are shown in Table I.

TABLE I

Average BER	1.25×10^{-8}, averaged over any 24 hours
Degraded Minutes (Minutes with BER $> 10^{-6}$)	1.44/day, averaged over any 30 days
Degraded Seconds (Seconds with BER $> 10^{-3}$)	0.86/day, averaged over any 30 days
Outage (> 10 Successive Degraded Seconds)	.005% (26 minutes), averaged over any one year

TABLE II

Transmitter/Optical Relay Output Power	-3.2 dBm
Regenerator Sensitivity at 10^{-9} BER	-34.2 dBm
Regenerator Overload at 10^{-9} BER	-19 dBm
Maximum Fiber & Component Loss (including buildout)	18.5 dB
Transmission Penalties (dispersion, etc.)	0.8 dB
Aging	3.9 dB
Repair Allowance	4.0 dB
Margin (includes margin for variation of above)	3.8 dB

Loss Budgets

The transmission performance requirements on bit error ratio are achieved by conservative engineering of the loss budgets for each of the cable-repeater sections in the system. In an undersea lightwave cable system, there is typically one standard loss budget for deep-water sections, one for shallow-water sections, one for land sections, etc. The differences result from many factors, including repair margins and allowances for additional splices (e.g., land sections). For a system as short as the Canary Islands system, each of the three original spans has its own loss budget. Once the repair trial has been completed, the two sections on either side of the repair repeater will have identical loss budgets. These sections, because they will be only 30 km long, will have line buildouts installed as part of the repair operation and will also have more margin than the original deep-water span. A high-level loss budget for these repair sections is shown in Table II.

It should be noted that the sensitivity and overload point of 10-9 BER in the above loss budget was chosen to ensure that all three error performance requirements (BER, degraded minutes, and degraded seconds) are met.

SYSTEM RELIABILITY

The system is designed to meet the reliability requirement of no more than three undersea repairs, caused by internal component failure, over the 25-year system life.

In order to ensure that the system meets its reliability requirements with margin, each component of the system is thoroughly tested at each stage of manufacture. In addition, redundancy is employed throughout the system design. The laser is the most critical component of the system and each regenerator, therefore, has at least two laser transmitters. To protect against failure of any other undersea component, one spare fiber and one spare regenerator protect each of the two working fibers and regenerators in a single direction of transmission. This regenerator-section redundancy is controllable on individual regenerator sections.

With an expected median laser lifetime of one million hours, the expected number of repairs for the Canary Islands system is 0.007. This represents a margin of 2.993 repairs. The corresponding expected number of laser switches in the system over the 25-year life is 1.7 and the expected number of regenerator-section switches is 0.2. Each switch results in an outage of less than 1 min. Thus a total of less than 5 min of outage over 25 years can be attributed to undersea failures. If the switches are made preemptively, this number will be even further reduced. The expected outage, per year, due to terminal failure is only 5.8 min. Therefore, the 26 min/year outage requirement is expected to be met with a significant margin.

It should be noted that there is less time during the manufacture of the Canary Islands System to test components than there is for the TAT-8 system. It is for this reason that the system is designed with very large margin. Even if the transmitters and all other repeater components missed their reliability objective by a factor of ten, the expected number of ship repairs would only be 1.6 over the system life and the outage per year due to undersea failures would be only a few minutes.

MAINTENANCE FEATURES

The SL system is designed with particular attention to maintenance features. Any failure of the undersea system or terminal equipment will automatically result in an office alarm. The majority of failure mechanisms will be handled automatically by the terminal equipment and will not require immediate operator intervention.

Undersea maintenance

As discussed earlier, the SCOUT system is responsible for monitoring the performance of the undersea system. If any failure occurs on the system, SCOUT will automatically locate the fault and switch in a redundant component to correct the fault. If the failure cannot be protected by redundancy, a repair will be necessary. A description of one type of deep-water repair follows in the discussion of the Test Programs.

Terminal Maintenance

The TTE has a central microprocessor-based control and display panel for maintenance of the terminal and end-to-end system. It has a 12-digit keypad entry pad and a 40 character display. The processor maintains records of the end-to-end error performance of the system, terminal alarm events, muldex protection switching events, etc. In addition, it is capable of isolating the majority of terminal failures down to a small group of individual circuit packs. Failure of a circuit pack will be indicated by a front-panel LED. This allows station personnel to easily replace any circuit pack that has failed with a spare.

The PFE has built-in alarms, meters, recorders, and a dummy load to assist personnel in maintenance of the power equipment. Since either terminal station can fully power the system, the PFE can be maintained without interrupting service.

The SCOUT system is designed with a built-in self-test program. In addition, the TTE routinely communicates with the SCOUT system. Failure of either the self-test program or TTE-SCOUT communication channel will automatically initiate a station alarm.

TEST PROGRAM

The Canary Islands system is being installed to prove-in SL technology. In addition, it will be used by CTNE to provide commercial service. The second commercial application of SL will be the next transatlantic cable system, TAT-8. This system will use the same basic design as the Canary Islands system, but will be significantly longer (approx. 6000 km) and will also include a branching repeater. The Canary Islands system is being manufactured, integrated, and installed in a manner as much like TAT-8 as possible. In order to demonstrate that those features incorporated in TAT-8 will meet design objectives, an extensive test program will be carried out on the shorter system. This test program will include both sea trials and transmission experiments.

Sea Trials

After the initial installations, a planned repair trial will be carried out to prove-in the procedures necessary to make a deep-water repair. A deep-water repair was chosen because it gives the most difficult recovery situation. In this trial, the 52-km cable section between repeaters 1 and 2 will be replaced with a repeater and two 30-km cable sections. This will allow the entire section to be thoroughly tested and evaluated after being subjected to all phases of the recovery and repair operation.

Branching Repeater and Armored Coupling Installation Trial

In the same time frame as the repair trial, a branching repeater with its associated main cable and two branch cables will be installed at a nearby site. Portions of one of the branches in the trial system will be armored and will contain a repeater. The purpose of this trial is to demonstrate the shipboard procedures associated with installing and recovering both the branching repeater and an armored cable-to-repeater coupling. Both the repeater and branching repeater will be totally passive, that is, they will contain only fiber feedthroughs and loopbacks. During the shipboard trial, the transmission of the system will be monitored through loss measurements.

System Tests

Additional tests on the Canary Islands system, over and above what will be done in TAT-8, will be conducted to characterize the transmission performance of the system. Extra tests will be run during manufacture, integration, and installation of the system, as well as after installation of the system.

During manufacture and integration of the Canary Islands System, testing will concentrate on transmission performance. Tests to precisely measure transmission margins of regenerator sections during assembly will be made. The jitter performance of each individual regenerator and the chain of regenerators in the assembled system will be characterized. Measurements of jitter transfer function, jitter tolerance, and jitter accumulation will all be made. In addition, long-term error-rate tests will be made on the assembled system.

During installation of an SL system, the end-to-end error performance of the system will be continuously monitored by the SALT computer and SALT test equipment aboard the cable ship. In addition, the supervisory parameters of the system will be periodically measured during the lay.

The installed system will be used as an SL testbed for several months following the final splice. During this test period, a number of tests will be executed. Most importantly, these tests will simulate the effects of a longer system and monitor the transmission performance (error rate and jitter) of the installed system and the extended systems.

To simulate the transmission performance of a long system, the regenerators in a number of additional repeaters will be concatenated with the undersea and terminal regenerators to form a chain of several dozen regenerators. These additional repeaters will be housed in an environmental chamber in Tenerife. Error statistics, jitter performance, and supervisory system performance will all be measured on this chain. Temperature, line frequency, regenerator sequence, and test pattern are only a few of the parameters that will be varied in these tests. A loop experiment [9], which loops a test pattern through a chain of regenerators a number of times, will be done to simulate the jitter accumulation of over 200 cascaded regenerators. The goal of this phase of the test program is to demonstrate SL performance for both short- and long-system applications.

SERVICE APPLICATION

Scope

The Canary Archipelago is made up of seven major islands organized in two administrative provinces: Las Palmas and Santa Cruz de Tenerife. Each of the two provinces has a population of about 700 000 inhabitants (most of them concentrated in the capitals Las Palmas and Santa Cruz de Tenerife) and about 210 000 telephones. Among their main economic activities, tourism and commerce are prominent. This fact, together with their strategic situation between Europe, Africa, and South and Central America, means that telephone traffic with the Spanish Mainland and between the Islands has had a constant growing rate in recent years and expectations are optimistic for the future. The cable system will provide significantly superior performance to an inter-island radio link, which is subject to fading.

Telephone Traffic

The planned undersea lightwave system will provide enough capacity for all the expected telephone traffic growth during its life, estimated as 25 years. Starting with perhaps 2 digital groups of 2 Mb/s from its inauguration, it is foreseen that telephone traffic (including other services like telegraphy and low-speed data transmission) through the system will reach about 200 2-Mb/s groups at the end of this century thus providing significant capacity for new services.

New Services

The development of high-speed optical digital transmission facilities provides possibilities of economically transmitting the many new services, together with the standard telephone traffic. These include: video-conference, telex, high-speed data transmission, and, most particularly, commercial television transmission. Currently, TV transmission between the Islands is via dedicated radiolinks that could be subject to fading. TV transmission demonstrations will be carried out over the system using video codecs during the system testing period.

ACKNOWLEDGMENT

The authors would like to acknowledge the following individuals from AT&T Bell Laboratories for their contributions: S. M. Abbott, J. L. Miller, Jr., M. D.Feinstein, S. E. McMeekin, and R. L. Easton. In addition, the combined efforts of all members of the Undersea Systems Laboratory at AT&T Bell Laboratories and the International Engineering Department at AT&T Communications are appreciated.

REFERENCES

[1] P. K. Runge and P. R. Trischitta, "The SL undersea lightwave system," this book, see ch. 4, p. 51.
[2] A. Adl, T-M. Chien, and T-C. Shu, "Design and testing of the SL cable," this book, ch. 16, p. 233.
[3] C. H. Gartside, III, and M. R. Santana, "High performance single mode lightguide media," presented at Global Telecommun. Conf. 1984, Atlanta, GA, Nov. 1984.
[4] S. R. Nagel, "Review of the depressed cladding fiber design and performance for the SL undersea system," this book, ch. 12, p. 157.
[5] M. W. Perry, G. A. Reinold, and P. A. Yeisley, "Physical design of the SL repeater," this book, ch. 21, p. 313.
[6] D. G. Ross *et al.*, "A highly integrated regenerator for 295.6 Mb/s undersea optical transmission," this book, ch. 26, p. 377.
[7] C. D. Anderson and D. L. Keller, "The SL supervisory system," this book, ch. 3, p. 39.
[8] J. L. Fromme and M. D. Tremblay, "Terminal transmission equipment (TTE) for the SL transmission terminal equipment," this book, ch. 43, p. 589.
[9] C. Chamzas and P. R. Trischitta, "Simulation of a chain of digital optoelectronic regenerators," presented at International Modelling & Simulation Conference, Athens, Greece, June 27–29, 1984.

9
The NL1 Submarine System

R. L. WILLIAMSON AND M. CHOWN

INTRODUCTION

The application of optical fibers in submarine cable systems is expected to lead to higher capacities, smaller diameter cable, and increased repeater spacing with, consequently, a lower cost per circuit·kilometer. As reported elsewhere in this book, development programs are now being carried out in several countries, and it is the purpose of this chapter to give details of the work being undertaken in the U.K. by STC, with the assistance of British Telecom Research Laboratories, Martlesham.

SYSTEM CONFIGURATION

The block diagram of the regenerator is shown in Fig. 1 and the mechanical arrangement of a pair of regenerators is shown in Fig. 2. Regenerators will be housed, normally in multiples of two for duplex operation, in steel repeater housings which will be shorter than, but otherwise similar to, STC's conventional repeater housings. Repeaters will be interconnected via a cable with a construction similar to that illustrated in Fig. 3. The cable can accommodate at least four pairs of fibers with a secondary coating of 0.85 mm OD. However, it is envisaged that many systems will need only one or two pairs of fibers and that the requirement for the first long-haul system will be for not more than three pairs giving a capacity equivalent to about 12,000 4 kHz voice channels (which is three times that of the largest submarine coaxial system in production). This capacity could, of course, be multiplied by a factor of around three by the use of digital speech interpolation.

SYSTEM PARAMETERS AND KEY COMPONENTS

The major parameters are summarized in Table I. Points worth noting are the following:

Wavelength

1.31 μm with fiber having a nominally zero dispersion at that wavelength has been chosen for initial systems. For the time being, operation at 1.55 μm where the fiber

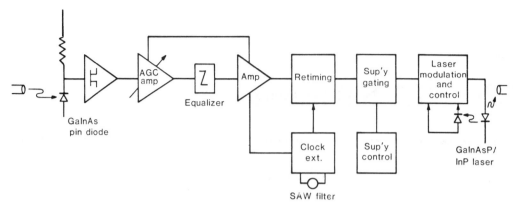

Fig. 1.   Block diagram of submarine regenerator.

Fig. 2.   A pair of regenerators with bulkhead and gland.

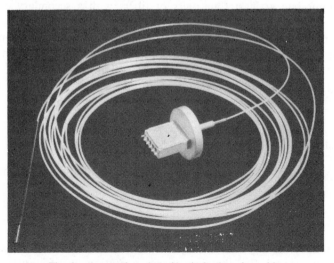

Fig. 3.   Armored version of optical submarine cable.

TABLE I
MAJOR SYSTEM PARAMETERS

Wavelength	:	1.31 µm
Bit-rate	:	280 Mbit/s
Line-code	:	7B/8B
Line-rate	:	325 Mbauds
Cable capacity	:	3 pairs of fiber(typ)
Transmitter	:	InGaAsP I.R.N. Laser
Receiver	:	InGaAs/InP PIN-GaAs FET
Timing Extraction	:	SAW filter
Electronics	:	Silicon bipolar IC's
Fiber dispersion	:	Zero at 1.31 µm
Fiber core diameter	:	8.5 µm

attenuation will be even lower is being avoided because optical components for this wavelength are not yet sufficiently advanced in their development.

Bit Rate

280 Mbits/s has been chosen to provide a simple high-level interface with a pair of 140 Mbit/s streams of the CCITT European hierarchy. Only minor modification is necessary to operate at the 274 Mbit/s rate of the North American hierarchy.

Line Code

A modified 7 bit/8 bit code giving a line rate of approximately 325 MBd has been chosen for its desirable properties of a high spectral density at the clock frequency (for easy timing extraction) and a low digital sum variation (to reduce dynamic range limitations in the integrating front end and to allow small value coupling capacitors). The version chosen permits engineering order wire information to be carried by spare code words and provides rapid recovery after loss of synchronization.

Transmitter

The module will use an inverted-rib waveguide (IRW) GaInAsP laser being developed at STC, Harlow, in a hermetic single-mode package (see Fig. 4) for manufacture by the ITT Components Group's Microwave and Opto-electronic Unit, Paignton, U.K.

The built-in waveguide of the IRW structure gives a stable zero-order mode near field distribution with a sufficiently narrow far field to allow high coupling efficiencies into

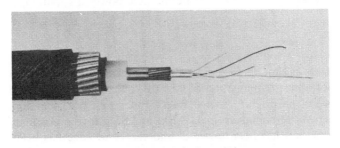

Fig. 4.   Packaged single-mode laser.

single-mode fiber. Launched powers greater than -4 dBm (mean) have been achieved with butt coupling and more than -1 dBm (mean) with lens coupling.

A major reliability qualification program on both laser and package is now underway and initial results are very encouraging. Threshold currents are 50–80 mA with no measurable change after 5000 hours at 20°C.

Receiver

The module uses an InGaAs/InP p-i-n diode combined with a GaAsFET mounted on a thick film substrate. The design is based on work carried out by BTRL and is being developed into a production item suitable for submarine systems' use by Plesey Caswell.

Timing Extraction

In exploratory development, experimental subsystems using phase-locked loops, LC resonant circuits, and helical resonators, as well as surface acoustic wave (SAW) filters were constructed and evaluated. The SAW appears to have a small margin of superiority, especially at the higher bit rates.

Electronics

The dominant factor in the choice of technology is that of reliability. Conventional coaxial submarine systems (such as STC's NG system which has a bandwidth of 45 MHz and a capacity equivalent to 4,140 4 kHz voice channels) have about one active semiconductor device per km. An optical system (with its much longer repeater spacing but greater circuit complexity) must have no more than about 15 active devices per one-way regenerator if their contribution to the overall reliability per circuit·kilometer is to be of the same order.

This makes the use of integrated circuits essential and, even then, each IC chip must have a reliability similar to that of an individual transistor in a conventional submarine system. However, it is reasonable to expect that this can be so because there is considerable evidence to show that semiconductor reliability depends on the active area of the chip rather than on its circuit complexity, and the semiconductor devices used on conventional systems have relatively large area. (The output transistor used on an NG system has, in fact, 32 separate emitter areas connected in parallel.)

Current development work is based on an uncommitted array using a silicon bipolar technology of proven reliability [1] and is expected to lead to a chip count, including supervisory and laser drive circuits, of 6–8 (which is around half the target figure given above). Chip count and circuit partitioning, which are dependent on the interaction of factors such as packaging, crosstalk, and permissible power dissipation, are not yet fully defined.

Fiber

The results shown at the 7th European Conference on Communications [2] described the achievement of a mean attenuation of 0.66 dB/km with a standard deviation of 0.08 dB/km for a total of 65 primary coated fibers. Since that time, development work has reduced the dispersion minimum from 1.37 μm to 1.31 μm (the nominal laser operating wavelength), attenuation has been lowered, and the bend edge has been pushed out to around 1.7 μm to give greater confidence that any cabling effects will be well away from the wavelengths of operation.

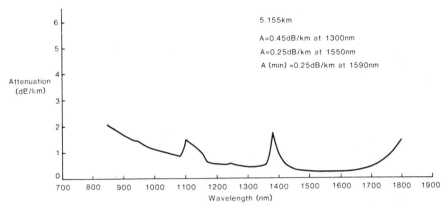

Fig. 5. Spectral attenuation of single-mode fiber.

Emphasis had been placed on reproducibility and uniformity so that spreads in attenuation have been reduced. Figure 5 shows a typical spectral attenuation curve.

POWER BUDGET

Tentative figures are given in Table II. Points worth noting are the following.

Source Level and Receiver Sensitivity

The target figures of −6 dBm and −3 dBm have been bettered in experimental units, but production values are not yet fully established.

System Design Penalties (SDP)

Allowances for jitter, for residual misequalization and mistiming, for dispersion and reflections, and for extinction ratios and turn on delays are included in the target figure of 2 dB. Its calculation involves the study of a large number of interactions.

For example, an increase in the penalty is produced by any timing offset of the decision point away from its optimum position in the "eye," and this can be caused in several ways.

1) The normal alignment jitter of all digital systems depending on the choice of code and of equalization and on the stability and Q-factor of the timing circuits.
2) Alignment jitter arising from some inservice supervisory and control techniques.

TABLE II
POWER BUDGET

Source Level	:	−6 dBm
Receiver Sensitivity	:	−36 dBm
System Design Penalties	:	−2 dB
Path Loss Capability	:	28 dB
Fiber attenuation	:	0.5 dB/km
Splice Loss	:	0.25 dB/splice
Path Loss	:	20 dB
System Margin	:	8 dB

3) Timing offsets arising from electronic limitations such as the setup and hold times of the flipflop.
4) The effect of laser partition noise in which any sudden changes in laser wavelength (which could result from mode jumps) are translated by chromatic dispersion of the fiber into a timing jitter at the next repeater.

These effects are not simply additive since the total penalty increases rapidly as the edge of the eye is approached. Furthermore the tradeoffs between system design penalties and production tolerances have to be considered. For example, production tolerances (and temperature variations) in both fiber and laser will lead to some mismatch between the laser operating wavelength and the wavelength of zero dispersion of the fiber.

Typical mismatches are likely to be a few tens of nm leading (for a wavelength jump of one mode which is approximately 1 nm) to a timing jump of several tens of picoseconds over a section length.

System Margin

The system margin (Mtot) is considered to be in three parts.

Mdes: An allowance for tolerances on source level, receiver sensitivity, path loss, and system design penalty.

Mdeg: An allowance for the effects of aging and for environmental effects.

Mrep: An allowance for repairs involving additional cable and/or additional splices.

Current calculations suggest these three contributions will be of approximately equal magnitude, although much work remains to be done to establish the minimum margins which can be routinely adopted with the confidence and reliability necessary for submarine systems.

RELIABILITY CONSIDERATIONS

The benign environment of the ocean bed with its nearly constant temperature and relative freedom from mechanical disturbance endows submarine cable systems with the potential of what is, for electronic systems, a very long operating life. At the same time, the inaccessibility of the submerged plant and the high cost of repair make it economically essential to invest heavily in efforts to establish that wear-out mechanisms are insignificant, that rogue failure levels are very small, and that infant mortality is negligible. With coaxial systems, it has been common to aim at a system life of at least 25 years and a submerged plant MTBF of over ten years. This has led to the design of special components, to accelerated life test programs (involving, for example, 6000 transistors on a 3000 hour test at temperatures ranging from 200 to 280°C equivalent to over one thousand years at normal operating temperature), and to manufacture on dedicated facilities where as much as 75 percent of the total output may be diverted to quality control testing leaving only 25 percent for system use.

Since the first long-haul submarine cable system was laid some 25 years ago and the first transistorized repeaters were put in service only 17 years ago, it is still too early to accurately judge just how successful these attempts have been. However, 800 million repeater hours of submarine operation have now been achieved, and the very small number of failures reported so far clearly indicates that the targets are close to being met and that the reliability which has been obtained is far greater than in any other branch of electronics.

With digital optical systems, the problems are, of course, very different, but the first indications are that similar figures should be achievable. For example:

1) The component count per circuit kilometer in the first-generation optical system is similar to that of the latest high capacity coaxial systems, in second-generation systems, which will probably have a higher level of integration and an even greater repeater spacing, it should be significantly less.

2) Recent results indicate that the new designs of optical components, intended for use at the longer wavelength of 1.3 μm, do not suffer from the wear-out mechanisms observed in some of the earlier short wavelength optical devices.

The work needed to establish the validity of these "first indications" is, however, far greater than it was for coaxial systems for the following reasons.

1) The variety of component types is far greater: Field effect transistors using GaAs, detector diodes using ternary semiconductor compounds, and laser diodes using quaternary material will all be needed (as well as bipolar integrated circuits using silicon which is the only semiconductor material now used in conventional repeaters).

2) Some of the above mentioned components are more novel and have a shorter development history than has usually been the case in submarine systems.

3) Silicon transistors will operate at temperatures as high as 200°C where many aging mechanisms are about 1000 times more rapid than they are at normal temperature so that accelerated testing at high temperature is a very powerful technique in establishing their reliability. In contrast, InGaAsP laser diodes will not lase at temperatures much above 90°C (although they continue to emit light at much higher temperatures). Therefore, the acceleration factor for those aging mechanisms which depend on the lasing process may be as low as ten. It follows that it may take far longer (and may cost far more) to establish, with a high degree of confidence, that lasers have the required reliability than is the case with other more easily tested devices.

This greater variety of more novel components, some of which are more difficult to test (although not necessarily less reliable) than the conventional ones, means that the early submarine optical systems may be committed to the use of some key components before their reliability has been demonstrated as fully as is usual in submarine cable applications. If this is so the use of remotely switched standby components to provide a measure of redundancy may be a worthwhile expedient, but a decision on this point will depend on the interim results of the test programs.

DEVELOPMENT PROGRAMS

Major events (past and future) of the development are shown in Table III and explained below.

TABLE III
PROGRAM MILESTONES

Cable Trial (Shallow Water)	Feb 1980 (Loch Fyne: 10 km)
Laboratory Demonstration	Feb 1981 (280 Mbit/s: 35 km)
Dry System	Mid 1983 (Preproduction components)
Cable Trial (Deep Water)	Autumn 1983
First short haul system	1984
Commercial short haul systems	1986 onwards (North Sea and Mediterranean)
Commercial long haul systems	1988 onwards

Cable Trial (Shallow Water) — February 1980

The prime purpose of the trial cable installed in Loch Fyne in February 1980 was to check that the design was suitable for handling and laying with a conventional cable ship [3]. The lay was satisfactorily accomplished with no significant changes in loss of any of the four multimode and two single-mode fibers of the cable. A repeater housing was also included and in the summer of 1980, this was recovered and two prototype regenerators (operating at 140 Mbits/s into multimode fibers) were installed and the cable relaid without incident.

A second purpose of the trial was to obtain long term data and the cable has been left in position so that measurements of fiber loss can be made from time to time. The repeater continues to function, and there is no significant change in the attenuation of the single-mode fibers at 1.3 μm (measurement accuracy is about ± 0.1 dB in a section loss of approximately 16 dB).

Laboratory Demonstration — February 1981

The figures given in Tables I and II have all been achieved or bettered in experimental setups and, in February 1981, a section length of 35 km at 325 MBd was demonstrated in the laboratory.

Dry Systems — Mid 1983

It is planned to construct a laboratory system consisting of a few sections (of regenerators and fiber but not cable or repeater housings). Its construction will serve initially, of course, as a further check on drawings and tolerances but it will also fulfill many of the functions of a reference set.

Cable Trial (Deep Water) — Autumn 1983

The design used in Loch Fyne (an armored version of the construction shown in Fig. 3) proved to be satisfactory in that trial and has successfully passed pressure and bending tests on the facilities of BTRL Martlesham. Theoretical calculations indicate that the unarmored version can be laid in deep water without problems, but the calculations indicate that recovery from deep water with typical ship motions could impose cable (and fiber) strains of about 0.4 percent. For safe recovery, this means that a fiber with a proof strain of greater than 1 percent would be needed.

Such fibers can be made in single lengths of several km, but it is not yet certain that production yields would be high enough to make this the most cost-effective approach. Consequently, an alternative design which could be recovered from the full operational depth of 7500 m using (if necessary) fiber of only 0.6 percent proof strain has been produced and will be evaluated in a deep water trial in the autumn of this year.

CONCLUSION

Most of the major parameters of STC's NL1 submarine system are now well established and development of the key components is well advanced. Emphasis is now being placed on the development of manufacturing, testing, and quality assurance procedures.

This first optical submarine system will have substantial advantages over its coaxial predecessors by virtue of its increased repeater spacing, smaller cable diameter, increased capacity, suitability for digital traffic, and its potential for the evolution of even greater improvements in later generations.

ACKNOWLEDGMENT

The authors express their thanks to the Directors of Standard Telephones and Cables plc, Greenwich, London, England, for permission to publish this chapter, to their many colleagues involved in this work, and, in particular, to G. J. Cannell, A. J. Jeal, P. A. Kirby, R. H. Murphy, G. Swanson, and J. G. Titchmarsh for their help in preparing this chapter.

REFERENCES

[1] D. Baker, "High reliability transistors for submarine systems" *Inst. Phys. Conf. Ser. 40*, pp. 87–105.
[2] J. Irven *et al.*, "Single mode fibre reproducibility" in *Proc. 7th Euro. Conf. Opt. Commun.*, Copenhagen, Denmark, Sept. 1981, pp. 2.4.1.–2.4.5.
[3] P. Worthington, in *Proc. 6th Euro. Conf. Opt. Commun.*, York, England, Sept. 1980, pp. 347–349.

10
The UK–Belgium No. 5 Optical Fiber Submarine System

GEORGE A. HEATH AND MARTIN CHOWN

INTRODUCTION

The design for this cable link is based on the advanced technology of the NL1 Submarine System [1] developed by Standard Telephones and Cables plc, London, in cooperation with the BT (British Telecom) Research Laboratories, Martlesham, and, therefore, is, generally, in accord with BT technology requirements [2]. The main points of interest are covered briefly in this chapter.

THE REQUIREMENTS

The co-owners of this system: Belgium, Netherlands, Germany, and the UK, require the provision of a large-capacity optical digital system between the UK and Europe to meet their forecast traffic needs.

Therefore, it has been decided to implement a fully engineered prototype system in June 1985 which will thereafter be ready for bringing progressively into commercial service. The usual submarine system design reliability requirements of no more than three failures in 25 years will apply to the full commercial system.

ROUTE DETAILS

The route will lie between terminal stations in Ostend (Belgium) and Broadstairs (United Kingdom) comprising some 122 km of suitably armored optical cable and three submarine repeaters. The proposed route of the system laid in 1985 is shown in Fig. 1.

SYSTEM CAPACITY

The optical submarine cable will be provided with three pairs of optical fibers each operating at 280 Mbit/s (traffic) giving a total capacity equivalent to 11 520 4-kHz voice channels, which could be increased using channel multiplication equipment as additional demand arises.

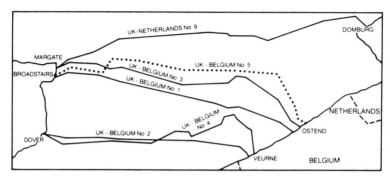

Fig. 1.   Proposed route of the world's first international undersea optical-fiber cable system.

Repeater Parameters and Supervision

The submarine repeater for the UK–Belgium system houses an assembly of six hermetically sealed regenerator modules. These are coupled in pairs and each Regenerator Pair is associated with a sealed Power Module.

The block diagram of an NL1 regenerator is shown in Fig. 2. Selected design details are as described below.

1) Operating Wavelength: 1.31 μm—with fibers having a nominally zero dispersion.

2) Operating Bit Rate: 280 Mbit/s per fiber pair. This has been chosen to interface with two 140-Mbit/s traffic streams, of the CCITT European hierarchy.

3) Line Code: The line code was chosen as a bounded 7B/8B code so that an integrating receiver design (explained below) could be used. It also gives a significant timing content in the data stream which greatly simplifies the timing extraction design.

For this code, a 7-bit data word is mapped into an 8-bit word for transmission, the redundancy allowing unwanted sequences to be forbidden. The digital sum variation (DSV) for this code is 6, and at the same time the code gives adequate clock frequency content (easing design requirements on clock recovery and the retiming circuit, particularly the limiting amplifier), bit parity redundancy (to allow error detection), and low frequency content (to allow ac coupling to be used). The line rate for the encoded signal is 324.315 MBd/s.

4) Transmitter: The transmitter block diagram is given in Fig. 3, in which binary input data is used to modulate the laser diode between current levels I_0 and I_1. The zero level I_0 is at the knee of the light power-current curve (Fig. 4) and the one-level I_1 is at a design value determined by the need for adequate optical power while operating within

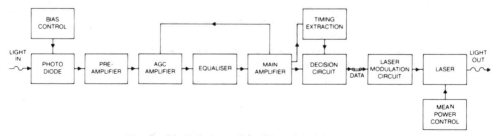

Fig. 2.   Block diagram of the NL-1 optical regenerator.

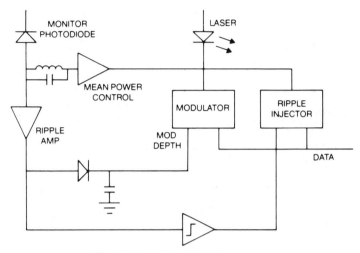

Fig. 3.   Transmitter block diagram.

safe margins for the components. Feedback is provided via the built-in monitor p-i-n photodiode to control both levels I_0 and I_1 against temperature or aging drifts. The modulation depth (dependent on $I_1 - I_0$) is controlled by injection of a ripple current at about 35 kHz. Fig. 4(a) shows the injected current as output optical power as data envelopes (since data is at a much higher frequency than ripple) for the case when I_0 is above threshold. In this case, the feedback circuit would recognize a relatively large ripple output from the monitor, and increase the modulation depth, tending now to the situation of Fig. 4(b). Here I_0 is below threshold and the detected ripple amplitude is now low, which forces a slight increase in modulation depth until the desired balance is achieved. A second feedback loop controls the bias current to maintain the desired mean optical power.

The semiconductor laser is a GaInAsP Inverted Rib Waveguide (IRW) type, chosen for its high reliability. Fig. 5 is a photograph of the laser package (with lid removed), including the sealed-in fiber tail, and Fig. 6 is a more detailed diagram covering the laser chip, the pin monitor photodetector, and lensed fiber end for efficient launching. The laser chip is Au/Su soldered onto a metallized diamond heatsink pedestal, and the fiber is than accurately aligned and held in position by means of an invar carrier (thermally expansion matched to the diamond pedestal) which is fixed to the copper submount by laser welding.

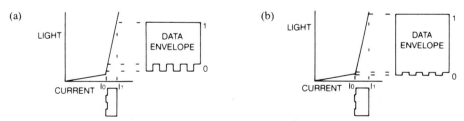

Fig. 4.   Ripple control of "zero" modulation level. (a) Zero's above threshold. (b) Zero's below threshold.

Fig. 5.   Photograph of the laser package.

A pin photodiode is mounted to receive light output from the rear facet of the laser for feedback purposes. The choice of oblique angle mounting is to avoid reflection problems (as secondary Fabry–Perot resonators lead to spectral broadening and instability), and this consideration also applied to the choice of lens launching into the fiber.

Lifetest data at room and elevated temperatures have amassed data to show that the IRW laser is a highly reliable device. Aging is found to fit a power law allowing us to predict an operating life well over 25 years at 50°C.

5) Receiver: The receiver front end consists of a GaInAs p-i-n photodiode, a GaAs FET having a high f_T, low-noise and low-gate capacitance, and a load resistor housed in a screened package, the p-i-n-FET node. See Figs. 7 and 8. This is the integrating type of detector in which a high value of load resistance is chosen to ensure good sensitivity. The consequence of a high RC product is that the incoming data stream is integrated, and has to be differentiated in a subsequent compensation network. An integrated sequence of data "1"'s would eventually be limited by supply rails, especially for a high received

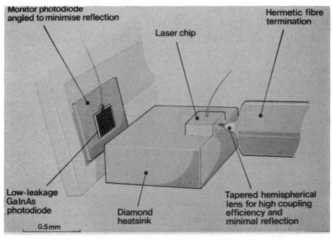

Fig. 6.   Diagram of the laser chip.

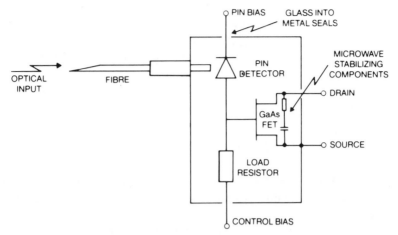

Fig. 7.   Schematic diagram of p-i-n FET node.

optical power, so much of the design is centred around dynamic range for the given code characteristics (especially DSV) and power rail voltages and sensitivity. With the chosen load resistor of 100K, a 20-dB dynamic range is achieved for the 7B/8B code. Contained in the same p-i-n-FET node package are the microwave stabilizing components to ensure unconditional stability over the band 1–20 GHz. The fiber tail, which is sealed into the package, is of a single-mode type to avoid modal noise. The central bias, is determined by an external feedback circuit to maintain the required drain voltage. The overall receiver block diagram is shown as Fig. 9, containing the bias controller and compensation network already referred to, and bipolar amplification stages.

6) Regeneration: This is accomplished in the following stages:

reamplification;
reshaping; and
retiming

as shown in the block diagram Fig. 2.

Fig. 8.   Photograph of p-i-n FET node.

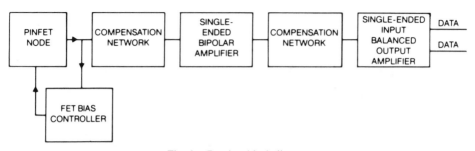

Fig. 9.   Receiver block diagram.

Employing this technique, the limitations of the timing extraction circuit give rise to jitter. For reasons of reliability no steps are taken to compensate for the accumulation of jitter effect in the repeaters.

The foremost factor in the choice of semiconductor technology for the regeneration electronics is that of reliability. The approach has, therefore, been to use integrated circuit chips which have a proven reliability pedigree inherited from their similarity with the individual transistors of a conventional coaxial analog submarine system.

The British Telecom ECL40 technology [3] has been used for many years in submarine systems and the integrated circuit chip count has been minimized by the use of an uncommitted array based on this ECL40 silicon bipolar technology of proven reliability.

7) Supervisory: The requirements of the system are

   a)   to locate, with the minimum ambiguity, a failed repeater; and
   b)   to locate a repeater giving a high error rate.

To fulfill the requirements the following supervisory facilities are provided

   i)   electrical "loop back" of the send and receive paths of either a failed repeater, or
         the preceding repeater;
   ii)   facilities for monitoring, in the terminal, the looped-back error rate;
   iii) measurement of the received light level of repeater; and
   iv)   measurement of laser status in a repeater.

Facilities i), ii), and iii) are provided on an out-of-service basis. Laser status associated with the input of any one fiber section, iv), is verified on an in-service basis.

Communication to a repeater is achieved by means of a tone which is frequency modulated onto the clock at the terminals at a level which does not degrade the system. In the repeater, a frequency demodulator is included in the clock extraction circuitry to recover the tone and use it for supervisory control.

The in-service laser health facility is operated by sending the address tone of a particular regenerator from the terminal at a level of less than ± 20-kHz deviation on the clock. This will not degrade the system, but will invoke the in-service laser health monitor. The regenerator will then return the address tone via phase modulation of the return clock at a level which is proportional to the laser threshold current of the addressed regenerator.

Fault location can be effected out-of-service by sending the address tone at a deviation level of greater than ± 20-kHz F.M. clock deviation. This will cause the addressed regeneration to go into electrical loop-back so that, by putting an error monitor on the terminal, the bit error ratio of the looped path can be established. If different regenerators

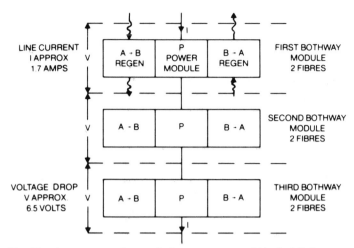

Fig. 10. Arrangement of power feed to regenerators within the NL-1 repeater.

are selected for loop-back the differing loops can be monitored and the offending repeater section can be established.

The receive light level of a regenerator is established by first sending its address tone to cause it to loop-back as described above. The data path is then used to send special 32-bit word patterns which have a 10 MHz frequency content. This 10-MHz content is filtered off within the regenerator, before the A.G.C. stage, and used to provide a dc level, which is compared with an internal reference. If the 10-MHz pattern content is now increased until the internal reference is exceeded then the loop-back command will be overridden. The received light level can now be computed by knowing the 10-MHz content of the pattern required to override the loop-back command.

Finally, this loop-back override can itself be overridden by sending the address tone at an F.M. deviation of greater than ±40 kHz.

8) Power Feeding: Each regenerator pair in a repeater receives stabilized power at about −6.5 V via its associated surge-protected power module within the repeater. Each of these both-way assemblies dissipates about 11 W via insulated thermal transfer assemblies to the repeater housing. The three both-way assemblies are connected in series as indicated by Fig. 10. Since the transmission and power feed paths are totally separate, power separation filters obviously are not required.

9) Housing: The repeater housing is basically similar to the well-proven construction used for analog coaxial cable submarine systems. Its features are as follows:

outer casing (cylindrical); bulkheads (high-to-low pressure): nickel–chrome–molybdenum high tensile steel;
outer casing protection: expoxy, coal tar, glass fiber reinforced compound;
cable termination: armor taper grip termination;
fiber termination; power feed termination: within low-pressure area splice chamber.

CABLE

Optical Core

The optical core consists of six secondary-coated single-mode fibers and two fillers of the same diameter. The fibers and fillers are bound to a central kingwire which consists of a nylon-jacketed copper-coated steel wire. The fibers are low-loss silica having nominal core and reference surface diameters of 8.7- and 125-μm, respectively. The fibers are protected by a primary coating of silicone rubber followed by a nylon secondary coating to an overall diameter of 0.85 mm.

All fibers are proof tested to 0.6-percent strain after secondary coating. Any splices that are included prior to cabling are proof tested at 1.0 percent.

Pressure-Resisting Housing

i) A split copper tube is formed around the optical package and drawn to a diameter of 5.32 mm. During this operation, the internal volume of the optical package and the copper tube is filled with a high-viscosity waterblocking material.

ii) A copper tube is formed and welded over this and drawn down to be a close fit over the inner copper tube. This forms a hermetic barrier to protect the fibers from ingress of water or hydrogen.

iii) The cable strength member consists of two layers of high-tensile steel wires. The first layer has ten 2.64-mm nominal diameter wires applied with a left-hand lay. The second layer has 32 1.19-mm nominal diameter wires applied with a right-hand lay.

The lay lengths are such that the strength member is torsionally balanced and does not generate any torsion in tension.

Note that i), ii), and iii) jointly form the power feed conductor for the submarine system. The dc resistance is less than 0.7 Ω/km.

Insulant

The power feed conductor is protected and insulated by a single-pass extrusion of natural-grade low-density polyethylene. The diameter over the polyethylene is nominally 26.2 mm for deep-water cables.

For shallow-water cables, where additional armoring is applied to the cable, the insulant diameter is increased to 44.5 mm.

Armoring

Polypropylene rove serving and two layers of steel armor wires coated with a coal tar compound are laid over the insulant to protect the cable on the seabed for this particular route. Further cable design details are given in [4].

TERMINAL EQUIPMENT

Power Feed

The power feed equipment is required to supply a total constant current of approximately 1.7 A to energize the repeaters.

The equipment is adapted from the well-established ultrasonic (20-kHz) power feed terminals used for analog coaxial cable systems. The features are shown in Fig. 11.

A duplicate power feed cubicle (PFC) is provided at each terminal with the standby cubicle working into a dummy load to provide switched hot transfer facilities (HTS).

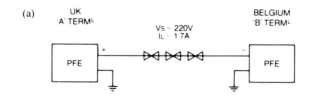

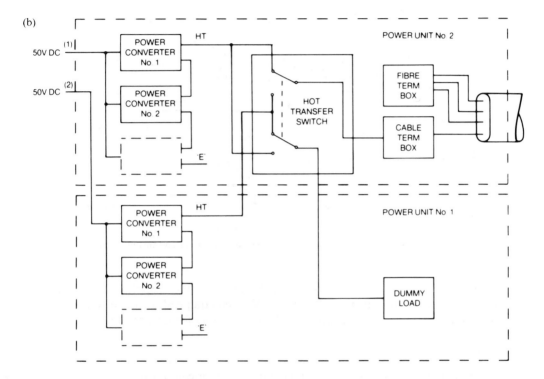

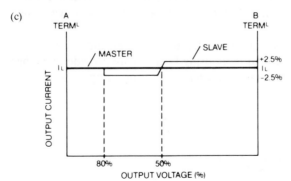

Fig. 11. Features of power feed for the UK-Belgium no. 5 Nl-1 optical system. (a) System schematic. (b) Block schematic of a terminal. (c) Master-slave operating characteristic for a double-end feeding system.

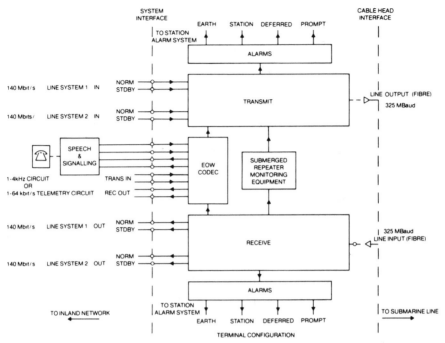

Fig. 12.   280-Mbit/s submarine transmission terminal—cablehead and inland system interfaces.

The equipment operates from the nominal station battery supply of 50 V and is conservatively used within its rated 2.0 A, 500 V when equipped with two converters.

Single end feeding with duplication and dummy loading for hot transfer is possible.

Transmission

The transmission terminals are constructed in a modern equipment practice to provide the interface between the submarine cable and the respective Ostend and Broadstairs terminal stations.

The interfaces of this equipment for the transmit and receive transmission paths are shown in Fig. 12. Alarm and monitoring facilities are provided together with access to the traffic path for the submerged repeater monitoring equipment.

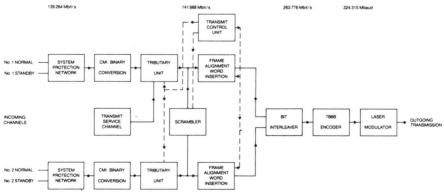

Fig. 13.   NL-1 optical submarine system transmit terminal.

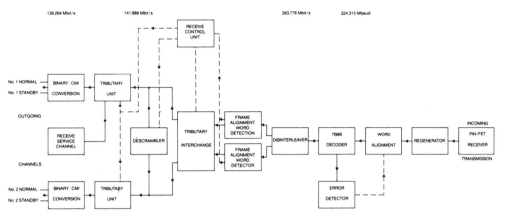

Fig. 14. NL-1 optical submarine system receive terminal.

Each 280-Mbit/s terminal transmission apparatus interfaces with two 140-Mbit/s CMI digital signal channels in accordance with CCITT Recommendation G703. These two input and output traffic ports, at 140 Mbit/s, are duplicated and a service protection network enables either normal or standby ports to be selected remotely.

A block diagram of the transmit terminal is shown in Fig. 13. After CMI to binary conversion the two working 140-Mbit/s traffic streams are synchronized and formed into a 280-Mbit/s frame structure. This elevated structure also provides for transmit service channels, scrambling, frame alignment, and transmit control timing. The two synchronized tributaries are then interleaved into a single data steam. Finally after 7B/8B encoding the transmission rate is raised to around 324.3 MB. Appropriate rates at the different stages are shown in Fig. 13. Optical transmission is effected via an IRW laser at a nominal mean output of −3.7 dBm.

Receive signals are detected at a level of −34 dBm (nominal) for a 10^{-9} BER, using a GaInAs p-i-n photodiode and integrally mounted FET amplifier. A block diagram of the receive terminal is shown in Fig. 14. For regeneration the 324.3-MB clock signal is extracted using a SAW filter, and this is followed by a word alignment process. The 7B/8B decoding operation, disinterleaver tributary interchange is a complementary process to that of the transmit terminal to ensure that the two halves of the data signal after CMI encoding are applied to the correct duplicate traffic output ports.

SYSTEM PLANNING

In an optical system, planning amounts to a study of the possible values of the power budget for a particular installation. For the UK–Belgium route the following values have been used.

A. Power Budget:

System Attenuation (dB):

splice losses	= 12
connector losses (terminals)	= 3
cabled fiber loss	= 55
total system attenuation (four sections)	= 70

B. Optical Power Margin:

source level	$= -3.7$ dBm
section loss for a three-repeater system	$=$ 17.5 dB
repair allowance	$=$ 3.5 dB
receive level	$= -24.7$ dB
receive sensitivity (for 10^{-9} BER)	$= -34$ dBm
system margin between receive sensitivity and receive level	$=$ 9.3 dB

This working margin is considered sufficient to cover device and measurement tolerances and degradation over the system life such as:

reduction in source level;
worsening in receive sensitivity;
increase in path loss; and
timing errors, jitter, etc.

C. Dispersion Budget:

Single-mode fibers are used in high-capacity systems such as NL1, because this choice avoids pulse distortion due to the range of propagating times which occurs in multimode fibers. However, there remain the limitations of material and waveguide dispersions, which combine to yield a variation in propagating time with wavelength, characterized by the parameter

$$M = \frac{d\tau}{d\lambda}$$

where τ is propagating time (ps/km) and λ is wavelength in nanometers. Now if the laser wavelength jumps (e.g., by a mode change) by $\delta\lambda$, there is a corresponding jump in time of arrival given by

$$t = M\delta\lambda L \qquad (1)$$

where L is the fiber path length in kilometers. This jump $\delta\lambda$ represents jitter, and, through the effect of displacing the received data pulses with respect to the steady clock stream, gives rise to a system penalty. We express this penalty as the decibel increase in optical power required to maintain specified maximum bit error rate.

In addition to the jitter, dispersion also gives rise to pulse spreading through a finite but constant spectral width of the laser. For the NL1 system, the effect is negligible.

Referring to (1), the system design approach is first to take a value of L somewhat above the maximum allowed by the power budget, and then to maintain low values both of dispersion M and laser spectral instability $\delta\lambda$.

The dispersion is kept low by operating near the zero-dispersion wavelength λ_0 of the fiber at which point material and waveguide components cancel and $M(\lambda_0) = 0$. The actual dispersion is finite, however, and limited by the fact that neither laser wavelength λ nor λ_0 can be perfectly controlled, so dispersion is proportional to $|\lambda_0 - \lambda|$. For NL1, we specify fiber and laser parameters to such that $\lambda_0 - \lambda < 40$ nm, to yield $ML < 120$ ps/nm over the working temperature range.

The next logical step is to determine the acceptable value of δt (clearly some fraction of the unit interval which is 3.1 ns at NL1 bit rate) and from that to determine the maximum δt to be specified.

This can be approached by drawing up a budget of all contributions to jitter and assigning a maximum value of δt, and hence determining a figure for $\delta \lambda$.

System experiments, however, showed a more complex situation, because the system penalty depends not on a single laser parameter $\delta \lambda$, but on the dynamic and statistical nature of these random mode changes. Also, the penalty is different for positive or negative values of M (i.e., $\lambda > \lambda_0$ or $\lambda < \lambda_0$), which has led to an asymmetrical specification on tolerances.

For the NL1 system, we have, therefore, had to develop a special laser test set, to demonstrate that each device has a spectral stability adequate in all respects to avoid performance degradation in the presence of the specified dispersion. This is described in [5].

Conclusion

This first optical system will give both designers and users an opportunity for hands-on experience of a commercially operating system. The lessons learned will be invaluable in assessing the requirements of future optical submarine systems and enable the evolution of this new technology to proceed on all fronts, with particular attention paid to reliability with economy.

Acknowledgment

The authors express their thanks to the Directors of Standard Telephones and Cables plc, London, England, for permission to publish, and to many colleagues who provided much of the material required for preparing this chapter, with particular appreciation due to S. Hill and P. Worthington for their assistance with the final draft.

References

[1] R. L. Williamson and M. Chown, "The NL1 submarine system," this book, ch. 9, p. 119.
[2] K. D. Fitchew, "Technology requirements for optical fiber submarine systems," this book, ch. 2, p. 23.
[3] D. Baker, "High reliability transistors for submarine systems," in ESSDERC, 1977.
[4] P. Worthington, "Cable design for optical submarine systems," this book, ch. 17, p. 251.
[5] P. J. Anslow, J. G. Farrington, I. J. Goddard, and W. R. Throssell, "System penalty effects caused by spectral variables and chromatic dispersion in single mode fiber optic systems," this book, ch. 32, p. 459.

11

A 150-km Repeaterless Undersea Lightwave System Operating at 1.55 μm

JOHN J. MCNULTY

INTRODUCTION

A repeaterless undersea lightwave system that will span 150 km is to be installed at an ocean depth of 2000 m during the first quarter of 1985. To transmit data through this distance without undersea regeneration, low-loss single-mode fiber and semiconductor lasers operating at 1.55 μm are required. The system line rate is 3.088 Mbit/s. This is double the input data rate of 1.544 Mbit/s because of the dipulse coding scheme used in the terminals.

This undersea lightwave system will be used as a data link from a shore terminal to a remote Air Combat Maneuvering Instrumentation (ACMI) range (see Fig. 1). ACMI ranges are used to train aircrews in Air Combat and Tactics Development using actual Jet Fighter Aircraft and Electronically Simulated Weaponry. This chapter describes the lightwave system architecture, design objectives, performance, and applications.

SYSTEM REQUIREMENT

To accommodate an ACMI range there exists a need for a highly reliable data transmission channel of 1.544-Mbit/s capacity between an on-shore control center and an off-shore floating platform some 150 km away.

A repeaterless lightwave system can span this distance using advanced lightwave technology available in early 1985. Important factors which need to be considered are operating wavelength and associated fiber loss and dispersion, data rate, laser reliability, output power, and receiver sensitivity.

In order to achieve this long length, the system would have to operate at the single-mode silica fiber loss minimum near 1.55 μm. However, at 1.55 μm, a typical single-mode fiber has 16 ps/km·nm of dispersion. The possibility of penalties associated with dispersion for transmission at 3.0 Mbit/s over long lengths of fiber and had already been evaluated [1], as shown in the Fig. 2 plot of the bit error rate versus received optical

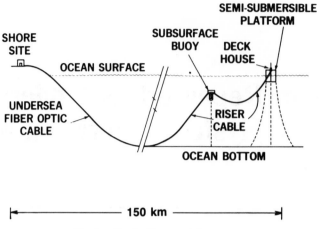

Fig. 1.   Physical layout of the system.

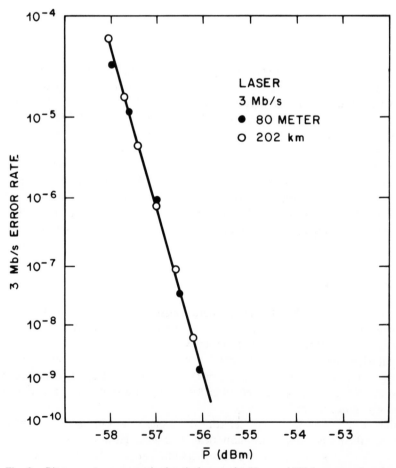

Fig. 2.   Bit error rate versus received optical power for 80-m and 202-km system lengths.

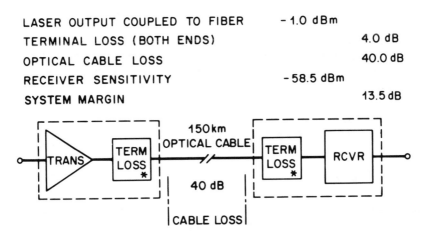

LASER OUTPUT COUPLED TO FIBER   − 1.0 dBm

TERMINAL LOSS (BOTH ENDS)   4.0 dB

OPTICAL CABLE LOSS   40.0 dB

RECEIVER SENSITIVITY   − 58.5 dBm

SYSTEM MARGIN   13.5 dB

*TERMINAL LOSS – CONSISTS OF CONNECTORS,
JUMPERS AND SPLICES

Fig. 3.   System optical loss budget.

power. As can be seen from the coincidence of the data corresponding to 80-m and 202-km fiber lengths, there is no measurable dispersion penalty. Therefore, at a data rate of 3.0 Mbit/s, the system would operate under a strictly loss-limited condition. The repeaterless span length is limited only by available coupled-in optical power, fiber loss at the operating wavelength, and receiver sensitivity.

This system, will be the first military undersea repeaterless optical system to be installed. Although this system pushes the technology, there still is a great degree of conservation built into the system. The optical loss budget given in Fig. 3, shows a substantial 13.5-dB margin. This margin should guarantee the required system performance throughout the life of the system.

Moreover, a repeaterless undersea system design does have the advantage, over a repeatered system, in that the optoelectronic devices are easily accessible at both ends. If repairs in the electronics are necessary, a ship is not needed for retrieval of a submerged repeater. Terrestrial system philosophy maintenance and improvement programs can be implemented on a repeaterless undersea system.

SYSTEM FEATURES

This system is a technically advanced fiber-optic system. The following shows some design highlights:

1) repeaterless span—150 km (approximately 93 mi);
2) transmission medium—single-mode depressed cladding optical fiber [2];
3) undersea cable design with high-strength fibers and metallic power conductors [3];
4) optical source–multilongitudinal mode InGaAsP semiconductor laser diode transmitters operating at 1.55 μm [4];
5) system capacity—1.544 Mbit/s with T1 interfaces (complies with CCITT recommendations);
6) line rate—3.088 Mbit/s using dipulse coding [5];

7) terminal regenerators—p-i-n receivers with, AGC, retiming and decision circuits optimized for 3.088-Mbit/s operation;

8) power transmission—Approximately 1/2 kW delivered to remote platform;

9) full duplex digital transmission over one pair of optical fibers;

10) 100-percent redundancy of optical transmission paths with automatic protection switching;

11) design error probability less than 10^{-7};

12) reliability—designed for a 25-year system life.

System Configuration

The basic system configuration is shown in Fig. 4. The input/output signals are bipolar at a rate of 1.544 Mbit/s (DS1 rate). A single full-duplex transmission channel is formed from the identical half-duplex channels, one in each direction (see Fig. 5). Since the input data is split before input to the transmit code converters, the data transmission channel is backed up with one-for-one redundant facilities. The output of each transmitter is coupled to a single-mode fiber. One optical circuit in each direction is considered the "active line," however, both lines are identical therefore, no preference is made.

Selection of the active circuit is made on the receive side of the system. The output of the receive code converters are connected to the protection switch circuit. This automatic switch circuit monitors both receive lines. Switching takes place and an alarm is activated when there is a loss of data or when excessive code violations are detected. (Criterion is one or more code violations during each of three consecutive 6.5-s time intervals. When these code violations are detected, the system bit error rate has exceeded the 10^{-7} requirement.)

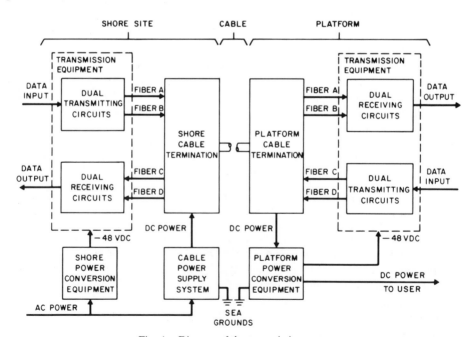

Fig. 4.   Diagram of the transmission system.

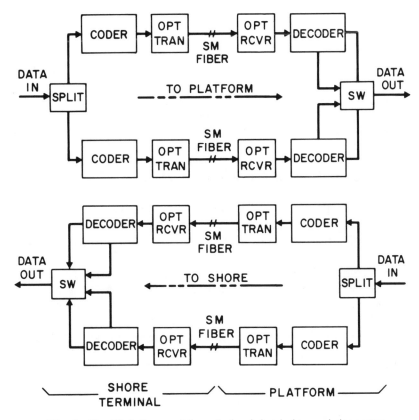

Fig. 5. Simplified diagram of the optical and electrical transmission system.

Data Transmission Link

Referring to the block diagram of Fig. 5, note that the terminal equipment at the shore and platform is identical. Automatic protection switching paths at the receive ends significantly increases the effective reliability of the data link because a minimum of two failures, one in each path in the same direction, are required to cause a long-term outage.

The input data to the system is bipolar return to zero (RZ) at the DS-1 rate. The bipolar RZ signal is then converted to binary nonreturn to zero (NRZ) data at 3.088 Mbit/s using dipulse coding.

An advantage of dipulse coding is that it results in a high density of transitions, thus providing simplified retiming. Also, framing at the receiver is trivial because there is no allowable "zero, one" sequence, making circuit implementation of encoding, decoding, and error monitoring simple. A disadvantage of dipulse coding is that its signal bandwidth is twice that of DS-1 which results in 3-dB signal-to-noise penalty at the receiver. For this application the price is acceptable. The code conversion and other circuitry in the electronic signal processors are implemented through use of small-scale integrated circuits.

SYSTEM—OPTICAL TRANSMITTER

The optical transmitter to be used in this system, which was initially designed for a higher bit rate, is shown in block form in Fig. 6. The laser is a highly index guided InGaAsP buried heterostructure design operating at 1.55 μm. It is packaged with a thermoelectric cooler and coupled to a lensed single-mode optical-fiber pigtail. The average power from the output of the fiber pigtail is -1.0 dBm. The temperature at which the laser operates is adjustable. For this application the laser stud temperature will be maintained at 25°C.

A back face photodiode is used with a feedback circuit to adjust the laser bias so as to maintain the constant average light output power. Laser aging and temperature effects are compensated for in this manner.

The input drive circuits are designed for higher speed operation but work well at 3.088 Mbit/s. Since a dipulse coded data stream has no long strings of ones or zeros, the low-frequency cutoff characteristics of the transmitter are adequate without circuit changes.

Two hybrid integrated circuits (HIC's) accomplish most of the external laser drive functions. The higher speed modulation function is implemented using microwave junction-isolated monolithic (MJIM) technology and the lower speed monitoring feedback function is implemented using complimentary bipolar integrated circuit (CBIC) technology. This transmitter is capable of operating at considerably higher speed than required by this application.

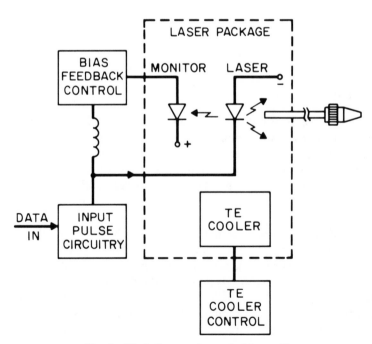

Fig. 6.   Block diagram of an optical transmitter.

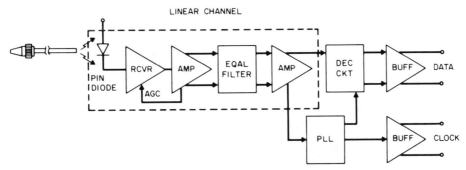

Fig. 7. Block diagram of an optical receiver.

The nominal characteristics of the transmitter are as follows:

average power coupled into the fiber	−1 dBm
wavelength	1.53 to 1.57 μm
spectral width	< 4 nm
maximum drive current (threshold plus modulation)	< 120 mA.

SYSTEM—OPTICAL RECEIVER UNIT

The optical receiver unit consists of InGaAs p-i-n photodetector, receiver front end, and regenerator circuits (see Fig. 7). The p-i-n diode and front-end integrated circuit are in a hermetic package with a multimode connectorized fiber pigtail coupled to the photodetector.

The receiver front end is a transimpedance feedback amplifier with an additional GaAs FET circuit which provides gain control. The receiver support circuitry consists of a buffer, a temperature compensated peak-to-peak detector, and a level shifting amplifier.

The regenerator portion of the receiver unit includes amplification, equalization, phase-locked loop retiming, and decision circuits.

The measured sensitivity of the receiver at 3.088 Mbit/s for 10^{-7} bit-error probability is less than −58.5 dBm. The optical dynamic range is greater than 24.0 dB.

UNDERSEA EQUIPMENT

The undersea fiber-optic cable connecting the shore terminal to the off-shore platform makes up the majority of underwater physical equipment. The physical layout of the system is shown in Fig. 1. The subsurface buoy is part of the attachment hardware for the cable to the platform.

The cable used for a majority of the distance (about 70 percent) in deep water, is similar to the SL system [3] lightweight cable, with the exception that it makes use of only four fibers (see Fig. 8). The fibers are embedded in an elastomer core and are helically wound around a central steel wire called the kingwire. The elastomer is used as a cushion for the fiber and to reduce cabling-induced microbending losses. The elastomer core is

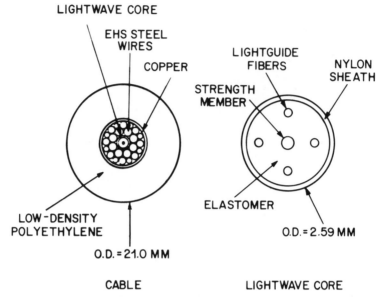

Fig. 8.   Lightweight cable (LW).

surrounded by a thin covering of nylon and then a series of steel strands which provide cable strength. Surrounding these strength members is a continuously welded copper cylinder which acts as a hermetic seal for the fiber core and as the conductor for delivering power to the platform. Surrounding the conductor is a layer of polyethylene which provides cable insulation and protection against abrasion.

Various forms of additional protection are provided for the cable in shallow water. The use of armor wires (single or double layers) (see Figs. 9 and 10) gives adequate protection up to the point where surf activity can cause further problems. In this region, "split pipe" will be used to cover the double armored cable.

A subsurface buoy is used to accommodate the connection of the riser cable to the platform. The main benefit afforded by the buoy is that it limits stress and motion on the

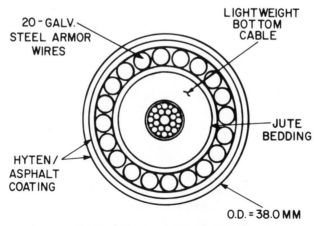

Fig. 9.   Lightwire armor cable (LWA).

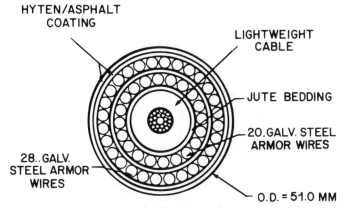

Fig. 10. Double armor cable (DA).

riser cable. The riser cable itself is a special design for this kind of application catering to surviving dynamic tension and flexure over its lifetime. With the exception of the fiber core previously described, the riser cable is completely different in its physical design. The steel strength strands are applied in counter directions so as to achieve a torque balance design. The welded copper cylinder is not used and a two-layer polyethylene jacketing system is used to protect against fish bite.

An Optical Fiber

A depressed cladding single-mode fiber is used in the cable. The refractive index profile is shown in Fig. 11. It is achieved through the use of the modified chemical vapor deposition (MCVD) process. At 1.5 μm, the fiber loss is at the minimum. The nominal characteristics of the fiber used in this system are as follows:

attenuation on the wavelength range 1.53 to 1.57 μm	$\leqslant 0.25$ dB/km
splice loss	$\leqslant 0.30$ dB
cutoff wavelength	~ 1.2 μm
fiber diameter	125 μm.

Laboratory System Demonstration

A successful laboratory demonstration of "error free" transmission through 150 km of cabled single-mode fiber was conducted in April 1984 (see Fig. 12). This repeaterless system experiment operated at a data rate of 3.088 Mbit/s and at a wavelength of 1.53 μm.

Cable retrieved from the SL System Sea Trial [6], which has twelve single-mode fibers embedded in the central cable core, was used as the transmission medium. By making eight passes through this 18.2-km length of cable plus additional fibers a total span length of 150 km was achieved. The loss of this span, including connectors, looping fiber jumpers, and fiber splices was 50.0 dB.

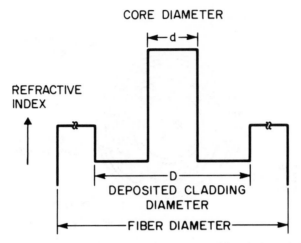

Fig. 11. Refractive-index profile of a depressed cladding single-mode fiber.

The optical transmitter used was a system prototype. This thermoelectrically cooled laser transmitter operated at 1.53 μm with average pigtail output power of -1.0 dBm.

The optical receiver-regenerator was also a system quality prototype. The p-i-n receiver had a BER of 10^{-7} for a received optical power of -57.8 dBm. This resulted in approximately 6.8 dB of margin. A plot of error probability versus received optical power is shown in Fig. 13.

The results of this demonstration using actual system hardware lends confidence to the expected design performance of the system. The transmitter and receiver performance in the system will be comparable to the demonstration. However, the cabled fiber loss is

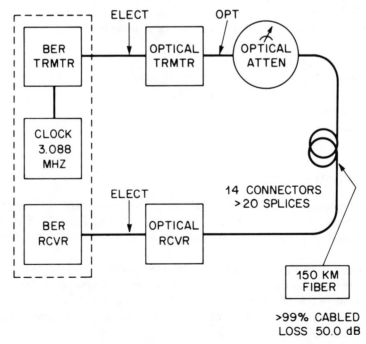

Fig. 12. Block diagram of a 150-km laboratory system demonstration.

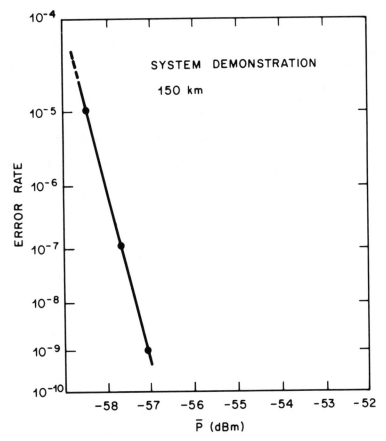

Fig. 13. Plot of error probability versus received optical power for laboratory system demonstration.

expected to be lower in the actual system. From this, one may infer that there will be a comfortable system margin.

CONCLUSION

The undersea lightwave system described in this chapter is made possible by the recent technical advances in lightwave technology. By taking advantage of low-loss single-mode fiber (at 1.5 μm), semiconductor lasers with narrow spectral width and receiver sensitivity performance approaching -60.0 dBm at 3.0 Mbit/s, a system spanning 150 km without regeneration, will be installed early in 1985. This is the first military application of a long-wavelength undersea system.

ACKNOWLEDGMENT

I would like to thank the many individuals from AT&T Bell Laboratories and AT&T Technologies who have contributed to the development of this system and provided information for inclusion in this chapter.

REFERENCES

[1] V. J. Mazurczyk, "202 km transmission spans at 1550 μm with multilongitudinal mode lasers," Post Deadline paper, Conf. Opt. Fiber Commun., New Orleans, LA, Jan. 1984.

[2] S. R. Nagel, "Review of depressed cladding fiber design and performance for the SL undersea system," this book, ch. 12, 155.

[3] A. Adl, T.-M. Chien, T. C. Chu, "Design and testing of SL cable," this book, ch. 16, p. 233.

[4] N. K. Dutta, R. B. Wilson, D. P. Wilt, P. Vesomi, R. L. Brown, R. J. Nelson, and R. W. Dixon, "Performance comparison of InGaAsP lasers emitting at 1.3 μm and 1.5 μm for lightwave system applications," *IEEE J. Quantum Electron.* (Special Issue), June 1985.

[5] J. S. Cook and S. D. Personick, "Optical communication system with bipolar input signal," U.S. Patent 4,001,578, Jan. 4, 1977.

[6] P. K. Runge, "Deep sea trial of an undersea lightwave system," in *Tech. Dig. Opt. Fiber Commun. Conf.*, vol. MD2, 8 (New Orleans, LA), 1983.

Part III
Undersea Fiber

Undersea fiber must be strong enough to survive cable laying and deep sea recovery operations. This has led to the development of high strength, low loss fiber for undersea use. The requirements imposed on the undersea use. The requirements imposed on the undersea fiber are examined with particular emphasis on the influence that hydrogen has on the loss of such fiber.

12

Review of the Depressed Cladding Single-Mode Fiber Design and Performance for the SL Undersea System Application

SUZANNE R. NAGEL, MEMBER, IEEE

INTRODUCTION

The proposal for high-capacity long-haul digital undersea lightwave cable systems presented many difficult challenges to the optimization and fabrication of the lightguide transmission media [1], [2]. System considerations for deployment of such an optical SL system in the late 1980's, based on cost, capacity, and reliability concerns, pointed to the use of single-mode fibers operating at 1.3 μm with injection laser light sources. The low dispersion of such lightguides was deemed suitable for 280-Mbit/s operation with relaxed requirements on the spectral purity of the laser source, while the potential for achieving cabled and spliced losses of 0.7–1.0 dB/km would allow repeater spacing of 27–54 km. Such considerations required not only a low-loss dispersion optimized fiber design, but one which was insensitive to microbending effects on cabling, deployment, and service, and which could be spliced with low loss. In addition, reliability considerations placed stringent requirements on the mechanical properties of the fiber, suitable for system lifetimes of 25 years. Long continuous lengths of fiber, capable of being cabled, being deployed, and should cable repair be necessary, withstanding recovery strains of up to 1 percent for 5 h, were essential. Such considerations dictated minimum initial strengths on the order of 200 ksi, underscoring the need for long-length high-strength fiber drawing as well as a high-strength fiber splicing technique. Finally, such fibers had to be capable of being produced in large quantity at reasonable cost. Formidable as these challenges appeared, the tremendous advances in research, development, and manufacture of the lightguide transmission media have resulted in fiber performance which has exceeded these initial goals using the depressed index cladding fiber design. This chapter will review how the results of research and development on the Modified Chemical Vapor Decomposition (MCVD) process; fiber drawing, coating, and splicing; and fiber design optimization have led to the simultaneous realization of stringent optical, dimensional, and mechanical properties, and culminated in the large-scale manufacture of the depressed index cladding fiber design by AT&T Technologies.

MCVD Preform Fabrication

The MCVD process [3], [4] has emerged as one of the major perform fabrication processes for high-quality optical communications fibers. The process uses vapor entrainment of halide reagents to form high-purity lightguide glasses by high-temperature oxidation. The chemicals are injected into a high-quality silica cladding tube which is mounted and rotated in a glass working lathe and heated by a traversing oxyhydrogen torch. Layer by layer of high-purity glass of controlled chemical composition is deposited to build up material of the desired refractive index distribution and thickness. Further heating is then used to collapse the composite tube to form a solid rod preform with the requisite core-cladding structure that then can be drawn down into fiber. Over the past decade, tremendous understanding of the basic physics and chemistry of this process has evolved, enabling a wide variety of fiber designs to be studied, as well as the economical large-scale manufacture of lightguide fiber [5]. The flexibility and versatility of the process have allowed the rapid transition from first generation multimode fiber fabrication to next generation single-mode fiber production by simple changes in the chemical delivery program.

In single-mode fiber preforms fabricated by MCVD, the silica starting tube provides the outer cladding in the lightguide fiber design. Sufficient low-loss cladding material is first deposited in order to ensure that the mode power traveling in the cladding at any wavelength is not attenuated by the lossy outer silica tube. SiO_2 cladding, typically modified with small amounts of P_2O_5 and/or F to control both the processing temperature and refractive index of the inner deposited cladding, is used. GeO_2-doped silica cores are most commonly used to achieve the desired refractive index structure, and both step index and graded index profiles can be made [6]. P_2O_5 is specifically avoided in the core to prevent deleterious P–OH absorption effects, especially at 1.55 μm. In actual MCVD fiber structures, deviations from an ideal step index profile can occur due to both diffusion of the dopants causing rounding of the profile from an ideal profile parameter of $\alpha = \infty$ and a central dip resulting from GeO volatization during the collapse step. However, it has been shown that dip widths of up to 45 percent of the core diameter have a negligible effect on the cutoff wavelength, microbending sensitivity, and spot size if equivalent step index parameters are used to predict these properties [7], and actual measured index profiles can be used to accurately compute the propagation characteristics as a function of wavelength for any fiber design [8].

Excellent single-mode fiber losses have been achieved in single-mode fibers made using MCVD since the original achievement of 0.2 dB/km by Miya *et al.* [9] in 1979. Control of the OH incorporation chemistry, particularly during the collapse step, has resulted in the achievement of very low OH levels [10] with OH peak heights at 1.39 μm as low as 0.05 dB/km reported [11].

Fluorine doping in MCVD was first reported by Abe [12] and subsequently used to make low-loss matched index cladding single-mode fiber designs where the deposited cladding was $F-P_2O_5-SiO_2$ composition whose index was equivalent to that of SiO_2 [13]. Ainslie *et al.* [14] first reported the use of large amounts of depressed index deposited cladding to fabricate structures based on this composition by increasing the fluorine/phosphorus ratio. The total percent index difference of the lightguide Δ was then given by

$$\Delta = \Delta^+ + \Delta^- \tag{1}$$

where Δ^+ is the index difference associated with GeO_2 doping, and Δ is the negative index for a given $F-P_2O-SiO_2$ composition. Increased absorption, scattering, and drawing-induced losses associated with GeO_2 were thus minimized and allowed low-loss fibers to be fabricated and a wide range of fiber designs to be explored. As precise control of the minimum dispersion wavelength λ_o became a concern for the SL system, the concept of depressed index cladding structures fabricated by MCVD was extended to allow simultaneous optimization of λ_o, minimum loss, cutoff wavelength, and microbending sensitivity of fibers leading to the depressed cladding index fiber design [15], [16].

All aspects of the chemistry and physics of MCVD, in particular in regard to single-mode preform fabrication, have been investigated. Definitive studies of GeO_2 incorporation chemistry allow controlled fiber core compositions to be fabricated [17]. The dependence of the achievable index depression with fluorine doping has been definitively related to the concentration of $[SiF_4]^{1/4}$ in the gas stream during high-temperature deposition. Thus, index depressions can be accurately predicted as a function of flow conditions, and use of SiF_4 as a starting dopant is preferred for maximum efficiency, especially for achieving deep index depressions or at high rates [18]. Use of chlorine during collapse is particularly important for achieving low OH single-mode structures. Optimized high-rate deposition with rates up to 2.3 g/min has been reported for single-mode preforms, having losses close to the theoretical limit and optimized geometry using controlled traverse velocity during deposition [19]. Detailed studies of the collapse rate and stability have allowed optimization of this step [20]. Thus, the understanding of the MCVD process has allowed the economical fabrication of single-mode fibers with theoretical losses and optimized process performance and rate.

PROPAGATION CHARACTERISTICS OF DEPRESSED INDEX SINGLE-MODE FIBER DESIGNS

The basic concept of making single-mode fibers by MCVD using an inner deposited cladding, whose index was lower than the outer silica substrate tube, and whose core, of refractive index n_c, was based on GeO_2-SiO_2 compositions, addressed a number of fabrication and design optimization issues [14]–[16], [21], [22], especially for fibers which were to operate at 1.3 μm. The cutoff wavelength λ_c, for single-mode operation, is achieved for ideal step index fibers when $V = 2.405$, where

$$V = \frac{2\pi a n_c}{\lambda}\sqrt{2\Delta}. \tag{2}$$

This placed limits on the allowed core radius a and percent index difference Δ since λ_c in the range of 1.20–1.27 μm was required. Value of the mode field radius ω or spot size, was another critical parameter since it is related to laser coupling characteristics, splice loss sensitivity, and microbending behavior. Compromises are necessary in optimizing this value since large ω increases laser coupling efficiency and decreases splice loss sensitivity, while small ω is favored to achieve good microbending performance. The mode field radius can be related to the V number of the fiber, most commonly by using the Marcuse analytical formula [23]

$$\omega = a[0.65 + 1.619/V^{3/2} + 2.879/V^6]. \tag{3}$$

The very strong dependence of microbending on ω in general favors operation at $V > 1.9$, although highly buffered coating and carefully designed cable structures can address such sensitivity. Additional concerns in choosing fiber design parameters relate to loss and dispersion properties. Since the intrinsic scattering and absorption increases as the GeO_2 doping level increases, the very lowest loss designs favor minimizing GeO_2 [9]. Moreover, as the GeO_2 level increases in step index single-mode fibers, greater sensitivity to draw-induced loss effects has been reported [14], [21]. When higher GeO_2 was used to increase Δ for microbending insensitivity, λ_o increased in wavelength, typically to ~ 1.35 μm due to the increased material dispersion [22], [24]. This can cause severe reduction in fiber bandwidth at 1.3 μm, especially if spectrally wide laser sources are used [25], [26].

Such considerations placed very practical constraints on single-mode fiber designs for use in the SL undersea system design which would operate at ~ 1.3 μm. The most common fiber design was the matched cladding single mode with $\Delta = 0.2$–0.3 percent and a core diameter of 9–10 μm to achieve the desired λ_c simultaneously with low loss and λ_o near 1.3 μm. However, this design was found to be quite sensitive to microbending, particularly at long wavelengths [22]. Decreases in a and increases in Δ decreased microbending sensitivity, but moved λ_o to unacceptably long wavelengths. By using the depressed index cladding design concept, such problems could be overcome. The core and the cladding index could be varied independently along with the core diameter. Thus, one could achieve the high Δ and small ω necessary for microbending insensitivity while keeping Δ^+ low enough to achieve the desired (λ_o as well as very low loss at 1.3 μm [15], [16].

High-bit-rate long-length SL undersea systems operating at 1.3 μm using commercial laser sources of finite spectral width require tight control on λ_o due to the dependence of the resultant bandwidth spectrum on the total chromatic dispersion as the spectral linewidth $\delta\lambda$ of the source increases [25], [26]. It has been shown that the peak bandwidth of the fiber using a laser operating at λ_o decreases as $(\delta\lambda)^2$. For wavelengths away from λ_o, the bandwidth falls away from λ_o causes the bandwidth to fall off proportionally to $\delta\lambda$. Thus, for a typical source with $\delta\lambda = 4$ nm, although the bandwidth is on the order of 1000 GHz·km near λ_o, the transmission bandwidth falls to 25 GHz·km at wavelengths ± 0.05 μm from λ_o [26]. Numerical parametric studies to calculate the dependence of λ_o on fiber design parameters in GeO_2–SiO_2 core fluorosilicate inner cladding fibers were shown to be an effective design tool for predicting λ_o. Figure 1 illustrates the dependence of λ_o on the total Δ for varying core diameters when an index depression $\Delta^- = 0.1$ percent is used [27]. The dashed curve indicates the limit for single-mode operation. Thus, such design tools are an effective guide for process optimization studies to control λ_o for fixed λ_c.

However, early fibers of the depressed index cladding design were shown to have higher losses at long wavelengths [14]–[16], attributed to a leaky mode loss mechanism which cause radiative losses of the fundamental mode [28]–[30]. The mode can be considered to be "cutoff" at wavelength λ_f where the effective index of the propagating mode equals that of the outer silica cladding. The λ_f moves to shorter wavelengths as Δ^+/Δ^- decreases. However, the magnitude of the leakage loss as a function of wavelength depends on the specific fiber design parameters used, as shown in Fig. 2 [30]. The wider the inner index well, the lower the leakage loss, and parameters can be chosen to simulate conventional step index structures. Additional modeling [30] determined the sensitivity of depressed index cladding designs to curvature induced radiative loss effects as a function of radius of curvature, thus further defining design parameters to minimize curvature loss values at any wavelength.

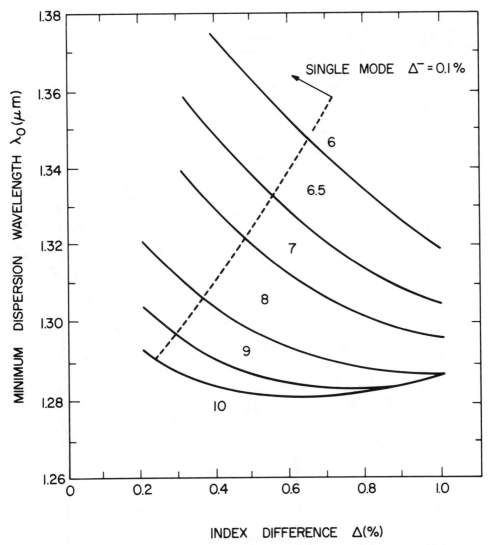

Fig. 1. Minimum dispersion wavelength versus total index difference Δ for depressed index cladding fibers with $\Delta^- = 0.1$ percent as core diameter varies from 6 to 10 μm. Dashed line indicates nominal limit for single-mode operation at 1.3 μm (after Cohen *et al.* [27]).

Experimental studies to examine a range of design parameters and their effect on fiber transmission properties were undertaken in order to explore optimal design parameters for MCVD fabrication of the depressed cladding design. Parameters which resulted in low loss in the 1.55-μm window while meeting the system goals for operation at 1.3 μm were determined [31], [32]. Based on values from such studies, fibers with nominal values of $\Delta^+ = 0.255$, $\Delta^- = 0.115$, core diameter $d = 8.3$ μm, deposited clad/core diameter ratios $D/d = 6.5$, and fiber $OD = 125$ μm were chosen as an optimized, practical, and manufacturable fiber design.

Figure 3 shows a typical MCVD refractive index profile achieved for such a fiber design. Despite the characteristic MCVD profile perturbations, excellent and reproducible

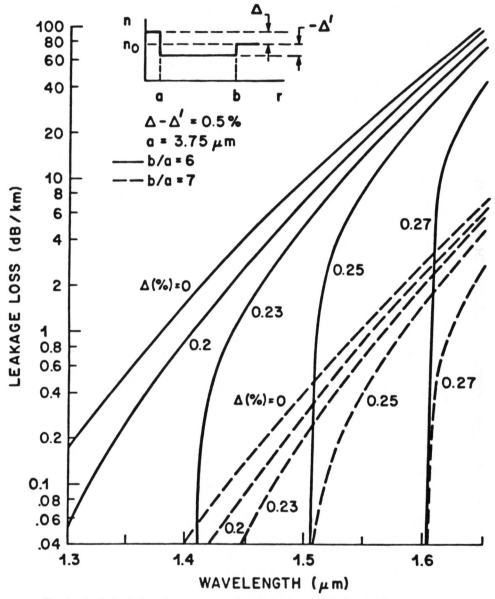

Fig. 2.   Radiative leakage loss versus wavelength for depressed index cladding designs as a function of varying index and clad/core ratio (after Cohen *et al.* [30]).

transmission results are achieved in such fibers. Figure 4 shows the typical dispersion versus wavelength curve for fibers of this nominal design where $\lambda_o = 1.312 \pm 0.002$ μm is achieved, as reported for fibers fabricated for use in prototype SL system experiments and the SL deep-water sea trial [33], [34]. Figure 5 represents a typical loss curve where losses of 0.35 dB/km at 1.3 μm and 0.20 dB/km at 1.55 μm can be achieved. Nominal mode field diameters of 8.7 μm are typical. Basket weave tests involving the deliberate crossover of successive layers on a reel were used to further characterize the microbending sensitivity

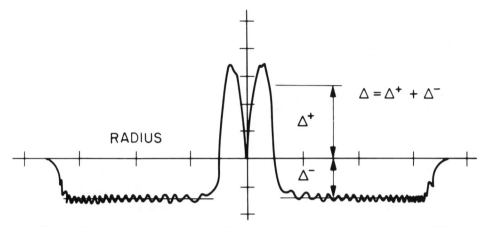

Fig. 3. Measured refractive index profile as a function of radial position for MCVD depressed index cladding design.

of fibers of this design, and no added loss at 1.3 μm with tensions up to 90 g was achieved [31], [35]. Although this test is a rather extreme measure of microbending sensitivity, it provides a rapid and valuable design tool to evaluate fiber performance. Such a microbend insensitive design greatly facilitates active monitoring of transmitted power during subsequent cabling and can be used to monitor any cabling problems. Excellent splice loss can be achieved as well, as discussed below. Based on such performance characteristics, the

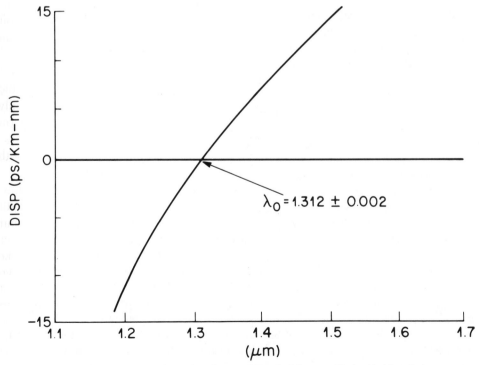

Fig. 4. Dispersion versus wavelength curve for typical depressed index cladding design.

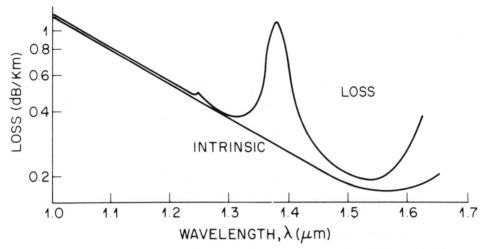

Fig. 5. Loss versus wavelength achieved for depressed index cladding relative to intrinsic material loss.

depressed index cladding fiber design met all the propagation requirements for its use in the proposed SL system.

MECHANICAL PROPERTIES OF SILICA FIBERS

Mechanical property optimization is also essential for the use of lightguides in undersea systems, and fundamental understanding of silica mechanical properties is critical to this end. The theoretical strength of fused silica, as reviewed by DiMarcello *et al.* [36], is on the order of 7–14 GPa ($1 - 2 \times 10^6$ lbf/in^2). However, the presence of defects or flaws can seriously degrade the intrinsic strength due to stress concentration at these defect sites. Such flaws are found to occur randomly along a given length of fiber, and the weakest link model predicts in fiber fracture where the most serious flaw occurs. In addition, the fracture stress that leads to failure is found to decrease with time, which is commonly referred to as static fatigue. Many studies have attempted to understand the long-term strength of silica fibers, and in general, the decreased strength is attributed to stress enhanced interaction of moisture at the defect site causing a subcritical flaw to grow with time to a critical size, resulting in fast fracture or failure.

Many studies have analyzed short gauge length strength data to examine the flaw distribution in fibers. In general, silica fibers show a very narrow unimodal strength distribution in short gauge lengths, with strength on the order of 700–900 ksi, but frequently a low-strength tail in the distribution is also observed, especially as longer gauge lengths are tested. This low-strength tail is attributed to extrinsic effects such as the creation of flaws on the fiber surface or bulk defects within the glass. It has been shown that the tensile strength statistics of high-strength fiber, when corrected for variation in fiber diameter, yield single values for the strength of silica, suggesting that in the absence of externally introduced flaws the fibers are flaw free as drawn. Thus the measured strength of 800 ksi in ambient conditions is a measure, under the test conditions, of the true theoretical strength.

However, the statistical nature of strength is such that as the desired length increases, the probability of an extrinsic flaw being present increases. Indeed, much of the work in fiber drawing has focused on the identification and elimination of all sources of such extrinsic effects and will be discussed in the following section. Because of this, proof-test screening is used to subject drawn fiber to a given strain so that any flaws greater than some critical size are eliminated in order to guarantee some minimum initial strength. For SL undersea cable application, a proof test of 200 ksi (1.4 GPa) corresponding to 2-percent strain is used to eliminate flaws 0.2 μm in size or greater. By using such total screening of fiber, a lifetime predictive diagram based on a fracture mechanics approach is used to take into account static fatigue effects [38]. This allows a very conservative estimate to be made of projected minimum time to failure for a given applied stress. Great care must be taken in fiber proof-test designs to ensure that significant crack growth does not occur during the testing itself, and such effects can be taken into account in deriving the lifetime predictive diagram. Although there are still many fundamental issues to be addressed in regard to the detailed mechanical properties of silica, tremendous progress has been made in optimizing these properties for lightguide applications.

FIBER DRAWING AND COATING

A critical second step in lightguide preparation is that of fiber drawing and coating, which can impact the optical, dimensional, and mechanical properties of the resultant lightguides. Many studies have been directed at the development of manufacturing technology capable of drawing fiber at practical line speeds with controlled diameter, applying in-line coatings which are concentric, uniform, and bubble free, and realizing the necessary control to yield very low transmission losses along with long lengths of reasonable strength (50–70-ksi) fiber. This has not only entailed a detailed understanding of the impact of process variations on strength, loss, and dimensions, but has required the development of sophisticated control and feedback techniques, the evolution of a compatible coating technology, and off-line studies of dynamic strength and static fatigue, fractography, and transmission characteristics [36], [39]. Use of optical fiber in undersea applications required even more detailed focus on factors affecting the strength of fiber during all processing steps since flaws on the order of 0.2 μm could lead to failure. Tremendous advances have been made in realizing high yield of high-strength MCVD single-mode fiber with excellent transmission properties and dimensional control, thus making the SL application feasible.

Figure 6 shows a state-of-the-art MCVD fiber drawing and coating apparatus. The preform is fed into a high-temperature furnace where it is heated in a controlled manner and drawn into fiber by means of a capstan to provide the pulling force. The fiber diameter is monitored and a signal is fed to the capstan to achieve a tightly controlled fiber outer diameter. In-line coatings are applied as liquids using a contactless coating applicator and cured by means of ultraviolet or thermal energy, and coating centering and diameter can be controlled using a feedback signal from a diameter monitor system. The cooling distance between where the fiber leaves the furnace and where it enters the coating cup is critical in determining achievable draw speeds as well as the ability to apply controlled uniform bubble-free coatings at a given draw speed. Clean Class-100 filtered air to purge the fiber path has been found to be critical in eliminating particles which could affect the fiber strength.

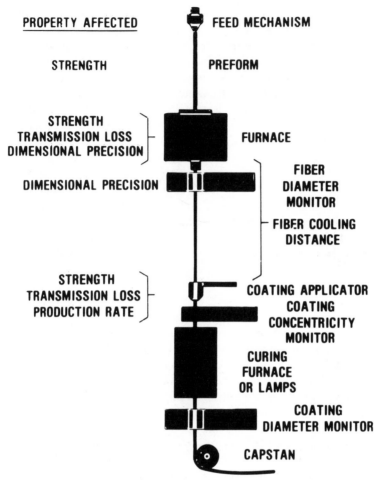

PROPERTY AFFECTED   FEED MECHANISM

STRENGTH   PREFORM

STRENGTH
TRANSMISSION LOSS   FURNACE
DIMENSIONAL PRECISION

DIMENSIONAL PRECISION   FIBER
DIAMETER
MONITOR

FIBER COOLING
DISTANCE

STRENGTH
TRANSMISSION LOSS   COATING APPLICATOR
PRODUCTION RATE   COATING
CONCENTRICITY
MONITOR

CURING
FURNACE
OR LAMPS

COATING
DIAMETER MONITOR

CAPSTAN

Fig. 6. Schematic of fiber drawing and coating apparatus.

Critical conditions for achieving long-length high-strength fibers from preforms drawn in practical high-temperature furnace environments were summarized and demonstrated by DiMarcello *et al.* [40]. An RF induction furnace with a yttria stabilized zirconia susceptor was used [41], capable of $\pm 1°C$ maintained temperature by means of feedback control from an optical pyrometer to the RF generator and not requiring a protective atmosphere. Significant factors for high strength included the quality of the initial starting tube and its subsequent handling during all steps of the preform and fiber drawing operation to avoid surface damage, using a clean heat source which does not generate contaminants that damage the glass surface, relatively high draw temperatures to smooth surface irregularities, a particulate-free coating material applied with a noncontacting applicator, and a clean particle-free environment during draw and coating. Subsequent work [42] defined the role of controlling convective flow on strength and diameter variations. In addition, lengths of 8.5 km at 300 ksi and 4 km at 500 ksi were reported: the first demonstration of the viability of long continuous lengths of proof-tested fiber. Successive proof testing at 200 ksi, followed by short gauge length testing which gave unimodal high-strength results, demonstrated that the proof test of single-coating UV curved acrylate fibers caused undetectable reduction of fiber strength by fatigue or coating degradation.

The first large database for high-performance high-strength 200-ksi fiber, providing an important manufacturability milestone, was reported for MCVD depressed index single-mode fibers prepared for SL undersea lightwave system experiments [43]. All aspects of high-strength drawing and coating technology were followed, including the use of a controlled convective flow zirconia furnace operated at 2200°C, carefully handled and firepolished preforms, prefiltered UV curable acrylate coating material applied with a double chamber, a bubble-stripper applicator, and a clean draw environment using filtered Class-100 air. A total of 54 preforms yielding 557 km of fiber was drawn and proof tested at 200 ksi; 74 percent passed proof test in 2.5 km or greater lengths, with a 50-percent yield in 6–12-km lengths and 6-km average length achieved. The resultant fiber had dimensional control to 0.25-μm standard deviation and average transmission loss of 0.38 dB/km at 1.3 μm. Thus, all elements of producing high-quality fiber for SL application were met.

FUSION SPLICING AND OVERCOATING

Despite the excellent results achieved for high-strength fiber in long lengths, as the length needed for a given application increases, the probability of achieving it decreases. This is shown in Fig. 7 [36] for the 557-km database previously discussed where the failure

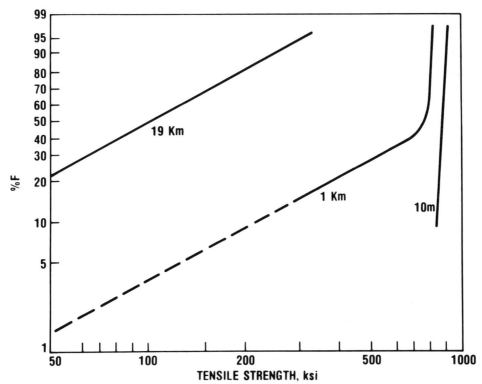

Fig. 7. Probability of a failure as a function of proof-test level for various unit lengths of fiber.

probability is plotted versus tensile strength for different fiber lengths. While short gauge length (10-m) data show characteristic unimodal high strength, the yield for gauge lengths of 1 km at 200 ksi drops to approximately 90 percent, and drops further to approximately 20 percent if 19-km unit lengths are required due to the low-strength tail typical for real fibers. Thus, for very long-length high-strength systems, a high-strength splicing technique is necessary to allow reasonable fiber manufacturing yields. Moreover, such a technique is critical for field assembly and repair, if necessary.

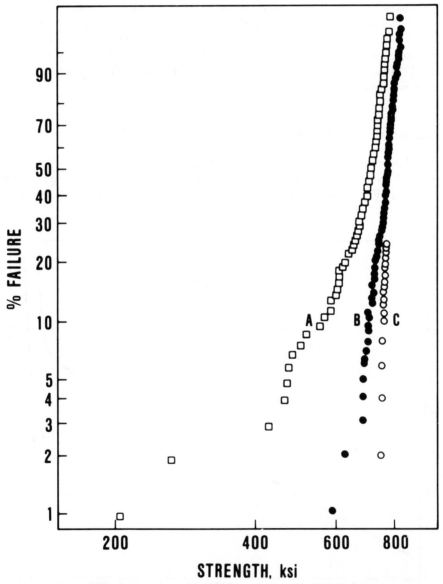

Fig. 8.   Strength distributions for A H_2/Cl_2 flame fusion splices, B H_2/Cl_2 flame fusion splices made with torch improvements, and C as-drawn fiber and splices shown in B when corrected for diameter and relative humidity variations (after Krause and Kurkjian [46]).

Work on fusion splicing was undertaken to address this need. Early work [44] demonstrated that the primary source of low strength in fusion splicing using a variety of heat sources was due to mechanical stripping of the coating and surface contamination degrading the fiber surface. By using chemical stripping techniques, median strengths in the range of 300–400 ksi (2.1–2.8 GPa) were obtained. Subsequent work [45] reported on the use of a Cl_2/H_2 torch to make flame fusion splices were median strengths of 570 ksi (246 Pa) were achieved. Such increases were attributed to reduction of water in the fusion atmosphere as well as to OH at the fiber surface, thus reducing thermally accelerated OH-enhanced corrosion at the high splicing temperatures. These strengths were further increased to 675 ksi by improved torch designs, as shown in Fig. 8, curve *A* [46]. Further improvements led to curve *B*, with median strengths of ~ 776 ksi, thus essentially retaining the strength of the pristine fiber, shown in Fig. 8, curve *C* [47]. Furthermore, the curve *B* data, when corrected for variations in fiber diameter and relative humidity, are equivalent to the data of curve *C*. This represents an unprecedented achievement in that it shows the ability to reheat a silica glass fiber to fusion temperatures within atmospheric

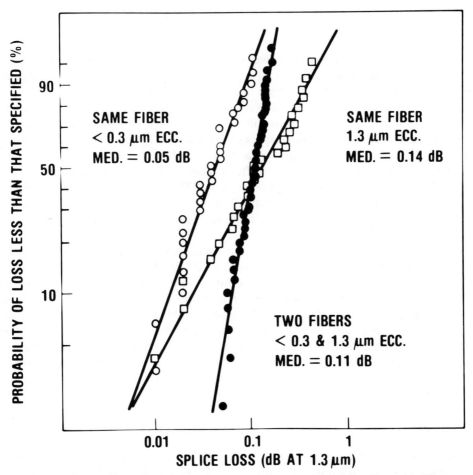

Fig. 9.   Flame fusion splice loss distribution for depressed index cladding fibers with different eccentricities (after Krause and Kurkjian [46]).

surroundings in a manner which prevents thermally induced corrosion. It also suggests that all OH at the surface is replaced before it causes corrosion and that Cl_2 or by-products in the flame itself do not cause strength degradation.

Splice losses achieved with the Cl_2/H_2 flame fusion technique are excellent, as shown in Fig. 9 for depressed index cladding single-mode fibers [47]. For fibers with low eccentricity (< 0.3 μm), median losses of 0.05 dB were achieved, while 1.3-μm eccentricity fibers showed median losses of 0.14 dB, thus demonstrating simultaneous achievement of very low loss with very high strength.

A further aspect of this work [46] demonstrated the ability to chemically strip the coating without causing swelling, thus rendering the fiber suitable for overcoating to the original coated diameter after splicing. An overcoating technique for rapidly coating fiber splices with UV curable coating materials has been developed [48] and allows the original coating diameter and concentricity to be maintained, along with high strength and low loss. Typically, when using such a procedure, the splice strength and loss are measured both before and after overcoating. This combination of low loss, high strength, and overcoating technology represents a critical achievement for realization of very long length undersea systems.

SYSTEM EXPERIMENTS

Two important undersea system experiments have been reported which demonstrate the feasibility of SL undersea cables using the MCVD depressed index cladding fiber design combined and high-strength fiber drawing and splicing.

The first [33] involved the preparation of a 101-km length of depressed index cladding fiber with a total of 12 high-strength splices, which was proof tested with a minimum tensile strength of 200 ksi. Individual splices were tested at 300 ksi. The total span loss at 1.3-μm was 0.38 dB/km and 0.29 dB/km at 1.5 μm, including splices. This span was used to demonstrate 1.3-μm repeaterless error-free transmission at 274 Mbits/s over 101 km and at 420 Mbits/s over 84 km. The fiber λ_o was 1.312 ± 0.002 μm with $\lambda_c = 1.23 \pm 0.02$ μm. This represented the first time that all elements of the technology were realized simultaneously to make long-haul high bit rate systems with high-strength fibers and splicing to meet undersea requirements.

The second experiment [34] was the SL deep-sea trial in September 1982, which used an 18.2-km cable containing twelve high-strength (200-ksi) fibers and high-strength splices (300 ksi). A number of important system transmission configurations were successfully demonstrated, with loss changes due to temperature, tension, and pressure being less than 0.1 dB under all conditions. A portion of the active cable (10.4 km) was laid, held 18 h, and recovered without any fiber breaks, with laying tensions as high as 6000 lb and recovery tensions up to 9300 lb. This first reported deep-sea trial of lightguide cable was a critical and important demonstration both of the ability to cable high-strength fiber and that the stringent mechanical property requirements could be realized in a deep-sea environment, coupled with optimal transmission properties.

H_2 STUDIES ON DEPRESSED INDEX CLADDING FIBERS

Considerable attention has recently been focused on the long-term reliability of fibers exposed to H_2 environments due to diffusion of H_2 into the lightguide with the potential for increased attenuation at wavelengths of system operation. Dissolved H_2 gives rise to a fundamental absorption band at 2.42 μm, which has its first overtone at 1.24 μm, and a number of other characteristic hydrogen peaks due to rotational and combination absorption bands [49], [50]. It is also possible for this H_2 to react with the glass network to form OH groups which could permanently increase the fiber loss. Studies of the depressed index cladding fiber design exposed to 3.5 atm of hydrogen for 86 h showed the characteristic H_2 absorption peaks, and on removal from the H_2 environment, exhibited complete outdiffusion and recovery to preexposure loss levels at ambient conditions after

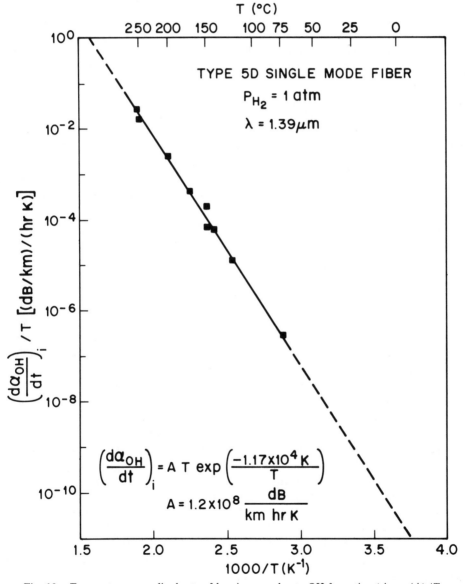

Fig. 10. Temperature normalized rate of loss increase due to OH formation $(d\alpha_{OH}/dt)/T$ versus reciprocal temperature for MCVD depressed index cladding fiber tested in 1 atm H_2 (after Lemaire and Tomita [51]).

60 days [51]. This lack of OH formation at room temperature is consistent with quantitative kinetic studies of hydrogen reactions with production MCVD depressed index cladding single-mode fibers [52]. Accelerated testing at $75°C < T < 250°C$ and varying pressures of H_2 allows predictions to be made for the long-term behavior of this fiber design at lower temperatures characteristic of ambient or undersea environments. Results of accelerated tests show that at a given temperature and hydrogen pressure, the rate of OH formation at 1.39 μm decreases steadily with time, and asymptotically approaches saturation value. A very conservative lifetime prediction can be made by plotting the initial rate of OH formation at 1.39 μm versus reciprocal temperature, as shown in Fig. 10 for the depressed cladding fiber design tested in 1 atm H_2. That the data all lie on a straight line indicates that a single reaction mechanism describes the OH formation over the temperature range investigated. For a 20-year lifetime, the predicted loss increase due to OH formation at 1.39 μm assuming $P_{H_2} = 1$ atm and $T = 4°C$, is only 0.0027 dB/km, corresponding to approximately five parts/billion (mole) OH. Thus, this fiber design is very resistant to OH formation.

Recent work [52] has also shown that an additional loss mechanism can occur in fibers exposed to H_2. The characteristic broad wavelength added loss due to this currently unknown H_2 interaction has an added loss signature which decreases weakly with wavelength in the range 1.0–1.7 μm. The magnitude of this loss has been shown to be the most significant loss mechanism associated with hydrogen in single-mode fibers, especially at wavelengths away from the OH and H_2 absorption maxima. However, the magnitude of this loss mechanism is such that the added losses at 1.3 and 1.55 μm are predicted to be less than 0.01 dB/km after 20 years at 4°C with $P_{H_2} = 1$ atm. Thus, the depressed index cladding fiber design has been shown to be very resistant to "permanent" loss increases due to H_2 effects for the SL undersea environment.

LARGE-SCALE MANUFACTURING RESULTS

A manufacturing process for the depressed index cladding fiber design was developed and introduced into large-scale manufacture by AT&T Technologies. The fiber design was optimized to provide minimal chromatic dispersion near 1.31 μm while maintaining a low-loss window at 1.55 μm. For the SL application, such a dispersion optimized design meets the criteria for dispersion less than 2.7 ps/nm·km at the system operating wavelength. Nominal design parameters of $\Delta^+ = 0.255$ percent, $\Delta^- = 0.115$ percent, $D/d = 6.5$, core diameter = 8.3 μm, fiber $OD = 125$ μm, and coating diameter = 250 μm are used, resulting in $\lambda_c < 1.27$ μm and mode field diameter of ~ 8.7 μm at 1.3 μm. Modified multimode MCVD stations are used to routinely produce preforms in 19×25 mm waveguide tubes by nontechnical hourly employees on multiple machine assignments on a five-day/three-shift basis. Data were initially reported for a 5000-km database [53] and more recently updated for 65 000 km of production fiber of this design [54]. Median losses of 0.38 dB/km at 1.3 μm and 0.21 dB/km at 1.55 μm were achieved, with the peak of the distribution having values of 0.35 and 0.21 dB/km at 1.3 and 1.55 μm, respectively, demonstrating the great capability of MCVD manufacture for high-quality single-mode fiber. Median core eccentricities of < 0.35 μm are typical [55], allowing very low loss splicing to be achieved.

More recently, high-strength depressed index fiber production has started at AT&T Technologies for the SL short-system experiment using MCVD preforms as described above [56]. All aspects of high-strength fiber fabrication were introduced successfully into the manufacturing facilities. Specific steps implemented to achieve high strength included the use of high-quality starting tubes, which were carefully cleaned and handled, then used in a clean preform fabrication environment, being particularly careful not to introduce torch contamination. Careful preform handling was used, taking care to remove any blemishes which might degrade fiber strength and to avoid surface contamination. A modified zirconia furnace to minimize particles in the furnace bore was used as a heat source to draw fiber in a clean draw environment. Single-layer clean UV curable acrylate coating material was applied in a clean application system with controlled fiber centering, bubble-free application, and proper fiber curing conditions. A proof-test apparatus designed to ensure uniform and smooth stress application at 200 ksi was used to test fibers. Initial results on 7500 fiber kilometers resulted in an average fiber length of 8 km after a 2-percent strain proof test. Such results are the largest database for highstrength lightguide fiber ever reported and indicate the suitability of MCVD large-scale manufacture of high-quality fiber for the high-strength applications. All fiber specifications required for use in the SL application have thus been realized in large-scale manufacture.

SUMMARY

The SL undersea lightwave cable system using single-mode fiber transmission media posed formidable fabrication challenges. Extensive fiber research and development effort by AT&T Bell Laboratories and AT&T Technologies has resulted in an extensive fiber technology base for the successful realization of the system goals. All key elements of lightguide technology have been demonstrated, and large-scale manufacture of depressed index cladding single-mode fibers has not only shown the economic viability of fabrication of this design, but has also achieved excellent optical and dimensional performance simultaneously with the achievement of very high strength. High-strength low-loss fiber splicing techniques have been developed in conjunction with overcoating technology, enabling the required long-length high-strength targets to be achieved. Cable manufacture, deployment, and recovery experiments have demonstrated the practicality of fiber design in regard to loss, dispersion, splicing, and strength requirements. Fiber reliability for the SL system lifetime has been addressed through static fatigue, proof test, and H_2 reactivity studies. Based on such results, the depressed index cladding designs have been chosen for use in the AT&T SL system to be used in TAT8 scheduled for deployment in 1988.

REFERENCES

[1] P. K. Runge, "High-capacity optical-fiber undersea cable system," in *Tech. Dig. CLEO*, 1980, p. 40.

[2] C. D. Anderson, R. F. Gleason, P. T. Hutchison, and P. K. Runge, "An undersea communication system using fiberguide cables," *Proc. IEEE*, vol. 68, pp. 1299–1303, Oct. 1980.

[3] J. B. MacChesney, P. B. O'Connor, F. V. DiMarcello, J. R. Simpson, and P. D. Lazay, "Preparation of low loss optical fibers using simultaneous vapor phase deposition and fusion," in *Proc. Xth Int. Congr. Glass*, 1974, pp. 6.40–6.44.

[4] J. B. MacChesney and P. B. O'Connor, "Optical fiber fabrication and resultant product," U.S. Patent 4 217 027, 1980.

[5] S. R. Nagel, J. B. MacChesney, and K. L. Walker, "An overview of the Modified Chemical Vapor Deposition (MCVD) process and performance," *IEEE J. Quantum Electron.*, vol. QE-18, pp. 459–476, Apr. 1982.

[6] M. A. Saifi, S. J. Jang, L. G. Cohen, and J. Stone, "Triangular profile single mode fiber," *Opt. Lett.*, vol. 7, no. 1, pp. 43–45, 1982.

[7] B. J. Ainslie, K. J. Beales, D. M. Cooper, and C. R. Day, "Fabrication and evaluation of MCVD single mode fibers with and without central index depression," *Electron. Lett.*, vol. 18, no. 19, pp. 809–811, 1982.

[8] U. C. Paek, G. E. Peterson, and A. Carnevale, "Effects of depressed cladding on the transmission characteristics of single mode fibers with graded index profiles," *Appl. Opt.*, vol. 21, no. 19, pp. 3430–3436, 1982.

[9] T. Miya, T. Terunama, T. Hosaka, and T. Miyashita, "Ultimate low loss single mode fibers at 1.55 μm," *Electron. Lett.*, vol. 15, pp. 106–108, 1979.

[10] K. L. Walker, J. B. MacChesney, and J. R. Simpson, "Reduction of hydroxl contamination in optical fiber preforms," in *Tech. Dig. 3rd IOOC*, San Francisco, CA, 1981, pp. 86–88.

[11] S. R. Nagel, S. G. Kosinski, and R. L. Barns, "Low OH MCVD optical fiber fabrication," *Ceram. Bull.*, vol. 61, no. 8, p. 822, 1982.

[12] K. Abe, "Fluorine doped silica for optical waveguides," in *Proc. 2nd ECOC*, Paris, France, 1976, pp. 59–61.

[13] B. J. Ainslie, C. R. Day, P. W. France, K. J. Beales, and G. R. Newns, "Preparation of long lengths of ultra low loss single mode fibre," *Electron. Lett.*, vol. 15, no. 14, pp. 411–413, 1979.

[14] B. J. Ainslie, K. J. Beales, C. R. Day, and J. D. Rush, "Interplay of design parameters and fabrication conditions on the performance of monomode fibers made by MCVD," *IEEE J. Quantum Electron.*, vol. QE-17, pp. 854–857, June 1981.

[15] P. D. Lazay, A. D. Pearson, W. A. Reed, and P. J. Lemaire, "An improved single mode fiber design, exhibiting low loss, high bandwidth and tight mode confinement simultaneously," in *Tech. Dig. Conf. Lasers Electroopt.* (Washington, DC), 1981, WG6.1–WG6.4.

[16] A. D. Pearson, P. D. Lazay, and W. A. Reed, "Fabrication and properties of single mode optical fiber exhibiting low dispersion, low loss, and tight mode confinement simultaneously," *Bell Syst. Tech. J.*, vol. 61, no. 2, pp. 262–266, 1982.

[17] D. L. Wood, K. L. Walker, J. R. Simpson, and J. B. MacChesney, "Reaction equilibrium and resultant glass compositions in the MCVD process," in *Tech. Dig. Opt. Fiber Commun.* (Phoenix, AZ), 1982, TUCC4, pp. 10–11.

[18] K. L. Walker, R. Csencsits, and D. L. Wood, "Chemistry of fluorine incorporation in the fabrication of optical fibers," in *Tech. Dig. Opt. Fiber Commun.* (New Orleans, LA), 1983, TUA7, pp. 36–37.

[19] K. L. Walker and R. Csencsits, "High rate fabrication of single mode fibers," in *Tech. Dig. Opt. Fiber Commun.* (Phoenix, AZ), 1982, PDI, pp. 1–2.

[20] K. L. Walker, F. T. Geyling, and R. Csencsits, "The collapse of MCVD optical preforms," in *Conf. Proc. 8th ECOC* (Cannes, France), 1982, pp. 61–65.

[21] B. J. Ainslie, K. J. Beales, C. R. Day, and J. D. Rush, "The design and fabrication of monomode fiber," *IEEE J. Quantum Electron.*, vol. QE-18, pp. 514–523, Apr. 1982.

[22] P. D. Lazay and A. D. Pearson, "Developments in single mode fiber design, materials and performance at Bell Laboratories," *IEEE J. Quantum Electron.*, vol. QE-18, pp. 504–510, Apr. 1982.

[23] D. Marcuse, "Loss analysis of single mode fiber splices," *Bell Syst. Tech. J.*, vol. 56, no. 5, pp. 703–718, 1977.

[24] L. G. Cohen, C. Lin, and W. G. French, "Tailoring zero chromatic dispersion into the 1.5–1.6 μm low loss spectral region of single mode fibres," *Electron. Lett.*, vol. 15, pp. 334–335, 1979.

[25] L. G. Cohen, W. L. Mammel, J. Stone, and A. D. Pearson, "Transmission studies of a long single mode fiber—Measurements and considerations for bandwidth optimization," *Bell Syst. Tech. J.*, vol. 60, no. 8, pp. 1713–1725, 1981.

[26] L. G. Cohen, W. L. Mammel, and S. Lumish, "Dispersion and bandwidth spectra in single mode fibers," *IEEE J. Quantum Electron.*, vol. QE-18, pp. 49–53, Jan. 1982.

[27] ——, "Numerical parametric studies for controlling the wavelength of minimum dispersion in Germanic fluoro-phosphosilicate single mode fibres," *Electron. Lett.*, vol. 18, no. 1, pp. 38–39, 1982.

[28] L. G. Cohen, D. Marcuse, and W. L. Mammel, "Controlling leaky-mode loss and dispersion in single mode lightguides with depressed index cladding," in *Tech. Dig. Opt. Fiber Commun.* (Phoenix, AZ) 1982, THCC1, pp. 52–53.

[29] P. D. Lazay, A. D. Pearson, and M. J. Saunders, "Control of the long wavelength loss edge in single mode fibers with depressed index cladding," in *Tech. Dig. Opt. Fiber Commun.* (Phoenix, AZ), 1982, THCC2, pp. 52–53.

[30] L. G. Cohen, D. Marcuse, and W. L. Mammel, "Radiating leaky-mode losses in single mode lightguides with depressed index claddings," *IEEE J. Quantum Electron.*, vol. QE-18, pp. 1467–1472, Oct. 1982.

[31] A. D. Pearson, P. D. Lazay, W. A. Reed, and M. J. Saunders, "Bandwidth optimization of depressed index single mode fibre by means of a parametric study," in *Conf. Proc. 8th ECOC* (Cannes, France), 1982, pp. 93–97.

[32] P. F. Glodis, W. T. Anderson, and J. S. Nobles, "Control of the zero chromatic dispersion wavelength in fluorine doped single mode optical fibers," in *Tech. Dig. Opt. Fiber Commun.* (New Orleans, LA), 1983, MF6, pp. 12–13.

[33] P. K. Runge, C. A. Brackett, R. F. Gleason, D. Kalish, P. D. Lazay, T. R. Meeker, D. G. Ross, G. B. Shawn, A. R. Wahl, R. E. Wagner, J. C. Williams, and D. P. Jablonowski, "101 km lightwave undersea system experiment at 274 Mb/s," in *Tech. Dig. Opt. Fiber Commun.* (Phoenix, AZ), 1982, PD7, pp. 1–4.

[34] P. K. Runge, "Deep-sea trial of an undersea lightwave system," in *Tech. Dig. Opt. Fiber Commun.* (New Orleans, LA), 1983, MD2, p. 8.

[35] A. Tomita, P. F. Glodis, D. Kalish, and P. Kaiser, "Characterization of the bend sensitivity of single mode fibers using the basket weave test," in *Tech. Dig. Symp. Opt. Fiber Meas.*, NBS Spec. Pub. 641, Boulder, CO, 1982, pp. 89–92.

[36] F. V. DiMarcello, C. R. Kurkjian, and J. C. Williams, "Fiber drawing and strength properties," in *Advances in Optical Fiber Communications*, T. Li, Ed. New York: Academic, ch. 4.

[37] C. R. Kurkjian and U. C. Paek, "Single-valued strength of perfect silica fibers," *Appl. Phys. Lett.*, vol. 42, no. 3, pp. 251–253, 1983.

[38] B. K. Tariyal, D. Kalish, and M. R. Santana, "Proof testing of long lengths optical fiber for a communications cable," *Ceram. Bull.*, vol. 56, no. 2, pp. 204–205, vol. 212, 1977.

[39] L. L. Blyler, Jr., and F. V. DiMarcello, "Fiber drawing, coating and jacketing," *Proc. IEEE*, vol. 68, pp. 1194–1197, Oct. 1980.

[40] F. V. DiMarcello, A. C. Hart Jr., J. C. Williams, and C. R. Kurkjian, "High strength furnace drawn fibers," in *Fiber Optics, Advances in R&D*, B. Bendow and S. S. Mitra, Eds. New York: Plenum, 1979, pp. 125–135.

[41] R. B. Runk, "A zirconia induction furnace for drawing precision silica waveguides," in *Tech. Dig. Opt. Fiber Commun.* (Williamsburg, VA), 1977, TuBS, pp. 22–24.

[42] F. V. DiMarcello, D. L. Brownlow, and D. S. Shenk, "Strength characterization of multikilometer silica fiber," in *Tech. Dig. IOOC* (San Francisco, CA), 1981, MG1, pp. 26–27.

[43] D. L. Brownlow, F. V. DiMarcello, A. C. Hart, and R. G. Huff, "High strength multikilometer lightguides for undersea applications," in *Conf. Proc. 8th ECOC* (Cannes, France), 1982.

[44] J. T. Krause, C. R. Kurkjian, and U. C. Paek, "Strength of fusion splices for fibre lightguides," *Electron. Lett.*, vol. 17, no. 6, pp. 232–233, 1981.

[45] ——, "Tensile strengths > 4 GPa for lightguide fusion splices," *Electron. Lett.*, vol. 17, no. 21, pp. 812–813, 1981.

[46] J. T. Krause and C. R. Kurkjian, "Improved high strength flame fusion single mode splices," in *Tech. Dig. 4th IOOC* (Tokyo, Japan), 1983, 29A, pp. 4–6, 96–97.

[47] ——, "Intrinsic glass strength achieved in fiber splices," in *Tech. Dig. Opt. Fiber Commun.* (New Orleans, LA), 1984, WI7.1–WI7.4.

[48] A. C. Hart, Jr. and J. T. Krause, "Coating technique for high strength fusion splices," *Appl. Opt.*, vol. 22, pp. 1731–1733, 1983.

[49] K. J. Beales, D. M. Cooper, and J. D. Rush, "Increase attenuation in optical fibres caused by diffusion of molecular hydrogen at room temperature," *Electron. Lett.*, vol. 19, no. 22, pp. 917–919, 1983.

[50] K. Noguchi, Y. Murakami, and K. Ishihara, "Infra-red absorption spectrum of hydrogen molecules in a silica fibre," *Electron. Lett.*, vol. 19, no. 24, pp. 1045–1046, 1983.

[51] E. W. Mies, D. L. Philen, W. D. Reents, and D. A. Meade, "Hydrogen susceptability studies pertaining to optical fiber cables," in *Tech. Dig. Opt. Fiber Commun.* (New Orleans, LA), 1984, WI3.1–WI3.4.

[52] P. J. Lemaire and A. Tomita, "Behavior of single mode MCVD fibers exposed to hydrogen," in *Conf. Proc. 10th ECOC* (Stuttgart, Germany), Sept. 1984, pp. 306–307.

[53] W. M. Flegal, E. A. Haney, R. S. Elliott, J. T. Kamino, and D. M. Ernst, "Mass production of depressed clad single mode fiber," in *Tech. Dig. Opt. Fiber Commun.*, Tu16 (New Orleans, LA), 1984, Tu16, pp. 56–57.

[54] D. P. Jablonowski, "MCVD fiber manufacture," *Advances in Optical Fiber Communications*, T. Li, Ed. New York: Academic, ch. 5.

[55] L. M. Boggs, "Optical measurements in the manufacture of optical fibers," in *Tech. Dig. Opt. Fiber Commun.*, WB1 (New Orleans, LA), 1984, pp. 88–89.

[56] F. Topalski, AT&T Technologies, PECC, private communication, 1984.

13
Influence of Hydrogen on Optical Fiber Loss in Submarine Cables

KIYOFUMI MOCHIZUKI, YOSHINORI NAMIHIRA, MEMBER, IEEE, MASAKUNI KUWAZURU, AND MAKOTO NUNOKAWA

INTRODUCTION

In submarine cables, it will be inevitable that some faults will occur which propagate seawater in the cables. In some coaxial cables, it has been known that hydrogen, which may be generated by electrochemical reaction between seawater and metals in the cables, has been accumulated in the cable [1]. Hence, in optical submarine cables, the necessity of the investigation of the long-term behavior of fibers in seawater and hydrogen environments had been proposed [1]. In terms of optical fibers themselves, there was no evidence of significant loss increase due to water. However, for the designer of the optical submarine cables, there has been a question whether the transmission loss in optical cables soaked in water is stable for the long term.

In these situations, it was found that hydrogen generated by electrochemical reaction between the metals and water in the cable causes the loss to increase due to H_2 vibration [2], [3]. The regions of the high additional loss lie in the wavelength range from 1.08 to 1.24 μm and in wavelengths longer than 1.59 μm. Meanwhile, at almost the same time, it was also reported that the transmission loss in phosphorus-doped fibers in a field-installed cable increased with time, especially at wavelengths longer than 1.2 μm, and it was identified in the aging tests that the loss increase results from OH formation due to hydrogen dissolved in optical fibers [4]. Since then, various investigations of the loss increase due to hydrogen permeation have been carried out [5]–[13], and a few kinds of countermeasures were reported [4], [14].

In this chapter, we describe the mechanism of the loss increase due to hydrogen permeation on the basis of previous studies, report the experimental results on the generation of hydrogen due to the electrochemical reaction, and finally, discuss the countermeasure against the problem, especially in optical submarine cables.

ABSORPTION LOSS INCREASE DUE TO HYDROGEN

Hydrogen permeation causes two different absorption losses in silica glass fibers. One is due to the vibration of the hydrogen molecule (hereafter we call it Type I absorption) [2]–[4], [8]. The other is due to OH formation (hereafter we call it Type II absorption) [4],

[15]. Here we describe the characteristics of these absorptions and discuss the relation between the two types of absorption.

Type I Absorption

The hydrogen molecule does not have a dipole moment in the gaseous state, and therefore it is not active for infrared absorption. However, once the hydrogen molecule diffuses into optical fibers, it will be polarized by the local field [16] or by the high electronegativity of oxygen in the defect centers [9] and will show infrared absorption. The absorption frequency (wavelength) coincides with the vibrational frequency of the hydrogen molecule, and it lies at 4132 cm^{-1} (2.42 μm) [4], [5], [8]. The first overtone lies at 1.24 μm. The absorption loss increases gradually with the diffusion of hydrogen molecules and reaches a saturation. The loss at the saturation ($L_{1.24}$) depends on ambient temperature [17], is proportional to the hydrogen pressure (P_{H_2} atm) [5], [18] and is approximately given by

$$L_{1.24} = 0.56 \cdot P_{H_2} \cdot \exp(1550/RT) \quad \text{(dB/km)} \tag{1}$$

where R is gas constant and T is absolute temperature. The diffusion speed of hydrogen molecule into fibers depends on the ambient temperature.

Calculation results on the time to achieve the saturation as a function of temperature are shown in Fig. 1. For the derivation of the results, we used the diffusion constant given by [18]

$$D(T) = 2.03 \cdot 10^{-7} \cdot T \cdot \exp(-E/RT) \quad \text{(cm}^2/\text{s)} \tag{2}$$

where E is activation energy and about 8.83×10^3 (cal/mol).

From Fig. 1, it is found that the saturation time at 3°C (temperature on the deep seabed) is about six times longer than that at 30°C.

In addition to the loss increase at 1.24 μm, many other peaks exist in the 1-μm wavelength region, which are due to the combinational vibrations with SiO$_4$ tetrahedral vibrations and rotation of hydrogen molecule [19]. The tails of the peaks affect the transmission loss in the wavelength range from 1.3 to 1.6 μm, which is the suitable region for optical fiber transmission systems. The loss increase at 1.3 ($L_{1.3}$) and 1.55 μm ($L_{1.55}$) is approximately given by

$$L_{1.3} = 0.033 L_{1.24} = 1.85 \times 10^{-2} \cdot P_{H_2} \cdot \exp(1550/RT) \quad \text{(dB/km)} \tag{3}$$

$$L_{1.55} = 0.083 L_{1.24} = 4.65 \times 10^{-2} \cdot P_{H_2} \cdot \exp(1550/RT) \quad \text{(dB/km)} \tag{4}$$

Type II Absorption

In addition to the Type I absorption, the Type II absorption occurs by the OH formation in fibers. The absorption peak arises at 1.41 μm for GeO$_2$-doped fibers and at wavelengths longer than 1.2 μm for P$_2$O$_5$-doped fibers [4]. In addition to the infrared absorption, it has been reported that broad wavelength dependent loss increase in the ultraviolet and visible wavelength dependent loss increase in the ultraviolet and visible wavelength region occurs simultaneously [17], [20]. In our experiment, both effects on the loss increase at 1.3 μm and 1.55 μm was measured without the distinction.

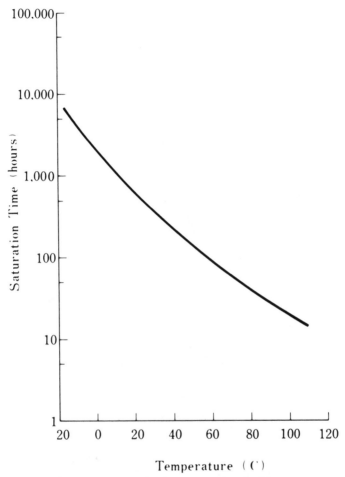

Fig. 1.   Saturation time of loss increase as a function of temperature.

The speed of the loss increase due to the OH formation is very slow compared to that of Type I absorption, and the formation is hard to occur below room temperature. However, it contains the danger that significant loss may arise for the long term.

Under high concentration of hydrogen and below room temperature, the loss at 1.3- and 1.55-μm wavelengths will be mainly affected by Type I absorption rather than Type II absorption. However, under very low concentration of hydrogen such that the loss increase due to Type I absorption does not much affect the loss at the wavelength to be used, it may occur that Type II absorption increases gradually with time, and the loss at 1.3 and 1.55 μm is affected by Type II absorption at last. The phenomenon has been observed in P_2O_5-doped optical fibers in a field-installed cable [4]. The loss increase speed is said to depend on the P_2O_5 concentration [13], and it has been considered that low P_2O_5 fibers will not suffer any significant degradation for operation at 1.3 μm for 20 years [13]. However, in terms of GeO_2-doped fibers without P_2O_5, the long-term magnitude of attenuation due to the OH formation is not well known. Hence, here we discuss the absorption loss due to the OH formation in GeO_2-doped fibers and estimate the additional loss for the long term on the basis of the aging test results.

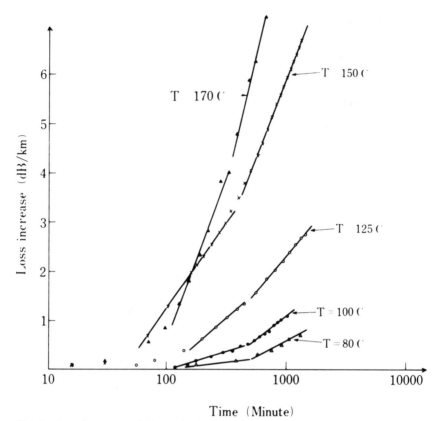

Fig. 2.   Loss increase at 1.41 μm in 2-atm hydrogen as a function of time for several temperatures.

Five pieces of GeO_2-doped multimode fibers (without P_2O_5) were used for the experiment. They were drawn from the same preform. The length of each fiber was about 500 m, respectively. Each fiber was placed in the vessel whose inner diameter was about 20 cm and whose height was 15 cm. Once the fiber was placed in the vessel, it was evacuated and pressurized in hydrogen at 2 atm. The vessel was left in the oven, whose temperature can be controlled from 30 to 200°C within the accuracy of 1°C. Both ends of the fiber were taken out from the oven in order to measure the loss variation in time. The loss increase of the five fibers at 1.41 μm was measured under the temperatures 80, 100, 125, 150, and 170°C, respectively. In order to avoid the effect of hydrogen diffusion on the time dependence of the loss increase, the fibers were saturated with hydrogen at room temperature in advance, and after that the tests were carried out at each temperature. The measurement results are shown in Fig. 2. The delay of the buildup time for the loss increase at 170°C is considered to be due to the imperfect saturation of hydrogen molecules in the core at the start. From Fig. 2, it is found that the loss increases almost linearly to a logarithm of time (t min). This indicates that the loss increase ($L_{1.41}$) follows the Elovich form [21] and is given by

$$L_{1.41} = A_0 \cdot \log(t + t_0) - B_0 \qquad (dB/km) \tag{5}$$

where t_0, A_0, and B_0 are constants depending on temperature.

From Fig. 2, it is also found that the slopes for each temperature change on the way. A similar change was also observed in the room temperature experiment in P_2O_5-doped fibers [13]. Since long term stability should be discussed by using the worst case, the loss increase for 25 years is estimated using the slope in the latter half. For the prediction of the loss increase for the long term, the following equation will be useful instead of (5).

$$L_{1.41} = A \cdot \log(t) - B. \tag{6}$$

The log value of A and B is obtained from the results in Fig. 2 and shown as a function of $1/T$ in Fig. 3. From Fig. 3, it is found that $\log(A)$ and $\log(B)$ are proportional to

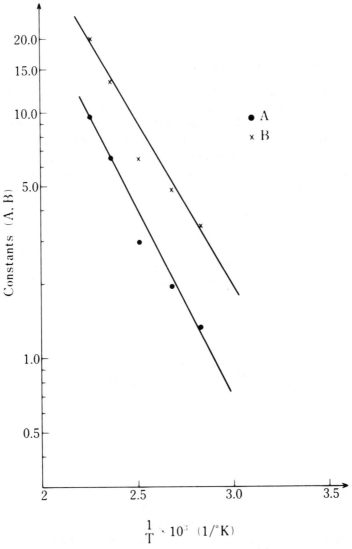

Fig. 3.   Log A and log B as a function of time.

$1/T$, and they can be approximately represented by

$$A = \exp\left(-3.55 \times 10^3 / T + 10.25\right), \tag{7}$$

$$B = \exp\left(-3.08 \times 10^3 / T + 9.95\right). \tag{8}$$

Meanwhile, it is said that the amount of the OH formation is proportional to the square root of the hydrogen pressure (22). If we follow the results, loss increase at 1.41 μm due to the OH formation can be given by

$$L_{1.41} = P_{H_2}^{1/2} \cdot \left(A_1 \cdot \log(t) - B_1\right) \tag{9}$$

where A_1 and B_1 is given by

$$A_1 = \exp\left(-3.55 \times 10^3 / T + 9.9\right) \tag{10}$$

$$B_1 = \exp\left(-3.08 \times 10^3 / T + 9.6\right). \tag{11}$$

Long-Term Predictions of Loss Increase

Using (3), (4), (9), (10) and (11), the loss increase at 1.3 μm and 1.55 μm due to H_2 permeation for 25 years are obtained. The contributions at 1.3 μm and 1.55 μm wavelengths from loss peak at 1.41 μm due to the OH formation are not exactly known. However, they are most unlikely to exceed 1/10 times the peak height. Hence, we calculated the loss increase at 1.3 μm and at 1.55 μm as $1/10 \cdot L_{1.41}$.

The results are shown in Figs. 4 and 5. If the tolerable loss increase in optical submarine cable is less than 0.01 dB/km, it is predicted from Fig. 5 that suppression of

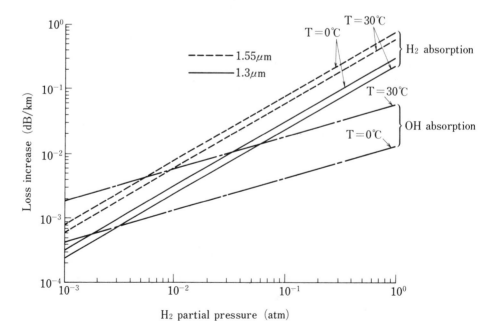

Fig. 4.   Predicted loss increase due to H_2 permeation for 25 years. The loss increase at 1.3 μm and 1.55 μm due to OH formation is considered to be $1/10\ L_{1.41}$.

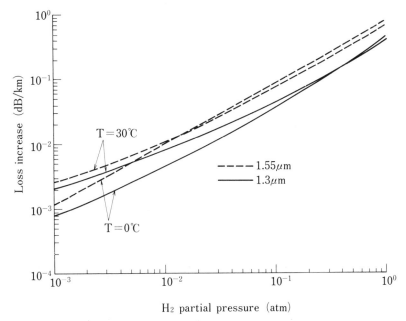

Fig. 5.   Predicted total loss increase due to H_2 permeation for 25 years.

hydrogen less than 2×10^{-2} atm for 1.3 μm transmission systems and 10^{-2} atm for 1.55 μm transmission systems is needed.

In terms of the long term prediction of the loss increase due to OH formation, several papers have been reported [20], [23]–[25]. The amount seems to depend on the fiber fabrication history and dopants. Hence, it will be important to examine the loss increase in the fiber to be used individually.

Hydrogen Generation Due to Electrochemical Reaction

It is well known that electrochemical reaction between metals and water generates hydrogen, and two possible water sources will be considered which introduce electrochemical reaction in submarine cables. One is the seawater which may propagate in the cables when the cables are cut under the sea, and the other is the vapor which may be confined in the cables during manufacturing. Here, we discuss hydrogen generation due to two possible water sources.

Effect of Seawater Propagation

In order to investigate the amount of hydrogen gas produced by the reaction between metals and seawater, two kinds of metals, out of copper (Cu), iron (Fe), and aluminum (Al), were immersed in seawater as shown in Fig. 6, and then the generated gas was analyzed by gas chromatography. The size of experimentally used metals was 50 mm $\times$ 10 mm $\times$ 2 mm. Two kinds of metals out of the three were separated in Type A, jointed without current supply in Type $B1$, and jointed with current supply in Type $B2$. Figure 7 shows the amount of hydrogen gas produced in Type A as a function of time, and Fig. 8 shows that in Type $B2$. In terms of Type $B1$, there was no evidence of hydrogen

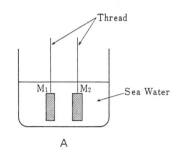

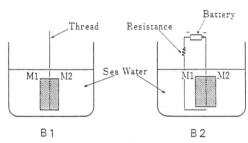

Fig. 6.   Schematic arrangement for the electrochemical reaction. Current (about 150 mA) is supplied in Type $B2$.

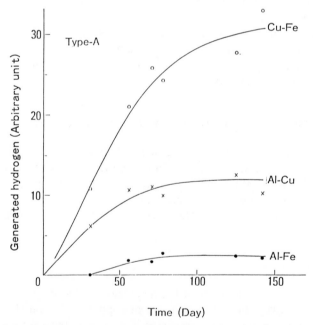

Fig. 7.   Relative hydrogen amount generated from Type A as a function of time for several kinds of the metal combinations.

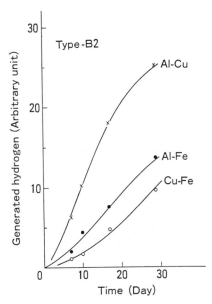

Fig. 8.   Relative hydrogen amount generated from Type $B2$ as a function of time for several kinds of the metal combinations.

generation, even after 150 days. It is interesting that there is no evidence of hydrogen generation from the jointed metals (Type $B1$). In order to clarify the reason, further investigation will be needed.

From Fig. 7, it is found that the generation amount of hydrogen is different among the three different combinations of metals, and the amount of hydrogen reaches a steady state, especially in the cases of Al–Cu and Al–Fe. As the reason of the steady state, it can be considered that the surface of Al was gradually covered by oxide. The hydrogen generation in Type $B2$ indicates that electric current has the influence on the generation of hydrogen and accelerates the reaction. The experimental results obtained here will not always be applied to the phenomena in the cables because the components in metals and their conditions will be different. However, it can be said that metals, whose ionization tendency is greater than that of hydrogen, have the tendency to produce hydrogen gas more or less when water immerses into the cable [2], and current supply will prompt the hydrogen generation.

Effect of Water Condensation in Cables

Although the water propagation in the cable will never happen in the normal operating condition, we should consider the case that the vapor confined in the cable may cause hydrogen generation even under the normal conditions. Then, next we consider the effect.

Assuming that the temperature in the cable factory is 30°C, and that on the seabed it is 0°C, water will condense in the cables after being laid on the seabed due to the difference of saturated vapor pressure, and it may cause hydrogen gas by electrochemical reaction with metals. Here, we consider the effect in the worst case.

When saturated vapor at 30°C (0.03037×10^{-3} g/cm^3) is cooled to 0°C (saturated vapor at 0°C: 0.00485×10^{-3} g/cm^3), the amount of condensed water per unit volume is given by

$$\text{(saturated vapor at 30°C)} - \text{(saturated vapor at 0°C)} = W_{\mathrm{H_2O}} \quad (\mathrm{g/cm^3})$$

$$= 0.026 \times 10^{-3} \quad (\mathrm{g/cm^3}).$$

If all amounts of condensed water are reduced to H_2 and O_2, the partial pressure of hydrogen gas is given by

$$W_{\mathrm{H_2O}} / \text{(molecular weight of } H_2O) \times 22.4 \times 10^3 = 0.032 \quad (\mathrm{atm}).$$

From Fig. 5, it is found that this partial pressure corresponds to about 0.01 dB/km at 1.3 μm and about 0.03 dB/km at 1.55 μm. The value obtained here considering the worst case will be tolerable for about 50-km repeater spacing systems using 1.3-μm wavelength. However, for the longer repeater spacing systems using the 1.5-μm wavelength region, careful investigations will be needed for this effect, even though the value obtained above is in the worst case.

Discussion

Here, we discuss the countermeasures on the basis of the above studies. Several kinds of countermeasures can be considered. However, they can be classified into three. The first one is not having the fiber itself introduce infrared absorption due to H_2, the second one is the coating method to prevent H_2 diffusion, and the last one is not having the cable structure introduce H_2.

In terms of the fiber itself, it has been observed that the fiber having OD formation by the high temperature treatment under deuterium will be more resistant to OH formation compared to the fiber without OD formation, although the absorption loss due to H_2 vibration is almost the same in both fibers [16]. From this fact, we see that the effect of OH formation can be overcome by diminishing the amount of P_2O_5 [4] or by filling the defect centers with some ion as OD formed fibers [16] and fluorine doped fibers [4]. However, in terms of the absorption loss due to H_2 vibration, it will be difficult to make the fibers which will not induce the infrared absorption to H_2 since the absorption due to H_2 itself originates from the induced dipole moment by the local field in the lattice interstices [16].

The practical barrier to H_2 diffusion into fibers was shown by Beales *et al.* [14]. They showed that the coating of silicon oxynitride is very effective for the prevention of H_2 diffusion. However, there are not many reports on the transmission characteristics in the fibers coated with silicon oxynitride, and we are anxious to see whether the transmission characteristics of the fibers coated with silicon oxynitride are similar to those of the conventional fibers being developed until now. For other coating materials to prevent the hydrogen diffusion, such as thin metals, it will also be difficult to expect the same transmission characteristics to the fibers coated with such materials. Hence, in the present situation, we require that the cable design must not introduce hydrogen, which is one of the effective countermeasures.

We showed that the electrochemical reaction causes the hydrogen generation more or less once water permeates the cables and current accelerates the reaction. In addition to that, it has been reported that some of the fiber coating materials produce hydrogen in much greater amounts than other materials, and it may cause significant loss increase for 20 years [26].

Hence, the hydrogen problem can be overcome by keeping out the materials to produce substantial amounts of hydrogen and preventing water filling in the cables. Furthermore, hydrogen generation due to confined condensed water in the cables can also be overcome by filling up the interstices of the cables with some filling or manufacturing the cables under dry circumstances.

CONCLUSION

Optical submarine cables are required to be kept stable for more than 20 years. In order to satisfy the requirement, it will be essential to make the factors which cause hydrogen generation as small as possible.

One of the factors is the materials used in the cables. Hence, it will be important to examine the hydrogen generation from the materials to be used and keep out the materials which generate large amounts of hydrogen.

When the cables are cut or broken, seawater will rush into the cables more or less. For these cases, the prevention of the water propagation will be most desirable. However, even if the prevention is not completely achieved, the effect of the water propagation will be overcome by the repairing methods.

Although further investigation will be needed on the effects of water condensation, the effect is estimated to be tolerable for 50-km repeater spacing systems using 1.3-μm wavelength, even in the worst case. However, the effects may become important as the repeater spacing gets longer and more pronounced by operating at 1.55 μm. In the present situation, it will be desirable to fill up the interstices of the cables with some fillings or to manufacture the cables under dry circumstances in order to avoid condensation effects.

ACKNOWLEDGMENT

The authors wish to thank Y. Iwamoto for his discussions. This acknowledgment is also extended to Dr. H. Kaji, Dr. K. Nosaka, Dr. K. Amano, Dr. C. Ota, and K. Furusawa of the Kokusai Denshin Denwa Company (KDD) Laboratories, and S. Mukasa of the KDD Ninomiya Cable Landing Station for their encouragement.

REFERENCES

[1] S. A. Taylor, "Mechanical testing and specification of submarine systems," in *Proc. Int. Conf. Submarine Telecommun. Syst., Inst. Elec. Eng. Conf. Pub.* (London, England), Feb. 1980, p. 183.
[2] K. Mochizuki, Y. Namihira, and H. Yamamoto, "Transmission loss increase in optical fibres due to hydrogen permeation," *Electron. Lett.*, vol. 19, pp. 743–745, 1983.
[3] N. Uesugi, Y. Murakami, C. Tanaka, Y. Ishida, Y. Mitsunaga, Y. Negishi, and N. Uchida, "Infra-red optical loss increase for silica fibre in cable filled with water," *Electron. Lett.*, vol. 19, pp. 762–764, 1983.
[4] N. Uchida, N. Uesugi, Y. Murakami, M. Nakahara, T. Tanifuji, and N. Inagaki, "Infrared loss increase in silica optical fiber due to chemical reaction of hydrogen," in *Proc. 9th ECOC*, Oct. 1983.
[5] K. J. Beales, D. M. Cooper, and J. D. Rush, "Increased attenuation in optical fibres caused by diffusion of molecular hydrogen at room temperature," *Electron. Lett.*, vol. 19, pp. 917–919, 1983.
[6] M. Fox and S. J. Stannard-Powell, "Attenuation change in optical fibres due to hydrogen," *Electron. Lett.*, vol. 19, pp. 916–917, 1983.
[7] Y. Namihira, K. Mochizuki, M. Kuwazuru, and Y. Iwamoto, "Effects of hydrogen diffusion on optical fibre loss increase," *Electron. Lett.*, vol. 19, pp. 1034–1035, 1983.

[8]   K. Mochizuki, Y. Namihira, M. Kuwazuru, and Y. Iwamoto, "Effects of hydrogen on infrared absorption characteristics in optical fibers," in *Proc. OFC '84*, WB2, Jan. 1984.

[9]   M. Kuwazuru, K. Mochizuki, Y. Namihira, and Y. Iwamoto, "Dopant effect on transmission loss increase due to hydrogen permeation," *Electron. Lett.*, vol. 20, pp. 115–116, 1984.

[10] K. Mochizuki, Y. Namihira, M. Kuwazuru, M. Nunokawa, and Y. Iwamoto, "Loss spectra of optical fibres under hydrogen and deuterium," *Electron. Lett.*, vol. 20, pp. 118–119, 1984.

[11] Y. Mitsunaga, T. Kuwabara, T. Abe, and Y. Ishihara, "Molecular hydrogen behavior for loss increase of silica fibre in cable filled with water," *Electron. Lett.*, vol. 20, pp. 76–78, 1984.

[12] H. Itoh, Y. Ohmori, and M. Nakahara, "Chemical change from diffused hydrogen gas to hydroxyl ion in silica glass optical fibres," *Electron. Lett.*, vol. 20, pp. 140–142, 1984.

[13] K. W. Plessner and S. J. Stannard-Powell, "Attenuation/time relation for OH formation in optical fibres exposed to H_2," *Electron. Lett.*, vol. 20, pp. 250–252, 1984.

[14] K. J. Beales, D. M. Cooper, W. J. Duncan, and J. D. Rush, "Practical barrier to hydrogen diffusion into optical fibres," in *Proc. OFC '84*, W15, Jan. 1984.

[15] K. Mochizuki, Y. Namihira, M. Kuwazuru, and Y. Iwamoto, "Behavior of hydrogen molecules adsorbed on silica optical fibers," *IEEE J. Quantum Electron.*, vol. QE-20, pp. 694–697, July 1984.

[16] K. Mochizuki, Y. Namihara, and M. Kuwazuru, "Absorption loss in optical fibres due to hydrogen," *Electron. Lett.*, vol. 20, pp. 550–552, June 1984.

[17] N. J. Pitt and A. Marshall, "Long-term loss stability of single-mode optical fibres exposed to hydrogen," *Electron. Lett.*, vol. 20, pp. 512–514, 1984.

[18] Y. Namihira, K. Mochizuki, M. Kuwazuru, and Y. Iwamoto, "Temperature dependence of hydrogen diffusion constant in optical fibers," *Opt. Lett.*, vol. 9, pp. 426–428, 1984.

[19] J. Stone, A. R. Chraplyuy, J. M. Wiesenfeld, and C. A. Burrus, "Overtone absorption and Raman spectra of H_2 and D_2 in silica optical fibers," *Bell Syst. Tech. J.*, vol. 63, pp. 991–1000, 1984.

[20] A. Tomita and P. J. Lemaire, "Hydrogen-induced loss increases in germanium-doped single-mode optical fibres: Long-term predictions," *Electron. Lett.*, vol. 21, pp. 71–72, 1985.

[21] A. Clark, *Theory of Adsorption and Catalysis.* New York: Academic Press, 1970.

[22] S. Tanaka, M. Kyoto, M. Watanabe, and H. Yokota, "Hydroxyl group formation caused by Hydrogen diffusion into optical glass fibre," *Electron. Lett.*, vol. 20, pp. 283–284, 1984.

[23] N. Uesugi, M. Tokuda, K. Noguchi, and Y. Negishi, "Loss increase characteristics due to hydrogen molecule diffused into optical fibers," presented at the 10th ECOC 1984, Stuttgart, Federal Republic of Germany.

[24] N. J. Pitt, A. Marshall, J. Irven, and S. Day, "Long term interactions of hydrogen with single-mode optical fibres," presented at the 10th ECOC 1984, Stuttgart, Federal Republic of Germany.

[25] J. D. Rush, K. J. Beales, D. M. Cooper, W. J. Duncan, and N. H. Rabone, "Hydrogen related degradation in optical fibres-system implications and practical solutions," *British Telecom Technol. J.*, vol. 2, pp. 84–93, 1984.

[26] E. W. Mies, D. L. Philen, W. D. Reents, and D. A. Meade, "Hydrogen susceptibility studies pertaining to optical fiber cables," presented at the OFC '84, Jan. 1984.

14
Drawing of High-Strength Long-Length Optical Fibers for Submarine Cables

SHIGEKI SAKAGUCHI

INTRODUCTION

Submarine cables are exposed to great mechanical stress during laying and recovering [1], and mechanically weak optical fibers often cannot endure the stress. In order to expand the application of optical fibers to submarine cable systems, the mechanical performance of the fibers should be further improved. Both the tensile strength and fatigue lifetime of an optical fiber at any length are fundamentally dependent on the maximum flaw present in the fiber. Therefore, to assure the long-term mechanical reliability of fibers, it is necessary to improve and to guarantee the minimum strength of the fibers.

Most of the macroscopic flaws, which result in the degradation of tensile strength, are formed in the fiber drawing process. Much effort has been made to minimize flaw formation during the drawing process. This includes preform surface treatment, a clean heating environment, the use of high-quality glass rods, and coating procedures. These techniques have led to the development of high-strength fibers [2]–[9]. However, present drawing techniques lack the reproducibility to draw sufficiently high-strength long-length fibers for submarine use. This is because flaws are formed by the complex interaction of many drawing factors, whose effects on tensile strength have not yet been sufficiently clarified. Among these factors, preform surface quality and drawing atmosphere cleanliness seem to be the two most dominant factors.

On the other hand, unfortunately, macroscopic flaws cannot be completely removed from the fibers in spite of strictly controlled drawing conditions. Thus, mechanical strength should be considered with the assumption that macroscopic flaws are inherent in the process. For this reason, it is necessary to evaluate the minimum strength of long fibers accurately. Proof testing is one effective method of confirming the minimum strength. In addition, nonbreak conditions for fibers would facilitate the analysis of cable lifetime.

This chapter describes a drawing process for high-strength long-length optical fibers intended for submarine use, focusing attention on the optimization of the preform surface treatment and clean heating atmosphere. Furthermore, the guaranteeing of the long-term reliability of optical fibers is discussed based on the fracture mechanics concept.

DRAWING OF HIGH-STRENGTH LONG-LENGTH FIBERS

Drawing Apparatus

Figure 1 is a schematic representation of the drawing apparatus using a carbon resistance furnace. Drawing factors affecting flaw formation are also shown at each drawing stage.

Surface Treatment

Cracks and impurity particles existing on the preform surface may remain on the fiber surface after drawing at temperatures higher than 2000°C, resulting in the formation of macroscopic defects such as flaws and microcrystals. To overcome this problem, surface treatment with hydrofluoric acid etching and oxyhydrogen flame polishing prior to drawing is effective in reducing flaw formations.

Figure 2 shows changes in the Weibull distribution of tensile strength for fibers drawn from fused silica rods with various surface treatments. Surface treatment procedures for these rods are as follows.

a) Without treatment, except for cleansing with acetone;

b) Oxyhydrogen flame polishing by traversing a silica glass burner for two passes at a peak temperature of about 1900°C; and

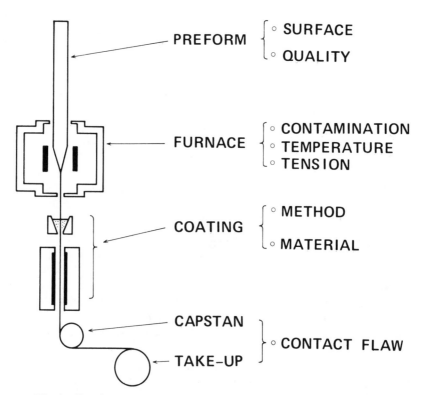

Fig. 1.   Drawing process parameters affecting flaw formation in optical fibers.

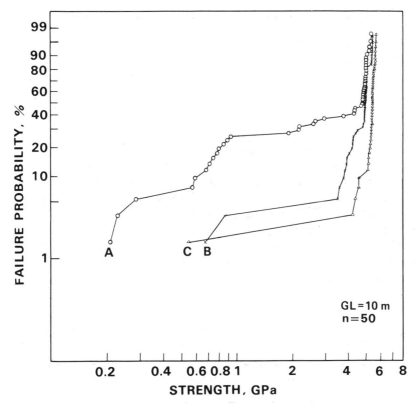

Fig. 2. Strength distributions for fibers drawn from fused silica rods with various surface treatments: A, without treatment; B, flame polishing; C, hydrofluoric acid etching and flame polishing.

c) Etching with 49 percent hydrofluoric acid for 15 min followed by oxyhydrogen flame polishing for two passes. Fibers with a 125 ± 1 μm diameter were drawn from fused silica rods having a 15 mm diameter at a fiber drawing tension of 10 ± 2 g and coated in-line with thermally curable silicone resin with a resulting diameter of 450 μm. Tensile tests were performed for 50 specimens of each fiber at a 10 m gauge length and a 0.05 min^{-1} strain rate.

As is clearly seen, strength distribution is greatly affected not only by whether the preforms were treated or not but also by the type of treatment. Distribution curve A for the fiber drawn from the preform without treatment shows a widely scattered strength distribution, in which a relatively large weak-strength tail and a high-strength region are seen. In contrast to curve A, the weak tails in curves B and C are remarkably reduced. In addition, the slope in the high-strength region for curve C is higher than that for curve B, indicating that uniformity in strength increases. This result indicates that the surface treatment combined with hydrofluoric acid etching and oxyhydrogen flame polishing is much more effective than flame polishing only.

Hydrofluoric acid etches the thin surface layer uniformly. As a result, the sharpness at the crack tip is moderated, and almost all impurity particles, which might cause crystallization due to locally deformed glass composition, are dissolved. A 49 percent hydrofluoric acid solution dissolves the silica at a rate of about 260 Å/s. Etching for 15 min removes the surface layer whose thickness is 2–3 μm. Oxyhydrogen flame polishing smoothes the preform surface. Flame polishing at 1900°C reduces the roughness to within 1 μm after only two passes, even though the initial preform surface roughness is more than 10 μm [10]. Since the depth of cracks existing on the glass surface is generally several microns, two pass flame polishing is adequate to remove them. Thus, treatment with hydrofluoric acid etching for 15 min and flame polishing for two passes seem to be the minimum required treatments.

Removal of Flaws Due to Dust Contamination

1) Effect of Dust Particle Size on Strength: In a carbon resistance furnace, dust particles are generated due to consumption of the heater at high operating temperatures and due to formation of SiC microcrystals derived from the reaction between carbon and silica. These particles are easily trapped on the fiber surface, which is activated at a high temperature, resulting in the formation of macroscopic flaws.

Figure 3 shows the effect of dust particle size on the tensile strength of silica fibers drawn in the dusty furnace and tensile tested at a 10 m gauge length. The dusty atmosphere was produced by forcibly feeding alumina and carbon particles with uniformly controlled diameters into the furnace by means of a purge gas flow. The controlled diameters of the alumina powders were 0.03, 0.3, and 1 μm while the diameter of the carbon particles was 20 μm.

As shown in Fig. 3, the plots of the reciprocal root of dust particle radius r versus the average tensile strength σ indicate an almost linear relationship, which is expressed by

$$\sigma = D/r^{1/2} \tag{1}$$

where D is a constant and determined to be 0.474 MPa m$^{1/2}$ in the present experiment. Therefore, in order to obtain fibers stronger than a required strength value of σ (MPa), it is necessary to remove dust particles with radii of larger than r (m) given by

$$r = (0.474/\sigma)^{2}.$$

This equation indicates the required cleanliness in the heating atmosphere.

A typical example of contamination on the glass surface due to carbon particles is shown in Fig. 4. This sample, which acted as a light scattering point, was obtained from the neckdown region in a drawn preform. The dust is identified by the intensity curve of the characteristic X-ray for carbon observed by XMA along the broken line. The carbon particle, having a diameter of about 20 μm and embedded in the glass, causes the flaw formation.

2) Elimination of Dust Particles: In order to eliminate flaws due to dust contamination, it is necessary to reduce the causes of dust particle generation and eliminate dust particles in the furnace. For this purpose, the furnace was improved by providing a single-piece muffle tube inside the heater as well as a gas flow port having a cross-sectional area of about 140 mm^2 on the top part of the furnace.

Figure 5 shows the distributions of tensile strengths obtained from tensile tests on a 20 m gauge length for silica fibers drawn with the port open and closed. When the gas port is closed, the distribution curve shows a transition from the high- to the weak-strength region at a failure probability of 30 percent. When the gas port is opened to ventilate the purge gas, the distribution curve is a unimodal straight line without a weak region. This is due to the removal of dust particles from the furnace by leaking a purge gas through the gas exhaust port [11].

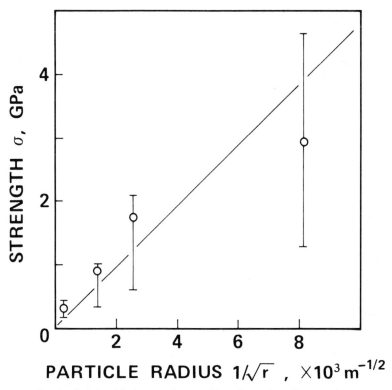

Fig. 3. Relationship between dust particle radius and tensile strength.

This indicates that using a carbon resistance furnace as a heat source yields high-strength fibers just as in the case of a CO_2 laser, which is thought to be the most clean heat source.

3) Time Dependent Changes in Dust Particles: The measurement results of an examination on how much dust is actually generated in the furnace are shown in Fig. 6. The figure shows time-dependent changes in dust particles having diameters of 0.3, 1, and 5 μm during furnace baking. For measuring the amount of dust, the purge gas in the furnace

⊢——⊣ 10 μm

Fig. 4. Typical example of flaw caused by carbon dust contamination. Characteristic X-ray intensity for carbon is shown along the broken line.

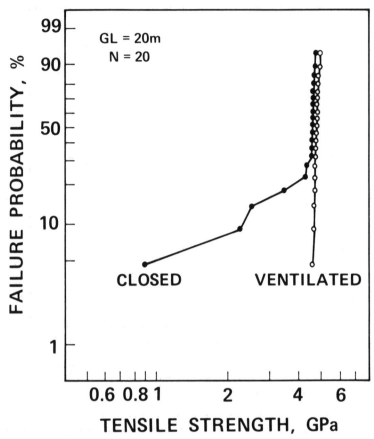

Fig. 5. Strength distributions for silica fibers drawn with the gas flow port both open and closed.

was guided into a particle counter through the top opening. At the same time, the oxygen content was measured using an oxygen analyzer. The percentage figures on the abscissa represent the ratios of supplied power to normal drawing operation power for the carbon heater.

Although no dust particles are generated when no power is supplied, they increase rapidly, as soon as a power of 32 percent is supplied, and then gradually decrease. This tendency is also observed as power rates of 64 and 100 percent are supplied. At 40–50 min after the supply of 100 percent power, dust particles are reduced to less than $10 \ 1^{-1}$ (higher than Class 300). Large particles having a diameter of 5 μm disappear at a relatively early stage in furnace baking. This suggests that sufficient furnace baking prior to drawing is needed to reduce dust particles. In this case, furnace cleanliness is kept higher than Class 300 as a result of baking for longer than 1 h.

On the other hand, the oxygen content, which affects dust generation by consuming the carbon in the furnace, cannot be clearly connected to the amount of dust, as shown in Fig. 6. The oxygen content shows merely a gradual reduction from 50 to 30 ppm. Thus, this oxygen content does not seem to greatly affect carbon consumption over relatively short periods of time.

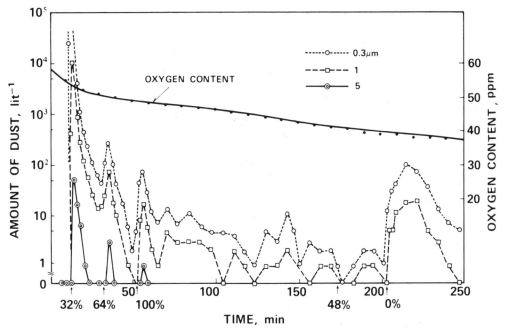

Fig. 6. Time dependence of the number of dust particles having various sizes and the oxygen content in the furnace during baking— ○: 0.3 μm μm, □: 1 μm, and ⊙: 5 μm. The percentage figures on the abscissa represent the ratio of the power supplied for the heater to normal operation power.

As described above, several steps are needed to improve the drawing process for the purpose of realizing high-strength fibers. These are summarized as follows.

1) Preform surface treatment combined with hydrofluoric acid etching and oxyhydrogen flame polishing;

2) Dust particle exhausting by means of ventilating purge gas; and

3) Furnace baking prior to drawing.

Proof Success Length of Long-Length Fibers

1) Effect of Glass Quality: The techniques mentioned above are closely related to the elimination of flaws formed extrinsically during the drawing process. On the other hand, preform rods contain originally internal macroscopic flaws, although such flaws occupy a small part of all macroscopic flaws. Thus, the effect of glass quality on mechanical performance strongly appears in the proof test data.

Based on the fundamentally improved drawing method, silica fibers having a length of about 6 km were drawn from fused silica, commercial synthetic silica, and VAD synthetic silica rods. These preforms, having diameters of 20–25 mm, were treated by hydrofluoric acid (49 percent content) etching for 30 min followed by oxyhydrogen flame polishing for five passes at 1900°C. Fibers with a 125 ± 1 μm diameter were drawn at a drawing tension of 10 ± 2 g and coated in-line with silicone with a resulting diameter of 450 μm. Their mechanical performance was examined by proof testing at various stress levels. Figure 7 schematically shows the proof test apparatus. The double driving rolls with soft rubber belts transport the fiber and provide a sufficient friction force so that the proof stress is

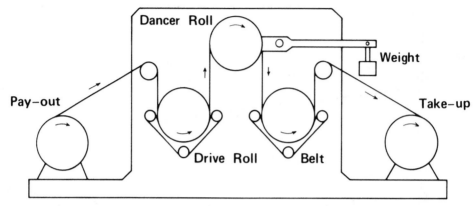

Fig. 7.   Schematic representation of proof test apparatus.

TABLE I
PROOF SUCCESS LENGTHS FOR FUSED SILICA FIBERS

Proof stress (GPa)	Av. success length (km)	Total length (km)	No. of fibers (n)	No. of breaks (N)	Yield > 2.5 km (%)
0.80	12.74	38.22	7	2	—
1.60	5.87	23.48	5	3	—
2.40	2.56	28.38	6	10	51.7

applied to the fiber through a dancer roll. The proof tests were performed at stress levels of 0.80–3.67 GPa, a proof time of 1 s, and an unloading rate of more than 1.5 GPa/s.

The results of fused silica fibers at proof stress levels of 0.80–2.40 GPa are listed in Table I, and those for synthetic silica fibers at 2.40 GPa are listed in Table II. The average success length L is determined using the formula $L = \hat{L}/(N+1)$ where $\hat{L}$ is the total length of the tested fibers and N is the number of breaks in the proof test.

The average proof success length values for fused silica fibers at 0.80 and 1.60 GPa proof stresses are 12.7 and 5.8 km, respectively (Table I). These values satisfy the mechanical performance required in a nonrepeatered system (maximum sea depth 1500 m) and in a repeatered system (maximum sea depth 5000 m) [1], respectively.

At a high proof stress level of 2.40 GPa, an average success length value of 8.7 km is obtained for VAD synthetic silica fibers. This suggests the feasibility of realizing high-strength long-length optical fibers for transocean submarine cables.

Figure 8 shows the plots of average success lengths versus proof stress. The average success length for fused silica fibers in a proof stress range from 0.80 to 2.40 GPa

TABLE II
PROOF SUCCESS LENGTHS FOR SYNTHETIC SILICA FIBERS AT A
PROOF STRESS OF 2.40 GPa

Fiber	Av. success length (km)	Total length (km)	No. of fibers (n)	No. of breaks (N)	Yield > 2.5 km (%)	> 5 km
VAD	8.71	52.28	8	5	95.7	66.0
commercial	4.42	35.37	6	7	81.4	48.2

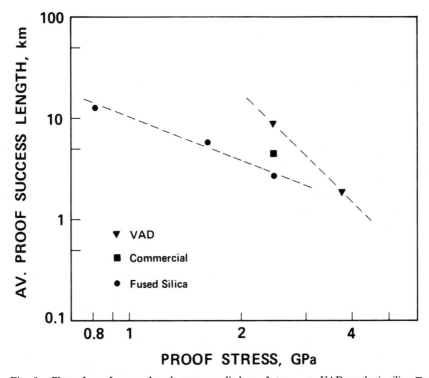

Fig. 8. Plots of proof success length versus applied proof stress—▼: VAD synthetic silica, ■: commercial synthetic silica, ●: fused silica.

decreases with increasing proof stress. The plots of the log of average success length versus the log of proof stress give the following equation:

$$L = C\sigma_p^{-a} \tag{2}$$

where C and a are constants and σ_p is the proof stress. The present success length data for fused silica fibers can be represented by

$$L = 9.80\sigma_p^{-1.42}$$

where L and σ_p are expressed in km and GPa, respectively.

The average proof success lengths of synthetic silica fibers drawn from commercial and VAD rods are also shown in Fig. 8. The average success lengths of the VAD and commercial synthetic silica fibers at a 2.40 GPa proof stress are 8.71 and 4.42 km, respectively. Two VAD fibers which successfully passed the 2.40 GPa proof test were again proof tested at 3.67 GPa for 11.36 km in total. The average success length was obtained as 1.89 km. Breakage occurred four times in the 3.67 GPa proof testing. This result is expressed by

$$L = 197\sigma_p^{-3.57}.$$

This clearly shows that the mechanical performance of the VAD synthetic silica fiber is appreciably higher than that of the fused silica fiber.

The relationship between the success length and the applied proof stress can be derived from the Weibull distribution of the tensile strength expressed by

$$F = 1 - \exp - L(\sigma/\sigma_0)^m \qquad (3)$$

where L is fiber length, and σ_0 and m are constants. From (3), the success length can be given as

$$L = \sigma_p^{-m}\sigma_0^m \ln\{1/(1-F)\}. \qquad (4)$$

The parameter a in (2) refers to the configuration parameter m in the Weibull distribution. Actually, they agree well with each other for fused silica fibers [12].

Generally, the proof success probability of synthetic silica fibers is superior to that of fused silica fibers. To examine the fracture origin, the fracture surfaces of fibers broken in the 2.40 GPa proof stress test were examined by SEM. However, the original fracture surfaces could not be observed because the fiber broke to pieces when the breakage occurred in the proof test.

The difference in proof success probabilities among the three kinds of silica fibers is primarily the result of the difference in the quality of silica glass, i.e., the concentration of internal defects such as impurity inclusions, voids, cracks, and veined structures. The synthetic silica contains very few defects, while the fused silica has a larger number of defects than the synthetic silica. Since most surface flaws formed during drawing can be eliminated by improving the drawing process, the effect of glass quality may appear in the mechanical performance of these fibers. The difference in the success length between VAD and commercial synthetic silica fibers may be attributed to the handling of the preform rods. The commercial rods go through many work processes such as polishing and sizing, and this results in a larger possibility for defects to occur than in the VAD rods.

Because the silica fibers which are drawn from whole synthetic preform rods prepared by the VAD method show very high proof success probabilities as described above, they are expected to be employed in transocean submarine cables.

2) Effect of Coating Material: Silicones used as a coating material strictly provide the mechanical performance of the fibers because of their very low elastic modulas. At the same time, they have a fear that the fibers are easily damaged by handling. In contrast to this, UV acrylates are expected to offer high efficiency in protecting the fiber surface due to their relatively high modulas.

Fig. 9 shows the plots of average proof success length of UV acrylate-coated fibers [10]. The fibers were drawn from the VAD preforms additionally jacketed with commercial synthetic silica tubes, and coated in-line with UV acrylates of dual layers, soft and hard layers, of a final diameter of 400 μm.

The average proof success length at a proof stress of 1.6 GPa is 19.0 km for 6 fibers of 57.0 km in total length, and 9.2 km at 2.4 GPa (4 fibers, 36.7 km in total). In Fig. 9, the solid rectangle represents the success length of silicone-coated commercial synthetic silica fibers (4.4 km, Table II). It is clear that the success length of UV acrylate-coated fibers is fairly high compared with that of silicone-coated fibers, even though the proof stress fraction, which is less than 8%, sustained by the coating layer is took account of. This is predominately attributed to high efficiency in keeping the fibers from being damaged.

A high modulas coating is preferred for the fabrication of high-strength fibers, so far as it does not bring microbending loss. Thus, it is possible to obtain high performance fibers in both mechanical and transmission by the optimization of coating.

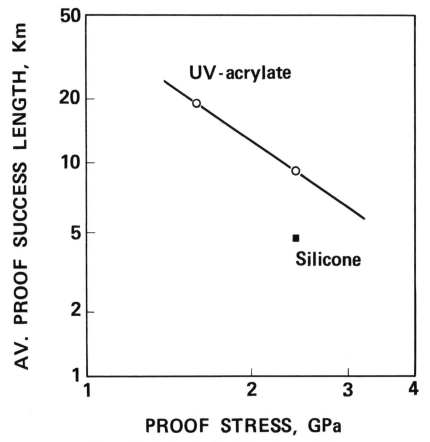

Fig. 9. Proof success length of UV acrylate-coated fibers.

STRENGTH EVALUATION FOR LONG FIBERS

Strength of Proofed Fiber

Even though long fibers are drawn under a strictly controlled process, it is difficult to completely eliminate macroscopic flaws, which degrade strength. Thus, it is necessary to guarantee the minimum strength level of fibers, especially those for submarine use.

Proof testing is a conventional method of assuring the minimum strength level of fibers proofed corresponding to the proof stress. It has been reported that, in the proof test, accuracy in guaranteeing the strength level is influenced by the unloading rate in the proof stressing cycle. This is because of the crack growth occurring during the unloading cycle due to the nature of corrosion cracking itself in glass.

Assuming that unloading is rapid enough that crack growth during unloading can be disregarded, the fiber fracture condition in the proof test is determined by using the basic fracture mechanics parameters

$$K_{IC} = Y\sigma_p\sqrt{c_p} \tag{5}$$

where K_{IC} is the fracture toughness, Y is a geometric parameter dependent on crack

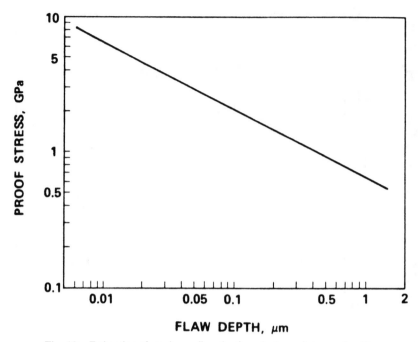

Fig. 10.   Estimation of maximum flaw depth against proof stress using (5).

length and configuration, σ_p is the proof stress, and c_p is the depth of the critical crack. This equation gives the maximum crack size which may remain in the proofed fiber. Fig. 10 shows plots for the maximum flaw which may remain in the fiber after proof testing, taking 0.80 MPa $m^{1/2}$ as the K_{IC} value [13]. Thus, in the tensile test following the proof test, the minimum strength corresponds to this maximum crack present in the fiber proofed at σ_p. Then, the minimum strength σ_m is given as

$$\sigma_m = \left\{ B(n+1)\dot{\sigma}\sigma_p^{n-2} \right\}^{1/(n+1)} \tag{6}$$

where $\dot{\sigma}$ is the loading rate, B is a constant, and n is the crack growth parameter.

Figure 11 shows the relationship between the minimum strength and the proof stress. The curves in Fig. 11 represent the minimum strength values for proofed fibers calculated using (6) [14]. Minimum strengths obtained experimentally are also shown. The vertical bar indicates the scattered range for minimum strengths obtained from 22 silicone–nylon-coated fibers, and the open circles represent those for 9 silicone-coated fused silica fibers. The total length of the 22 fibers, which were commercially available, was 106 km (individual fibers varied from 2 to 7 km). The fibers proofed at a 0.75 GPa stress for 1 s and a 1.5 GPa/s unloading rate were tensile tested at a 100 m gauge length and a 0.1 min^{-1} strain rate. The 9 fibers (drawn from fused silica rods), having a total length of 45 km (individual fiber length was about 5 km), were proof tested at various stresses ranging from 0.8 to 3.67 GPa for 1 s and at an unloading rate higher than 1.5 GPa/s. Tensile tests were performed at a 20 m gauge length and a 0.025 min^{-1} strain rate.

All the experimental minimum strengths (scattered range of minimum strengths for 22 silicone–nylon-coated fibers and open circles for 9 silicone-coated fibers) clearly fall in

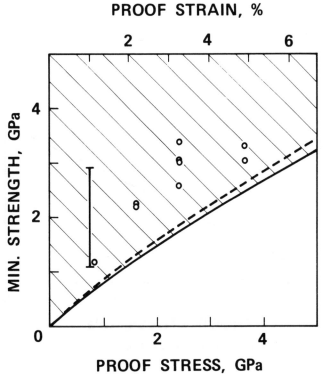

Fig. 11. Relationship between proof stress and minimum strength. Solid line and dashed line represent values calculated using (6) for tensile test at strain rates of 0.025 and 0.1 min^{-1}, respectively. Vertical bar and open circles represent experimental data for 22 silicone–nylon-coated fibers and 9 silicone-coated fibers, respectively.

the hatched region, whose lower limit is given by curves calculated using (6). These results confirm the truncation of the minimum strength of the fibers proofed not considering crack growth during unloading and reflect the reliability of the proof test.

Allowable Stress Condition

Although the minimum strengths of fibers are guaranteed by the proof test, it is important to clarify the allowable stress conditions in a fail safe manner. It is well known that when the applied stress condition is lower than the critical stress intensity for stress corrosion cracking K_{ISCC}, no stable crack growth occurs. In order to apply this concept to the assurance of the long-term mechanical reliability of fibers, it is necessary to clarify fatigue behavior in low-strength fibers containing macroscopic flaws. This is done by comparing the fatigue to that in a plate type specimen. A detailed examination of fatigue characteristics for low-strength fibers indicates that the fatigue behavior is quite the same as that for plate specimens. This is seen by comparing the crack growth parameters between them [15]. This means that a lower limit for the crack growth of macroscopic flaws in a low-strength fiber also exists, i.e., the critical condition is given by K_{ISCC}, which cannot be measured directly on fibers. Based on this assumption, the allowable applied stress σ_a, at which no growth will occur for the maximum crack c_p, which is guaranteed

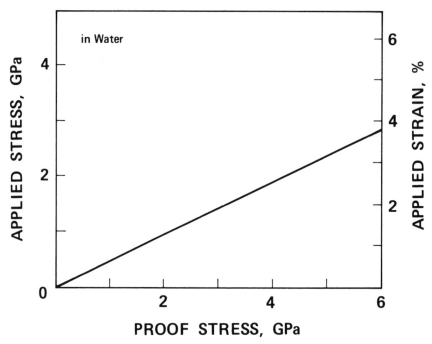

Fig. 12.   Estimation of required proof stress against applied stress for silica fibers in water, using (8).

by the proof test, is given by

$$K_{\text{ISCC}} = \sigma_a Y \sqrt{c_p} .$$ (7)

Then, from (5) and (7), the allowable stress is expressed by

$$\sigma_a = \sigma_p K_{\text{ISCC}} / K_{\text{IC}}.$$ (8)

In other words, when the stress condition is given considering factors such as the cable laying process and the cable structure, the required proof stress can be determined from (8).

For example, in water the K_{ISCC} value is evaluated at 0.37 MPa $m^{1/2}$ [16], while the K_{IC} is obtained as 0.80 MPa M^{12} [13]. Thus, the fiber must pass the proof test at the stress given by

$$\sigma_p = 2.1\sigma_a.$$

Figure 12 shows required proof stress against applied stress for silica fibers in water.

The above discussion gives the nonbreak conditions for optical fibers using some basic fracture mechanics parameters measured on a plate specimen. This is confirmed by the fact that the crack growth process for macroscopic flaws is quite the same as for plate specimens.

SUMMARY

A drawing process for high-strength long-length optical fibers has been developed using a carbon resistance furnace. Preform surface treatment conditions are explored in order to heighten surface quality. The influence of dust particles on the tensile strength of the fibers is clarified quantitatively as a function of particle radius, giving the required cleanliness for the heating atmosphere. These improvements in the drawing process have led to an average proof success length of more than 8 km at a 2.4 GPa proof stress for VAD synthetic silica fibers.

An evaluation of the mechanical strength of long fibers has also been described. The reliability for the proof test in determining minimum strength was examined by tensile testing for long fibers proofed at various stress levels. The allowable stress conditions, at which no breaks occur, were examined on the basis of the fracture mechanics concept. The required proof stress was determined to be a function of applied stress using fracture toughness and critical stress intensity for stress corrosion cracking.

ACKNOWLEDGMENT

The author would like to thank N. Inagaki for encouragement and is also indebted to M. Nakahara for useful discussions.

REFERENCES

[1] N. Kojima, Y. Miyajima, Y. Murakami, T. Yabuta, O. Kawata, K. Yamashita, and N. Yoshizawa, "Studies on designing of submarine optical fiber cable," *IEEE J. Quantum Electron.*, vol. QE-18 no. 4, pp. 733–740, 1982.

[2] R. D. Maurer, "Strength of fiber optical waveguides," *Appl. Phys. Lett.*, vol. 27, no. 4, pp. 220–221, 1975.

[3] H. Schonhorn, C. R. Kurkjian, R. E. Jaeger, H. N. Vazirani, R. V. Albarino, and F. V. DiMarcello, "Epoxy-acrylate-coated fused silica fibers with tensile strengths > 500 ksi (3.5 GN/m²) in 1-km gauge lengths," *Appl. Phys. Lett.*, vol. 29, no. 11, pp. 712–714, 1976.

[4] T. J. Miller, A. C. Hart, Jr., W. I. Vroom, Jr., and M. J. Bowden, "Silicone- and ethylene-vinyl-acetate-coated laser drawn silica fibers with tensile strengths > 3.5 GN/m² (500 kpsi) in > 3 km lengths," *Electron. Lett.*, vol. 14, no. 18, pp. 603–605, 1978.

[5] H. Schonhorn, H. N. Vazirani, and H. L. Frisch, "Relationship between fiber tension and drawing velocity and their influence on the ultimate strength of laser-drawn silica fibers," *J. Appl. Phys.*, vol. 49, no. 7, pp. 3703–3706, 1978.

[6] T. Yamanishi, K. Yoshimura, S. Suzuki, S. Seikai, and N. Uchida, "Modified silicone as new type of primary coat for optical fibre," *Electron. Lett.*, vol. 16, no. 3, pp. 100–101, 1980.

[7] T. T. Wang and H. M. Zupko, "Strengths and diameter variations of fused silica fibers prepared in oxy-hydrogen flames," *Fiber Integrated Opt.*, vol. 3, no. 1, pp. 73–87, 1980.

[8] U. C. Paek, C. D. Spainhour, C. M. Schroeder, and C. R. Kurkjian, "Tensile strength of 50-m-long silica fibers drawn with a laser galvanometer scanning system," *Bull. Amer. Ceram. Soc.*, vol. 59, no. 6, pp. 630–634, 1980.

[9] R. D. Maurer, "Effect of dust on glass fiber strength," *Appl. Phys. Lett.*, vol. 30, no. 2, pp. 82–84, 1977.

[10] H. Hanafusa, S. Sakaguchi, Y. Tajima, and Y. Hibino, to be published in *ECL Tech. J.*, NTT, in Japanese.

[11] F. V. DiMarcello, D. L. Brownlow, and D. S. Shenk, "Strength characterization of multi-kilometer silica fibers," in *Tech. Dig. MG6, IOOC '81*, San Francisco, CA, Apr. 27–29, 1981.

[12] S. Sakaguchi, M. Nakahara, and Y. Tajima, "Drawing of high-strength long-length optical fibers," in *Tech. Dig., IOOC '83*, Z9A4, Tokyo, Japan, June 27–30, 1983.

[13] S. M. Wiederhorn, "Fracture surface energy of glass," *J. Amer. Ceram. Soc.*, vol. 52, no. 2, pp. 99–105, 1969.

[14] S. Sakaguchi and M. Nakahara, "Strength of proof-tested optical fibers," *J. Amer. Ceram. Soc.*, vol. 66, no. 3, pp. C-46–C-47, 1983.

[15] S. Sakaguchi and Y. Hibino, "Fatigue in low-strength silica optical fibers," *J. Mater. Sci.*, vol. 19, no. 10, pp. 3416–3420, 1984.

[16] S. Sakaguchi, Y. Sawaki, Y. Abe, and T. Kawasaki, "Delayed failure in silica glass," *J. Mater. Sci.*, vol. 17, no. 9, pp. 2878–2886, 1982.

15

Polarization-Maintaining Optical Fibers Used for a Laser Diode Redundancy System in a Submarine Optical Repeater

YUTAKA SASAKI, TOSHIHITO HOSAKA, AND JUICHI NODA

INTRODUCTION

A laser diode (LD) redundancy system which improves reliability by implementing a sparing scheme for LD's in a submarine optical repeater has been proposed [1]–[4]. This redundancy system consists mainly of LD's and a light-switching coupler. The system is required to have high reliability and low optical loss and to be fabricated easily. Therefore, to realize long-term reliability, it would be desirable for this redundancy system to have no mechanical parts and to also utilize LD polarization characteristics to reduce optical loss [5], [6]. Furthermore, if linear polarization-maintaining optical fibers were used to couple the LD's to the light-switching coupler, such a redundancy system could be fabricated easily and would be highly reliable, so the LD's could be installed anywhere without regard to the location of the light-switching coupler [7].

Many types of linear polarization-maintaining optical fibers have already been reported [8]–[17]. Among these, stress-induced birefringent fibers [8]–[13] have the highest linear polarization-maintaining ability. The structural parameters of PANDA optical fiber [11], [13], [18], [19], which is a stress-induced birefringent fiber, have been shown to vary only slightly over an 11-km length [19]. PANDA optical fibers have been adopted as the polarization-maintaining optical fiber in an LD redundancy system [20].

This chapter describes the polarization stability and mechanical strength of short-length PANDA fibers and shows that the fibers satisfy requirements for use in an LD redundancy system in a submarine optical repeater.

REQUIREMENTS FOR POLARIZATION-MAINTAINING OPTICAL FIBERS

An LD redundancy system utilizing laser diode polarization characteristics [7] is shown in Fig. 1. This redundancy system is composed of two laser diodes ($LD_{1,2}$), two polarization-maintaining optical fibers ($PMF_{1,2}$), a light-switching coupler, that is, a

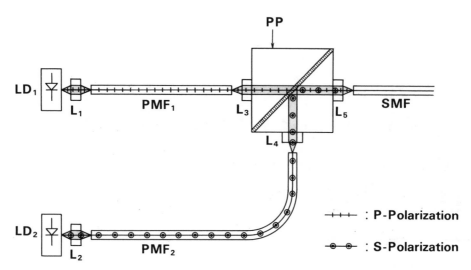

Fig. 1. An LD redundancy system utilizing LD polarization characteristics. $LD_{1,2}$: laser diodes; L_{1-5}: GRIN rod lenses; $PMF_{1,2}$: polarization-maintaining optical fibers; SMF: single-mode optical fiber; PP: polarization prism.

polarization prism (PP) consisting of two glass prisms and a dielectric thin film filter, five GRIN spherical rod lenses (L_{1-5}), and a single-mode fiber (SMF). Two 1.3-μm wavelength channels are installed in the redundancy system: one transmits only P-polarized light and the other only S-polarized light.

It has been shown that crosstalk or extinction ratio in an LD is about -20 dB, and that in the polarization prism is less than -25 dB [6]. Therefore, crosstalk in the polarization-maintaining fibers must be -20 dB or less.

The requirements for the polarization-maintaining fibers that couple the two LD's to the polarization prism in a submarine optical repeater are listed in Table I [7]. The fiber diameter and the spot size of light emerging from the fiber ends should be 125 and 10 μm, respectively, to make the parameters of the polarization-maintaining fibers match those of conventional single-mode fibers.

The crosstalk degradation induced by external factors, which include bends with radii of more than 25 mm, twists of 1 turn/m or less, and ambient temperatures of 5–40°C, must also be -20 dB or less. The maximum allowable transmission loss requirement for

TABLE I
REQUIREMENTS FOR POLARIZATION-MAINTAINING OPTICAL
FIBERS IN A SUBMARINE OPTICAL REPEATER

	ITEMS	REQUIREMENTS
	Fiber Diameter	125 μm
	Spot Size	10 μm
Crosstalk	Bending Radius: 25 mm	-20 dB
	Number of Twists: 1 turn/m	
	Temperature Range: 5–40°	
Reliability	Crosstalk: -20 dB	25 years
	Strength	
	Loss	10 dB/km

polarization-maintaining fibers is sufficiently satisfied by a loss value of less than 10 dB/km. The submarine optical repeater is required to maintain its transmission characteristics for a period of more than 25 years, which is the maintenance period required for submarine optical fiber cables under high pressure. Therefore, the polarization-maintaining fibers will also be required to maintain their polarization characteristics and mechanical strength for a period of more than 25 years.

FIBER PARAMETERS AND CROSSTALK MEASUREMENT SYSTEM

Fiber Design

A cross-sectional view of the PANDA optical fiber is shown in Fig. 2. The diameters of the core and the cladding are represented $2a$ and $2b$, respectively, and the relative refractive index difference between the core and cladding is represented by Δ. The half distance between the stress-applying parts is represented by r, and the diameter of the stress-applying parts is represented by t. The dopant concentration for the stress-applying parts is ρ, while, projection angle θ is the angle subtended by each stress-applying part at the core center. High modal birefringence B can be obtained by choosing $\theta \approx 90°$ [13]. As

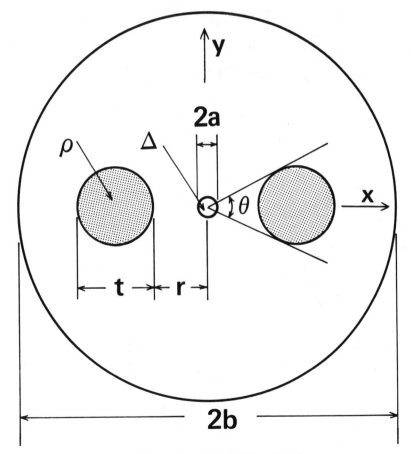

Fig. 2.   Cross section of a PANDA optical fiber.

for the fiber's principal axes, the axis in the direction of the stress-applying parts is denoted as the x-axis, and that normal to it is denoted as the y-axis.

In order to achieve low transmission loss, losses due to the introduction of stress-applying parts must be reduced. Origins which increase the transmission loss are

1) Absorption loss due to OH-ion in the stress-applying parts.
2) Infrared absorption loss due to B_2O_3 in the stress-applying parts.
3) Scattering loss due to the refractive-index mismatch between the cladding and the stress-applying parts.

If normalized distance r/a is set to be more than two, OH-ion contributes mainly to transmission loss increase, which is less than 5 dB/km at 1.3 μm wavelength.

Figure 3 shows the relationship between normalized stress-applying part diameter t/b and modal birefringence B. When the normalized distance r/a is 2, 3, and 4, the

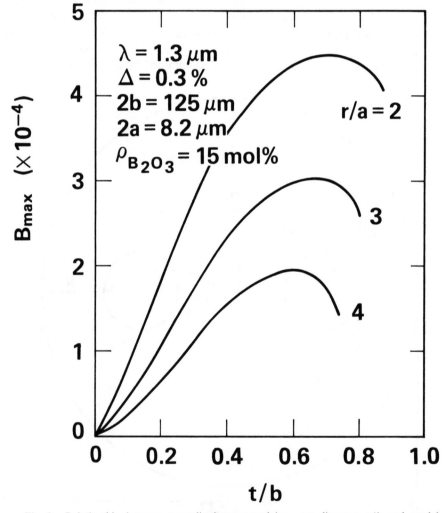

Fig. 3. Relationship between normalized stress-applying part diameter t/b and modal birefringence B for $r/a = 2$, 3, and 4.

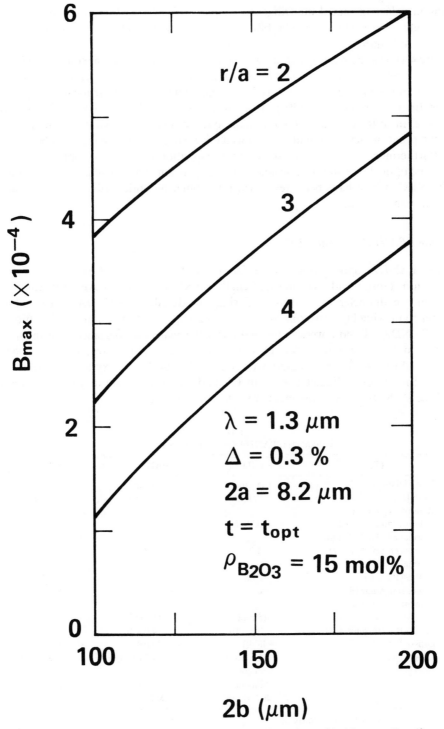

Fig. 4. Relationship between outer diameter $2b$ and maximum birefringence B_{max} for $r/a = 2$, 3, and 4.

maximum values of modal birefringence B_{max} are 4.5×10^{-4}, 3.0×10^{-4}, and 2.0×10^{-4}, respectively. These values of modal birefringence enable to realize crosstalk less than -30 dB in a 1 m length.

Figure 4 shows the relationship between outer diameter $2b$ and maximum birefringence B_{max}. As outer diameter $2b$ increases, maximum modal birefringence B_{max} increases. When the normalized distance r/a is 3, the maximum modal birefringence (B_{max}) values for the fibers with 125 and 150 μm of $2b$ are 3.7×10^{-4} and 3.0×10^{-4}, respectively. Therefore, in order to realize low crosstalk, outer diameter is desirable to be larger. But from the point of view of handling and mechanical strength, it is required to be 200 μm or less. Furthermore, the outer diameter of polarization-maintaining optical fibers in a submarine optical repeater, is desirable to be 125 μm, since highly reliable fiber tools, such as NO-NIK's which have been developed for submarine optical fiber with 125 μm outer diameter, can be used.

Fabricated PANDA Optical Fibers

Three PANDA optical fibers were fabricated to compare to each other regarding polarization stability and mechanical strength. PANDA's 1 and 3 have a 125 μm diameter to match the diameter of conventional single-mode fibers. PANDA 2 has a 150 μm diameter to provide high birefringence, which was expected to maintain polarization more stably. PANDA's 1 and 2 were coated with only thermally curable silicone resin (silicone), resulting in a 400 μm diameter, or both silicone and nylon, resulting in a 900 μm diameter. PANDA 3 were coated with only UV curable epoxy-acrylate resin (UV), resulting in a 400 μm diameter, or both UV and nylon, resulting in a 900 μm diameter. The three PANDA fiber parameters are listed in Table II.

TABLE II
FABRICATED PANDA OPTICAL FIBER PARAMETERS

ITEMS	PANDA 1	PANDA 2	PANDA 3
Fiber Diameter ($2b$)	125 μm	150 μm	125 μm
Core Diameter ($2a$)	8.0 μm	8.6 μm	8.5 μm
Relative Refractive Index Difference (Δ)	0.29%	0.26%	0.28%
Normalized Half-Distance between SAP*[1] (r/a)	3.0	2.3	2.0
Normalized Diameter of SAP (t/b)	0.66	0.64	0.62
Projection Angle (θ)	$\sim 80°$	$\sim 90°$	$\sim 90°$
Dopant Concentration in SAP (ρ)	15 mol%	15 mol%	15 mol%
Modal Birefringence (B)	3.4×10^{-4}	4.0×10^{-4}	3.5×10^{-4}
Crosstalk*[2] (CT)	< -40 dB	< -40 dB	< -40 dB
Loss*[3] (α)	1.5 dB/km	4.2 dB/km	4.8 dB/km
Coat	Silicon Silicone & Nylon	Silicone Silicone & Nylon	UV UV&Nylon

*[1] SAP: Stress-Applying Parts
*[2] $\lambda = 1.3$ μm, $l = 1$ m
*[3] $\lambda = 1.3$ μm

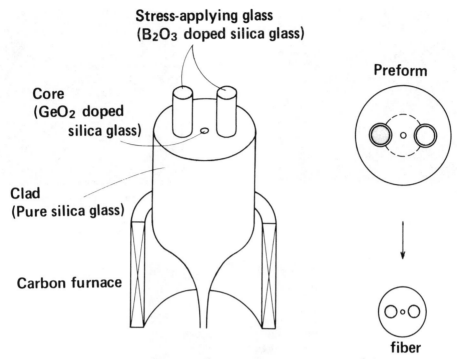

Fig. 5.   PANDA optical fiber fabrication by pit-in-jacket method.

These PANDA fibers were fabricated using a new method called the pit-in-jacket method [18] as shown in Fig. 5. In that method, VAD synthesized preforms are used as the core-cladding preforms, and MCVD preforms are used as the stress-applying preforms. The relative refractive index difference (Δ) values for PANDA's 1, 2, and 3 were set at 2.6, 2.9, and 2.8 percent, respectively, to make the spot sizes of light emerging from the fibers match those of conventional single-mode fibers. The cores of the fibers were GeO_2-doped silica glass, having $2a = 8.0$, 8.6, and 8.5 μm, respectively. The claddings were pure-silica glass. The effective cutoff wavelength, λ_c, was in the range for single-mode operation at 1.3 μm. The stress-applying parts consisted of B_2O_3-doped silica glass. Dopant concentration ρ is 15 mol%. The normalized half distance between stress-applying parts, r/a, was 3.0, 2.3, and 2.0 for PANDA 1, 2, and 3, respectively. The normalized diameter of stress-applying parts t/b for these fibers were about 0.65. The projection angle (θ) values for the fibers were set at optimum value, $\sim 90°$. Modal birefringence B for each of the fibers was greater than 3×10^{-4}. According to the order of PANDA's 3, 2, and 1, the losses increase because the ratio r/a values decrease. These increased losses in the fibers were mainly due to OH-ion absorption. Losses in the fibers were less than 5 dB/km at 1.3 μm. Therefore, these fibers satisfied the maximum allowable loss requirement for polarization-maintaining fibers in a submarine optical repeater.

Figure 6 shows end-face photographs of PANDA's 1, 2, and 3. The shapes and dimensions of the stress-applying parts in these fibers varied only slightly between the input and output fiber ends. The distance ($2r$) variations and the diameter (t) variations in the stress-applying parts of the fibers were within 1 μm over a 1 km length for the fibers.

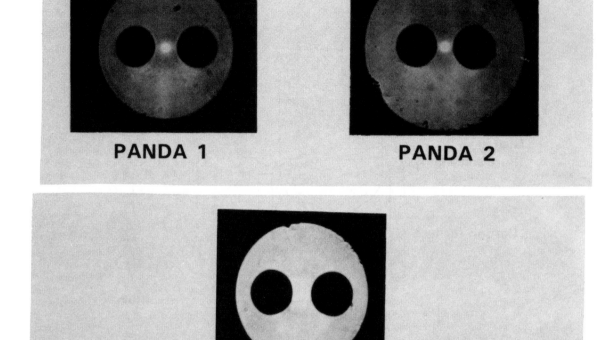

Fig. 6.   Cross sectional photographs of fabricated PANDA optical fibers.

Crosstalk Measurement System

The experimental arrangement used to measure the polarization characteristics of three PANDA optical fibers, is shown in Fig. 7. Linearly polarized light emitted from a 1.3-μm wavelength InGaAsP/InP LD is focused into the test fiber through a fixed Glan–Thompson polarizer (P_1), a fixed quarter-wave plate ($\lambda/4$) and a rotatable Glan–Thompson polarizer (P_2). At the fiber output, the light is focused through a rotatable Glan–Thompson analyzer (P_3) onto a germanium photodiode (PD) and synchronously detected by a lock-in amplifier (L.A.). The LD is driven by a pulse generator with 1 kHz frequency in order to make it operate with broad-band spectrum. The fiber axes are located by rotating both the input polarizer (P_2) and the output analyzer (P_3) for minimum detected power. In this condition, the input polarizer is parallel to the HE_{11}^{x} mode and the output analyzer is perpendicular to that mode. Rotation of the analyzer through $\pi/2$ allows measurement of the crosstalk or extinction ratio, $CT = 10 \cdot \log(\eta)$, where $\eta = P_y/P_x$ and P_x and P_y represent the output power in the two polarization modes. If the input polarizer is set parallel to the HE_{11}^{y} mode, η can be expressed as P_x/P_y. In this experimental arrangement, crosstalk can be accurately measured and found to be -45 dB in a 1 m fiber. The crosstalk measurement accuracy degrades at $CT = -48$

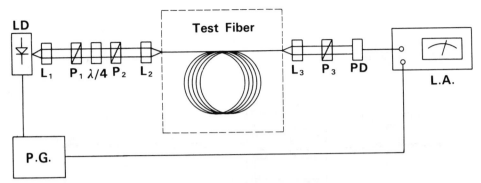

Fig. 7. Crosstalk Measurement system. LD: Laser diode ($\lambda = 1.3$ μm); L_{1-3}: Microscope objectives lenses, P_{1-3}: Glan–Thompson polarizer; $\lambda/4$: Quarter-wave plate; PD: Ge photodiode; L.A.: Lock-in amplifier.

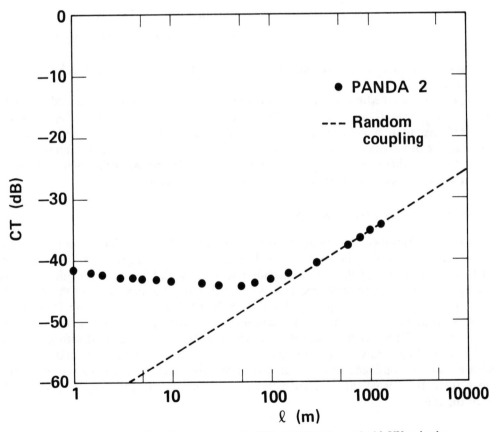

Fig. 8. Fiber length dependence of crosstalk CT for PANDA 2 coated with UV and nylon.

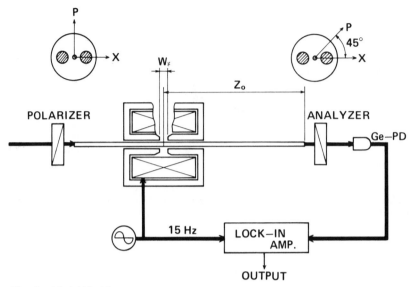

Fig. 9. Modal birefringence measurement system by magneto-optic modulation method.

dB because the detected power onto the germanium photodiode tends to reach maximum and saturate.

Figure 8 shows the fiber length dependence of crosstalk for PANDA 2 coated with UV and nylon. Crosstalk of 1.3 km long fiber was -34 dB at 1.30 μm wavelength. When crosstalk degradation obeys mode coupling yielded random perturbations, the fiber length dependence of crosstalk is shown as the dashed line in Fig. 8. Crosstalk saturation in a short length is due to propagation of the cladding mode excited at the input fiber end. This cladding mode can be dissipated by lossy primary coating.

Modal Birefringence Measurement System

Modal birefringence B is given by $B = \lambda_0/L$ where L is beat length and λ_0 is the free-space wavelength. The beat length L is measured by magneto-optic modulation method as shown in Fig. 9. A DFB injection laser (LD) with 1.3 μm wavelength is used as the light source. Linearly polarized light is coupled into a 2-m long piece of the PANDA fiber, parallel to one of the principal axies. At the fiber output an analyzer is set at 45 degrees azimuth. The fiber passes through a 0.8 mm diameter bore in the center of an electromagnet. The pole gap width W_g is 1 mm, and the magnet is excited with sinusoidal AC at 15 Hz. The analyzer output signal is detected with a Ge photodiode and processes with a current amplifier and a lock-in amplifier. When the magnet is moved along the fiber, the output signal varies periodically as shown in Fig. 10, indicating the beat length L of the fiber.

POLARIZATION STABILITY REGARDING EXTERNAL FACTORS

The polarization state in polarization-maintaining fibers in a submarine optical repeater can be modified by environmental influences that vary in an unpredictable manner. Therefore, it is necessary to understand the polarization-maintaining ability in

PANDA 1
Wave Length : 1.3 μm

L = 3.8 mm

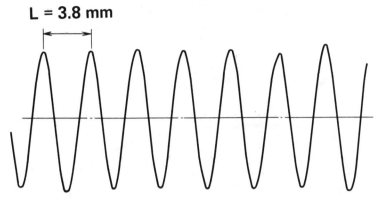

Fig. 10.   Output light intensity variation by magneto-optic modulation method.

Bending Stress σ_b (kg/mm^2)

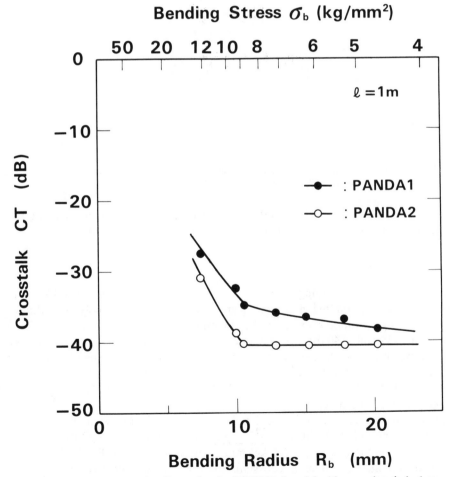

Fig. 11.   Crosstalk versus bending radius for PANDA's 1 and 2 with gauge length $l = 1$ m. Bending stress is shown only for the fibers with $2b = 125 \mu$m.

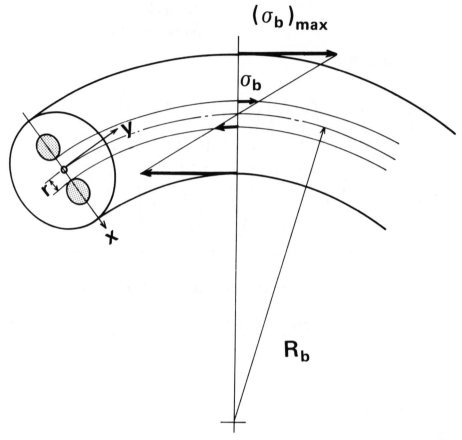

Fig. 12. Schematic view of bending experiments. R_b is bending radius and σ_b is bending stress at $x = \pm r$. $(\sigma_b)_{max} = \sigma_b \cdot (b/r)$, where b is fiber radius.

PANDA fibers. The polarization stability in PANDA fibers versus external factors, including bend, twist, tension, and temperature change, was examined.

Crosstalk Due to Bending

Crosstalk CT versus bending radius R_b is shown in Fig. 11 for PANDA's 1 and 2 coated with silicone and nylon. Fibers 1 m long were coiled loosely on cylinders of various diameters without respect to the direction of the x and y axes in the fiber. Crosstalk of -38 and -40 dB was maintained for bending radii of 20 mm for PANDA's 1 and 2, respectively. Bending stress σ_b in Fig. 11 corresponds to the bending radius for a fiber having 125-μm fiber diameter, that is, PANDA 1, and is the value at the locations $x = \pm r$ in Fig. 12. The σ_b values were calculated from $\sigma_b = r \cdot E/(R_b + r) \approx r \cdot E/R_b$, where E is Young's modulus.

As shown in Fig. 11, the crosstalk CT for PANDA 1 increased monotonically with decreased bending radius R_b in the 10–20 mm range and reached -32 dB with $R_b = 10$ mm, which corresponds to $\sigma_b = 9.4$ kg/mm^2. The crosstalk for PANDA 2 remained almost constant at $R_b = 10$ mm. Crosstalk for both fibers increased rapidly with R_b reduced below 10 mm. The incident polarization power loss also increased with R_b

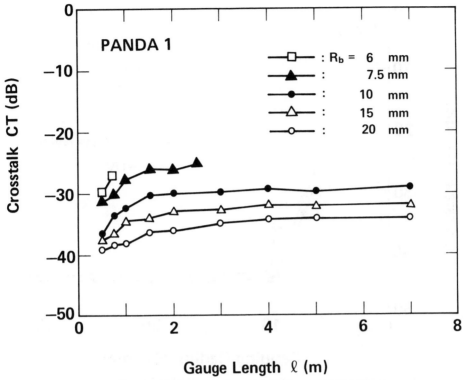

Fig. 13.   Crosstalk *CT* versus gauge length 1 for PANDA 1.

reduced below 10 mm, and crosstalk measurement at $R_b = 6$ mm was impossible for both fibers due to abrupt bending loss increase. What is evident on comparison of the two curves in Fig. 11 is that, as for polarization stability under bending, PANDA 2 is better than PANDA 1.

Figure 13 shows crosstalk *CT* versus gauge length 1 in PANDA 1 coated with silicone and nylon for five different bending radii: $R_b = 6$, 7.5, 10, 15, and 20 mm. Crosstalk for these bending radii, except for $R_b = 6$ mm, saturated at lengths of more than 2 m. For $R_b = 6$ mm, incident polarization power was very lossy and crosstalk could not be measured in a 1 m fiber. It is clear from Fig. 13 that if PANDA 1 were installed with a bending radius of 10 mm or more, crosstalk of −30 dB could be preserved in the fiber length range of 1 m to several meters.

Figure 14 shows crosstalk *CT* versus bending radius R_b for PANDA 3 coated with only UV and both UV and nylon. Crosstalk of about −45 dB for the fibers was maintained stably up to bending radius of 10 mm. Crosstalk increased rapidly with R_b reduced below 10 mm.

It was confirmed from the results in Figs. 11 and 13 that short-length (1 m) PANDA fibers, whose diameters are even 125 μm as well as 150 μm, could sufficiently maintain less than −20-dB crosstalk for a more than 25-mm bending radius, which satisfies the bending requirement for PANDA fibers in a submarine optical repeater.

It was confirmed from Figs. 11 and 14 that the PANDA fiber coated with only UV and both UV and nylon could sufficiently maintain also less than −20 dB crosstalk for a more than 25-mm bending radius regardless of coating materials.

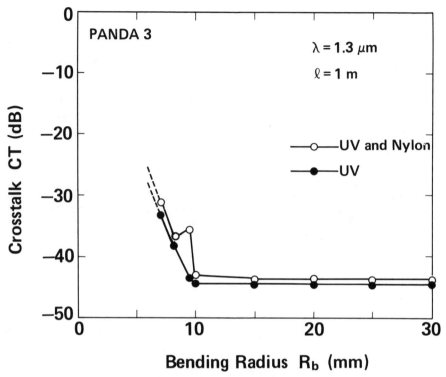

Fig. 14. Crosstalk CT versus bending radius R_b for PANDA 3 coated with only UV and both UV and nylon.

Crosstalk Due to Twisting

To investigate crosstalk degradation due to twisting, both ends of 1-m long fibers were fixed to a rotating jig with epoxy adhesive. Crosstalk CT versus number of twists N is shown in Fig. 15 for PANDA's 1 and 2 coated with silicone and nylon, both with and without sag. Shearing stress σ_s in Fig. 15 corresponds to the number of twists for the fiber with a 125-μm fiber diameter, that is, PANDA 1, and is the value at the locations $x = \pm r$ in Fig. 16. The σ_s values were calculated from $\sigma_s = \pi \cdot r \cdot E(N/1)/(1+\nu)$, where ν is Poisson's ratio. Crosstalk CT remained at less than -30 dB up to 60 turns for both fibers, regardless of sag, where $N = 60$ turns corresponds to $\sigma_s = 15.0$ kg/mm^2 and 12.4 kg/mm^2 for PANDA's 1 and 2, respectively. With sag, crosstalk degraded rapidly with increased number of twists above 60 turns. In the range from 75 to 85 turns, crosstalk exceeded -20 dB for both fibers. The rapid crosstalk degradation in this range results from several bends having extremely small curvature. Because the number of localized bends for both fibers increased with increased twisting above 85 turns and, as a result, the incident polarization power loss also increased greatly, crosstalk could not be measured with N of more than 100 turns.

In the case without sag, when about 50 g tension was applied to the fibers, crosstalk increased gradually with increased number of twists. However, CT of -20 dB was maintained even with $N = 160$ turns for both fibers. Crosstalk for PANDA 1 was measurable even with 200 turns and was about -21 dB. Because PANDA 2 broke down

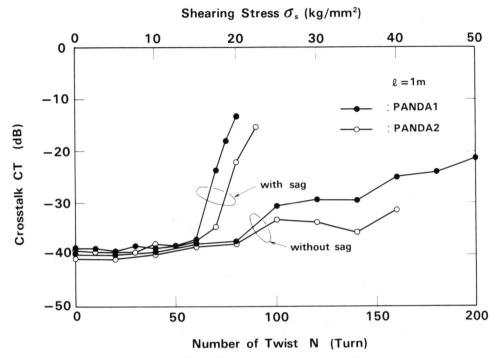

Fig. 15.   Crosstalk CT versus number of twists N.

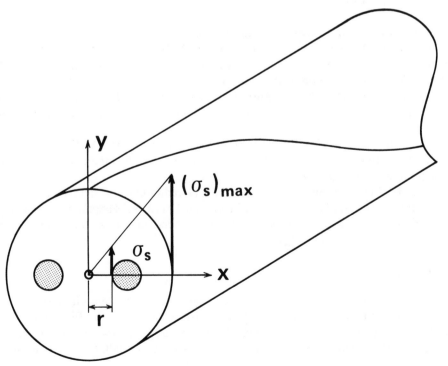

Fig. 16.   Schematic view of shearing experiments. σ_s is shearing stress. $(\sigma_s)_{max} = \sigma_s \cdot (b/r)$, where b is fiber radius. Shearing stress is shown only for the fibers with $2b = 125$ μm.

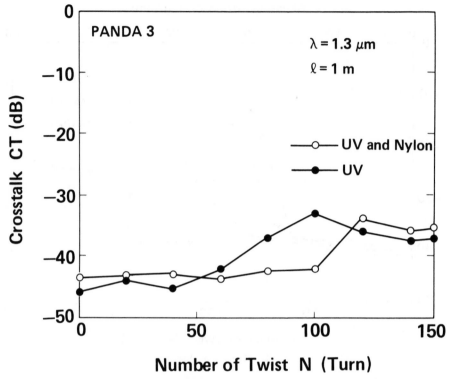

Fig. 17. Crosstalk CT versus number of twists N for PANDA 3 coated with only UV and both UV and nylon.

after more than 160 turns, crosstalk measurement for PANDA 2 was impossible. From the case without sag in Fig. 15, it is clear that PANDA 2, with its higher birefringence ($B = 4.0 \times 10^{-4}$), preserves polarization better than PANDA 1, which had lower birefringence ($B = 3.4 \times 10^{-4}$). However, it is also evident that, for twisting strength, PANDA 1, with its smaller diameter ($2b = 125$ μm), is stronger than PANDA 2, which has a larger diameter ($2b = 150$ μm). This will be discussed further in the section on mechanical strength.

Figure 17 shows crosstalk CT versus number of twists N for PANDA 3 coated with only UV and both UV and nylon. Crosstalk CT remained at less than -30 dB up to 150 turns without sag for both fibers.

The polarization-maintaining fibers for an LD redundancy system in a submarine optical repeater are installed with an N of less than 1 turn/m. Therefore, when PANDA fibers are used as the polarization-maintaining fibers, crosstalk degradation due to twisting will be negligibly low regardless of sag.

Crosstalk Due to Tension

To carry out accurate crosstalk measurement for tension, only tension must be applied along test fibers and, especially, side pressure must be removed. The fiber setup for the tenile strength experiment is shown in Fig. 18. In this setup, the fiber is fixed to the inside of steel pipes ($SP_{1,2}$) with epoxy adhesive. Figure 19 shows crosstalk CT versus tension W.

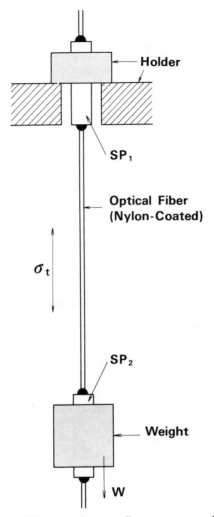

Fig. 18.   Fiber setup for crosstalk measurement under tension.

Tensile stress σ_t in Fig. 19 corresponds to the tension for a fiber with $2b = 125$ μm, that is, PANDA 1. Crosstalk of about -40 dB or less was maintained for PANDA's 1 and 2 coated with silicone and nylon up to a tension of $W = 2.5$ kg. Crosstalk for PANDA 1 remained at -40 dB up to $W = 3$ kg, which corresponds to $\sigma_t = 245$ kg/mm². Crosstalk increased gradually with tension above 3 kg and reached -36.5 dB at 3.5 kg. Crosstalk for PANDA 2 also remained at -45 dB up to $W = 2.5$ kg and then degraded from -45 to -36 dB with tension increased from 2.5 to 3.5 kg. Although tension was increased to the range of 4–5 kg in this experiment, accurate curves could not be obtained because the silicone buffer coating in the fibers slipped from the nylon jacket and the nylon jacket creeped, and as a result, bends of extremely small curvature occurred in the vicinity of the holders. From Fig. 19, it was confirmed that the tensile requirement for polarization-maintaining fibers in a submarine optical repeater is sufficiently satisfied.

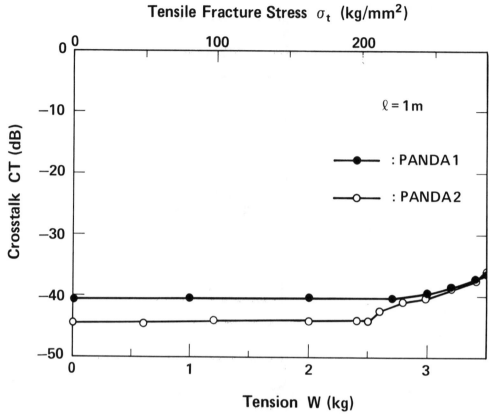

Fig. 19. Crosstalk CT versus tension W for PANDA's 1 and 2 with gauge length $l = 1$ m. Tensile stress is shown only for fibers with $2b = 125$ μm.

Crosstalk Due to Temperature

The test fibers were formed into 300- or 50-mm diameter coils and put into an oven. Then, temperature was gradiently changed in the $-40-80°$C range for PANDA's 1 and 2 or $-80-100°$C range for PANDA 3. The length of fiber parts inside the oven was about 6 m, and that outside the oven was 4 m : 2 m each on the input and output ends.

The temperature dependence of crosstalk is shown in Fig. 20. Crosstalk for PANDA 1 with silicone and nylon in the temperature range 20–80°C was found to be -38 dB or less. Crosstalk for PANDA 1 with $R_b = 150$ mm increased gradually with temperature T, decreased below 20°C, and reached -24 dB at $-40°$C. Crosstalk for PANDA 2 with silicone and nylon with $R_b = 150$ mm remained an almost constant value, that is, CT of -45 dB, in the temperature range $-40-80°$C. The temperature dependence of crosstalk for PANDA 2 was found to be better than that of PANDA 1.

Figure 21 shows the temperature dependence of crosstalk with $R_b = 150$ mm for PANDA 3 coated with only UV and both UV and nylon. Crosstalks for the fibers remained almost constant values, which were less than -40 dB, in the temperature range $-20-100°$C and were less than -35 dB even in the temperature range $-80- -20°$C.

Comparison of the curve for $R_b = 25$ mm to that for $R_b = 150$ mm for PANDA 1 in Fig. 20 shows that there is no significant difference between the crosstalk values with both

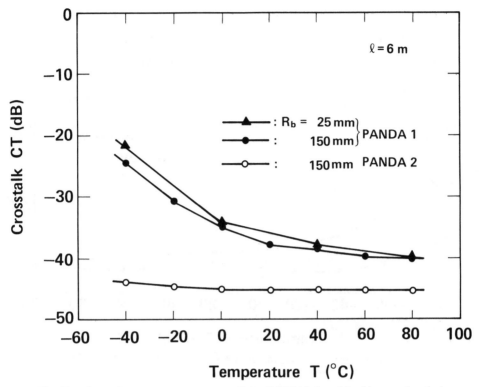

Fig. 20. Crosstalk CT versus temperature T for PANDA's 1 and 2 with gauge length $l = 6$ m.

bending radii, and the bending at $R_b = 25$ mm causes hardly any crosstalk degradation. Therefore, it was confirmed that there is no problem of the crosstalk requirement for PANDA fibers with respect to temperature in a submarine optical repeater because the temperature is in the range 5–40°C in the repeater.

MECHANICAL STRENGTH

Fiber strength is one of the most important long-term reliability factors for an LD redundancy system in a submarine optical repeater. Therefore, fiber fracture strength was determined by bending, twisting, and tensile tests for PANDA's 1 and 2 and compared to that for a conventional single-mode fiber with a 125-μm fiber diameter and GeO_2-doped core. A 125-μm diameter silica-only fiber was also used in the bending fracture test. The fibers used in the bending fracture test were coated with only silicone. The fibers used in tests other than the bending fracture test were coated with both silicone and nylon.

Bending Fracture

Bending fracture time was measured by the mandrel method. PANDA fibers 1 m in length were coiled on cylinders with diameters of 2.5, 3.0, 3.5, and 4.0 mm. The bending strength distributions were plotted on a Weibull probability chart. The bending fracture time for 1 percent failure probability was obtained by extrapolating the fitted curve.

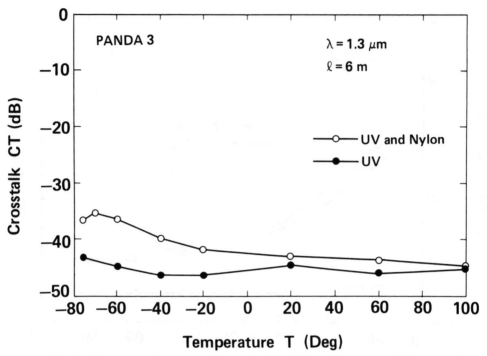

Fig. 21.   Temperature dependence of crosstalk CT with $R_b = 150$ mm for PANDA 3 coated with only UV and both UV and nylon.

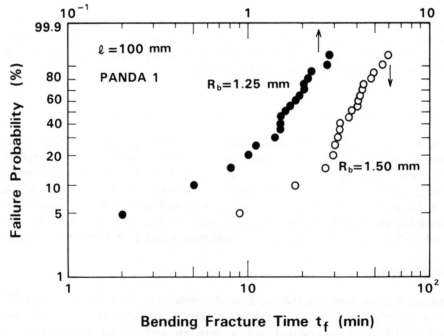

Fig. 22.   Weibull probability plot for bending fracture time t_f with $R_b = 1.25$ and 1.5 mm. Gauge length l is 100 mm.

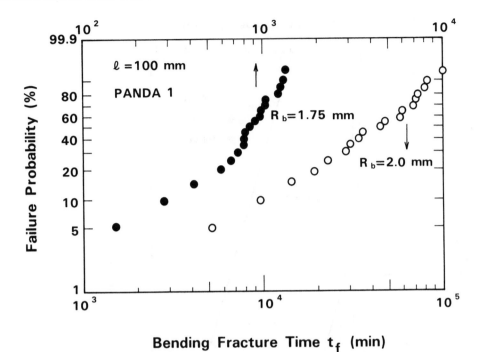

Fig. 23. Weibull probability plot for bending fracture time t_f with $R_b = 1.75$ and 2.0 mm. Gauge length l is 100 mm.

Figures 22 and 23 show the failure probability distributions for PANDA 1. From Fig. 22, obtained bending fracture time (t_f) values at bending radius $R_b = 1.25$ and 1.50 mm for PANDA 1 with gauge length $l = 100$ mm are 3×10^{-2} and 1.5 min, respectively, which correspond to 1.9×10^{-3} and 10^{-1} min, respectively, for a 1-m gauge length. Similarly, from Fig. 23, obtained bending fracture times at $R_b = 1.75$ and 2.0 mm for PANDA 1 with gauge length $l = 1$ m are 5 and 2×10^2 min, respectively. Bending fracture times for a conventional single-mode fiber and a silica-only fiber were also measured.

The obtained relationship between fracture time, $\log(t_f)$, and maximum bending stress $(\sigma_b)_{max}$ is as shown in Fig. 24. The fibers used in these experiments had 125 μm diameters. Maximum bending stress $(\sigma_b)_{max}$ where $(\sigma_b)_{max} = \sigma_b \cdot (b/r)$ was as shown in Fig. 12, and corresponds to the bending radius for the 125-μm diameter fibers shown in Fig. 24. From Fig. 24, the bending fracture time t_f for PANDA 1 at bending radius $R_b = 25$ mm, that is, the bending stress, $(\sigma_b)_{max} = 3.8$ kg/mm^2, was $10^{4.8}$ years/m. It was confirmed by these results that the curve for PANDA 1 is the same as that for a conventional single-mode fiber and a silica-only fiber. In other words, the life against bend in PANDA fibers is the same as that of other fibers.

Twisting Fracture

Twisting fracture tests were carried out by fixing both ends of 100 mm long fibers to a rotating jig with epoxy adhesive. Twisting strength distributions for PANDA's 1 and 2, and a conventional single-mode fiber were plotted on a Weibull probability chart, as shown in Fig. 25. Shearing fracture stress $(\sigma_s)_{max}$ where $(\sigma_s)_{max} = \sigma_s \cdot (b/r)$ as shown in Fig. 16, corresponds to the number of twists N for fibers with $2b = 125$ μm, that is, PANDA 1 and the single-mode fiber.

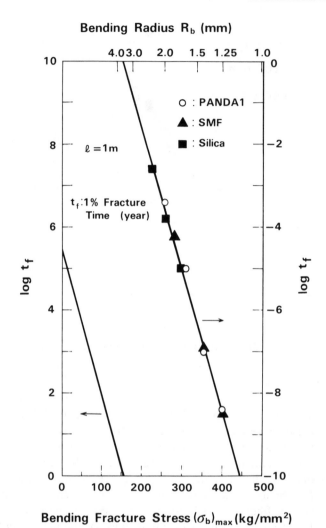

Fig. 24. Relationship between bending fracture time $\log(t_f)$ and bending fracture stress $(\sigma_b)_{max}$ with gauge-length $l = 1$ m. The fibers used in this experiment had 125 μm diameter. Bending radius R_b is shown for $2b = 125$ μm.

From Fig. 25, the greatest number of twists for PANDA 1 is 30.5 turns, and that for PANDA 2 is the same as that for the single-mode fiber, i.e., 24 turns. This number of twists corresponds to 39.4, 37.2, and 31.0 kg/mm^2 of shearing fracture stress $((\sigma_s)_{max})$ values for PANDA's 1, 2 and the single-mode fiber, respectively. Comparing PANDA 1 to the single-mode fiber regarding the greatest number of twists shows that the twisting strength of the PANDA fiber is stronger than that of the single-mode fiber. Comparison of the shearing fracture stresses of both PANDA fibers also shows that the smaller diameter ($2b = 125$ μm) PANDA fiber is slightly stronger than the greater diameter ($2b = 150$ μm) one.

Although there were not sufficient experimental data for the twisting fracture tests, it has been found that the twisting strength of PANDA fibers is the same as that of conventional single-mode fibers.

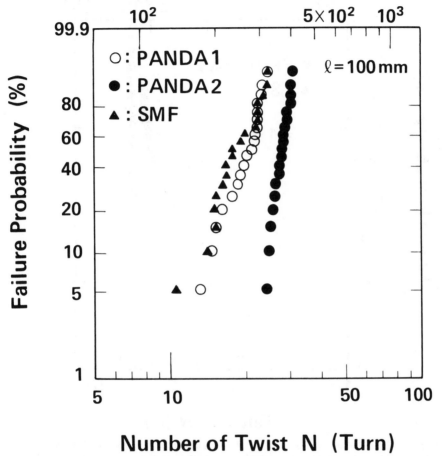

Shearing Fracture Stress $(\sigma_s)_{max}$ (kg/mm^2)

Fig. 25. Weibull probability plot for number of twists N with gauge length $l = 100$ mm. Shearing fracture stress $(\sigma_s)_{max}$ is shown only for fibers with $2b = 125$ μm.

Tensile Fracture

Tensile strength tests were carried out for a PANDA 1 fiber and a conventional single-mode fiber with 0.3-m and 1-m gauge lengths, respectively, using a universal tensile test machine at a constant strain rate of 10 mm/s at room temperature.

Figure 26 shows the failure probability distributions for PANDA 1 and those for the single-mode fiber with the failure probability for PANDA 1 modified to that for a 1-m gauge length. Tensile fracture stress σ_t in Fig. 26 corresponds to tension W for fibers with $2b = 125$ μm.

From Fig. 26, the greatest tension for PANDA 1 and the single-mode fiber is 7.4 and 6.8 kg, respectively, and corresponds to 601 and 552 kg/mm^2 of tensile fracture stress (σ_t) values, respectively. Averaged tension values were 7.2 and 6.2 kg, which correspond to $\sigma_t = 587$ and 505 kg/mm^2 for PANDA 1 and the single-mode fiber, respectively. Aver-

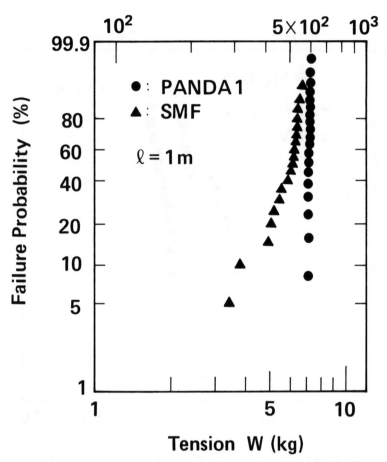

Fig. 26. Weibull probability plot for tension W with gauge length $l = 1$ m. The fibers used in this experiment had 125 μm diameters. Tensile fracture stress σ_t is shown only for fibers with $2b = 125$ μm.

aged tensile fracture stress for both fibers was in the high tensile strength range. These results show that the tensile strength of PANDA fibers is the same as that for single-mode fibers.

Conclusion

Short-length PANDA optical fibers have been investigated for use as polarization-maintaining optical fibers for an LD redundancy system in a submarine optical repeater. Such investigation centered on polarization stability and mechanical strength. The investigation showed the following.

1) Crosstalk of less than -30 dB in a 1-m PANDA optical fiber has been achieved with a bending radius greater than 10 mm, twists of less than 60 turns, and tension of up to 3 kg, and an ambient temperatures of -20–$80°$C.

2) Mechanical strength of PANDA optical fibers is the same as that of conventional single-mode fibers.

PANDA optical fibers are, therefore, applicable for use in a submarine optical repeater.

Acknowledgement

The authors wish to thank K. Takata, N. Inagaki, and T. Miyashita for their continuous encouragement. They are also grateful to K. Okamoto for his discussion of PANDA fibers, to T. Ito and Y. Mitsunaga for discussions on submarine optical repeater requirements, to H. Nagai for supplying semiconductor lasers, and to Y. Miyajima for the tensile test equipment.

References

[1] C. D. Anderson, R. F. Gleason, P. T. Hutchison, and R. K. Runge, "An undersea communication system using fiberguide cables," *Proc. IEEE*, vol. 68, pp. 1299–1303, 1980.

[2] Y. Niiro, H. Wakabayashi, and H. Tokiwa, "Design and experimental results of optical submarine repeater circuits," in *Proc. 7th ECOC Conf.*, 1981, pp. 15.1.1–15.1.4.

[3] C. M. Miller, R. B. Kummer, S. C. Metter, and D. N. Ridgway, "Single-mode optical fiber switch," *Electron. Lett.*, vol. 16, pp. 783–784, 1980.

[4] W. C. Young and L. Curtis, "Cascaded multiple switches for single-mode and multimode optical fibers," *Electron. Lett.*, vol. 17, pp. 571–573, 1981.

[5] R. Kishimoto, "Optical coupler for laser redundancy system," *Electron. Lett.*, vol. 18, pp. 140–141, 1982.

[6] ____, "A consideration of an optical coupler for optical submarine transmission laser redundancy systems," *IEEE Trans. Commun.*, vol. COM-31, pp. 232–244, 1983.

[7] S. Tsutsumi, Y. Ichihashi, M. Sumida, and H. Kano, "LD redundant system using polarization components for a submarine optical transmission system," in this book, ch. 35.

[8] R. H. Stolen, V. Ramaswamy, P. Kaiser, and W. Pliebel, "Linear polarization in birefringent single-mode fibers," *Appl. Phys. Lett.*, vol. 33, pp. 699–701, 1978.

[9] T. Katsuyama, H. Matsumura, and T. Suganuma, "Low-loss single-polarization fibers," *Electron. Lett.*, vol. 17, pp. 473–474, 1981.

[10] T. Hosaka, K. Okamoto, T. Miya, Y. Sasaki, and T. Edahiro, "Low-loss single-polarization fibers with asymmetrical strain birefringence," *Electron. Lett.*, vol. 17, pp. 530–531, 1981.

[11] Y. Sasaki, K. Takada, T. Hosaka, and N. Shibata, "Polarization-maintaining and absorption-reducing fibers," in *Proc. 5th OFC*, 1982, pp. 54–56.

[12] R. D. Birch, D. N. Payne, and M. P. Varnham, "Fabrication of polarization-maintaining fibers using gas-phase etching," *Electron. Lett.*, vol. 18, pp. 1036–1038, 1982.

[13] N. Shibata, Y. Sasaki, K. Okamoto, and T. Hosaka, "Fabrication of polarization-maintaining and absorption-reducing fibers," *IEEE J. Lightwave Technol.*, vol. LT-1, pp. 38–43, 1983.

[14] S. Machida, J. Sakai, and T. Kimura, "Polarization conservation in single-mode fibers," *Electron. Lett.*, vol. 17, pp. 494–495, 1981.

[15] R. B. Dyott, J. R. Cozens, and D. G. Morris, "Preservation of polarization in optical-fiber waveguides with elliptical cores," *Electron. Lett.*, vol. 15, pp. 380–382, 1979.

[16] T. Okoshi and K. Oyamada, "Single-polarization single-mode fiber with refractive-index pits on both sides of core," *Electron. Lett.*, vol. 16, pp. 712–713, 1980.

[17] T. Hosaka, K. Okamoto, Y. Sasaki, and T. Edahiro, "Single-mode fibers with asymmetrical refractive-index pits on both sides of core," *Electron. Lett.*, vol. 17, pp. 191–193, 1981.

[18] Y. Sasaki, T. Hosaka, K. Takada, and J. Noda, "8 km-long polarization-maintaining fiber with highly stable polarization state," *Electron. Lett.*, vol. 19, pp. 792–794, 1983.

[19] Y. Sasaki, T. Hosaka, and J. Noda, "Low crosstalk polarization-maintaining optical fiber with an 11 km length," *Electron. Lett.*, vol. 20, pp. 784–785, 1984.

[20] ____, ____, "Polarization-maintaining optical fibers used for a laser diode redundancy system in a submarine optical repeater," *IEEE J. Lightwave Technol.*, Special Issue, 1985.

Part IV
Undersea Fiber Cables

Undersea fiber cables must be designed to protect the undersea fibers from excessive strains during deep-sea laying and recovery operations as well as shield the fibers from possible damage due to abrasion and pressure on the ocean bottom. In this part we will examine five different undersea cable designs with emphasis on the requirements each design places on fiber strength.

16
Design and Testing of the SL Cable

ALI ADL, TA-MU CHIEN, MEMBER, IEEE, TEK-CHE CHU

DESIGN OF SL CABLE

Design Objectives and Constraints

The design of an undersea lightguide cable, as for any engineering project, should meet certain goals and satisfy certain constraints. The design goals for the SL cable are as follows:

protect singlemode lightguides
  —from excessive strain during laying and recovery,
  —from the pressure of the deep ocean,
  —from external aggression;
provide a powering path for repeaters;
provide at least 25-year service life; and
be manufacturable at a competitive price.

The design constraints for the SL cable have been described and explained previously by Gleason *et al.* [1]. They are

minimum horizontal sinking speed > 0.97 km/h;
minimum strength >19 km of cable weight in water;
maximum strength <180 kN; and
maximum weight in water < 8.4 kN/km.

From the above constraints, a cable design window can be constructed on strength-weight axes as shown in Fig. 1. The equation

$$S = 30.6 \times W - 11.5$$

describes the maximum strength attainable from an all-steel cable. The minimum sinking speed requirement results in a minimum weight requirement for a particular cable outer diameter. The vertical line at $w = 2$ kN/km in Fig. 1 shows the minimum weight for 21-mm-diameter SL cable. The unshaded region is then the design window for the cable.

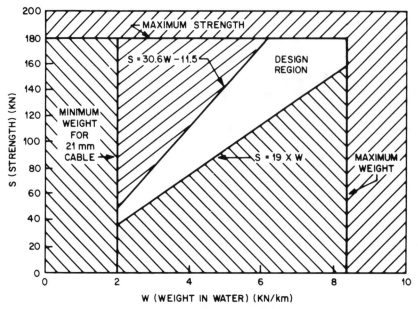

Fig. 1.   Cable design diagram.

Physical Design

The SL cable, design to meet the above goals and constraints, is illustrated in Fig. 2. The cable consists of a lightguide core protected by two layers of high-strength steel stranded wires, a copper sheath, and low-density polyethylene for high-voltage insulation and environmental protection. The lightguide core (or unit fiber structure) consists of a center kingwire, up to 12 helically wound fibers embedded in elastomer, and a thin outer nylon sheath. Some important features of this cable design are

low fiber curvature (1500-mm radius);
steel strand package acts as pressure vessel as well as strength member; and
fibers and steel strand package move together, with no relative motion between them, simplifying deep-water repair.

The tensile and torsional properties of the cable are plotted in Figs. 3 and 4, respectively. The theoretical results were obtained using a method previously published

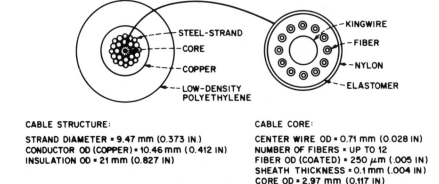

CABLE STRUCTURE:

STRAND DIAMETER = 9.47 mm (0.373 IN.)
CONDUCTOR OD (COPPER) = 10.46 mm (0.412 IN)
INSULATION OD = 21 mm (0.827 IN)

CABLE CORE:

CENTER WIRE OD = 0.71 mm (0.028 IN)
NUMBER OF FIBERS = UP TO 12
FIBER OD (COATED) = 250 μm (.005 IN)
SHEATH THICKNESS = 0.1 mm (.004 IN)
CORE OD = 2.97 mm (0.117 IN)

Fig. 2.   Cable constructions and dimensions.

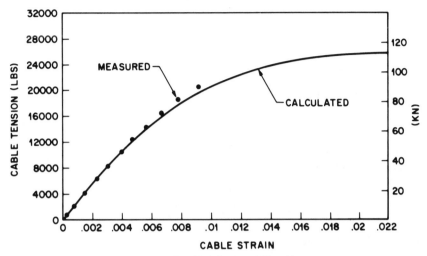

Fig. 3.   Tensile behavior of SL cable.

[2]. Other cable properties are listed in Table I. More than 200 km of this type of cable have been manufactured. Typical fiber loss at each manufacturing stage is shown in Fig. 5. The loss increase due to cabling is typically less than 0.02 dB/km.

The cable tensions during laying, recovery, and holding for repair can be evaluated according to Zajac's work [3]. The results are the following.

Laying:

Where
    laying speed is 13 km/h,
    ocean depth is 5.5 km,
    angle between cable and horizontal is 6.22°,
    cable touchdown time is 3.93 h.

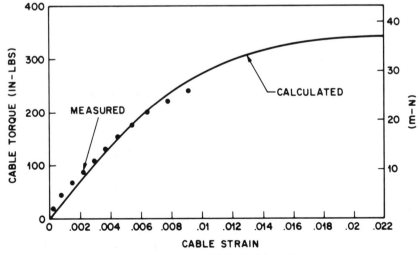

Fig. 4.   Torsional behavior of the cable.

TABLE I
PROPERTIES OF SL CABLE

Weight in air	8.4 kN/km
Weight in sea water	4.9 kN/km
Minimum breaking strength	107 kN
DC resistance	0.72 ohm/km at 20°C
Cable hydrodynamic constant	43.5 degree-knots

Then
   maximum tension during laying is 26 kN, and
   maximum cable strain (from Fig. 3) is 0.21 percent.

Recovery:

Where
   recovery speed is 1.85 km/h,
   recovery angle is 85°,
   maximum ship vertical velocity due to wave motion is 5.2 m/s (sea state 6),
   ocean depth is 5.5 km,
   repeater weight in water is 1.34 kN,
   suspended cable length is 9.1 km (see [3]),
   recovery time is 4.89 h.
Then
   maximum recovery tension is 78 kN, and
   maximum cable strain is 0.78 percent.

Holding:

Where
   ocean depth is 5.5 km,
   cable angle during holding is 75°,

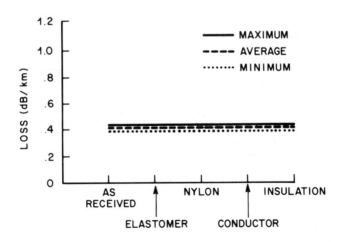

PROCESSING STEP

Fig. 5.   Typical fiber loss versus processing step in cable manufacture.

holding time is 48 h,
   maximum ship vertical velocity due to wave motion is 5.2 m/s, and
   period of wave motion is 7.3 s.
Then
   maximum holding tension is 53 kN, and
   maximum cable strain is 0.46 percent.
The minimum breaking strength of SL cable is about 107 kN. Hence, the operating
tensions are well below the breaking strength.

Required Fiber Proof Stress

From the above operating tensions, strains, and time durations, the required fiber
proof-test stress can be evaluated [4]. Using fracture mechanics theory and assuming a
power law relationship for crack growth, it can be shown that the largest crack size a_p,
after fiber proof testing, is

$$a_p = \left(\sum_1^q \left(1 - \frac{n}{2}\right) A(YE)^n \int_0^{t_i} [\epsilon_i(t)]^n \, dt \right)^{2/(2-n)}$$

where A and n are constants depending on the material properties and environment (see
[4]), Y is a constant determined by the crack geometry (1.241 for a semicircular
part-through crack), E is the Young's modulus of the fiber (71.9 GPa), $\epsilon_i(t)$ is the fiber
strain as a function of time, and q is the number of stress loadings.
   By knowing the crack size a_p, the required proof-test stress can be determined from the
fracture mechanics relationship

$$\sigma_p = \frac{K_{IC}}{Ya^{1/2}}$$

where K_{IC} is the fracture toughness of the material. For fused silica fibers, $K_{IC} = 0.789$
MN/m$^{3/2}$.
   Adopting a conservative approach, the maximum operating tension or strain is assumed
for the duration of the operation. The value $n = 17$ is selected for a high-humidity
environment. Substitution of these values into the above equations yields a required
proof-test stress of 163 kpsi. However, a proof-test stress of 200 kpsi ($\approx$ 2-percent strain)
is specified for the transatlantic cable to ensure that adequate margin is provided to offset
any effects associated with the cable manufacturing and handling processes that are not
covered by the model.
   Data on the behavior of this cable during testing will be given in the following sections.

Armored Cable

Armored cable for use in the SL system has been constructed by armoring the basic
deep-water cable described above. The double armored version is shown in Fig. 6. The
single-armored version is similar, but with the intermediate jute layer and the outer wire
layer absent.
   For special applications there is also available a somewhat heavier single-armored cable
design using 18 armor wires of 5.16 mm diameter.

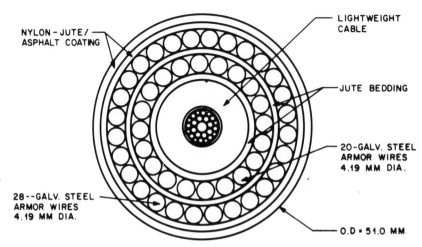

Fig. 6. SL double-armored cable.

MECHANICAL TESTS AND FIELD TRIALS

One of the main requirements for the SL undersea cable is to protect the transmission media, the optical fibers, from the harsh undersea environment. To accomplish this, the cable must maintain its mechanical integrity throughout cable ship handling and undersea operations. Extensive testing has been performed which has demonstrated that the SL cable maintains its integrity under low temperature, high pressure, and high tension. Mechanical results from factory tests and sea trials are discussed below. Specific optical results will be presented in later sections.

Factory tests which simulate cable ship handling during installation and recovery include:

tension to 80 kN;
50 reverse bends with a 1-m radius, no tension;
tension/torque tests;
tension/twist tests; and
extended-range temperature tests ($-20°$ to $+40°C$).

Results from all of these tests indicate that there is no significant permanent mechanical or optical degradation induced.

Actual deep-water sea trials have also been beneficial in proving SL cable reliability. Two sea trials have been performed which were highly successful. The first sea trial used an 18-km length of cable with a repeater at the end providing optical loopback path. The cable was deployed and recovered twice in 5.5-km ocean depth. The combined effects of

two cable and repeater laying and recovery cycles in 5.5-km ocean depth,
18-h cable suspension over the ship's bow,
transient bends at a 1.5-m radius under full recovery load, and
4°C operating temperature

resulted in only a 0.005-dB/km average increase in the optical attenuation and no mechanical degradation.

The second sea trial used a 4.5-km cable with optical loopbacks at a cable-to-cable joint. The cable was deployed in 4.5 km of water and held for several hours. A handling operation used during cable installation, known as a bow-to-stern transfer, was performed on the cable. During these operations, the optical transmission was monitored at a wavelength of 1.52 μm (the fiber is somewhat more sensitive to microbending at this wavelength than at the SL transmission wavelength of 1.31 μm). No measurable change in optical transmission was observed. The cable was recovered and no handling-induced mechanical degradation was found.

OPTICAL TESTS, CABLING, AND FIELD TRIAL RESULTS—MANUFACTURE OF SEA TRIAL CABLE

Twelve lengths of spliced fiber, approximately 20 km long, were produced from November, 1981 to April, 1982 for use in the cable for the first SL sea trial. The characteristics of these fibers are described elsewhere [5]. Six of the fibers have matched-index deposited claddings and six have depressed-index claddings. They are identified by the letters *M* and *D*, respectively.

The losses of these fibers were measured at various stages of manufacture, from fiber spool to finished cable. These values are summarized in Tables II and III.

The final loss in the cable of the spliced *D*-type fibers was lower than that of the spliced *M*-type fibers on average. The *M*-type fibers generaly exhibit slightly higher loss wound on a spool than stranded in a cable, probably due to their higher bending sensitivity. The *D*-type fiber design, with its tighter mode confinement, has been selected as the design for the SL system based on its superior performance.

The minimum dispersion wavelengths were also measured on the unspliced constituent fiber samples at AT&T Bell Laboratories, Murray Hill, NJ. The values ranged from 1.305 to 1.320 μm.

TABLE II
SEA TRIAL CABLE TYPE *M* FIBER LOSS AT 24°C, 18.25 km LONG
(AVERAGE TYPE *M* CABLED FIBER LOSS IS 0.45 dB/km)

Column	1	2	3	4
Fiber ID	Loss (dB) on Spool at MH*	Loss (dB) on Spool at SWC*	Loss (dB) in Cable at SWC (18.25 km)	Cabling Loss Change From 2 to 3 (dB/km)
1M	8.01	7.75	7.35	−0.02
2M	9.50	9.81	9.58	−0.01
3M	8.02	7.98	7.95	0.00
4M	8.60	8.58	8.13	−0.03
5M	N.A.	N.A.	9.22	N.A.
6M	7.64	7.62	7.37	−0.01

N.A.: Data not available.
*Fiber length ranged from 19.0 to 21.0 km; all losses have been normalized to 18.25 km finished cable length.
MH: AT&T Bell Labs location in Murray Hill, NJ; SWC: Simplex Wire and Cable Co. in Newington, NH.

TABLE III
SEA TRIAL CABLE, TYPE D FIBER LOSS AT 24°C, 18.25 km LONG
(AVERAGE TYPE D CABLED FIBER LOSS IS 0.40 dB/km.)

Column	1	2	3	4
Fiber ID	Loss (dB) on Spool at MH*	Loss (dB) on Spool at SWC*	Loss (dB) in Cable at SWC (18.25 km)	Cabling Loss Change from 2 to 3 (dB/km)
10D	7.19	7.26	7.45	+0.01
11D	7.48	7.09	7.11	0.00
12D	7.01	7.08	7.03	0.00
13D	7.57	7.59	7.771	+0.01
14D	N.A.	7.35	7.78	+0.02
15D	7.08	7.06	7.25	+0.01

N.A.: Data not available.
*Fiber lengths ranged from 19.0 to 21.0 km; all losses have been normalized to 18.25 km finished cable length.

After the 18.25-km cable containing the twelve fibers was manufactured at Simplex Wire & Cable Co. in Newington, NH, it was moved into a temperature-controlled chamber. Measurements of spectral loss and dispersion were performed at 24° and at 7°C. No loss changes exceeding measurement reproducibility (~ 0.02 dB/km) were observed in the twelve fibers in the 1.3-μm region. Also, no changes in dispersion exceeding measurement accuracy ($\sim \pm 1$ nm) were observed in the four fibers measured at the two temperature extremes.

It is interesting to look at the cabled fiber loss at 1.55 μm since this is the wavelength being considered for future systems. The splice loss was measured at 1.3 μm, so the fiber loss can legitimately be corrected for splice loss only at this wavelength. However, even without this correction, the loss at 1.52 μm of fiber 11D was less than 0.2 dB/km with three splices. Table IV gives a summary of these results. The behavior of this cable during the sea trial is discussed in the section "Mechanical Tests and Field Trials".

TABLE IV
SEA TRIAL CABLED FIBER AND SPLICE LOSS

Fiber ID	Loss (dB) at 1.30 μm with Splices	# of Splices	Corrected Fiber Loss (dB/km) with no Splices at 1.30 μm	Loss (dB/km) at 1.55 μm with Splices
1M	7.35	2	0.37	0.26
2M	9.58	3	0.47	0.48
3M	7.95	1	0.41	0.27
4M	8.13	1	0.43	0.26
6M	7.37	2	0.36	0.35
10D	7.45	3	0.36	0.59
11D	7.11	3	0.34	0.16
12D	7.03	2	0.35	0.21
13D	7.71	3	0.37	0.19
14D	7.78	2	0.42	0.23
15D	7.25	2	0.39	0.25

TABLE V
OSF TEST CONDITIONS

Parameter	Range
Pressure	0–69 MPa
Temperature	3°–30°C
Tension	0–80 kN

TESTING IN THE OCEAN SIMULATING FACILITY

Introduction

In addition to trial installation and measurement work at sea, the Ocean Simulating Facility at the Holmdel Laboratory has been used to model deep-sea laying and operating conditions.

Test Capability

In the ocean simulator, cable samples of 100-m length are used. Extremely sensitive and accurate loss measurement methods were developed using balanced bridge techniques and temperature control. The sensitivity of the loss measurement sets is 0.0001 dB. The short-term (less than 2 h) error is less than ± 0.002 dB, and for long-term (more than 2 h), it is less than ± 0.010 dB.

The deep-sea environmental conditions that can be simulated are shown in Table V. A typical test program is shown in Fig. 7; it takes up to two weeks to complete.

Results

1) Fiber loss at 1.3 μm is insensitive to pressure variation (measured changes are less than the equipment stability values noted above and are not correlated to pressure) and there is no indication of a "first-time" effect. Moreover, response to other environmental conditions is independent of pressure.

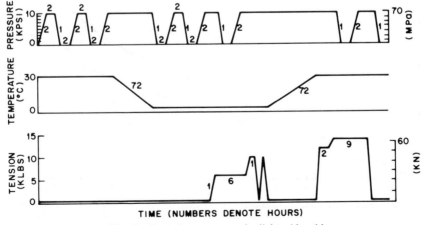

Fig. 7. Typical test program for lightguide cable.

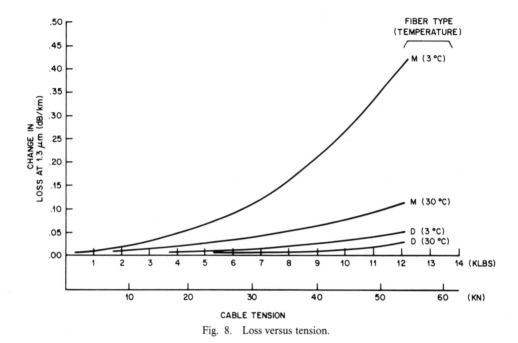

Fig. 8.   Loss versus tension.

2) Fiber loss change as a function of temperature is very small, within the measurement uncertainty range noted above.

3) Fiber loss change as a function of tension is shown in Fig. 8. As indicated in the figure, there is a considerable amount of coupling between temperature and tension. This is believed due to the hardening at low temperatures of the elastomer which surrounds the fibers.

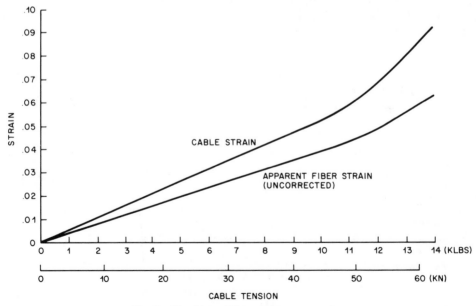

Fig. 9.   SL cable and fiber strain versus tension.

4) Cable and fiber strain as a function of tension are shown in Fig. 9. The fiber strain was measured by a transit time technique [6]. If one applies the proper correction factor for fused silica fiber [7], the actual fiber strain is essentially the same as that of the cable.

5) Strain due to pressure has been meausrd. As the cable is pressurized, its length increases since the power conductor is exposed to the atmosphere at each end. The amount of strain observed is

$$\epsilon(p) = [5.2 \times 10^{-6}/\text{MPa}] \times p.$$

6) Length change due to temperature change has been measured. The thermal strain is given by

$$\epsilon(T) = [8.7 \times 10^{-6}/°\text{C}] \times \Delta T.$$

Conclusion

With the OSF, a complete characterization of a cable design can readily be performed. The results found in the OSF have been fully consistent with those determined from sea trails.

Control of Hydrogen Generation in SL Cable

Effect of H_2 on Silica Fiber

Free gaseous hydrogen will diffuse into silica-based optical fibers and produce substantial loss increases in the near-infrared transmission bands. This poses a potential problem to the long-term stability of optical-fiber cables, where free hydrogen could be generated either from the corrosion of constituent metals or through the breakdown of certain organic components.

There are three loss mechanisms of concern for fibers exposed to a hydrogen environment. The primary loss mechanism is due to the absorption spectra which is generated by hydrogen molecules freely diffusing into the fiber. The magnitude of this loss is approximately linear with ambient hydrogen pressure, and has an added loss coefficient on the order of 0.2 dB/km·atm at 1.31 μm and 0.5 dB/km·atm at 1.54 μm.

A secondary loss mechanism which can occur for fibers exposed to a hydrogen environment is that of the diffused hydrogen chemically reacting with the glass structure to form new infrared absorbing compounds. This has been observed primarily at elevated temperatures ($>150°$C) in the form of irreversible hydroxyl (OH) bonding. A third irreversible mechanism has also been observed at high temperature, producing a broad spectrum loss increase. The details of this mechanism are under active investigation.

To maintain the long-term transmission stability required in undersea cable installations, it is therefore necessary that a cable design not generate significant amounts of hydrogen. A number of tests have been performed on SL cable to ensure that this is the case.

Tests on SL Cable

High-temperature (200°C) aging tests were performed on each of the organic materials used within SL cable. Based on mass spectrometry measurements of the evolved hydrogen, an estimate was made of the maximum pressure of hydrogen which could accumulate within the cable over a 20-year period at ocean-bottom temperature. Using this estimate, the cumulative level of hydrogen generation within SL cable would be less than 0.010 atm in 20 years, which would account for an added loss of no more than 0.002 dB/km at 1.31 μm. Even based on this worst case estimate, the organic constituents of SL cable will not degrade transmission performance due to hydrogen evolution.

To determine the susceptibility of SL cable to corrosion-induced hydrogen generation, a series of tests was conducted. One test made use of a nominally dry cable, which would model the deployed in-service condition of the system. A second modeled a cable break, in which the cable core was flooded with seawater. The cable break test represents a far more severe condition for corrosion, since the seawater provides the electrolyte necessary to complete a galvanic cell between the steel strength wires and the copper power conductor. After 16 months of monitoring the nominally dry cable, and 6 months of monitoring the soaked cable, neither of these samples has shown any sign of transmission degradation.

From the quantitative results of these and other tests, and from analytical studies, it is concluded that SL cable is not subject to a hydrogen problem.

CABLE-TO-CABLE SPLICING

Introduction

In SL systems, it is necessary to be able to join cable sections together for system assembly, installation, and repair. Like the SL cable, joints must protect the fibers during handling, installation, and service, and provide for power transmission.

Requirements

Since joints connect cables and must perform wherever cables perform, the requirements are similar to those for the cable. That is, joints must provide optical, mechanical, and electrical continuity while satisfying the following requirements:

usable in ocean depths to 7.5 km;
capable of being assembled easily and safely;
compatible with tension, torsion, and bending of the cable;
capable of being handled with normal cable machinery;
capable of being stored in cable tanks; and
reliable for 25 years.

Architecture

1) Armorless Cable: Cable joints consist of five basic components:

strength terminations;
splice shelf;
pressure housing;
electrical insulation; and
bend limiters.

The assembly is illustrated in Fig. 10. The strength terminations are assembled onto the cable ends to be joined. These terminations transfer tension and torsion from each cable

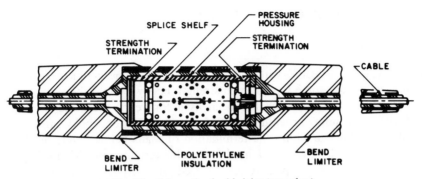

Fig. 10. Schematic of cable joint (armorless).

to the pressure housing. They also establish electrical contact between cable power conductors and both the splice shelf and the pressure housing. The shelf and housing form the power transmission path between terminations.

The splice shelf is fastened to each termination and provides storage for excess fiber. It also protects the fiber splices from being damaged during handling and installation. The shelf contains numerous guide pins to ensure acceptable low-level bending strain in the stored fibers.

The housing forms a pressure vessel that isolates the splice shelf and fibers from the sea. The housing and terminations are overmolded with polyethylene which forms the high-voltage insulation from sea ground and also inhibits water ingress to the fiber.

Finally, elastromeric bend limiters are attached at each cable-to-joint interface to protect the cable from being bent severely during handling. These bend limiters or "boots" also provide a smooth transition from the SL cable to the 120-mm-diameter joint. The length of the housing and termination assembly is about 305 mm; with bend limiters, the overall assembly is about 2.4 m long.

2) Armored Cable: When armored cables are joined, the outer armor wires are unlaid to expose the core cable. Then a cable-to-cable joint is installed as described above to connect the core cables. The armor wires are relaid so that the wires from one cable lay over the joint and overlay the armor wires on the other cable. This technique establishes tension and torsion coupling between the armored cables.

Testing

Reliability and performance of joints are checked by subjecting them to a wide range of tests. These include mechanical, optical, and electrical testing independently and in combination, including:

tension-bending-torsion;
pressure-temperature;
shock-vibration;
optical transmission;
high voltage; and
handling-assembly.

The results of these tests and field tests in deep-ocean environments during sea trials in the Atlantic Ocean in 1982 and 1983 show that cable-to-cable joints exhibit stable behavior under all operating conditions.

ARMORLESS CABLE-TO-REPEATER COUPLING

Cable Termination

Traditionally, undersea cables with high tensile strength properties have been terminated in assemblies which use some type of high-strength epoxy to hold the strength members of the cable. While this approach has been used for years, there are some problems associated with it, including providing a path for epoxy injection, epoxy curing, and small amounts of cable slippage within the termination of loads below failure level. With the development of SL cable, there was a need for a cable termination with no slippage. Such a termination has been developed; it can be used with any cable having high-strength wires as strength members.

This termination consists of three major parts, as shown in Fig. 11. These are

1) a tapered socket, which is the body of the termination, made out of a high strength alloy such as beryllium copper or steel;
2) a mating cone, also made out of a high strength alloy, with a center through-hole; and
3) a conical copper sleeve which has the same angle as the cone and fits over it with close tolerance.

The strength members of the cable (steel wires) are flared out to form a conical shape. They are then elastically reformed into a strand, and pushed into the socket where the wires elastically return to the conical surface of the socket. Next, the copper sleeve is placed over the steel cone and this combination is threaded over the fiber core and pressed inside the wire array. The strength members of the cable are thus trapped and held between the socket wall and the steel cone. Due to the high stresses which result from the injection force, the copper sleeve cold flows around the steel wires of the cable and partially fills the gaps between the steel wires. Work hardens during this process.

Since the steel wires in the cable are of various sizes for compactness, the cold flow of the copper transfers shear to all wires, in contrast to holding only the largest diameter wires which would be the case if no copper sleeve were used. The assembly is held together by friction.

As tension is applied to the cable, the force seats the cone more firmly into the socket, increasing the available friction force on the wires. Analysis shows, and experiments

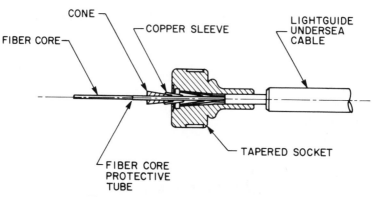

Fig. 11.   SL strength termination assembly.

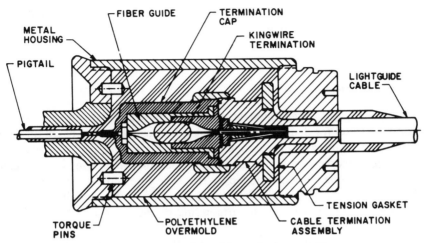

Fig. 12. SL armorless cable termination assembly.

verify, that the available restraining friction force exceeds the tension on the cable when

$$\tan \alpha < \mu$$

where α is the cone half-angle and μ is the combined coefficient of friction between the steel wires and the socket wall and the steel wires and the cone-copper sleeve combination.

A large number of these cable terminations, for different size cables, have been tested under the most severe conditions with no failure. The cold-flow self-energizing cable termination has replaced the epoxy cable termination for undersea lightguide cable due to its reliability and ease of assembly.

Coupling Assembly

The lightguide core passes through the cable termination freely, contained in a protective tube as shown in Fig. 11. As shown in Fig. 12 the kingwire is terminated adjacent to the strength termination. The glass fibers are flared around the kingwire termination in such a way that the required fiber bending is maintained and a small amount of free fiber is accumulated. The kingwire termination and the fiber guide assemblies are protected by a pressure housing which is attached to the cable termination and the pigtail.

The pigtail is an insulated flexible metallic pressure housing which carries the fibers and power. A water block assembly is provided between the termination and splice box to prevent water ingress into the splice box in the event of a cable break near the repeater. The pigtail terminates in a seal assembly which is pressed into the splice box during cable-to-repeater integration. This whole system is protected by a continuous insulating polyethylene jacket which connects to the cable polyethylene jacket for high-voltage insulation from the sea ground.

The overmolded strength and kingwire termination assembly is contained in a metallic housing. As shown in Fig. 12, tension is transferred to this housing through a polyethylene gasket, and torque is transferred through two metallic pins embedded in the molded polyethylene. This metallic housing attaches to the repeater through a gimbal arrangement similar to that used in coaxial systems. The pigtail permits the assembly to flex freely at the gimbal while protecting the fibers.

ARMORED-CABLE COUPLING

Two coupling designs to accommodate armored cables are planned, one for single-armored and the second for double-armored cable. Both couplings are based on the most recently used coaxial cable coupling design. The singe armored version is shown in Fig. 13. The double-armored version is similar, having a second armore ring to terminate the second armor layer. Cable tension is shared by the cable core and the armor wires in mid span. However, the armor wires alone carry the tension at the couplings. Tension is transferred from the armor wires to the armor ring by means of crimped ferrules. The tension components of these couplings have been tested to 440 kN. The torque produced by the cable, when under tension, is also carried to the coupling by the armor wires.

Isolation between the steel armor wires and the copper-beryllium components of the coupling assembly is required to prevent galvanic corrosion in sea water. This is accomplished by use of dielectric spacers which have adequate mechanical strength to transfer tension and torsion loads.

The coupling provides 50° of free gimbaled movement in any angular direction from the cable centerline. Fairing devices, called boots, are used to guide the couplings and their respective repeater into and out of cable handling machinery. Repeaters, couplings, and cable are easily accommodated on a 3 meter diameter sheave.

Transition from the cable core to the pigtail is accomplished in an overmolded chamber which provides for power continuity, some fiber slack storage, and attachment of the pigtail. The pigtail carries the power and the fibers from the transition to the splice box through a waterblock. The cable core and the overmolded transition are allowed freedom of movement axially within the armor assembly. This feature is provided to accommodate the motion of the cable core with respect to the armor wires which occurs when armored cable is tensioned. The pigtail carries system power and fibers to the splice box as in the armorless coupling. The coupling is mechanically connected to the repeater by means of a threaded clamp ring.

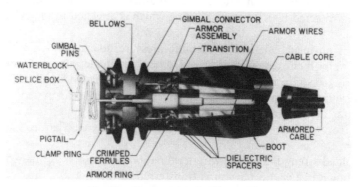

Fig. 13.   Single-armored-cable coupling.

CONCLUSION

The designs of SL cables, armored and armorless, and the associated joints and couplings have been described. In addition, test results which demonstrate required performance and reliability have been discussed.

REFERENCES

[1] R. F. Gleason, R. C. Mondello, B. W. Fellows, and D. A. Hatfield, "Design and manufacture of an experimental lightguide cable undersea transmission system," in *Proc. Int. Wire and Cable Symp.* (Cherry Hill, NJ), 1978.
[2] T. C. Chu, "A method to characterize the mechanical properties of undersea cables," *Bell Syst. Tech. J.*, vol. 62, no. 3, Mar. 1983.
[3] E. E. Zajac, "Dynamics and kinematics of the laying and recovery of submarine cable," *Bell Syst. Tech. J.*, vol. 35, no. 5, Sept. 1957.
[4] D. Kalish, B. K. Tariyal, and H. C. Chandan, "Effect of moisture on the strength of optical fibers," in *Proc. Int. Wire and Cable Symp.* (Cherry Hill, NJ) 1978, pp. 331–341.
[5] A. D. Pearson, P. D. Lazay, W. A. Reed, and M. J. Saunders, "Bandwidth optimization of depressed index single mode fiber by means of a parametric study," in *Proc. European Conf. Optical Commun.*, paper AIV-3, p. 93, Sept. 1982.
[6] M. Johnson and R. Ulrich, "Fiber-optical strain gauge," *Electron. Lett.*, vol. 14, pp. 433–437, 1978.
[7] W. Primak and D. Post, "Photoelastic constants of vitreous silica and its elastic coefficient of refractive index," *J. Appl. Phys.*, vol. 30, no. 5, May 1959.

17
Cable Design for Optical Submarine Systems

PETER WORTHINGTON

INTRODUCTION

Optical cables for transoceanic cable systems have been under development for several years and sea trials have been reported by all the major manufacturers of submarine cable systems.

The first international commercial systems are to be installed during 1985 and the next major transatlantic system (TAT-8) will be an optical system installed in 1988.

Optical cables offer the major benefits of long repeater spacings and smaller cables with increased capacity compared to coaxial cables. The last coaxial transatlantic cable (TAT-7) has a capacity of 5000 circuits and uses cable of over 50 mm diameter with a repeater spacing of 9 km.

In contrast TAT-8 will have a basic capacity of 8000 circuits using 25-mm-diameter cable and repeater spacing of about 40 km.

Because of these obvious benefits there has been intense activity to develop fibers, cables, and repeaters for long-haul optical cable systems. The main concern with the design of cables is to provide a very high reliability product that ensures the stability and integrity of the fiber transmission path for a system design life of 25 years. This chapter describes the design and development of cables proposed for the first commercial systems.

CABLE DESIGN CONSIDERATIONS

Cable Requirements

The cable must meet several important design requirements for satisfactory system operation. They can be summarized as follows:

to provide a pressure and moisture-free environment for the fibers;
to ensure minimum strain in the fibers during all manufacturing and service conditions;
to provide a low-resistance insulated power feed conductor for the repeaters; and
to be capable of being laid and recovered in depths of 5.5 km using conventional shipboard handling methods.

In addition to these it is necessary to ensure that the fiber environment has negligible hydrogen contamination. The effect of hydrogen on the attenuation of fiber was first reported in 1983 [1] and a great deal of investigation has taken place since then.

Hydrogen can be generated by several mechanisms both externally and internally in the cable. It is important, therefore, that the cable design provides a barrier against diffusion of externally generated hydrogen and that there is negligible hydrogen generated within the cable environment.

Service Conditions

Cable for submarine systems must be designed for the appropriate seabed conditions. Most of the length of a long-haul system is in deep water, greater than 2 km. Here the cable is subjected to high hydrostatic pressures, but enjoys a generally stable and cold environment (2–3°C). There is a little risk of damage from external agencies. The depth of deployment does mean, however, that tensions in the cable during laying and recovery of the cable can be high. It is important that the cable is designed so that the strains in the fiber are sufficiently low that fiber failure will not occur. In the next section the strain in the cable in deep water is derived.

For the shallow-water continental shelf regions of submarine cable routes the major hazard to cables is from attack by fishing vessels and other shipping.

There are several regions where there is a high demand for submerged cable systems and a high level of fishing activity. For instance, the North Sea between the U.K. and Europe has many operational cable systems and is also heavily fished. Some of the fishing vessels use Dutch beam trawl gear which uses a large steel beam dragged across the seabed and presents a major hazard to cables.

It will, therefore, be necessary to protect future optical cables from breakage or excessive strain caused by attack from such trawlers. Where possible, cable burial is increasingly favored as a method of protection. However, 100-percent burial is not normally possible and it is necessary to have designs for trawler resistant cable for those areas where the cable is exposed on the seabed.

Deep-Water Cable Design

Description of Design

Figure 1 shows a cross-sectional view of the cable for the deep-water region of the U.K. section of TAT-8. (This is referred to as a lightweight cable in line with current coaxial cable terminology in order to distinguish it from externally armored cables.) At the center of the cable is a package of eight fibers (or fibers plus fillers) laid around a central kingwire and held together by a light Kevlar yarn.

A high-viscosity water-blocking material is contained, with the fibers, in a thick-walled composite copper tube. This composite tube is made in two stages. Firstly a "C" shaped section is closed and drawn into a tube around the fiber package. Secondly, a copper tape is formed around this, longitudinally seam welded and drawn to form a close-fitting hermetic tube. It is this tube that provides the hydrogen barrier for the fiber package. It also protects the fiber package from external hydrostatic pressure and is the main power feed path for the system line current.

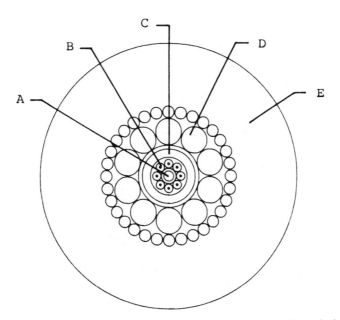

Fig. 1. Deep-water cable. *A* kingwire, *B* eight fibers, *C* copper tube, *D* steel wires and *E* polyethylene.

The water-blocking material has a very high viscosity at seabed temperatures so that in the event of the cable being cut on the seabed, axial penetration of water is limited to a few meters. It also provides mechanical coupling of the fiber package to the copper tube and ensures negligible relative movement between fibers and cable when the cable is subjected to strain.

The strength member of the lightweight cable consists of two layers of high tensile steel wires applied with opposite directions of lay. Their interstices are also filled with a water-blocking compound, in order to limit axial water penetration in the event of cable damage. Insulation is provided by an extrusion of low-density polyethylene giving a finished cable with an overall diameter of 26 mm.

Mechanical Performance

Figure 2 shows the tensile load versus elongation characteristic of the cable. The cable has a high strength for a small diameter cable. This ensures that the strain in the cable in deep water is low. The service strains are considered in greater detail next.

The strength member of the cable is torsionally balanced so that there is negligible twist generated when the cable is in tension. This means that there is no elongation of the cable as a result of untwisting, and there will be little tendency for the cable to be thrown into loops on the seabed as a result of torsional instability. The combination of the steel wires and copper tube provides a highly pressure resistant structure to protect the fibers from external hydrostatic stresses.

Strains in Service

In order to establish a proof test for the fiber that will guarantee the integrity of the transmission path, it is necessary to know the cable strains and time of application of the

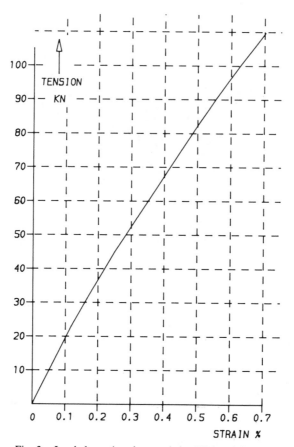

Fig. 2. Load-elongation characteristic of lightweight cable.

strain. For lightweight cable in deep water the highest strains occur during a repair operation when long lengths of cable are in suspension.

Strains during initial laying of the system are lower, but the whole system cable length is exposed to this strain so that a high degree of assurance of survival is required.

1) Cable Recovery and Repair: Deep-water cable repairs in lightweight cable are infrequent events because of the high reliability of the submerged plant and the generally stable environment provided by the seabed.

However, in the event that a repair is necessary it is essential that the cable can be recovered, held in suspension for the duration of the repair, and re-laid reliably.

The cable tension is a maximum at the ship and is a combination of the static weight of the suspended cable plus dynamic loading caused by ship motion.

a) Static tension: If the cable in suspension is vertical then the static tension at the ship is simply the weight per unit length of the cable multiplied by depth. The tension rises if the ship departs from the vertical station. Figure 3 shows the relationship between cable tension and ship position in a depth of 5.5 km. Generally, the ship will be able to maintain the cable to within about 15° of the vertical, resulting in maximum tensions of about 54 kN. This corresponds to a strain of 0.3 percent.

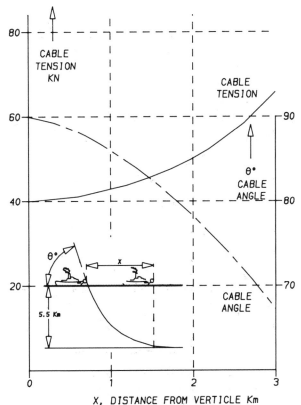

X, DISTANCE FROM VERTICLE Km
Fig. 3.   Tension in LW cable in deep water.

The tension also rises with increasing speed of recovery (winding in) of the cable, and the method of calculating this is derived by Zajac [2]. Figure 4 shows the relationship calculated for the deep-water cable in a depth of 5.5 km.

It can be seen that the tension rises rapidly with increasing speed of recovery and it is, therefore, necessary to recover at moderate speed to avoid excessive tension in the cable.

b) *Dynamic loading:* The wave-induced motion of the cable ship can cause high dynamic tensile forces superimposed on the static loading. In shallow water, the whole suspended cable length will be subjected to uniform motion in phase with the motion of the bow of the ship and the tension can be calculated from inertial and drag forces acting on the cable.

In deep water, the time of propagation of the tension wave becomes significant and the motion of the cable is no longer uniform.

The calculation of the dynamic tension assuming the suspended cable is of semi-infinite length is given by Zajac [2]

$$T = (EA \cdot m)^{1/2} \cdot V \qquad (1)$$

where

T cable tension,
EA cable tensile stiffness,
m mass/unit length, and
V velocity of the ship.

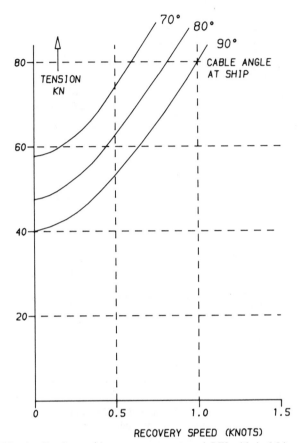

Fig. 4.   Tension at ship versus recovery speed. LW cable in 5.5 km.

However, for small, lightly damped cables there may be significant reflection of the tension wave at the seabed that will produce a longitudinal resonance. This can increase the tension in the cable at the ship.

It is possible for the bow motion of the ship to be approximately sinusoidal for several cycles which will induce the resonance condition. The worst case condition will arise if the depth of water is equal to a quarter wavelength of the tension wave in the cable at the frequency of the bow motion.

For typical wave-induced vertical bow motion, the periodicity is about 6–7 s (0.15 Hz).

For the deep-water cable, the quarter wavelength resonant length at this frequency is about 6 km.

Figure 5 shows the dynamic tension in the cable at the ship as a function of depth for a sinusoidal bow acceleration of 1.8 m·s^{-2} at a period of 6.5 s (1.8 m·s^{-2} is an estimate of the maximum significant bow acceleration under which a cable ship would continue cable operations. It corresponds approximately to the worst case motion that would be experienced in sea state 6 or low 7.)

It can be seen from Fig. 5 that the dynamic tension at the ship can be greatly increased in the event of resonance. Since the duration of a deep-water repair may be quite prolonged (48 h or more), it is quite possible for one or more periods of resonance to occur. It is, therefore, necessary to use this pessimistic estimate of cable tension in estimating the maximum strains in the cable and fiber.

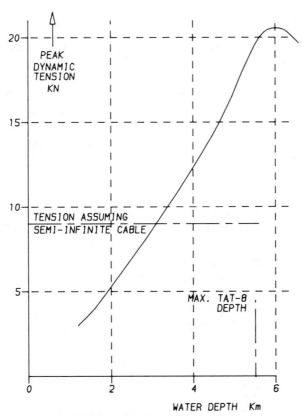

Fig. 5.   Dynamic tension in cables caused by ship motion.

From Fig. 5, the maximum dynamic tension under resonant conditions in a depth of 5.5 km is about ±20 kN. This is superimposed on the maximum static tension of 54 kN to give a peak tension of 74 kN. This will cause a peak strain in the cable of approximately 0.45 percent. The proof level of the fiber must be high enough to withstand this peak strain during the recovery operation.

2) Cable Laying: Although the highest service tensions occur in recovery in deep water, a repair operation only affects a small fraction of the total system length. In the laying operation the whole system cable is subjected to tension so it is important to establish that the strains involved do not present a significant hazard to the fibers.

In steady-state laying, the cable is paid out from the ship at a constant speed of about 6 knots. The system is laid with slack (typically 2 percent for the lightweight cable) so that the tension at the seabed is zero. Under these circumstances the cable assumes a straight line configuration between ship and seabed and the tension in the cable is given by

$$T = mh$$

where

m weight per unit length in water
h water depth.

For the lightweight cable in 5.5 km, the tension at the ship is about 40 kN corresponding to a strain of only about 0.23 percent. This strain falls with distance from the ship as the cable sinks to the seabed, which takes about 4 h in a depth of 5.5 km.

Cable Protection in Shallow Water

Shallow water cables require additional protection from external damage by anchors or trawlers. Traditionally, coaxial cables have been protected by one or more layers of mild steel armoring. The center conductor of the cable is normally made of solid copper or copper and mild steel. This construction is extremely tolerant to high elongation and can still continue to function after being pulled almost to breaking point. For optical cables, high elongation and resulting high residual strain could lead to subsequent failure of the fibers from static fatigue. It is therefore necessary to provide protection of the fibers against excessive residual strain as well as providing protection from impact damage.

Armored Cable

For the U.K. section of TAT-8, a double-armored cable has been proposed for areas where the cable is exposed on the seabed. The center of the cable uses the same construction as the deep-water lightweight cable (Fig. 1).

The protection and strength is provided by two layers of medium tensile grade 65 steel armoring, which give the cable very high breaking strength (greater than 1000 kN) and good linearity of the load-elongation characteristic up to high strains.

This very high breaking strength ensures the cable will resist attack by small vessels. However on infrequent occasions it may be pulled to such high tension that breaks at the

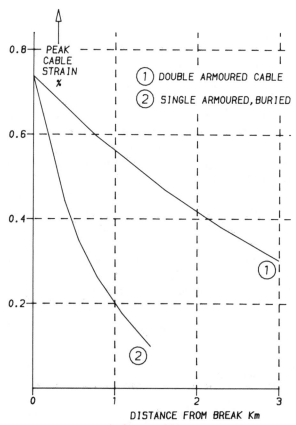

Fig. 6. Armored cable-localization of strain.

point of hooking. Tests on similar coaxial armored cables have shown that failure occurs at about 75–80 percent of the ultimate tensile strength of the parent cable, producing peak strains in the cable immediately adjacent to the break of about 0.75 percent.

Figure 6 shows the estimated peak strain in cable adjacent to a break. In the next section it is shown that under these circumstances fiber breaks will be limited to within 1 km of the cable breakpoint.

Buried Cable

Cable burial has been used increasingly in recent years. Cables can be buried either during the initial lay using a plough towed behind the ship or post-lay buried using a jetting in technique. Cable burial has been shown to be effective in reducing cable damage in many routes. However, the cables must still be sufficiently strong and robust to be laid and recovered without excessive strain. Experience of burying lightweight coaxial cables has not been particularly successful because of the difficulty of recovering a relatively weak buried cable.

With optical cable it is even more important that no excessive strain is imparted to the cable during burial or subsequent recovery. For this reason, the proposed buried cable for the U.K. section of TAT-8 is a high-strength single-armored cable. It can withstand tensions of up to 400 kN with less than 0.1-percent residual strain.

In the unlikely event that this cable is hooked and pulled to break, the very high axial drag forces acting on the buried cable ensure that the tension falls away rapidly on each

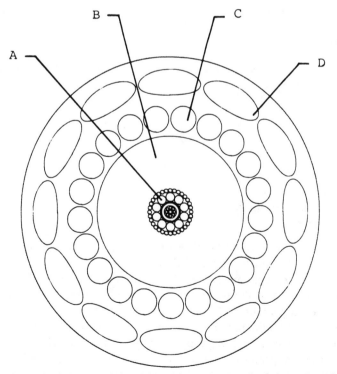

Fig. 7. Rock armor cable. *A* lightweight cable center, *B* polyethylene, *C* medium tensile steel wires, and *D* rock armoring.

side of the break, so that only short lengths of cable adjacent to the break are exposed to high strain (see Fig. 6).

Protection from Beam Trawls

As mentioned in the section on cable design considerations some areas, particularly the North Sea between U.K. and Europe are heavily fished by ships using Dutch Beam Trawling gear. Cables in this area can be subjected to repeated heavy impact from this type of trawl which can lead to shunt faults (penetration of the insulation).

For the proposed U.K, to Belgium optical cable, a rock armor cable design has been developed (Fig. 7). This is a double-armored cable in which the outer layer has a very short lay length. Externally, this armoring looks like a close spaced coil spring and it has been shown to be extremely resistant to penetration and crushing and will withstand repeated impact by a beam trawl. The insulation thickness has been increased to an overall diameter of 45 mm. In order to improve the penetration resistance of the cable and also to enable the number of wires in the first layer of armoring to be increased. This inner layer of medium tensile armor wires can then provide strength of more than 800 kN.

Fiber Reliability in Service

All fiber is proof tested prior to cabling at a sufficiently high strain to ensure that it will not fail when subsequently exposed to the service strains in the cable.

In order to determine the required proof-test level, fracture mechanics theory has been used together with experimental data from tests on long lengths of representative fibers.

Fracture Mechanics Theory

If a fiber is subjected to a strain $\epsilon(t)$ and fails after time T, then the relationship of strain and time can be expressed as

$$\epsilon_i^{N-2} = \frac{E^2}{10^4 B} \int_0^T (\epsilon(t))^N \, dt \tag{2}$$

where B, N are experimentally determined constants

    E Youngs modulus for silica,

    $\epsilon(t)$ applied strain, percent, and

    ϵ_i inert breaking strain.

ϵ_i is the breaking strain that would be observed if the tensile test were done under inert conditions with no crack growth prior to failure.

A proof-test level of ϵ_p guarantees that the minimum inert strength of fiber is ϵ_p. If service strains are low so that the required inert strength for survival given by (2) is less then ϵ_p, then we have a guarantee of fiber survival.

If service strains are high the required proof test for guaranteed survival may be higher than is necessary to give a very high probability of fiber survival. In such cases a substantial reduction in proof-test level can be achieved by accepting a very small probability of failure rather than zero. To do this it is necessary to know the inert breaking strain probability distribution of the fiber after proof test in order to predict the survival probability in service. This can be obtained by performing dynamic tensile tests on a large sample of the fiber population.

From this, together with a knowledge of the strains in the fiber and time of application, the survival probability can be estimated.

Deep-Water Cable

The strains in the cable during laying, recovery, and repair of the deep-water cable in a depth of 5.5 km were derived previously. Maximum strains for laying and repair were 0.23 and 0.45 percent, respectively.

Proof-test levels for the fiber have been based on the following requirements for fiber survival probability:

Cable Laying	100 percent
Deep-Water Repair	99 percent

In order to meet the laying requirement, a proof-test level of 0.75 percent is adequate. However, to meet the repair criterion, a proof-test level of 1.2 percent is required.

Shallow-Water Cable

The criterion for the proof-test level for fibers in shallow-water armored cable has been based on the localization of possible fiber breaks close to a cable tensile break caused by a ship trawl or anchor.

In the previous section, the fall of strain in the cable with distance from the break was derived. From this, the survival probability of the fiber in the cable adjacent to the break can be estimated. This is shown in Fig. 8 for both the double-armored cable (on the seabed) and the buried single-armored cable for a fiber proof test of 1 percent. It can be seen that at distances greater than 1 km there is negligible probability of fiber failure in the double-armored cable. For the buried cable, the corresponding distance is less than 0.5 km.

HYDROGEN

The cables described in the preceding sections have been designed so that there will be minimal adverse effect of hydrogen on the fiber.

Hydrogen can be generated as a corrosion by-product where metals in contact and moisture are present. Many materials also absorb hydrogen which can subsequently out-gas and cause contamination of the fiber environment.

Chemical changes in some polymer materials can also lead to hydrogen generation.

The features of the cables that have been included to minimize these effects are as follows:

A welded copper tube around the optical core prevents ingress of externally generated hydrogen.

Water and gas blocking, in both the steel strength member and inside the hydrogen barrier to inhibit axial penetration of water and hydrogen in the event of cable damage on the seabed.

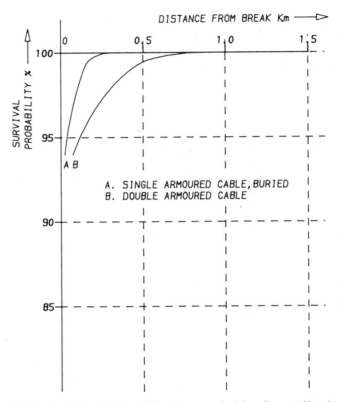

Fig. 8.   Survival probability of fibers in armored cables adjacent of break.

No mixture of metals inside the hydrogen barrier which could give rise to electrolytic generation of hydrogen.

All materials inside the hydrogen barrier have been assessed and shown to exhibit negligible out-gassing or hydrogen generation from chemical changes.

The results of an extensive investigation of the materials in the cable lead to the conclusion that the effect of hydrogen on the fiber will be negligible for the system life of 25 years.

References

[1] Uchida *et. al.* "Infrared loss increase in silica optical fiber due to chemical reaction of hydrogen," presented at 9th European Conf. Optical Commun., Oct. 1983.

[2] E. E. Zajac, "Dynamics and kinematics of the laying and recovery of submarine cable," *Bell Syst. Tech. J.*, Sept. 1957.

18
Design of Deep-Sea Submarine Optical Fiber Cable

YUKIYASU NEGISHI, KOUSHI ISHIHARA, YASUJI MURAKAMI,
MEMBER, IEEE, AND NOBUYUKI YOSHIZAWA

INTRODUCTION

By applying the excellent characteristics of optical fibers, such as low loss, small diameter, and wide bandwidth, there is a possibility of realizing high-performance submarine optical fiber cables with wider spacing of repeaters in a system. Therefore, research on the application of optical fibers to submarine cables is being carried out actively in many countries, and various reports have been prepared on this subject [1]–[3].

The field of optical fibers has already developed so much that low-loss 0.2-dB/km single-mode fibers [4], [5] can be produced. The advantages of submarine optical fiber cables are increasing.

Submarine optical fiber cables have to withstand great elongation during laying and recovery. Fiber proof test strain is determined by cable elongation during laying and recovery [6]. Therefore, high tensile proof tests are required for deep submarine optical fiber cables. In a previous report [6], it was clarified that about 2.5-percent proof test strain is required for application in depths of 6000 m and more than 3-percent proof test strain is required for 8000-m water depths.

This chapter describes a newly developed submarine optical fiber cable which is designed to demonstrate low elongation under high tension and both lateral and hydraulic pressure-resistant characteristics and also describes an optical unit which is able to withstand lateral pressure.

II. UNIT STRUCTURE DESIGN

A cross section of the newly developed submarine optical fiber cable [6] is shown in Fig. 1. As this figure shows, the submarine optical fiber is composed of an optical unit, an inner pipe, a tension member, an outer pipe, a dielectric layer, and a jacket. Among these parts of the submarine optical fiber cable, the inner pipe, the tension member, and the outer pipe are collectively called a composite tension member.

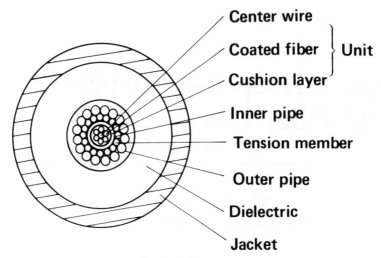

Fig. 1.   Cable structure.

This structure is adopted from the viewpoint of elongation characteristics and high lateral and hydraulic pressure resistances. This chapter is concerned with studies on unit structure considering lateral pressure in cabling processes.

The unit structure, which consists of six-coated fibers located around a center wire and filled by a buffer resin, is shown in Fig. 2. The buffer layer is enclosed by a jacket. The outer diameter of the unit, determined by the inner pipe diameter, is 2.5 mm. The coated fiber diameter was set at 0.4 mm in order to accommodate up to 12 fibers. Coated fibers are single-mode fibers to be used at the 1.3-μm wavelength, and the fiber and coated fiber outer diameters are 0.125 and 0.4 mm, respectively.

The unit model for the submarine optical fiber cable is shown in Fig. 3. R_1 is the unit radius, R_0 is the center wire radius, P and R are the stranding pitch and stranded fiber radius, respectively, and η_{R1} is the change of unit outer diameter which is caused by the lateral pressure applied during the cabling process. The change of unit diameter is expressed as [7]

$$\eta = \frac{R^2 - R_0^2}{R_1^2 - R_0^2}\left(\frac{R_1}{R}\right)\eta_{R1}. \tag{1}$$

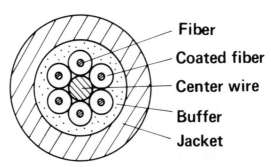

Fig. 2.   Unit model.

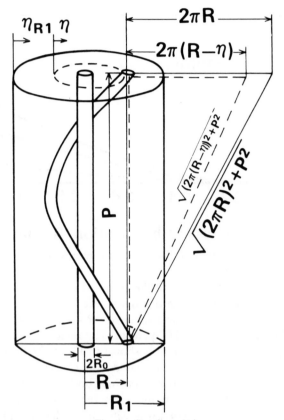

Fig. 3.   Standing pitch.

Fiber length l in the unit, as shown in Fig. 3, is given by the following expression:

$$l = \sqrt{(2\pi R)^2 + P^2}.$$ (2)

Fiber length l' due to added lateral pressure is given by

$$l' = \sqrt{(2\pi(R - \eta))^2 + P^2}.$$ (3)

Using (2) and (3), fiber compressive strain ϵ is expressed as

$$\epsilon = \frac{l' - l}{l} = 1 - \sqrt{\frac{(2\pi(R - \eta))^2 + P^2}{(2\pi R)^2 + P^2}}.$$ (4)

Fiber buckling, which causes the loss increase, takes place when the coated fiber is cooled and the compressive strain exceeds the critical value ϵ_{cr} [8]

$$\epsilon_{cr} = \sqrt{\frac{E_1}{\pi E}}$$ (5)

where E_1 and E are Young's modulus of the buffer in the coated fiber and the fiber. The

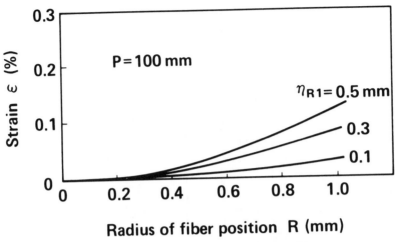

Fig. 4.   Relation between strain and fiber position.

relationship between the fiber strain ϵ and the stranded fiber radius changing parameter of η_{R1} given by (4) is shown in Fig. 4. In this figure, the fiber strain increases as the fiber stranded radius R increases. Substituting values of $R = 1$ and 0.4 mm, respectively, at $\eta_{R1} = 0.5$ mm, we obtain $\epsilon = 0.12$ and 0.02 percent. The relation between the compressive strain and stranded pitch at $\eta_{R1} = 0.3$ mm is shown in Fig. 5. The dotted line shows the critical strain of fiber buckling, which is calculated from (5). The compressive strain must be smaller than the critical strain in order to prevent both loss increase and fiber failure caused by lateral pressure. Fiber strain increases remarkably as the stranding pitch becomes shorter than 100 mm. Therefore, it is determined that an appropriate stranding pitch is 100 mm for practical use.

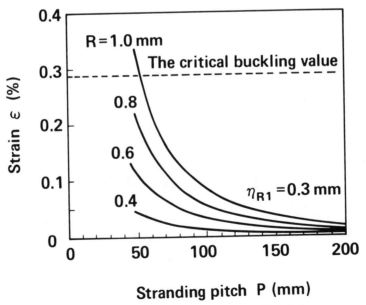

Fig. 5.   Relation between strain and stranding pitch.

Design of Cable Structure

Cable Elongation

Submarine optical fiber cables are required to endure mechanical force applied on them during laying and recovery as well as subsequent damage caused by fishing boats and trawling nets. However, the designing of cable elongation for deep-sea submarine optical fiber cables is done considering the cable tension incurred during laying and recovery operations because damage caused by fishing boats or trawling nets rarely occurs in the deep sea.

The elongation of submarine optical fiber cables during laying and recovery is nearly given by the following equation [9]:

$$\epsilon_c \simeq \alpha \rho h (1 + A'\rho'/A\rho)/\beta E_t \qquad (6)$$

where α is the ratio of cable weight in water corresponding to the sea depth to cable tension during laying and recovery, estimated from 1.0 to 1.5 during laying and from 2.5 to 3.0 during recovery. β is the corrective coefficient for tensile rigidity caused by the structure of the tension member. h is water depth, E_t is the secant modulus of the tension member, A' is the cross-sectional area of the pressure-resistant pipe, and ρ' is the density in water. A and ρ are the cross-sectional area of the tension member and the density in water, respectively.

In order to decrease cable elongation during laying and recovery operations, it is necessary to study four areas:

1) laying and recovery methods, concerning parameter α;
2) tension member structure, concerning parameter β;
3) tension member materials, concerning parameters ρ and E_t; and
4) pressure-resistant material concerning parameter ρ'.

The areas 2), 3), and 4) are connected with the cable structure design. As for 2), both the rewinding and the displacement in the radial direction of the tension member increase cable elongation because the tension member consists of standard wires. When the tension member is sunk between the inner and outer pipe, the tension member rewind and radial direction displacement can be restrained, so the elongation caused by structural factors can be suppressed. The tensile rigidity of the tension member material can be closely maintained and β can be kept close to 1.

As for 3), a low density ρ and a high Young's modulus of the tension member decrease cable elongation. Therefore, steel wire was selected as the tension member because of its small ρ/E_t value.

As for 4), the material of the pressure-resistant pipe is affected by the relationship $A'\rho'/A\rho$. Therefore, reducing the density and the section area would be effective in decreasing cable elongation. For this purpose, aluminum is selected instead of copper as a low-density metal, and the thinnest dimension was selected that could be manufactured for the wall of the pipe.

The stress–strain characteristics of the steel wire which is used in the cable tension member are shown in Fig. 6. Steel wire is within the elastic region of up to 0.6-percent elongation and shows plastic distortion after more than 0.6-percent elongation. The stress–elongation characteristics of the cable shown in Fig. 1 depend on those of steel

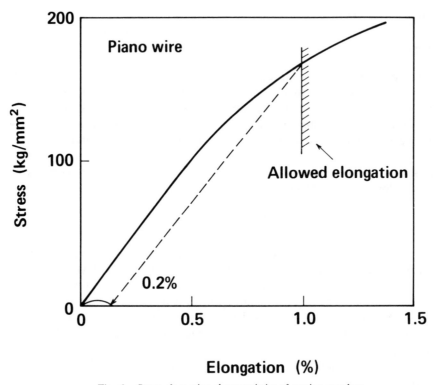

Fig. 6.   Stress-elongation characteristics of tension member.

wire [9]. Designing the allowed residual elongation to be less than 0.2 percent, the maximum allowed cable elongation is about 1 percent. The relation between cable modulus and applicable maximum water depth in the case of case recovery is shown in Fig. 7. In this figure, the cable tension 3 W·h: cable weight in water is estimated for the cable raising method without cutting it, and 2 W·h is for the cable recovery method, which is to cut the cable on the bottom and take it up. The definition of cable moduli is as follows:

$$\text{Cable modulus (km)} = \frac{\text{Cable tension corresponding to 1-percent elongation (kg)}}{\text{Unit length cable weight in water (kg/km)}}. \quad (7)$$

Cable moduli of copper pipe cable and aluminum pipe cable are shown in Fig. 7.

The maximum applicable water depth is about 6000 m for copper pipe and about 8000 m for aluminum pipe cable when the cable tension during The recovery process is 2.5 W·h.

The elongation characteristics of both aluminum and copper pipe cables with nearly the same dimensions are shown in Fig. 8. The elongation during laying and recovery is affected by the cable weight in water corresponding to the water depth. From the figure, it is proved that if aluminum is used for the pressure-resistant pipe, cable elongation can be reduced by 3.0 percent compared to a pressure-resistant pipe which uses copper.

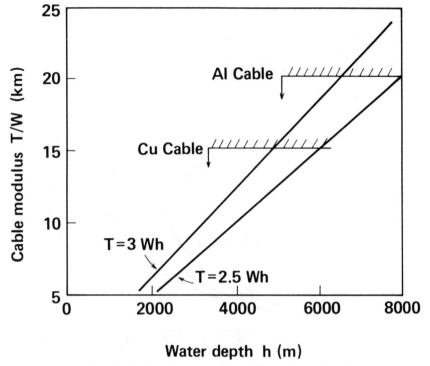

Fig. 7.   Relation between cable modulus and water depth.

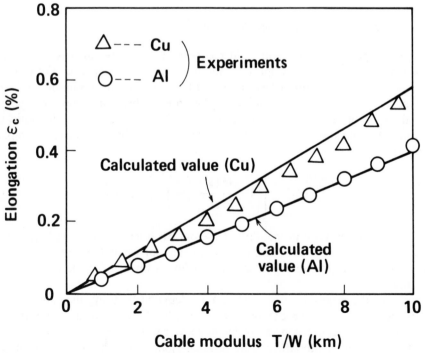

Fig. 8.   Cable elongation characteristics.

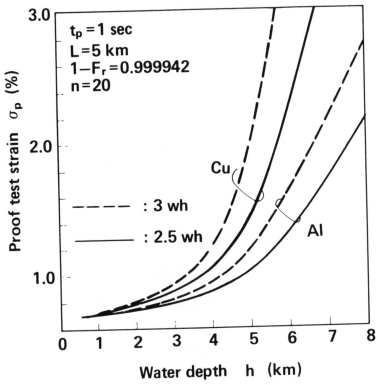

Fig. 9.   Necessary proof test strain.

Fiber Proof Test Value for Aluminum Pipe Cable

The cable elongation during laying and recovery in a deep-sea area is of a scale that cannot be ignored because it may approach the fiber breaking elongation. For this reason, it is necessary to conduct a proof test on optical fibers for use in submarine optical fiber cables. Assuming the strain to be applied to the optical fiber under the proof test to be strain σ_p, the time of test t_p and the sum of the strain to be applied to the optical fiber and the time of application for the cable manufacturing, laying, recovery, and repair are explained [6]

$$\sum_{i=1}^{i=M} \sigma_{ri}^{q} t_{ri} = \left[\left(1 + \frac{\ln(1-F_r)}{\ln(1-F_p)} \right)^{1/b} - 1 \right] \sigma_p^{q} t_p \tag{8}$$

where $(1-F_p)$ is the survival probability of the optical fiber of fiber length L when it is subjected to the proof test with σ_p and t_p. q is constant corresponding to the static fatigue. $(1-F_r)$ is the survival probability of the optical fiber over a full cable length when the strain is applied.

The results of the necessary proof test strain calculated by using the cable elongation characteristics shown in Fig. 8 can be seen in Fig. 9 [10]. As shown in this figure, the proof test value approaches a constant value in a lower water depth region; this is why the value in this region is mainly affected by residual elongation. The results are shown for two cases of the recovery method coefficient, $\alpha = 2.5$ and $\alpha = 3$.

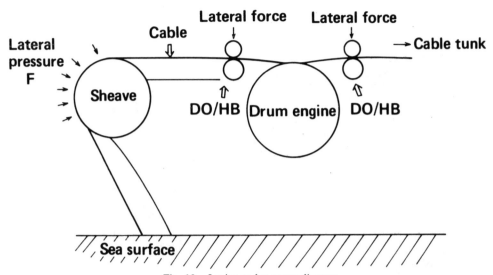

Fig. 10.   Laying and recovery diagram.

A proof test value of approximately 2 percent is difficult to achieve with an optical fiber as long as 5 km in length. With copper pipe, the proof test value will reach the 2-percent level when the cable is applied to 5000-m depth. Therefore, it is impossible to obtain a cable which can be applied in deep-sea areas up to 5000 m in depth. However, if the elongation suppressing structure is combined with the use of aluminum for the pressure-resistant pipe, the necessary proof test value can be greatly reduced.

Cable Design Based on Lateral Pressure

Lateral pressure is applied to the cable both at the sheave and under the DO/HB (draw-off/hold-back gear) of the laying ship, as shown in Fig. 10. DO/HB is the equipment which holds the cable at a maximum lateral load of almost 1 ton during laying and recovery. Lateral pressure is also applied to the cable at the sheave due to cable tension T during laying and recovery. The amount of lateral pressure is calculated as follows [13]:

$$F = T/R_K \tag{9}$$

where R_K is the radius of the sheave.

Cable tension during recovery is greater than during laying. It is estimated that about 10 tons of tension are incurred when a cable such as the one shown in Fig. 1 is recovered from a maximum water depth of 8000 m. If the sheave radius is 1.25 m, lateral pressure F comes out at 80 kg/cm from (9).

As mentioned above, a composite tension member can give a large lateral pressure resistance. Moreover, it is expected that the cable which consists of a composite tension member with a jacket is stronger than one which consists of a composite tension member under lateral pressure. In this section, the cable deformation under lateral pressure is experimentally studied.

A diagram of the cable under lateral pressure on the sheave is shown in Fig. 11 (a) and (b). In this figure, F represents lateral pressure, R_a and t are the mean radius and pipe

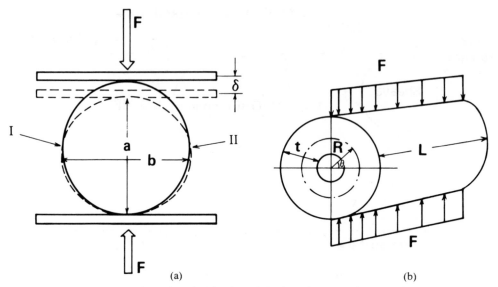

Fig. 11.   Experimental method for determining lateral pressure characteristics.

thickness, and δ is the displacement of cable diameter, respectively. Curves I and II indicate the before and after cable outer diameters before and after the lateral pressure test. The test cable structure details, including diameter, material of inner/outer pipe materials, material constant σ^*, and stress hardening index m are shown in Table I. Every cable consists of a double pipe such as that shown in Fig. 1. Deformation is evaluated from the elliptical rate e, which is defined by the following equation, using the elliptical model shown in Fig. 11(a):

$$e = \frac{a-b}{a+b} \times 100 \text{ percent.} \tag{10}$$

The elliptical rate for cables under lateral pressure is shown in Fig. 12. As this figure shows, little elliptic deformation is caused in the inner pipe, compared to one in the jacket.

An elastic deformation δ of the pipe subjected to lateral pressure F/L is given by [11]

$$\delta = (3\pi - 24/\pi)(1 - \nu^2)(R_a/t)^3 P/EL \tag{11}$$

TABLE I

Type	Diameter (mm)			Pipe material	σ^* kg/mm^2	m
	Inner pipe	Outerpipe	Jacket			
A	5.6 (0.85)	13.1 (0.75)	26	Al/Al	7.5	0.1
B	5.6 (0.85)	12.8 (0.75)	26	Al/Al	7.5	0.1
C	5.2 (0.8)	9.8 (0.6)	25	Cu/Cu	27	0.1
D	5.2 (0.7)	11.8 (0.6)	24	Cu/Cu	27	0.1
	(thickness)			$\sigma^*_{PE} = 1.5$ kg/mm^2		

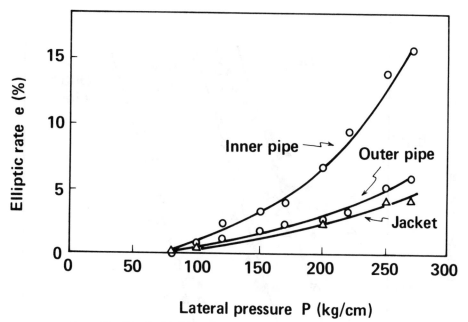

Fig. 12. Residual elliptic rate for cable under lateral pressure.

where ν is Poisson's ratio. This relationship is based on the elastic theory for curved bars. The strain of plastic materials can be expressed as [12]

$$\epsilon = \sigma/E + (\sigma/\sigma^*)^{1/m}. \tag{12}$$

In a large deformation region, neglecting the first term of (11), stress σ is approximately given by

$$\sigma \simeq \epsilon^m \sigma^*. \tag{13}$$

When the strain exceeds the elastic region, plastic deformation occurs. The equation for the lateral deformation in the plastic region has been derived as [11]

$$\delta = K(P^*/\sigma^*)^{1/m}$$
$$P^* = (R^{1+2m}/t^{2+m})(P/L) \tag{14}$$

based on (11) in the same way as the derivation of (10). Here K is a constant. We assume that a and b are equal to $(R - \delta)$ and $(R + \delta)$, respectively. Then the elliptical rate e which is given by (10) is also expressed by

$$e \propto (P^*/\sigma^*)^{\ell} \tag{15}$$

where ℓ is the constant. The relationship between the elliptical rate e and P^*/σ^* for each cable in Table I is shown in Fig. 13. The elliptical rate e and P^*/σ^* of the jacket, inner pipe, and outer pipe are given as follows:

$$\begin{aligned} \text{jacket} \quad & e = 0.024(P^*/\sigma^*)^3 \\ \text{inner pipe} \quad & e = 0.0025(P^*/\sigma^*)^3 \\ \text{outer pipe} \quad & e = 0.000036(P^*/\sigma^*)^3. \end{aligned} \tag{16}$$

The elliptical rate e of each part is approximately equal to three times P^*/σ^*. The elliptical rate e of the outer pipe is smaller than that of both the jacket and the inner pipe.

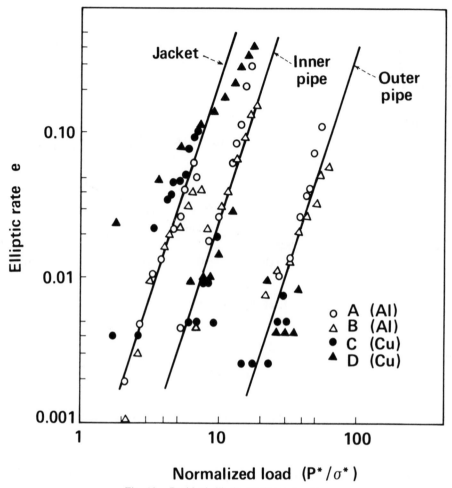

Fig. 13.   Residual elliptic rate for each cable.

The reason is thought to be that resistance to lateral pressure is reduced by the wire's torsional rigidity [13].

Cable Design in Consideration of Hydraulic Pressure

Excess loss occurs when the optical fiber receives hydraulic pressure [14]. Hydraulic pressure of 800 kg/cm^2 is applied on the cable at the 8000-m water depth. Therefore, the submarine optical fiber cable employs a pressure protection pipe in order to protect optical fibers from hydraulic pressure.

It is a well-known phenomenon that buckling stress for a cylindrical shell with initial deformation decreases considerably compared to buckling stress without initial deformation. This section deals with the effect of the initial elliptic deformation on hydraulic buckling stress, which is caused by lateral pressure during laying.

The initial deformation effect of the inner or outer pipe deformation on hydraulic buckling pressure is shown in Fig. 14. It is apparent that buckling pressure decreases in proportion to $e^{2/3}$ [14], the same as it does for a cylindrical shell. These results also

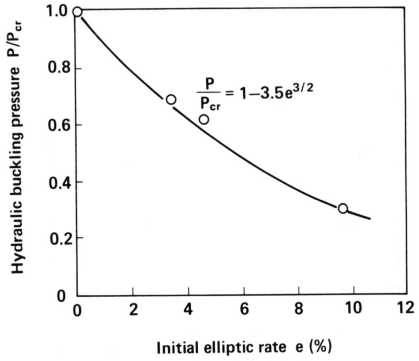

Fig. 14.   Initial residual effect on hydraulic buckling stress.

indicate that hydraulic pressure resistance for the pipe decreases if it has initial deformation. Although this analysis investigates hydraulic buckling, which occurs immediately under hydraulic pressure, creep buckling has to be considered since submarine optical fiber cable is used for a long period of time at the sea bottom. Therefore, it is preferable for the pipe deformation after laying to have no elliptic deformation. The allowable P^*/σ^* of the pipe on the hydraulic buckling pressure is shown in Fig. 15. From this figure, it can be seen that the submarine optical fiber cable structure should be designed at $P^*/\sigma^* \leqslant 4$.

CONCLUSION

This chapter presents the results of a study on the unit structure, cable elongation, and pressure pipe deformation caused by the sea bottom during the lateral and hydraulic pressure at cabling, laying, and recovery operations at the sea bottom. The results are as follows.

1) The coated fiber diameter and the stranding pitch are set at 0.4 and 100 mm, respectively, in order to withstand the lateral pressure which is incurred during the cabling process.

2) A maximum applicable water depth as deep as 8000 m is attained by decreasing the cable weight using aluminum pressure-resistant pipes. The proof test strain is 2.2 percent for 8000-m water depth under the condition that only one fiber breakage occurs in 20 years for a 1000-km long stationary line.

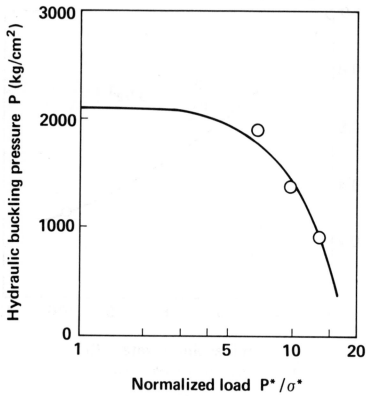

Fig. 15.   Normalized load effect on hydraulic buckling stress.

3) The elliptical rate e of the submarine optical fiber cable under lateral load is in proportion to $(P^*/\sigma^*)^3$. The allowable elliptical rate of the pressure-resistant pipe is found to be under 0.3 percent considering the lateral pressure on the sheave and the hydraulic pressure at the sea bottom.

ACKNOWLEDGMENT

The authors would like to thank N. Kojima and N. Uchida for their valuable discussions and H. Fukutomi for his encouragement.

REFERENCES

[1] N. Kojima, Y. Negishi, M. Kawase, and T. Matsumoto, "Studies on optical fiber cable for undersea transmission systems," IECE Japan, Tech. Rep. CS78-217, 1979 (in Japanese).
[2] R. F. Gleason, R. C. Mondello, B. W. Fellows, and D. A. Hadfield, "Design and manufacture of an experimental lightguide cable for undersea transmission systems," in Proc. 27th IWCS, 1978.
[3] P. Worthington, "Application of optical fiber system in underwater service," in Proc. Int. Conf. Submarine Telecommun. Syst., Feb. 26–29, 1980.
[4] N. Niizeki, "Single mode fiber at zero-dispersion wavelength," in Proc. Topical Meet. Integrated Guided Wave Opt., Jan. 16–18, 1978.
[5] T. Miya, T. Terunuma, T. Hosaka, and T. Miyashita, "An ultimate low loss single mode fiber at 1.55 μm," Electron. Lett., vol. 15, pp. 106–108, Mar. 1979.

[6] N. Kojima, Y. Miyajima, Y. Murakami, T. Yabuta, O. Kawata, K. Yamashita, and N. Yoshizawa, "Studies on designing of submarine optical fiber cable," *IEEE J. Quantum Electron.*, vol. QE-18, pp. 733–740, Apr. 1982.

[7] S. P. Timoshenko, *Strength of Materials.* Princeton, NJ: Van Nostrand, 1956.

[8] Y. Katsuyama, Y. Mitsunaga, Y. Ishida, and K. Ishihara, "Transmission loss of coated single-mode fiber at low temperature," *Appl. Opt.*, vol. 19, pp. 4200–4205, 1980.

[9] T. Yabuta, N. Kojima, Y. Miyajima, N. Yoshizawa, and K. Ishihara, "Studies on designing of submarine optical fiber cable elongation," *Trans. IECE Japan*, vol. J65-B, pp. 695–772, 1982 (in Japanese).

[10] T. Yabuta, K. Ishihara, and Y. Negishi, "Submarine optical-fiber cable design considering low elongation under tension," *Electron. Lett.*, vol. 18, no. 22, pp. 943–944, 1982.

[11] Y. Mitsunaga, Y. Katsuyama, and Y. Ishida, "Optical cable deformation characteristics under lateral load," *Trans. IECE Japan*, vol. J64-B, p. 142, 1981 (in Japanese).

[12] J. V. Schmitz, *Testing of Polymers*, Vol. 1. New York: Wiley, 1965.

[13] T. Yabuta, K. Hoshino, and N. Yoshizawa, "Lateral and hydraulic pressure characteristics of submarine optical fiber cable," *ASME J. Energy Resources Technol.*, to be published.

[14] M. Kawase, K. Yamashita, M. Nishimura, T. Yamanishi, and Y. Sugawara, "Optical loss change caused by hydraulic pressure in multimode optical fiber," *Electron. Lett.*, vol. 15, no. 7, pp. 208–229, 1979.

19

The Submarine Optical Cable of the Submarcom S 280 System

JEAN-PIERRE TREZEGUET, PIERRE OLLION, PIERRE FRANCO, AND JEAN THIENNOT

INTRODUCTION

Submarine telecommunication cables, coaxial or optical, must satisfy multiple requirements resulting from the following:

1) their objective: transmission of information across the ocean for 25 years;
2) their environment: the bottom of the sea with its stresses and risks;
3) their deployment and eventual repair: cable ship loading, handling, laying, and recovering.

We have provided over 40 000 km of coaxial submarine cable all around the world, and have accepted the challenge of optical fiber technology. The cables which are described below are the result of several years of development.

SPECIFICATIONS

Transmission

The cables have to transmit in each direction two 295.6 Mbd streams of information on two pairs of optical fibers guiding light at 1.3 μm. In order to minimize the number of repeaters, the optical loss and the dispersion of the fibers must be as low as possible. Also, the cable structure must keep the optical performance of the fibers from degrading during the 25 years of the system life.

Electrical

The repeaters have to be powered through the cable. The dc resistance of the cable and its high-voltage insulation shall be consistent with 10 000 km of cable length and the associated repeaters.

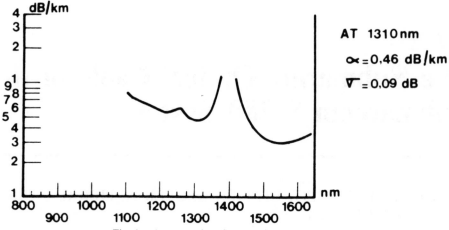

Fig. 1. Average value of attenuation on 600 km.

Mechanical

The cable must protect the fibers against excessive mechanical stress during deployment, eventual anchor or fishing tools aggression, and subsequent repair. The cable must protect the fibers against seawater and its associated depth pressure, even after a cable break on the seabed.

The cable must sustain temperature variation during storage handling and on the sea bottom or along the terrestrial underground path.

The cable must be laid and recovered by existing cable ships.

Maintenance

The cable shall be easily repaired by a cable ship. The damage to the cable and its fibers after a cable break must be localized at the vicinity of the break.

GENERAL DESIGN OF CDL CABLES

Optical Fibers

For long-span high bit rate transmission, single-mode fibers are necessary. They are produced using the MCVD process. Figure 1 shows the average-loss-versus-wavelength relationship on a 600-km production run. Typical figures for production to date are listed in Table I.

Now controlled in the laboratory and scheduled for mass production for 1986, optical fiber performances are listed in Table II.

TABLE I

	$\lambda = 1290$ nm	$\lambda = 1330$ nm
Optical loss (dB/km)	0.5	0.5
Chromatic dispersion (ps/nm × km)	5	2

TABLE II

	$\lambda = 1290$ nm	$\lambda = 1330$ nm
Optical loss (dB/km)	0.42	0.42
Chromatic dispersion (ps/nm × km)	3.5	3.5

TABLE III

Outside diameter	(μm)	210
Cladding outside diameter	(μm)	125
Beam diameter	(μm)	8
Core to cladding eccentricity	(μm)	1
Proof test value including splices ($t = 0.5$ s)	(percent)	0.9

TABLE IV

Number of splices tested		100
Mean splice loss	(dB)	0.09
Standard deviation	(dB)	0.08

Fiber elongation resistance is a basic problem for optical submarine telecommunication cables. The specification of the mechanical properties of the fibers is closely dependent on the cable design that will be described below. As a result, the fibers shall be proof tested at 0.9-percent elongation ratio for 0.5 s before being used in the S 280 cable.

Table III lists the essential physical properties of the fibers.

For identification in the cable, different colors may be applied on the coating.

The fiber splicing is performed by the arc fusion technique. Table IV gives the characteristics of the splices.

Cable

1) General Design of Cables: The main stresses to be encountered by the cable designer are the following.

a) For the 25-year system life:

- seawater with its pressure due to the sea depth;
- high electrical voltage;
- sea bottom roughness and profile;
- temperature variations;
- tidal or submarine water current;
- anchor or fishing tool aggression including cable break.

b) Due to its deployment or repair subsequent to aggression:

- abrasion;
- bending;
- elongation.

All these stresses are well known by the coaxial cable designer, and very reliable cables have been designed, manufactured, laid, and successfully worked for many years.

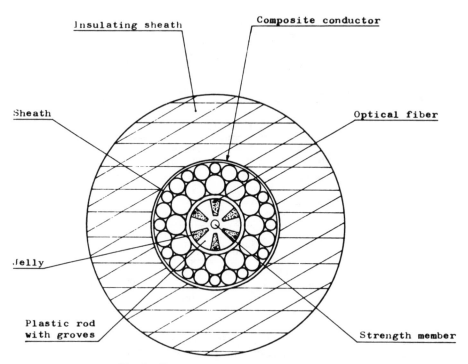

Fig. 2.   Cross section of type D4 deep-sea cable.

The new challenge is due to the sensitivity of the optical fiber to stress corrosion (or stress aging).

During their deployment or recovery, submarine cables are exposed to elongation induced at least by their own weight, eventually increased by the dynamic strengths caused by ship and cable motions. Moreover, aggression or breaking of the cable may induce instantaneously very high elongation strain on the cable (>1.5–1.9 percent) with high remanent elongation (> 0.5 percent).

To overcome this problem and meet the requirements of the specification, a strong cable with slack on the fibers has been designed. The deep-sea cable (Fig. 2) is the basic design of all other types of cables. Armored cables are manufactured by adding on the deep-sea cable several lays of galvanized steel wires and synthetic yarn impregnated by a bituminous compound.

The deep-sea cable is composed of a fiber unit structure, a steel vault copper cladded and insulated by polyethylene. The steel vault is longitudinally waterproofed by periodic stopping blocks.

The fiber unit structure is composed of a strength member, a plastic rod with six U-grooves filled with special hydrophobic jelly and closed by plastic tapes and a plastic sheath.

2) Elongation of the Cable: Fig. 3 demonstrates the basic principle that avoids elongation of the fibers when the cable is elongated during the normal operations of deployment and repair of the system. Moreover, in case of breaking, thanks to the cable structure, the elongation at the break of the cable induces on the fibers an elongation which is still lower than the proof test value.

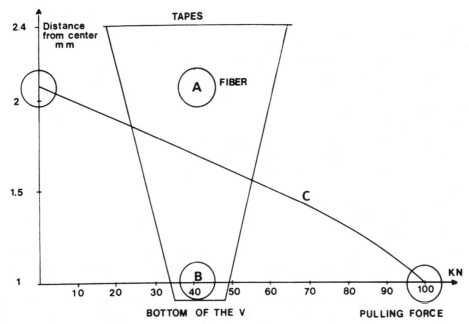

Fig. 3.   Fiber position in V versus pulling force in the deep-sea cable.

During manufacturing, fibers are placed in the U-grooves in the *A* position (Fig. 3) corresponding to a radial distance from the center of the rod shown on the vertical axis. When the cable is elongated under an increasing tension, shown on the horizontal axis, the position of the fibers in the U-grooves follows the curve *C*.

Beyond 100-kN pulling force (deep-sea cable) the fibers are on the bottom of the grooves and some elongation stress appears on the fibers.

When the cable is hooked by a grapnel, an anchor, or a fishing device, the cable break occurs typically at about 80 percent of the in-line breaking strength corresponding to 1.3 percent of elongation. In such a situation the elongation stress on the fibers themselves is only 0.2 percent. It can be assumed that fibers' breaks are located at the cable break point.

3) Seawater and Pressure Resistance:

   a) Seawater resistance: Two different situations may be encountered.

The first and the most general is the situation when the cable is operating on the seabed. The fibers are protected against water or moisture ingress by successive radial barriers.

From the sea to the fiber are, successively: an insulating sheath; a copper tube; a plastic tube; and hydrophobic jelly.

A second situation occurs after a cable breakage. The cross section of the cable is exposed to water under very high pressure corresponding to the sea depth.

In such a situation, the water ingress along the steel wires of the strand induces a high-pressure stress on the fiber unit structure and consequently on the fibers. The subsequent recovery and repair process does not wear out this stress, and moreover, pressurized water is left for the lifetime of the system around the fiber unit structure and the fiber. To overcome this difficulty, our cable design includes water blocking areas periodically placed along the steel strand.

The water ingress in the cable along the steel wires is first restrained, then stopped at the water blocking areas in the steel strand on each side of the break point. Less than a few hundred meters of the cable are so exposed to seawater and sea pressure. The fiber unit structure is itself protected against water ingress by the filling product, and the humidity content of the fiber unit structure increases very slightly on a few meters close to the break point.

b) Pressure resistance: The tensile wires are stranded in order to obtain a vault effect preventing the fiber unit structure from any pressure effect. This vault, covered by a copper tube and a polyethylene sheath, is designed to withstand pressure at least equal to 70 MPa.

Evolution of the cable has been checked in a pressure vessel. The cable is wound on a 270 mm diam and volume variation is measured at the outer part of the vessel. Cycles up to 70 MPa are achieved.

Under this severe bending condition the vault perfectly resists pressure and comes back to the initial dimension. Maximum volume variation is 0.5 percent, which corresponds to 1.5-μm-diameter variation; fibers are thus protected against external pressure.

4) High Electrical Voltage Resistance: For the power feeding purpose, the cable has been designed to withstand a 12-kV tension over 25 years. Polyethylene used in coaxial cable has been selected for its 20-year trouble-free service.

5) Abrasion Resistance — Aggression Resistance: The polyethylene has exhibited a good resistance to abrasion. Where necessary, steel armored cables are used. Burying increases the protection.

6) Bending Resistance: Cables have been bent on a 0.30-m radius. After 100 reverse bendings no degradation was observed on cable components.

7) Temperature Variations:

a) Fibers: Intrinsic attenuation of the fiber was measured with respect to temperature conditions. A 2.7-km fiber was uncoiled and submitted to temperatures between $-30°C$ and $+45°C$. The results are the following: no loss increase in the $+0°$ to $+45°$ range; loss increase less than 0.05 dB/km at $-10°C$ (reversible); and loss increase of 0.15 dB/km at $-30°C$ (reversible).

In the submarine (portion of a link, no loss variation is expected. In the land portion, climatic conditions margins for temperature variation have to be calculated.

b) Cable: The maximum temperature range is to be found in storage conditions, i.e., from $-20°C$ to $+40°C$. Thus, the maximum temperature difference to be encountered is 60°C. Calculations of temperature effects regarding slack variation or pressure variations show that the magnitude of these effects is negligible.

Reliability

Because optical fibers are submitted to stress cracking mechanisms, their introduction in submarine telecommunication cables makes the cable designer as well as the system operator somewhat anxious. In fact, polyethylene is also sensitive to stress cracking, but the choice of a special grade with severe incoming tests has pushed the critical conditions far away from operational conditions, making needless systematic lifetime calculation. We shall examine below the fibers' lifetime evaluation in CDL cable.

The minimum lifetime t_f under a constant stress is given by the relation

$$t_f > B \frac{\sigma_p^{n-2}}{\sigma_a^n}$$

where σ_p is proof test stress, and σ_a is the constant stress applied to the fiber.

Experiments of breaking strength were conducted under atmospheric conditions between 2 and 4 GPa. Plotting the time-to-failure data gave values for n between 25 and 31. In dry conditions, assuming 10–20 ppm of humidity content, values of n would certainly increase, and a program has been undertaken to give an evaluation of n in the filling compound. To give a conservative evaluation of lifetime, and assuming that water is not present in the cable, values of B and n have been taken as the following:

$$B = 10^{-2} \text{ GPa}^2 \text{ s}$$
$$n = 20.$$

σ is expressed in gigaPascals.

A proof test gives an upper limit for flaw size, but does not imply that lower sized flaws have not grown close to the critical value so that fiber could be made fragile. To get ensurance it must be qualified. Experiments on randomly chosen fibers are conducted by repeating the proof test several times. It was experienced that fibers having successfully resisted one proof test also resisted three proof tests so we concluded that the proof test does not weaken the fiber and that we can apply the lifetime evaluation with relative levels of screen test and strain in the cable.

1) Application to the Cable:

a) Maximum permanent stress in the cable: Evaluations of slack and maximum permanent stress due to the bending radius are made according to the following formula:

$$S = \left[1 + \frac{\pi^2}{p^2} \left(\phi_1^2 - \phi_0^2 \right) \right]^{1/2} - 1$$

where S is the available slack, ϕ_1 and ϕ_0 are, respectively, the maximum and minimum winding diameters of the fiber, and p is the pitch.

The maximum stress applied to the fiber is expressed as follows:

$$\sigma_a = E_g \frac{d}{\phi_1} k \sin^2 \left[\text{arctg} \left(\frac{\pi \phi_1}{p} \right) \right].$$

In this formula, E_g is the glass young modulus and d is the cladding diameter, i.e., 125 μm. With our cable design the following values are obtained:

<table>
<tr><td>minimum available slack</td><td>$S = 1.1$ percent</td></tr>
<tr><td>maximum permanent stress</td><td>$\sigma_a = 70$ MPa.</td></tr>
</table>

The maximum calculated stress σ_a is only applied on the external part of the fiber. However, in a conservative purpose, we shall consider that this stress is applied on the whole fiber.

When reporting permanent stress in our lifetime formula, and assuming a 0.9-percent elongation proof test, the lifetime of the whole fiber is about $1.5 \cdot 10^{11}$ years.

TABLE V

Plastic rod diameter (mm)	5
Fiber unit structure diameter (mm)	5.7
Number of optical fibers	up to ɔ
Slack on fibers (percent)	1.1
Steel vault diameter (mm)	11.9
Number of wires in the first lay (left hand)	12
Diameter of the wire (mm)	1.98
Number of wires in the second lay (left hand)	24
Diameter of the wires (mm)	1.18
	1.60
Composite conductor diameter (mm)	12.8
Outside diameter (mm)	25
Weight in air (kN/km)	11.50
Weight in water (kN/km)	6.32

We can assume that stress corrosion aging due to the bending radius is negligible for well over 25 years.

In the case of splicing fibers before cabling, the same proof test is applied to the splice to ensure that no lifetime reduction could occur.

b) Maximum permanent stress in junction boxes: In a junction box the minimum local bending radius is 40 mm. Consequently, and with the same remark as above, the maximum local permanent stress is 125 MPa, and the related lifetime is 10^6 years.

We can conclude that even with a pessimistic estimation of n values, there will not be any effect on the fiber in our cable because the stress on fiber is known and constant in all circumstances during system life.

DEEP-SEA CABLE

Description

The deep-sea cable which is the basic cable of the S 280 system has been described in the previous section and Fig. 2.

The deep-sea cable is referenced as type D4 (four fibers) or D6 (six fibers).

The main characteristics are listed in Table V.

Performances

The main performances are listed in Table VI.

Hydrodynamic Constant

The sinking angle α of a cable related to the horizontal axis follows the relationship

$$\tfrac{1}{2} C_D \rho \, d V_s^2 \sin^2 \alpha = w \cos \alpha$$

where

V_S cable ship velocity related to ground in meters per second,
C_D transverse drag coefficient,
w weight of the cable in water in newtons,
ρ mass density of the seawater,
d diameter of the cable in meters.

TABLE VI
DEEP-SEA CABLE PERFORMANCES

Breaking load	(kN)	140
Load canceling slack on fibers	(kN)	100
Modulus	(km)	22
Full safe modulus (no stress on fibers)	(km)	14.5
Torque	(m×N/kN)	0.6
Gyration under 100 kN load	(t/m)	1
Reverse bend test resistance on 0.3 m radius	(number of cycles)	100
Pressure resistance	(MPa)	70
Hydrodynamic constant	(m/s)	0.696
Ohmic resistance	(Ω/km)	0.6
dc insulation strength (3 mn)	(kV)	50
Mean fiber attenuation on cable including splices in the operational 1.3 nm wavelength range	(dB/km)	0.44
Maximum fiber attenuation in cable in the operational 1.3 μm wavelength range	(dB/km)	0.46
Total maximum dispersion	(ps/nm×km)	3.5

For small values of α, $\sin \alpha \# \alpha$ and $\cos \alpha \# 1$ and the hydrodynamic constant H may be calculated

$$H = \left(\frac{2w}{C_D \rho d} \right)^{1/2}$$ expressed in radians, meters per second, or degree knots

and

$$\alpha = \frac{H}{V_s}.$$

For greater values of α

$$\alpha = \arccos \frac{-H^2 + \sqrt{H^4 + 4V_s^4}}{2V_S^2}.$$

In general α is too great for the simplified formula, and the last formula must be used. Unfortunately, it must be noted that α is not proportional to V_s.

SHALLOW WATER CABLE

Generality

As said before, shallow water cables are manufactured by adding suitable protection on deep-sea cables.

The use of high-strength galvanized steel wires with long pitch for armoring allows very high tension on the cable without cancelling the slack on fibers and inducing additional stress on them. So, armored cables are consistent with the general design of our cable. Moreover, because of the high strength of the inside deep-sea cable, the armored cables break simultaneously with the inside deep-sea cable, i.e., at less than 2-percent elongation.

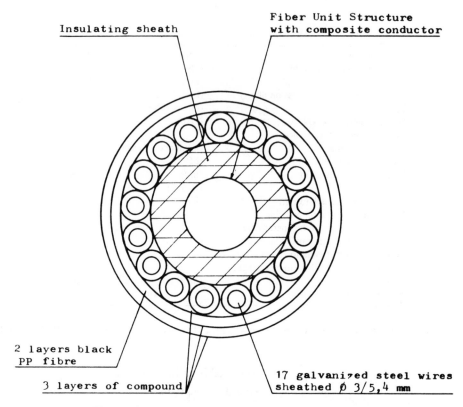

Insulating sheath

Fiber Unit Structure
with composite conductor

2 layers black
PP fibre

3 layers of compound

17 galvanized steel wires
sheathed ⌀ 3/5,4 mm

Fig. 4.   Cross section of type D4 AL lightweight armored cable.

Three types of armored cables are available, corresponding to different seabed conditions.

When the cable is laid between 500 and 1500 m of sea depth, it may be protected by a light armor if local fishing activity appears to be a threat for the cable.

When the cable is laid between 100 and 800 m of sea depth, it may be protected by a single armor.

Between the shore end and 100 m of sea depth, the cable is protected by double armor.

The final choice between the three types of armored cable and their distribution along the cable route is made after the survey. For example, where the cable will be buried, lightweight armor cable may be used in place of single armor.

Lightweight Armored Cable

1) Description: Lightweight armored cable is manufactured from deep-sea cable over which a layer of high-strength galvanized steel wire is wound with long pitch. These wires are previously polyethylene coated and the layer is covered by two compound wetted layers of polypropylene yarn (PP) (see Fig. 4).

The lightweight armored cable is referenced as type D4 AL. The main characteristics and performances of the D4 AL cable are listed in Table VII.

TABLE VII
ARMORED CABLE CHARACTERISTICS AND PERFORMANCES

Characteristics	Unit	Light-weight armor D4 AL	Single armor D4 A	Double armor D4 AR	Land cable D4 AT
Deep-sea cable diameter	mm	25	25	25	25
First lay steel wire diameter	mm	3	5	5	3
Coated steel wire diameter	mm	5.4	—	—	5.4
Number of steel wires in the lay (left hand)		17	19	19	16
Pitch	mm	760	620	620	300
Second lay mild steel wire diameter	mm	—	—	7.62	—
Number of steel wires in the lay (left hand)		—	—	12	—
Pitch	mm	—	—	125	—
Outside diameter	mm	43	43.8	64	35.8
Weight in air	kN/km	28.6	46	116.5	23.2
Weight in water	Kn/km	13.8	30.4	83.3	—
Performances					
Breaking load	kN	> 270	> 680	> 750	170
Load canceling the slack on fibers	kN	> 200	> 500	> 550	110
Modulus	km	21	21	9	—
Full safe modulus (no stress on fibers)	km	15	16.7	6.6	—
Torque	mN/kN	1.2	2.1	5	
Gyration under 100 kN load	t/m	0.5	0.2	0.1	
Reverse bend test resistance on 0.5 m radius	cycles	120	120	120	100
Pressure resistance	MPa	70	70	70	
Hydrodynamic constant	m/s	0.76	1.16	1.59	
Electrooptical performances		same as deep-sea cable			

Single-Armored Cable

1) Description (Fig. 5): The single-armored cable, referenced as type D4 A, is manufactured from deep-sea cable on which is laid successively one lay of serving polypropylene yarn wetted with bituminous compound, one lay of high-strength galvanized steel wires, and two lays of polypropylene serving yarn wetted with bituminous compound.

The main characteristics and performances of the D4 A cable are listed in Table VII.

Double-Armored Cable

1) Description (Fig. 6): The double-armored cable, referenced as type D4 AR, is manufactured like single-armored cable, but a second lay of galvanized steel wires with short pitch is cabled, protected by two lays of PP serving yarn wetted with bituminous compound.

The purpose of the second steel lay is not to increase the strength of the cable but to increase its resistance against anchors, trawlers, abrasion, and perforation.

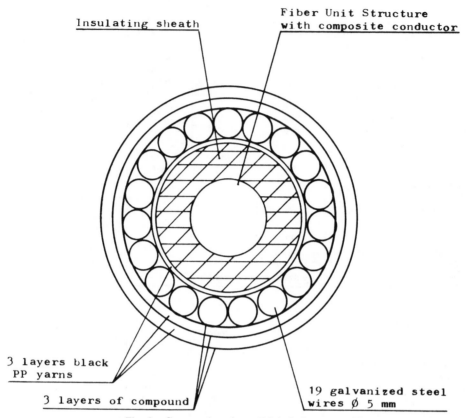

Fig. 5.   Cross section of type D4 A single-armor cable.

Roc armor is especially well adapted to the protection of cable against fishing tool threats, even beam trawlers.

The main characteristics and performances of the D4 AR cable are listed in Table VII.

Transitions Between Cables

The torque of each type of cable is very different, so if no special arrangement is used, strong gyration may be induced on the cable during laying and recovering. In order to minimize these gyrations we use progressive transitions between each type of cable.

For example, between single armor and light weight armor cable the transition is as long as the sea depth (500 m), and a pair of 3-mm steel wires are substituted for a pair of 5-mm steel wires every 55 m.

In this way the torque changes progressively, and there is no risk of local looping on the sea bottom or excessive twist on a cable with a lower inertia modulus.

Land Cable

Description (Fig. 7)

The land cable is manufactured from deep-sea cable on which is laid an armoring made of mild steel galvanized and polyethylene sheathed wires. If necessary, after the study of

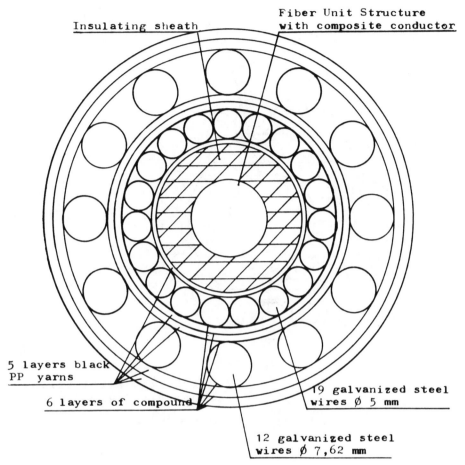

Fig. 6.   Cross section of type D4 AR double-armor cable.

the local conditions, additional protection may be used (e.g., copper tapes associated with mild steel tapes). The land cable is referenced as D4 AT type.

The main characteristics of the land cable are listed in Table VII.

COUPLINGS

Deep-Sea Coupling

The couplings are designed for the following purposes: optical continuity between the line cable and repeater; electrical continuity between the line cable and repeater; allowing remote power feeding; mechanical continuity between the line cable and repeater; and watertightness necessary for correct operation.

A waterproof polyethylene molded steel box is used for this purpose.

The design of the deep-sea couplings is shown in Fig. 8.

On the repeater side an access cable (pigtail) containing optical fibers is used. It consists of an insulated stainless steel tube covered with electrolytic copper cladding. This access cable ensures optical and electrical continuity between the box and repeater.

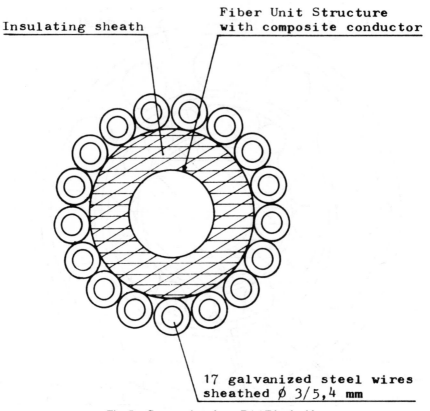

Fig. 7.   Cross section of type D4 AT land cable.

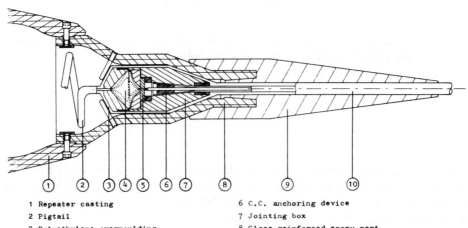

1 Repeater casting   6 C.C. anchoring device
2 Pigtail   7 Jointing box
3 Polyethylene overmoulding   8 Glass reinforced epoxy part
4 Coiling tank for fiber slack   9 Bend limiter
5 Sealing plate with fiber's pass through   10 D4 cable

Fig. 8.   Deep-sea coupling.

TABLE VIII
DEEP-SEA COUPLING PERFORMANCES

Minimum breaking load	(kN)	140
Load canceling the slack on fiber	(kN)	100
Maximum load on cable without any electrooptical or mechanical impairment when the repeater and the coupling are engaged on a 3 m diameter sheave	(kN)	80
dc resistance of the coupling including pigtail	(Ω)	0.06
dc insulation strength (3 mn)	(kV)	50

The box contains an excess of fiber to allow the cable-to-repeater splicing.

A longitudinal water barrier is provided to avoid water penetration on stored fibers in case of cable break in the vicinity of a repeater.

The mechanical continuity with the repeater is performed by a glass reinforced epoxy part. This part is of a conical shape. Its inside shape is adapted to that of the conical anchoring which bears on it when tension is applied to the cable. This part is fixed to the casing of the repeater by cuproberyllium screws. The torque transmission is performed by two reliefs on the polyethylene sheath of the box, which are fitted to corresponding shapes in the glass reinforced part.

A specially shaped bend limiter is fitted on the cable to prevent excessive bending when the repeater passes around the sheave of the cable ship.

The main performances of the deep-sea coupling are listed in Table VIII.

Armored Cable Coupling

The description of the armored cable coupling is shown in Fig. 9.

The connection between the inside deep-sea cable of the armored cable and the repeater pigtail is exactly the same as for deep-sea coupling.

The mechanical continuity is different. We use a gimbal technology to allow a high load on the cable when a repeater passes around the bow sheave of the cable ship. The armor wires are clamped between the glass reinforced epoxy part and a jamming ring of the same material.

The main characteristics of the armored cable coupling are listed in Table IX.

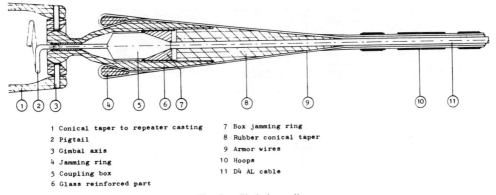

1 Conical taper to repeater casting
2 Pigtail
3 Gimbal axis
4 Jamming ring
5 Coupling box
6 Glass reinforced part
7 Box jamming ring
8 Rubber conical taper
9 Armor wires
10 Hoops
11 D4 AL cable

Fig. 9.   Gimbal coupling.

TABLE IX
CHARACTERISTICS OF ARMORED CABLE COUPLING

Maximum breaking load	(kN)	300 depending on the cable type
Load canceling the slack on fibers		same as armored cable
Maximum load on cable without any electrooptical or mechanical impairment when the repeater is engaged on a 3 m diameter sheave	(kN)	200

CABLE REPAIR

Because it is necessary to get slack on the fiber to perform the splicing of the fibers, junction or repair with operation on the fiber unit structure needs a "junction box" in which the excess length of the fibers is stored.

The junction box is designed for the following purposes: optical continuity; electrical continuity; mechanical continuity; and waterproofness.

Waterproofness is provided by a polyethylene overmolded steel box.

The mechanical continuity is performed through an anchoring of the steel wires of the composite conductor of each cable and a steel tube screwed on each steel anchoring device. This metallic junction between the composite conductor of the cable and the box provides the electrical continuity.

The optical continuity is performed by arc fusion splicing on the fibers. The excess of fiber lengths is coiled in special tanks placed on each side of the box.

Figure 10 shows the junction box design for deep-sea cable.

Such a repair is able to sustain the same load as deep-sea cable and to be handled without additional care by the cable ship.

For armored cables the rebuilding of the armoring after repair of the deep-sea cable induces additional slack on the steel wires.

When the cable is submitted to a further elongation, the slack on wires induces a local and strong elongation on the repair area. To avoid this problem we make two armored cable couplings jointed together through a cylindrical brace made of glass reinforced epoxy resin (Fig. 11).

Performances of such a repair are those of armored cable coupling.

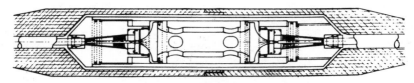

Fig. 10.   Repair box.

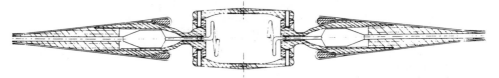

Fig. 11.   Armor cables repair.

Sea Trials and Experimental Links

Sea Trials

The development steps of the submarine optical cables have been marked with several sea trials.

April 1981:   Vault effect efficiency under elongation demonstrated.

January 1982:   Validation of fiber with slack in plastic grooves associated with the steel vault.

September 1982:   107 kN on the cable without additional stress on the fibers.
12 h station on the bow sheave.
Fulfillment of a cable repair on board the cable ship.

April 1984:   Deep-sea couplings and repeater laid and recovered.
Water ingress limitation demonstrated by 2220 m sea depth.

Experimental Links (Fig. 12)

1) First Experimental Link, September 15, 1982: This link has a length of 20 km and connects two telephone exchanges located at Juan-les-Pins and Cagnes-sur-Mer on the French Mediterranean Coast. 9 km of the link was laid at more than 500-m sea depth. The maximum depth was 1070 m.

The cable has six fibers, four multimodes, and two monomodes. Their attenuations were measured at the factory, on board the cable ship, after laying, and periodically from this time. No additional attenuation has been observed after 19 months.

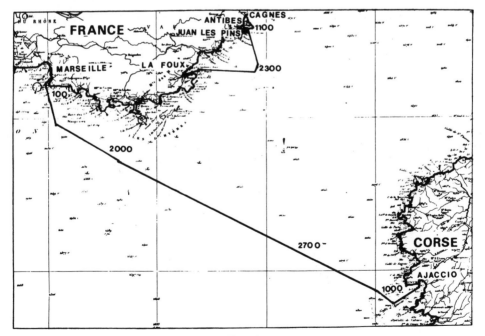

Fig. 12.

TABLE X
PERFORMANCES OF FIBERS USED IN THE ANTIBES–PORT GRIMAUD LINK

Fibers' attenuation in cable	(dB/km)	0.53
Operational wavelength range	(nm)	1290–1330
Chromatic dispersion in the operational wavelength range	(ps/nm × km)	5
Fiber splice attenuation mean value	(dB)	0.15
Standard deviation	(dB)	0.09

The four multimode fibers are used for transmission of two CIT-ALCATEL 34 Mbit/s systems which are under monitoring.

This link is a validation of the suitability of the cable design to be manufactured, handled, and laid by a cable ship.

2) Second Experimental Link, April 9, 1984: The link has a length of 80 km with two submerged repeaters.

The link connects the terminal station of Port-Grimaud/La Foux and the telephone exchange of Antibes on the French Mediterranean Coast.

The cable contains four monomode fibers. The deep-sea cable is 53 km long, laid beneath 500-m sea depth with a maximum depth of 1850 m.

Between the shore ends and 500 m of sea depth the cable is of the D4 A type. 1 km is buried in the St. Tropez bay.

The two repeaters are laid, respectively, in 1700 and 1000 m of sea depth.

The system is a 2×280 Mbit/s system that will be put on traffic after two years of experimentation. It will be used to carry a segment of long-distance terrestrial traffic between Nice and Marseille, avoiding difficult terrestrial works in this mountainous area with overcrowded roads.

Each regenerator is fitted with a supervision circuit, shore-end-controlled electronic looping devices, and lightguide commutators associated with four lasers (three on cold standby).

The performances of fibers used in the link are listed in Table X.

CONCLUSION

Cables de Lyon/Submarcom has developed a first generation of submarine optical cables which is fitted to the S 280 system. This system will be laid for the first commercial use between Marseille and Ajaccio on a 420-km link at the end of 1985. At the end of 1987, it is planned to be laid on the TAT 8 link between the French shore end and the deep-sea submerged interface with the AT&T SL system and the STC NL2 system.

20

Development of Optical Fiber/Power Line Composite Tether Cable for Remotely Operated Unmanned Submersible

YUICHI SHIRASAKI, KENICHI ASAKAWA, JUNICHI KOJIMA, YOSHINAO IWAMOTO, AND MASANORI OHKUBO

INTRODUCTION

Various types of remotely operated unmanned submersibles have been used for inspection and repair of underwater structures and the exploration of ocean environments.

The unmanned submersible is connected to surface control equipment with a tether cable which is composed of power lines, signal transmission lines, and tension members. Ideal characteristics which the tether cable must have are small diameter, lightweight, flexibility, high load strength, low signal attenuation, very high signal bandwidth, and freedom from electromagnetic interference and crosstalk.

The performance of the unmanned submersible system is very dependent on the performance of the tether cable. However, it is much more difficult for conventional metal tether cables to have these required characteristics because the metal conventional tether cable uses coaxial cables and twisted pair cables for the signal transmission line.

The optical fiber/power line composite tether cable (fiber optic tether cable), which adopts optical fibers for signal transmission lines instead of metal conductors, seems to satisfy these required characteristics.

The design of the fiber optic tether cable is considerably different from that of the conventional metal tether cables because the transmission characteristics of the optical fibers change easily by external force, and the elongation of the optical fiber against breaking stress is much smaller than that of the metal cable.

We have studied the fiber optic tether cable [1], [2] for the remotely operated unmanned submersible MARCAS (Marine Cable Search System) since 1979 and developed several types of fiber optic tether cables. Most of the fiber optic tether cables which were reported were not yet in practical use, and design consideration was not reported [8], [9], [10], [11].

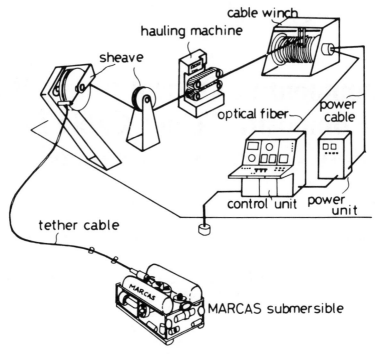

Fig. 1.   Configuration of MARCAS.

This chapter initially considers a number of fundamental design requirements and describes cable design, evaluation results, and the termination for the tether cable which has been in practical use for MARCAS.

DESIGN REQUIREMENTS

Figure 1 shows the configuration of MARCAS [2]. Table I is specifications of MARCAS. The MARCAS submersible moves with four propulsion thrusters to inspect the seabed and submarine cables by underwater TV cameras and to locate the buried cables and the

TABLE I
SPECIFICATION OF MARCAS

Operating Depth	200 m
Size	$1.4W \times 2.0L \times 1.1H$ (m)
Weight	650 kg (air), -15 kg (water)
Performance	Forward Speed Max. 2.5 Knots
	Descending Rate Max. 20m/min
	Turning Rate Max. $35°/s$
Propulsion	Three 3HP Thrusters
	(2 axial, 1 vertical)
	One 2HP Thruster
	(Lateral)
Control	Manual Control
	Automatic Control (Altitude/Depth,
	Direction Keeping)

faulty points of the buried cable by magnetometers and to bury the cables by water jetting tool. One of the optical fibers is used for video signal and data transmission in which a four wavelength division multiplexed bidirectional transmission system [2] is adopted.

The tether cable is subject to various significant loads such as tension, repeated bending, twisting, lateral force, water pressure, and heating from power lines. It is most important for designing the fiber optic tether cable to maintain the optical fiber performance from such various loads. In order to determine the design requirements, we will consider the above mentioned loads.

Tension Load

Tension applied to the tether cable is a function of the propulsion thrust, drag force of submersible, and diving depth (D) of the submersible, length (L), diameter (d), weight (W) of tether cable, and sea current speed (Vs).

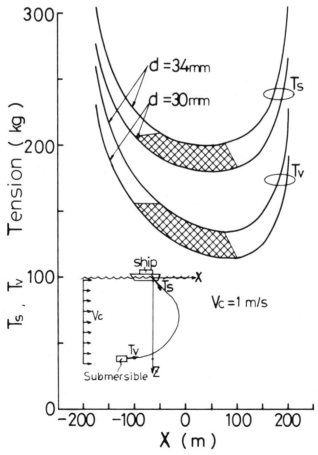

Fig. 2.   Tension applied to tether cable.

Figure 2 shows the tension Tv, Ts at the tether cable end of the mother ship and at the submersible, respectively, calculated under the following condition.

$$D = 200 \text{ m} \quad W = 0.2 \text{ kg/m} \quad L = 300 \text{ m} \quad Vc = l \text{ m/s.}$$

The shaded portion in Fig. 2 corresponds to the MARCAS submersible movable area, which is determined by the maximum propulsion thrusts.

It is seen from Fig. 2 that the maximum tension applied to the tether cable is about 200 kg under static conditions. Under dynamic conditions, however, the maximum tension will increase due to ship motion. The maximum tension will be estimated at about 300 kg.

On the other hand, the tether cable should tolerate an emergency tension when the submersible is lifted by only the tether cable in emergency conditions. This emergency tension is estimated at about 1000 kg.

TWISTING LOAD

The tether cable is twisted when the submersible turns round. Many turning to one direction will accumulate the twists in the tether cable. Excessive cable twisting will cause buckling in cable helical components such as power lines, optical fibers, and tension members. The buckling causes microbending loss increase in optical fibers and conductor failure in power lines.

To avoid this undesired accumulation of cable twisting, the number of cable twists is usually monitored on the surface control equipment, and the submersible is operated not to accumulate the cable twists.

Bending Load

The tether cable is bent on the cable winch drum and sheaves. The cable in the vicinity of the cable terminator is bent severely when the submersible turns rapidly. The required bend diameter is more than 20 times the tether cable diameter from the view of decreasing the cable fatigue [4].

Lateral Force

The tether cable is subject to lateral force on the sheaves, the hauling machine, and the cable winch drum. The maximum lateral force is 10 kg/cm on the hauling machine for the MARCAS system. The lateral force (F) on the sheave is given in the following equation:

$$F = (\text{cable tension})/(\text{radius of sheave}) \tag{1}$$

the radius of the sheave is usually designed to be more than 20 times the radius of the tether cable. As the radius of the sheave is 45 cm for the MARCAS system, the maximum lateral force is about 6.7 kg/cm under usual operation and 22.2 kg/cm under emergency conditions.

Water Pressure

As the maximum operating depth of the MARCAS submersible is 200 m, the maximum water pressure is 20 kg/cm^2.

TABLE II
DESIGN REQUIREMENTS OF TETHER CABLE

Length	500 m
Maximum Operating Strength	1000 kg
Breaking Strength	9000 kg
Transmission line	
Power	more than 20 metal conductors
	(600 V, 10 A)
Signal	more than 6 optical fibers
	(GI silicone 50/125 μm)
Weight	less than 1 kg/m (air)
	less than 0.2 kg/m (water)
Diameter	less than 34 mm
Operating Condition	
Temperature	$-30°C–70°C$
Maximum water pressure	20 kg/cm^2
Maximum lateral force	25 kg/cm
Minimum bending radius	40 cm

Cable Heating Problem

The tether cable is heated by the electrical power dissipation. The tether cable operating temperature is expected to rise because of the conflicting requirements such as small diameter, light weight, and high power capacity. The heat dissipation of the tether cable in water is sufficient to keep the internal cable temperature within acceptable limits. But the tether cable rolled up on the cable winch drum is heated up to severe temperatures.

In the case of the tether cable of MARCAS, the joule loss of 20 w/m raises the internal temperature of the tether cable in air at 20°C up to 60°C. As the quality of the plastics on the optical fiber will change at high temperatures, more than 80° C, the tether cable on the cable winch drum should be cooled by a sprinkler.

Design Requirements

We determined the design requirements for the MARCAS tether cable as shown in Table II from the above discussion and the MARCAS system requirements.

Twenty power cables are for 4 thrusters and circuits. The number of the optical fiber is more than 6. The tension member is Kevlar 49, which is required to realize lightweight and flexible tether cable. The outer jacket is made of polyurethane, which has superior durability. The weight of the tether cable in water is slightly positive for copper power conductors and is neutral for aluminum power conductors.

DESIGN

Figure 3 (a) and (b) shows the structure of two types of the tether cable which were designed according to above mentioned requirements. The optical fibers are stranded in the optical fiber unit in both tether cables. For the tether cable [2] as shown in Fig. 3(a), the optical fiber is a tightly jacketed type. The optical fiber unit is stranded on the core of the central power cable, and two layer Kevlar tension members are wrapped helically around the inner jacket. On the other hand, for the tether cable as shown in Fig. 3(b), the optical fiber is the loose tube type. The optical fiber unit is in the center of the cable, the

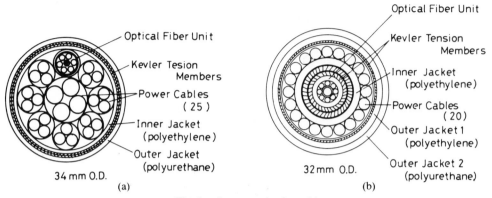

Fig. 3.   Structure of tether cable.

two layer Kevlar tension members are stranded on the optical fiber unit, and power cables are stranded on the Kevlar tension members.

Laboratory and field test results of the tether cable as shown in Fig. 3(a) indicated the following serious problems.

1) When the tether cable was twisted in a direction to unlay the optical fiber unit, the optical fiber unit was buckled easily. Consequently, the transmission loss of the optical fibers increased.

2) When the tether cable was bent and twisted repeatedly, the arrangement of the Kevlar tension members became irregular. Consequently, the tether cable was deformed. This deformation not only reduces the cable breaking strength, but also reduces the cable fatigue performance.

3) Internal temperature rising in the cable introduced stretch of the power cables, the optical fiber unit, and the jacket, and shrink of the Kevlar tension members whose thermal expansion coefficient is slightly negative. Consequently, the optical fiber unit and power cables subjected to the shrink force are easily buckled when other external forces are added to the tether cable.

These problems were solved in designing the tether cable as shown in Fig. 3(b). This type of tether cable has been in practical use for the MARCAS. We will discuss the tether cable as shown in Fig. 3(b) in the following section.

The principal feature of the tether cable is the following.

1) The optical fiber unit is put in the center of the cable to avoid the buckling due to bending and twisting.

2) Kevlar tension members are put between the optical fiber unit and the power cables to avoid the deformation of tension members, and to decrease the heat stresses to the optical fiber unit because of the low thermal conductivity of Kevlar.

3) A loose tube optical fiber which has a strain-relaxing effect [3] is adopted to decrease the strain of the optical fiber against the tension loading.

Optical Fiber Unit

Figure 4 shows the structure of the optical fiber unit. This unit is composed of six loose tube optical fibers, two dummy fibers, an FRP tension member, and a polyethylene jacket.

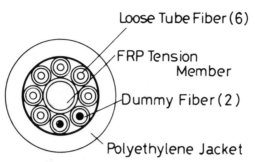

Fig. 4.   Structure of optical fiber unit.

Design parameters of the optical unit shown in Table III were determined from the following considerations.

1) Tension-Strain Characteristics: The tether cable with the Kevlar tension members has initial strain characteristics; its strain is usually less than 0.2 percent. We will design the excess strain of the loose tube fiber to more than the initial strain. When the optical fiber unit is stretched, the optical fiber in the loose tube is not subject to stress until the elongation of the unit becomes the following value:

$$\epsilon_s = (d_{pi} - d_f)(d_t + d_{po})(\pi/P)^2. \tag{2}$$

The stranding pitch P is usually selected to be more than 100 mm to decrease the microbending loss due to the small bend radius of optical fiber. As the excessive optical fiber is usually installed in the loose tube, the optical fiber is not subject to more than the strain of (2). ϵ_s is designed to be 0.2 percent.

2) Twisting Characteristics: The optical fiber in the loose tube will move to the inside or the outside and will contact the tube wall when the optical fiber unit is twisted. This contact induces the increase of the loss of the optical fiber.

The torsion angle θ when the optical fiber contacts the loose tube wall is expressed as the following equation under the condition that the unit length does not change by its twisting.

$$\theta = \pm \frac{d_{pi} - d_f}{2(d_t + d_{po}) - (d_{pi} - d_f)}. \tag{3}$$

This maximum torsion angle is designed to be 300°/m.

TABLE III
DESIGN PARAMETERS OF OPTICAL FIBER UNIT

Optical Fiber with Primary-Coat	
Outer diameter d_f	0.4 mm
Nylon-Loose Tube	
Inner diameter d_{pi}	1.0 mm
Outer diameter d_{po}	1.4 mm
FRP Tension Member	
Outer diameter d_t	2.5 mm
Optical Fiber Unit Jacket	
Inner diameter D_i	5.4 mm
Outer diameter D_o	8.0 mm
Stranding Pitch P	100 mm

3) Mechanical Strength Against Water Pressure: Although the water pressure to the optical fiber unit is less than that to the tether cable, we will assume the maximum water pressure for designing the mechanical strength of the unit jacket operational water pressure of 20 kg/cm^2.

Allowable pressure to the thin cyclidrical tube Pw (kg/cm^2) is

$$Pw = \frac{2E}{1-\nu^2}\left(\frac{D_o - D_i}{2Di}\right)^3 \tag{4}$$

where

- E Young modulus of polyethylene (40 kg/cm^2),
- t thickness of unit jacket (1.3 mm),
- ν Poisson ratio of polyethylene (0.3).

The length of the loose tube shrinks as the optical unit jacket is shrunk by the water pressure. This excess shrinking of the loose tube will induce the microbending loss of the optical fibers. The relation between shrinking strain ϵ_l of the loose tube and water pressure Pw is

$$\epsilon_l = 1 - \sqrt{\frac{\pi^2(1-\epsilon_u)^2 + \left(\dfrac{P}{d_t + d_{po}}\right)^2}{\pi^2 + \left(\dfrac{P}{d_t + d_{po}}\right)^2}} \tag{5}$$

$$\epsilon_u = \frac{2}{E} \cdot \frac{P_w}{1 - \left(\dfrac{D_i}{D_i + 2t}\right)^2} \cdot \tag{6}$$

Pw is 47.2 kg/cm^2 and ϵ_l is 0.02 percent at Pw of 20 kg/cm^2.

4) Lateral Force Characteristics: The transmission loss of the loose tube fiber does not increase until the optical fiber contacts the loose tube wall by lateral force. This maximum deformation Δ_l is

$$\Delta_l = \frac{d_{pi} - d_f}{d_{po}} . \tag{7}$$

But the optical fiber unit should be designed to deform in an elastic limit of the loose tube by the lateral force. The relation between the unit deformation Δ_u and the loose tube deformation Δ_l is

$$\Delta_u = \frac{2(t + \Delta_l \cdot d_{po}) + d_t}{D_o} . \tag{8}$$

Because the tether cable is a complex multishell structure as shown in Fig. 3(b), the relation between the lateral force to the tether cable and the lateral force to the optical fiber unit cannot be shown in a simple equation. The preliminary test indicated that the lateral force to the optical fiber unit is about one half of that to the tether cable. The maximum lateral force to the optical fiber unit is designed to be 12 kg/cm.

The deformation δ of the optical fiber unit jacket under the lateral force P_l [5] is

$$\delta = 3\left(\pi - \frac{8}{\pi}\right)\frac{1-v^2}{E}\left(\frac{D_i+t}{2t}\right)^3 \cdot P_l. \tag{9}$$

5) Temperature Characteristics: As the thermal expansion coefficient of the loose tube [7] is larger than that of the optical fiber, the transmission loss of the loose tube optical fiber varies with temperature. In order to improve the temperature characteristics of the optical fiber unit, the loose tube fiber is stranded on the FRP tension member, which has a low thermal expansion coefficient.

The effective thermal expansion coefficient α_e of the unit [6] is

$$\alpha_e = \frac{\sum_i \alpha_i A_i E_i}{\sum_i A_i E_i}, \tag{10}$$

where

A_i cross section area,
E_i Young modulus,
α_i Thermal expansion coefficient,
i 1) Nylon loose tube; 2) polyethylene unit jacket; 3) FRP member.

Temperature characteristics improvement can be achieved by selecting the cross section area of an FRP member. We designed to be $1 \times 10^{-5} \,^\circ\text{C}^{-1}$.

Cable Tension – Elongation Design

The larger the ratio of the breaking tension to the operational tension is, the smaller is the fatigue of the tether cable. But the large ratio implies a large diameter of the tether cable. We designed the maximum operational tension as 10 percent of the breaking strength, and the elongation of the optical fiber as less than 0.2 percent against maximum operating tension.

The relation between the elongation ϵ_l of the tether cable and tension T is shown in (11) under the assumption that the elongation of the tether cable depends on only Kevlar tension members.

$$\epsilon_l = \frac{1}{E_k}\frac{T}{K_1\sin^2\theta_1 + K_2\sin^2\theta_2}, \tag{11}$$

where

E_k Young modulus of Kevlar tension members (8,000 kg/mm),
K_i Number of Kevlar tension members,
θ_i Pitch angle of Kevlar tension members,
$i=1$ Inner tension member,
$i=2$ outer tension member.

K_i and θ_i are determined to satisfy two characteristics: one is breaking tension, another is torsionless characteristics.

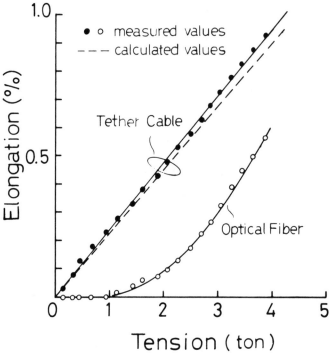

Fig. 5. Tension–elongation test.

Evaluation Results

Tension Elongation Characteristics

Figure 5 shows the elongation of the tether cable and the optical fiber against the tension applied to the tether cable. The calculated value is due to (11) after due consideration of the outer braid Kevlar tension members.

Initial elongation of the tether cable was 0.08 percent after an initial tension load of 2000 kg. The fiber elongation was measured using a phase-delay measuring method.

The strain relaxing of the optical fiber showed more than 0.2 percent which is larger than the value designed for the loose tube fiber. That is why the strain relaxing contains excess length of the optical fiber in the loose tube and excess length due to shrink of the unit by Kevlar tension member stranding on the unit.

This result shows that no strain of the optical fiber is under maximum operation tension of 1000 kg.

Cable Twisting Characteristics

When the tether cable of 1 m length was twisted up to 360, the transmission loss of the optical fiber did not increase. But the electric power cables were buckled when the tether cable was twisted to about 200°/m in a direction to unlay the power lines.

This undesired buckling depends on the twisting speed and the existence of lateral force, and will occur at the angle of more than 120°/m.

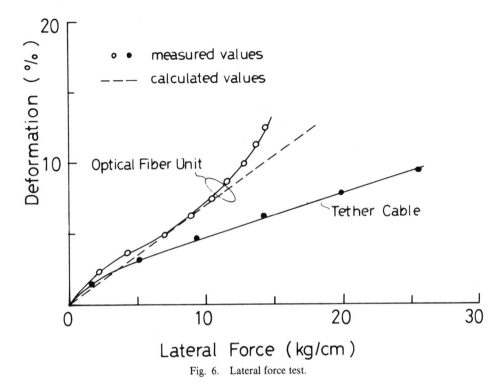

Fig. 6. Lateral force test.

Lateral Force Characteristics

Figure 6 shows the relation between deformation of the tether cable and the optical fiber unit under the lateral force. The results indicate that the optical fiber did not change under the maximum lateral force of 25 kg/cm. The calculated value is due to (9).

Water Proof Test

Maximum water pressure of 50 kg/cm² was applied to the tether cable of 15 m length. Increase of the transmission loss of the optical fiber was less than 0.01 dB/km.

Heat Cycle Test

When the electric current of 10 A was supplied to the 18 power cables of the tether cable of 30 m length in air, the internal cable temperature rose up to 50°C from 20°C. The temperature rose up to 70°C with a current of 15 A. However, the transmission loss of the optical fibers did not increase in either case.

TERMINATION FOR TETHER CABLE

The tether cable terminator has the following functions: 1) connection of the tether cable with the submersible; 2) separation of the power cable and the optical fibers; 3) protection of the power cable and optical fiber against immersion in seawater through a broken outer cable sheath.

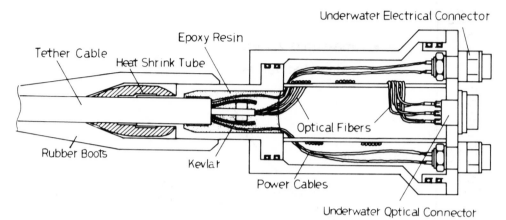

Fig. 7.   Structure of cable terminator.

Figure 7 shows the structure of the terminator. Kevlar tension members are fixed to the pressure case by the epoxy potting compound. The broken strength of this terminator is about 7 t which is about 80 percent of the cable breaking strength.

Both the heat shrink tube and the resin filled assemblage are used for the required tension strength and the waterproof characteristics, and enable field repairability.

Power cables and optical fibers are connected to underwater electrical connectors and a newly developed underwater optical connector [12]. Six optical fibers are wound on the spooling frame to obtain extra length required for splicing of the optical fiber with the optical connector.

Quick turning of the submersible will cause the tether cable to buckle in the vicinity of the terminator, which increases the transmission loss of the optical fibers. Rubber protection boots of 1.5 m length cover the tether cable in the vicinity of the terminator to avoid the above troubles. The flexural rigidity of the tether cable with the protection boots is designed to be more than 50 cm bending radius when the tether cable with the protection boots is bent laterally with a tension of 120 kg. Repeated bend tests of 100 times under tensions of 100–150 kg showed much more stable performance in which the fluctuation of transmission loss of the optical fibers was less than 0.01 dB.

CONCLUSION

We developed the optical fiber/power line composite tether cable which has the following characteristics.

1) The tether cable is composed of 6 loose tube optical fibers, 20 power lines, and Kevlar tension members.

2) The transmission characteristics of the optical fiber are maintained under such loads as tension of 1000 kg, lateral force of 25 kg/cm, water pressure of 50 kg/cm^2, twisting of 360°/m, and temperature up to 75°C.

Many of the optical fiber/power line composite tether cables will be in use in the field of remotely operated unmanned submersibles. This new type of tether cable is expected to be required for the development of an advanced submersible robot which can do more sophisticated various offshore works.

ACKNOWLEDGMENT

The authors thank Dr. Y. Nakagome, Dr. H. Kaji, Dr. K. Nosaka, Mr. T. Nakai, and Mr. H. Ishihara, for their guidance during this work. The authors also grateful to Mr. Y. Tajika, Mr. M. Ogai, and Mr. U. Kawazoe for their useful discussions.

REFERENCES

[1] Y. Iwamoto, S. Suzuki, Y. Shirasaki, and K. Asakawa, "Fiber optic tether cable for submersible" *IECE Japan* (in Japanese), Tech. Rep. CS80-142, 1980

[2] Y. Iwamoto, Y. Shirasaki, and K. Asakawa, "Fiber optic tethered unmanned submersible for searching submarine cables," in *Proc. of OCEANS'82 Conf.*, 1982, p. 65.

[3] P. R. Bark, U. Oestreich, and G. Zeidler, "Stress-strain behavior of optical fiber cable," in *Proc. of 28th IWCS*, 1979, p. 385.

[4] J. A. Walter and P. T. Gibson, "Performance characteristics of ROV tether cables," in *Proc. ROV'83 Conf.*, 1983, p. 37.

[5] Y. Mitsunaga, Y. Katsuyama, and Y. Ishida, "Optical cable deformation characteristics under lateral load," *Trans. IECE Japan*, vol. J64-B, no. 2, p. 142, 1981 (in Japanese).

[6] Y. Sugiwara, "Attenuation increase mechanism of jacketed and cabled fiber at low temperature," *Trans. IECE Japan*, vol. J62-C, p. 864, 1979 (in Japanese).

[7] S. Stueflotten, "Low temperature excess loss of loose tube fiber cables," *Appl. Opt.*, vol. 21, no. 23, p. 4300, 1982.

[8] H. R. Talkington, "Fiber optic data links in vehicle tethers—An update," in *Proc. SUBTECH'83 Conf.*, 1983, p. 419.

[9] A. A. Sadler, "Design and manufacture of strain umbilicals incorporating optical fibre," in *Underwater Technol.*, vol. 10, no. 4, p. 16, 1984.

[10] T. Aoki, M. Hattori, and K. Takahashi, "Fiber–optic–tethered vehicle HORNET-500," in *Proc. ROV'84 Conf.*, 1984, p. 59.

[11] E. W. Hughes and J. F. Wadsworth, "Component, system and sea testing of the oceaneering DUAL HYDRA 2500 remotely operated vehicle system," in *Proc. ROV'84 Conf.*, 1984, p. 155.

[12] K. Asakawa and Y. Shirasaki, "Underwater optical connector for remotely operated submersible," in *Electron. Lett.*, vol. 20, no. 25/26, p. 1031, 1984.

Part V
Undersea Lightwave Repeater Design

Since for long distance undersea lightwave communications the optical signal must be regenerated periodically along the ocean bottom, the physical design of undersea lightwave repeaters is an important aspect of the overall system design and reliability. In the part, the mechanical design of undersea lightwave repeaters is discussed with emphasis on technology requirements of the feedthrough of the fibers into the repeater housing. Also, a chapter on an undersea branching repeater design is included. Undersea branching allows for multiple landing points for an undersea lightwave communications system and introduces the possibility of future undersea networking capabilities.

21
Physical Design of the SL Repeater

MICHAEL W. PERRY, GORDON A. REINOLD,
AND PAUL A. YEISLEY

INTRODUCTION

We have developed an undersea repeater for a single-mode optical transmission system operating at 295.6 Mbit/s. This repeater (Fig. 1) consists of six regenerators (each capable of carrying up to four remotely switchable laser transmitters for reliability), a power supply network, and a supervisory network for in-service monitoring, out of service fault location, and restoration, and switching control. Figure 2 shows a block diagram of this repeater.

Our goals in this design were

1) to achieve a temperature at the critical points on the chassis no higher than 6°C above the external ambient;
2) to be capable of withstanding up to 50-g shocks and 3-g vibrations inherent in the shipping and emplacement process; and
3) to arrive at an inherently manufacturable structure which could provide modularity, and, therefore, an ability to provide subsequent flexibility to meet differing traffic needs and requirements.

It is well known, for electronic components, that reliability increases with decreasing temperatures [1]. Therefore, our prime thrust in this design was to keep critical component temperatures as low as possible. In previous undersea cable repeaters the inner chassis was spring mounted within the pressure vessel [2]. That arrangement, while providing shock isolation and adequate thermal performance, was not the most efficient heat-transfer design. To obtain a better passive heat-transfer design, we mounted the internal unit against a thin epoxy layer on the inside wall of the pressure vessel (see Fig. 3). The epoxy serves two purposes.

First it isolates the high-voltage (~ 7.5-kV) chassis potential from the sea, and second it supports the inner unit and, thereby, becomes a main element in the heat-conducting path to the ocean.

The decision to put the inner unit against the inside wall necessitated many other considerations.

1) A method was needed to place a controlled layer of epoxy on the inside wall of the pressure vessel.

Fig. 1. SL repeater.

2) A means of spreading the heat flow over a large area of the epoxy was needed.

3) The above had to be done in a cylinder which could change internal dimension by as much as 0.023 inches due to the deep ocean pressure.

4) A means was needed to force the inner chassis assembly against the internal cylinder wall while still allowing the necessary "breathing" space to account for the contraction of the cylinder.

5) Because the electronics package was no longer spring mounted the electronic components had to be designed and qualified to be able to withstand the handling shocks and vibrations encountered during installation.

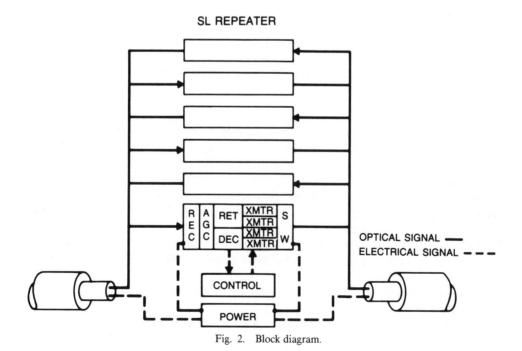

Fig. 2. Block diagram.

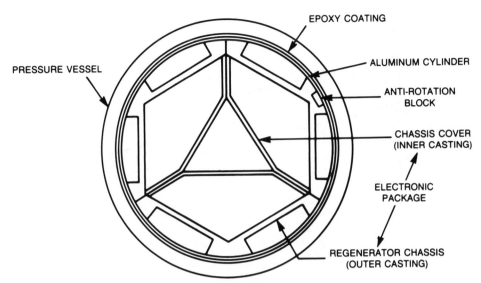

EPOXY COATING

PRESSURE VESSEL

ALUMINUM CYLINDER

ANTI-ROTATION
BLOCK

CHASSIS COVER
(INNER CASTING)

ELECTRONIC
PACKAGE

REGENERATOR CHASSIS
(OUTER CASTING)

Fig. 3.   Repeater cross section.

In addition to all the above, it was necessary to find a convenient means to pass the optical fibers into the pressure vessel while allowing ready access to the individual fiber ends for joining them to the cable while maintaining the low-loss characteristics of the fibers. The following paragraphs describe the resultant design, some of the experiments performed to support the design, and much of the motivation of the design features.

THERMAL DESIGN

Since it is a well-known fact that the reliability of electronic components and devices increases with decreasing temperatures, effort was started on evaluating techniques in which the repeater physical design could be modified or supplemented to reduce the temperature at the laser and thus improve laser reliability. Figure 4 is a hypothetical plot which illustrates the potential savings in spares which are realizable by lowering the operating temperature of the laser.

The curves are based on the "rule of thumb" estimate of an order of magnitude enhancement in laser life per 20°C reduction in operating temperature [3]. Lasers which are currently on long-term aging are expected to provide the data to establish this relationship.

The potential savings, along with consideration of the fact that the SL Repeater was expected to dissipate three to four times more heat than its analog predecessor, SG, made the thermal design the focal point of the physical design.

Physical Design Approach

Four different approaches were considered for the physical design of the SL Repeater. They were as follows:

1) use of the existing analog/coaxial repeater structure;
2) use of the existing analog/coaxial repeater structure with supplementary cooling aids;

UNDERSEA LIGHTWAVE COMMUNICATIONS

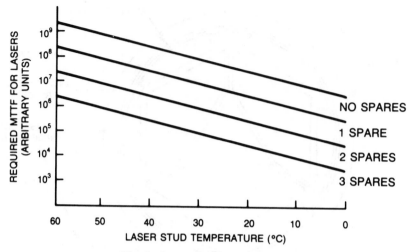

Fig. 4.   Laser reliability requirements.

3) modification of the existing analog/coaxial structure; and

4) modification of the existing analog/coaxial repeater structure with supplementary cooling aids.

The idea of using the existing design, although it had the obvious advantage of proven reliability, was quickly abandoned because of the expected three to four times increase in heat that had to be dissipated. Use of it would fail to meet our goal of achieving a temperature rise of no greater than 6°C. With an expected dissipation of 30 W, the previous design would have resulted in a temperature rise as high as 20°C above ambient.

The next approach was to consider various ways in which we could reduce the effective thermal impedance by supplementary cooling methods. The methods considered were liquid evaporation cooling, heat pipes, and thermoelectric coolers. The first two were quickly dismissed because of reliability concerns and implementation difficulties.

Thermoelectric coolers merited further considerations because, if necessary, they could provide local cooling for the laser or laser transmitter. These devices, however, could not solve the problem of the overall dissipation of 30 W in the repeater. In addition to not providing a solution to this problem, there were several concerns about thermoelectric coolers themselves. The concerns included:

a) questionable long-term reliability;

b) possible condensation internal to the repeater, and its effects; and

c) extra power and, therefore, extra system voltage.

Since thermoelectric coolers do not use moving parts, they are expected to be highly reliable devices. However, to be used in undersea cable repeaters, this reliability has to be verified. Since there was little useful data available, verification would require an extensive long-term testing program.

Also, we were concerned with the possibility of internal condensation on the coolers and/or the cooled parts and its effect. At steady-state temperatures, this posed no real problems in that the thermoelectric cooler could always be maintained at a temperature higher than the internal dew point of the repeater. The problem arises during the transient period when the repeater is being rapidly cooled in the powered state. Albeit, during the 25-year life of a repeater, this particular occurrence of events happens infrequently, it is still an area of concern which would need to be addressed.

The power concern manifests itself in three ways. They are as follows:

a) the additional system power required;
b) the dissipation of power in the repeater; and
c) the effect of the additional dissipation on the other components and devices in the repeater.

With the concerns of using thermoelectric coolers in mind, the next approach was to look at ways of modifying the structure of the existing repeater such that it would be more efficient in dissipating heat. For the thermal model shown in Fig. 5 it was found that the largest thermal impedance was that of the spring and air layer between the chassis and the high-pressure housing. There were at least two possible solutions to this problem: eliminate the springs or modify the springs to reduce their thermal impedance. The second alternative was briefly considered but dismissed because of the design and development effort required. It was decided to eliminate the springs and go with a unit which was not shock isolated. This necessitated requalifying component and subassemblies to a new shock level.

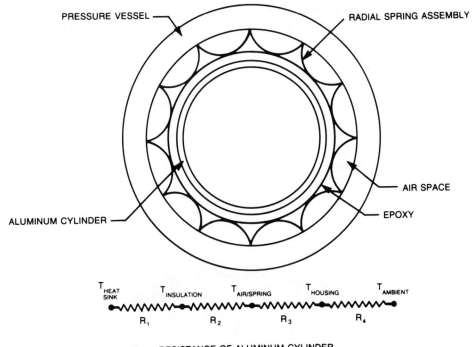

Fig. 5. Thermal model of the SG repeater.

In conjunction with the approach of eliminating the springs, the possible use of thermoelectric coolers was still being considered. However, as it became more obvious that the passive approach alone might be sufficient to reach our goal their need became less. This, coupled with the previously stated concerns, and the realization that the consequence of a thermoelectric cooler failing would result in the laser operating temperatures being hotter than they would without a cooler, led us to remove coolers from further consideration.

Description

Figure 6 shows a cutaway view of a section of the repeater and illustrates the thermal path from the laser transmitter to the high-pressure housing which serves as the heatsink. As illustrated, the path goes from the laser transmitter case, through the chassis, through the legs of the chassis, to an aluminum sleeve, and, finally, through the epoxy to the high-pressure housing.

To minimize the thermal contact resistance, the laser transmitter case is bolted to the chassis. The chassis, being aluminum, serves as an intermediate heatsink. The heat from the chassis is carried to the aluminum sleeve via the legs as shown. The purpose of the aluminum sleeve is to serve as a heat spreader. It is shrink fitted into the epoxy-coated housing assembly.

By providing a spreading force between the chassis castings, the springs serve to minimize the thermal contact resistance between the chassis and the sleeve, and to allow the electronics unit to deflect under the external loads that it would be subjected to at sea-bottom pressures.

Results

A model of the SL Repeater was tested in the laboratory over a range of power dissipations. The results of these tests are illustrated in Fig. 7. As shown, the average

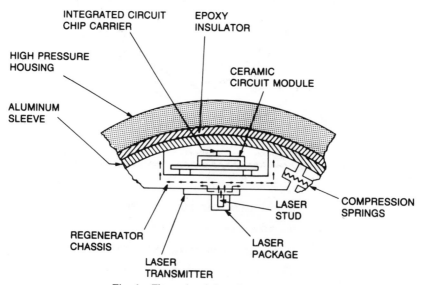

Fig. 6.   Thermal path for a laser transmitter.

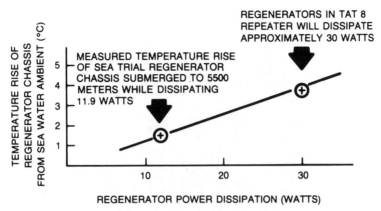

Fig. 7.   Thermal measurements of the SL repeater.

thermal impedance as measured from the position on the aluminum chassis to the external ambient is approximately 0.13°C/W. At an expected dissipation of 30 W this translates into a temperature rise of 3.9°C. This value meets our stated objective of a temperature rise of less than 6°C with margin.

Most repeaters will operate in deep water where the temperature is approximately 4°C, resulting in a chassis temperature of approximately 8°C. In shallow water, the repeater chassis will be approximately 12°C higher, which implies a 20°C operating temperature.

In addition to the laboratory models which were tested, measurements were also made during the 1982 Sea Trial on a partially equipped repeater containing actual regenerators. The results confirmed those obtained in the laboratory.

Design Consequences

One of the primary consequences of the approach of eliminating the radial springs was the loss of shock and vibration isolation of the electronics unit from the high-pressure housing. This required that the internal units and components be designed and qualified to withstand greater dynamic loads. One area of concern were the ceramic circuits. Besides the shock and vibration consequences of the selected design approach, there were other consequences as well. They include the overall repeater assembly, as well as the design of the internal unit. The resultant designs are described in the following sections. A design to accommodate ceramic modules is described and explored in more detail.

REPEATER ASSEMBLY

Pressure Vessel

The pressure vessel (enclosure) for the repeater is similar to that used in previous systems, except for the material system used. Over the past five years we have had an active program of developing a spinodally hardened copper nickel tin alloy (10-percent, Ni, 8-percent Sn) for use in our pressure vessel. This program has led us to the conclusion that this alloy is in all ways compatible to the previously used beryllium copper alloy (1.65-percent Be). Similarity in mechanical properties and excellent compatibility in a submerged seawater environment allows for a mix and match of these materials as

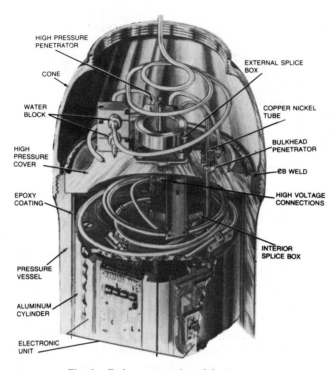

Fig. 8.   End cut-a-way view of the SL repeater.

generally determined by the lowest cost. However, fastener hardware will be made from the spinodal alloy because of its superior stress corrosion characteristics.

Bulkhead penetrators are of an existing, proven design incorporating a copper nickel (70-percent Cu, 30-percent Ni) central tube to maintain the integrity of the glass fibers (see Fig. 8). With these penetrators the repeaters are capable of withstanding pressures up to 12,000 psi (37.2 MPa.) with helium leak rates of less than 0.00000005 std. cc/s. Water blocks are strategically located on both sides of the external splice box. These devices insure the seawater integrity of the repeater in the event of a cable break.

Epoxy Insulator

The electronic package is nestled into an extruded aluminum cylinder located coaxially within the pressure hull (Fig. 3). An insulating layer is required between these two cylinders to insulate the high voltage used to power the system (which is carried by the aluminum sleeve) from the sea ground (which is carried by the pressure vessel). A centrifugal casting technique was developed to create this layer of epoxy. By spinning the pressure hull while pouring in a metered quantity of epoxy, and maintaining centrifugal force on the epoxy until it is cured, an internal epoxy cylinder is formed. The resultant inside diameter has a mirror finish and, by careful metering of the amount of epoxy injected, is held to within 0.002 in of its required dimension. The aluminum cylinder is then shrunk fit into the housing. This assembly is coherent through all of the required pressure and temperature cycles.

Cylinder Dimension Change Under Pressure

The pressure vessel diameter will change due to the external sea pressure and temperature. Experimental verification of calculations has shown that the pressure hull may change as much as 0.023 in (0.906 mm), when subjected to deep ocean conditions of 12,000 psi. (37.2 MPa) and 4°C.

Unit Design

The assembled optoelectronic network consists of three pairs of castings (see Fig. 3). Each pair consists of an inner and outer casting. The outer casting has two completed regenerators (one for each direction of transmission) mounted externally. Thus three outer castings hold the six regenerators in a repeater. The inner castings provide mounting for power and supervisory circuits. To provide economical pore-free castings, the low pressure permanent mold [4] process was used. Inserts were used instead of tapped holes throughout the castings. This allows for multiple assemblies and disassemblies while minimizing the threat of galling of fasteners within the aluminum.

Spring and Spring Release

Because of the necessary clearances between the unit and the cylinder, a method was established to contract and expand the unit. This was accomplished by locating springs between the adjacent external castings, and by compressing these springs with a block moving along an inclined plane on each end of every casting (see Fig. 9). Each pair of blocks is activated using a threaded rod with left- and right-handed threads. The springs are compressed to give the assembly minimum diameter for insertion, and released to provide an intimate fit with the inner aluminum cylinder. This fit provides adequate

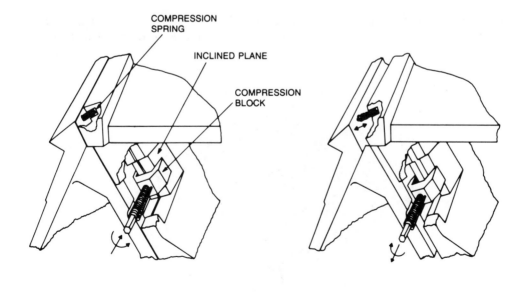

Fig. 9.   Spring release.

electrical and thermal conductivity, and the springs allow the assembly to breath with dimensional changes brought about by pressure and temperature changes. Rotation of the assembly is controlled by a block mounted on the inner cylinder and nestled in the cover plate of the unit.

Flexible Interconnections

Because of the relative motion of the internal castings, interconnections between them must be made with flexible leads. Because of the potential of inadvertently losing a fine strand of wire, stranded wire cannot be used in this high-reliability product. We use individual wire leads.

Ceramic Circuit Modules

Description

The ceramic circuit modules are the functional circuit units. They include the AGC, the retiming circuit, the decision circuit, the supervisory linear/logic circuit, and the supervisory interface circuit. A typical circuit is shown in Fig. 10.

Physically, they consist of a metallized ceramic with thin film resistors (i.e., a film integrated circuit), ceramic capacitors, and one to two silicon integrated circuits. The silicon integrated circuits are mounted in hermetically sealed chip carriers which are attached to their respective film integrated circuits (FIC's) by reflow of solder "bumps." The mounting structure consists of a lead frame which is inserted into a printed wiring board.

Environmental Considerations

Ceramic circuits have never been used previously in undersea repeaters. This, coupled with the selection of the "springless" physical design approach, required that the design

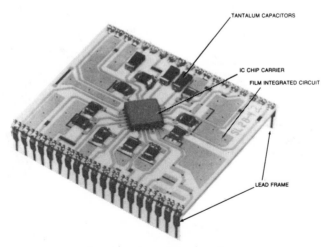

Fig. 10.   Ceramic circuit module.

be carefully and extensively evaluated with respect to whether it could perform in the expected SL Repeater environment. The two primary items of concern were

a) the maximum chip temperature of the integrated circuits; and
b) the shock and vibration capability of ceramic circuits.

The maximum chip temperature was a concern because of the additional thermal impedances in the path from the chip to the printed wiring board. Shock and vibration were a concern because of the suspected fragility of the ceramic substrate.

1) Maximum Chip Temperature: To ensure the long-term reliability of the integrated circuits used in the repeater, it was decided that the maximum chip temperature should be less than 65°C. This temperature corresponds to the maximum allowable junction temperature for transistors used in the SG Repeater. Using the thermal models as shown in Fig. 11, the maximum allowable chip temperature is expressed by

$$T_{chip}(max) = \Delta T_{chip/fic} + \Delta T_{fic/pwb} + \Delta T_{pwb/rc} + \Delta T_{rc/amb} + T_{amb} \leqslant 65°C$$

where

T_{chip} temperature at the surface of the chip,
$\Delta T_{chip/fic}$ temperature rise from the chip to the film integrated circuit,
$\Delta T_{fic/pwb}$ temperature rise from the integrated circuit to the regenerator printed wiring board,
$\Delta T_{pwb/rc}$ temperature rise from the regenerator printed wiring board to the regenerator chassis,
$\Delta T_{rc/amb}$ temperature rise from the regenerator chassis to ambient, and
T_{amb} ambient temperature of environment surrounding repeater.

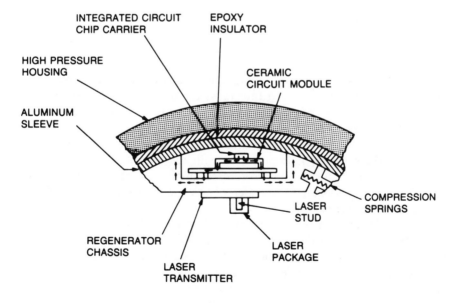

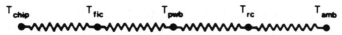

Fig. 11.   Thermal path for integrated circuits.

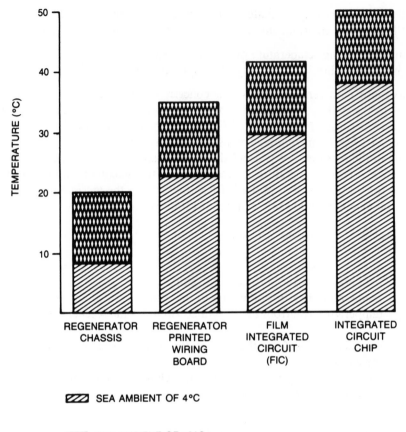

SEA AMBIENT OF 4°C

SEA AMBIENT OF 16°C

Fig. 12.   Maximum repeater temperatures.

We used an AT&T Bell Laboratories finite element analysis computer program known as Thermal Analysis of Substrates and Integrated Circuits (TASIC) in conjunction with an infrared television system to determine that the expected worst case values for $\Delta T_{\text{chip/fic}}$ and $\Delta T_{\text{fic/pwb}}$ are 7.9°C and 6.6°C, respectively.

Similarly, the temperature rise on the regenerator printed wiring board is not expected to exceed 15°C. These values, in conjunction with the previously mentioned value of 3.9°C for $\Delta T_{\text{rc/amb}}$ and a maximum worst case ambient, suggest that the maximum integrated circuit will be around 50°C under worst conditions (see Fig. 12). This is well within our stated objective of 65°C or less.

2) Shock and Vibration Capability: The decision to use a "springless" physical design necessitated that the SL Repeater unit, its subassemblies, and components be capable of withstanding a more rugged environment than its analog predecessors. On a repeater level, our goal was that it be capable of withstanding up to 50-g shocks and 3-g vibration which are expected during shipment and emplacement. To provide margin, the amplitude requirement on the internal subassemblies and components was upped to 100 g for shock.

To evaluate ceramic circuit modules in this regard, shock tests were performed: first on a qualification basis at 100 g, then on a destructive basis up to 5000 g. Of the units tested, none failed.

Ceramic circuit modules were also vibration tested per the SL Repeater requirements. The requirements are such that these modules be capable of

a) withstanding one-half inch p-p vibration levels from 5 to 11 Hz; and
b) withstanding 3-g maximum sinusoidal vibration from 11 to 500 Hz.

All the units that were tested met, and in some cases, surpassed these requirements. Some units were also tested at a level of 3 g from 500 to 1000 Hz without failure.

FIBER JOINTING

Jointing of Transmitters to Optical Relay

PVC coated fibers originating from three individual optical relays and three receivers are brought through the cover plate at both ends of the repeater in molded tubes. They are coiled and clamped until final assembly. Up to four transmitters are connected to a 4X1 optical relay (see Fig. 13).

The joint is organized and stored in a storage tray located between the two devices. These fibers are cleaved, fusion spliced, recoated, and coiled, and located in this tray.

Jointing of Cover (and Seals) to Unit

After locating the electronics unit inside the pressure hull, the power connections to the bulkhead penetrator on the cover are made and encapsulated. Fibers from the high-pressure seal are removed from the splice box attached to the inboard side of the high-pressure cover (i.e., the interior splice box) and joined to the fibers from the unit in the same manner as the joints made between optical relays and transmitters. The joined fibers are then individually coiled and located in the splice box. After closing the splice box, it is mounted on the high-pressure cover. Covers are then located in the pressure hull, as shown in Fig. 8, and sealed by electron beam (eB) welding. The assembled unit is vacuum dried and leak tested in the same manner as previous designs [5].

Jointing of Repeater to Coupling

Assembled repeaters are shipped to the cable factory or ship for joining to the cable. This integration is performed in much the same manner as closing and sealing the

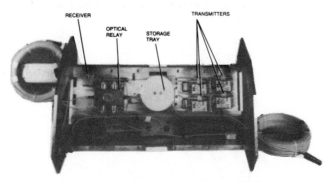

Fig. 13. Transmitter and optical relay assembly.

repeater. The splice box (in this case, the external splice box) is removed from its mount and located outside the cone. A high-pressure penetrator originating from the coupling assembly is mounted in the splice box and tested for pressure integrity. This penetrator carries the fiber and power from the coupling. After carefully locating the fiber in the splice box, the power connection is made and encapsulated. The fibers from the repeater and the coupling are then color-code matched and fusion spliced. These joined fibers are recoated and individually replaced into the box. The box is then sealed, evacuated, leak tested, and back-filled with dry nitrogen. Water blocks are located between the high-pressure cover (outboard side) and the splice box, and between the splice box and the cable termination in the coupling. After completing all procedures on the splice box, it is returned to its original position on the high-pressure cover and the coupling integration procedures are completed.

CONCLUSION

This repeater structure was used successfully in our sea trial demonstration in September of 1982. It has been refined since then through redesign of the fiber seal feedthroughs, redesign of the cable to repeater coupling, and the large-scale integration of the high-speed circuits. The internal structure is, however, the same. The thermal, shock, and vibration performance of this structure met all requirements during the aforementioned sea trial. As a result we now consider the SL optical repeater to be ready for initial manufacture. The first application of this repeater will be in a short system placed between the islands of Gran Canaria and Tenerife in the Canary Islands. This installation will be done as a joint effort of the AT&T Company and CTNE (Compania Telefonica Nacional de Espana). Experience gained during this installation will allow us to provide any necessary refinements for the planned transatlantic cable three years later.

REFERENCES

[1] Bell Telephone Laboratories, "Physical design of electronic systems," vol. IV, in *Design Process*. Englewood Cliffs, NJ: Prentice Hall, 1972.
[2] "Repeaters and equalizers for the SD submarine cable system," *Bell Syst. Tech. J.*, July 1964 or "Repeater and equalizer design," *Bell Syst. Tech. J.*, May–June 1970.
[3] Bell Telephone Laboratories, "Physical design of electronic systems," vol. IV, in *Design Process*. Englewood Cliffs, NJ: Prentice Hall, 1972, pp. 337–340.
[4] J. V. Milos and P. A. Yeisley, "Manufacturing aluminum castings and extrusions for use in SG submarine cable repeaters," *Western Electric Engineer*, July 1975.
[5] "SG undersea cable system: Repeater and equalizer design and manufacture," *Bell Syst. Tech. J.*, Sept. 1978, p. 2361.

22
FS-400M Submarine Optical Repeater Housing

IWAO KITAZAWA, HARUO OKAMURA, AND SHINJI NAKAMURA

INTRODUCTION

A submarine optical repeater housing is an essential part of the fundamental technologies which make it possible to construct optical transmission networks for use in severe deep-sea environments.

Repeater housings are required to protect the repeater units from mechanical forces during cable laying and recovery operations and from high hydraulic pressures and corrosion for more than 25 years in the deep-sea environment.

Experience in designing submarine repeater housings has been gained as a result of having developed coaxial submarine transmission systems [1].

Given this extensive back ground, development efforts have concentrated on some key components. The fundamental study was begun in 1979 [2] and after several laboratory and ocean evaluation tests [3], a prototype repeater housing performed successfully in the first field trial [4] from 1982 to 1984.

This chapter describes the advanced design and characteristics of S-400M submarine optical repeater housing, with a heat-dispersive and shock-absorbing structure using cushion of rubber mixed with Al powder and metal fins, an optical-and-electrical feedthrough using epoxy resin and solder, a caulking type cable termination using a multihole sleeve, and a completely insulated small joint chamber located in the cable coupling.

DESIGN OBJECTIVES

The design objectives are listed in Table I. The housing must be compatible with present NTT cable ship equipment and must be designed to keep a repeater maintenence-free for more than 25 years in the deep-sea environment at a maximum depth of 8000 m. The required tensile strength is 10 tF and the maximum shock is 50 G. The aim of shock absorption is that the shock ratio between the inner housing and pressure housing should be less than 1.0. Thermal design objectives require that the temperature difference between the laser diode (LD) stem and seawater should be less than 10°C. The housing must also be gas tight so that the relative humidity of the inner space of pressure

TABLE 1
DESIGN OBJECTIVES

Capacity	3 systems (6 repeater circuits) maximum
Hydro pressure proof	800 kgf/cm^2 (Water depth of 8000 m)
Tensile strength	More than 10 tf
Shock	50 G
Shock absorbing design	Less than 1.0 (Shock ratio between inner and outer housing)
Thermal design	Less than 10°C between LD stem and sea water
Gas tight	Less than 20% (Relative humidity in inner housing)
Environmental temperature	$-2 \sim 30$°C
Optical loss	Feedthrough: less than 0.1 dB Cable coupling: less than 0.1 dB
Life	More than 25 years

housing remains less than 20 percent for at least 25 years. Finally, the housing must function in an environment where the ambient temperature is from -2 to 30°C. Optical losses for satisfying the system design [5] are determined.

Technical design considerations are as follows.

1) Performance, reliability and ease of manufacturing must be considered in determining the basic structure design and component allocations.

2) A pressure housing with an effective heat-dispersive and shock-absorbing structure is needed.

Optical repeater units, which dissipate about 5 times as much power as coaxial submarine repeater units, are composed of many heat-sensitive and finely structured devices such as laser diode (LD), avalanche-photodiode (APD), etc.

3) Highly reliable, pressure-proof, gas-tight, and low-loss feedthroughs are required so that fibers and electric power line can enter the pressure housing.

4) Highly reliable and low-loss cable coupling with pressure proof and adequate tensile strength is needed.

5) Cable to repeater joint structure.

During the repairing process, jointing work is carried out on board ship where the cable is suspended under a tensile load, so a highly reliable joint structure which can be conveniently constructed is required.

BASIC STRUCTURE AND FEATURES

Figure 1 shows the schematic structure of the FS-400M submarine optical repeater housing. This repeater housing is made of Cu–Be alloy having high tensile strength and superior anticorrosion characteristics. The basic structure is a gimbal type which passes easily through cable laying equipments such as drum-type cable engines, sheeves, etc.

As shown in Fig. 1, the repeater units are inserted in a pressure housing with an effective heat-dispersive and shock-absorbing structure and submarine optical cable is terminated in a cable coupling. Fibers and electrical power line are led to the repeater unity by a tail cable passing through a joint chamber and a feedthrough. The tail cable has a small diameter polyethylene insulated pipe which serves as a fiber protecter and an electrical conductor. A feedthrough located on an end plate allows fibers and an electrical

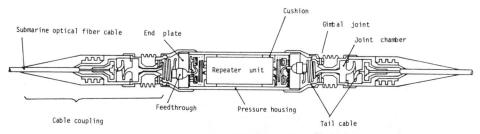

Fig. 1. Schematic structure of FS-400M submarine optical repeater housing.

conductor to enter the pressure housing. The joint chamber is a part of jointing tail cables mutually, where reinforced fiber splices and excess fibers are encapsulated. The pressure case is brazed to the tail cable pipe conductor and polyethylene sheath is moulded with that of tail cable, too. As a result, fibers are protected by the pressure pipe or case which are completely insulated by a polyethylene sheath from submarine optical cable to feedthrough.

Figure 2 shows the typical arrangement of the repeater on a cable engine drum or a sheeve. It is evident from this figure that the distance between gimbal joints and maximum gimbal angle are important. The former is determined to be 1180 and 1410 mm for 2 and 3 systems (4 and 6 repeater circuits), respectively, and the latter is 52 deg. Maximum diameter is 265 mm and total unit length is about 3 m.

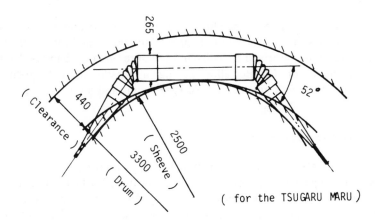

(for the TSUGARU MARU)

Cable ship	Drum type cable engine		Bow sheeve
	Diameter	Clearance	Diameter
KUROSHIO MARU	3800 mm	510 mm	3000 mm
TSUGARU MARU	3300 mm	440 mm	2500 mm

Fig. 2. Typical scheme of the repeater housing on a cable engine or a sheeve.

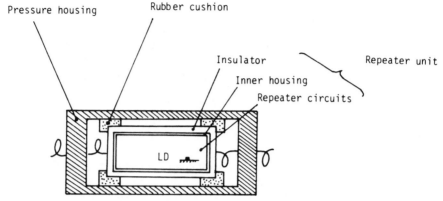

Fig. 3.   Heat-dispersive and shock-absorbing structure.

It is, therefore, possible for repeaters with 2 and 3 systems to be wound on 2.5- and 3.0-m-diam curvatures, respectively. Locating the joint chamber inside the cable coupling contributes greatly to reducing the distance between gimbal joints.

Design and Characteristics of Key Components

Pressure Housing

The pressure housing is designed to be almost similar to the repeater housing presently used in the coaxial cable systems. The Finite Element Method is sued for a detailed stress analysis and the housing is confirmed to withstand pressures greater than 800 kg·F/cm².

The housing is made of Cu–Be alloy having a tensile strength greater than 100 kg·F/mm², a yield strength greater than 90 kg·F/mm² and a corrosion less than 20 μm per year.

Shock Absorption and Thermal Design

Figure 3 shows the heat-dispersive and shock-absorbing structure.

Heat generated in a repeater unit is dispersed into seawater after being transfered to the insulator, shock absorbing space, and pressure housing. LD stem temperature is slightly higher than the repeater unit average temperature because of heat distribution in the repeater unit. By an experimental result, this temperature difference was about 4°C in our repeater. As can be seen in Fig. 4, there is also temperature difference of about 0.5°C between the seawater and the pressure housing. The design objective of 10°C is the sum of these values and the temperature difference between the inner housing and the pressure housing. Thus the shock-absorbing structure and polyethylene insulator accounts for the remaining 5.5°C.

Almost all of the heat resistance between the inner housing and the pressure housing is due to the shock absorber and its air space. Therefore, a cushion of rubber mixed with Al powder is used and metal fins are inserted in the space between cushions to reduce the heat resistance.

The relationship between metal fins and temperature difference between the inner and pressure housing is shown in Fig. 5. With 20 fins in addition to the rubber cushion, the

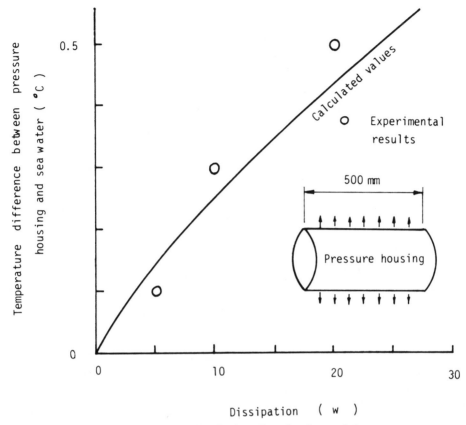

Fig. 4. Pressure-housing heat-dispersion characteristics.

temperature difference is about 3°C for both 2 and 3 systems. Here, a 1°C difference is due to the polyethylene insulator. The new rubber cushion alone reduces the temperature difference to by about 1/2 compared with the conventional structure. Together with the metal fins, the difference becomes 1/5–1/6. Figure 6 shows the heat distribution in this new structure.

As a result, the highest LD stem temperature is 37.5°C where the seawater temperature is 30°C. This satisfies the design objective for LD reliability.

This structure also shows such an exceeding shock-absorbing characteristics. It limits the shock ratio between the inner housing and the pressure housing from 0.7 to 1.0 in both cases of 2 and 3 systems against the shock of 50 G whose acting time is 5 ms. Here, the rubber hardness is about 60 deg and the spring stiffness of the metal fins is designed to be negligibly small compared with that of the rubber cushion in order not to interfere with the shock-absorbing performance of rubber cushion.

Feedthrough

Figure 7 shows the schematic structure of a feedthrough. A maximum of 6 fibers is inserted in the cylindrical conductor in which the mechanism for hydraulic pressure proof and gas tight around fibers are designed. The conductor has a stopper disk to guard against pressure and is moulded with a polyethylene insulator. This disk-type feedthrough

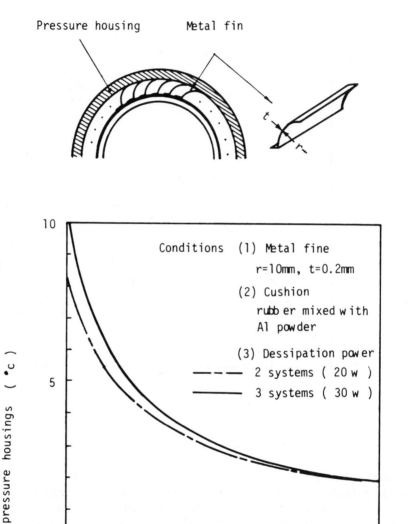

Fig. 5.   Effect of heat dispersion by a cushion of rubber with Al powder and metal fins.

is confirmed to have hydraulic pressure strength of 800 kg·F/cm^2 and dielectrical strength of more than 15 kV. The gas tight between the feedthrough body and the end plate is achieved by a cone seal. In this structure there is no leakage path except in the polyethylene layer which satisfies the design objective adequately.

The mechanical design surrounding the fibers is the most important part, where fibers are set in axial grooves, respectively, and are fixed by epoxy resin and solder. Solder is useful for gas-tight applications but has creeping characteristics. On the other hand, some kins of epoxy resin have high adhesion strength with almost no creep but have a tendency

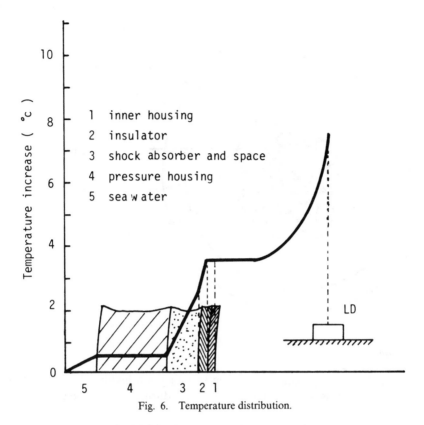

Fig. 6.   Temperature distribution.

to degrade in the presence of water vapor. Thus epoxy resin is used to withstand the hydraulic pressure and solder located at the sea side of the resin is used to keep the resin from coming into contact with water vapor. Solder is anticorrosion coated by another resin which is resistant to water vapor. In order to make the solder seal complete and the assembling process simpler, a metal-coated fiber is applied.

It was empirically confirmed that all of the design objectives are satisfied. The design and detail characteristics are reported in Chapter 18.

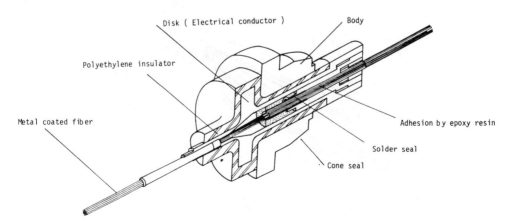

Fig. 7.   Schematic structure of feedthrough (optical and electrical composite feedthrough).

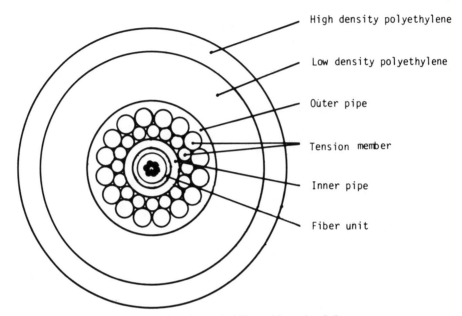

Fig. 8.   Submarine optical fiber cable sectional view.

Cable Coupling

The most important requirement for cable coupling is a highly reliable cable termination being able to withstand a large tensile load during cable laying and recovering operations.

Figure 8 shows a sectional view of submarine optical fiber cable [7]. Fibers are located at the center and protected by a pressure pipe. Tension member consists of two layers of piano wire and has a breaking strength of about 12 tF.

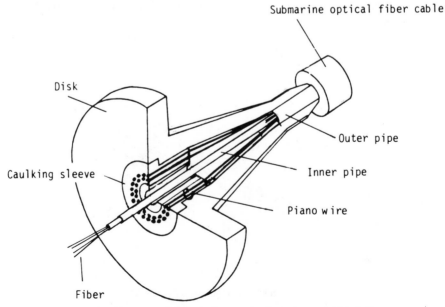

Fig. 9.   Caulking-type tension member termination using a multi-hole sleeve.

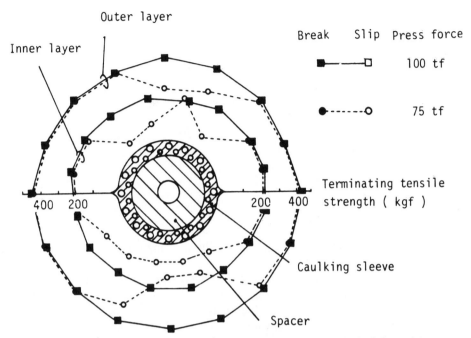

Fig. 10.   Relation between press force and terminating tensile strength of piano wires.

Several effective tension member terminating techniques, such as adhesion with epoxy resin and fixing by means of caulking with a depressing tapered pin [8] have been already proposed. However, resin has the tendency to degrade in high-humid environment and careful control of curing time, temperature, and surface treatment of piano wires are needed. Caulking with a tapered pin is not thought to be suitable for a multilayer tension member, because crossed piano wire could possibly cause termination instability.

Figure 9 shows a type of new caulking using a sleeve with multiple holes. Each piano wire is inserted in a hole and termination is achieved by pressing the sleeve from the radial

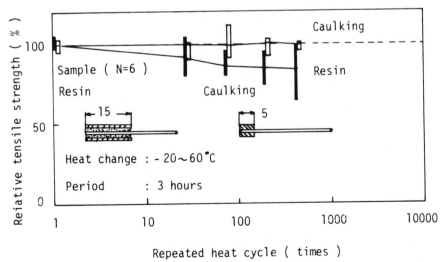

Fig. 11.   Terminating tensile strength stability under heat cycles.

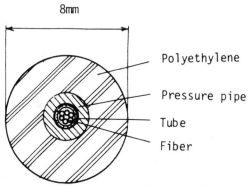

Fig. 12.   Tail cable structure.

direction. Fibers pass through in the center hole without being influenced by the pressing. Effective caulking length is about 20 mm.

Figure 10 shows the relation between caulking press force and individual piano wire terminating tensile strength. From these results, it is evident that caulking press force must be more than 100 tF. At this point, all of piano wires break before they slip. The distribution of terminating strength shown in the case of 75 tF is due to the compressive stress distribution in the sleeve. Terminating tensile strength is greater than 90 percent of the original tensile strength of the piano wire. Total terminating strength of the whole cable is greater than 11 tF, which exceeds the specification.

Figure 11 shows the stability of a single piano wire termination after repeated heat cycle tests. No degradation is seen even after 800 cycles where temperature change ranged from −20 to 60°C over 3-H periods. In addition, in further tests, no degradation was seen where terminated cable samples were exposed to severe environmental conditions with temperatures of 115°C and at a relative humidity of 95 percent for 2.5 months.

Tail Cable and Joint Chamber

Figure 12 shows a sectional view of a tail cable. Fibers are inserted in a small diameter pressure pipe insulated with polyethylene sheath. Dimensions are determined to have adequate hydraulic pressure strength and flexibility withstanding the stress of handling work during assembly and the bending action of the gimbal joint. The inner to outer diameter ratio of the pressure pipe is about 0.5 and the thickness of the polyethylene sheath is 2.5 mm. It was experimentally confirmed that this cable can withstand hydraulic pressures of up to 1000 kg·F/cm^2 and that the polyethylene sheach can withstand a voltage of 20 kV after gimbal joint bending tests of 200 times in two directions crossing each other at right angles.

Figure 13 shows the basic structure of joint chamber. Spliced and reinforced fibers are installed in the ring space surrounding the center core. The pressure case doubles as electrical conductor. At the end of the joint operation, both sides of the polyethylene sheaths are molded together. The ring space structure produces a very small chamber because the center core can effectively support the hydraulic pressure. This completely insulating structure using the polyethylene molding technique results in high reliability during long-term use.

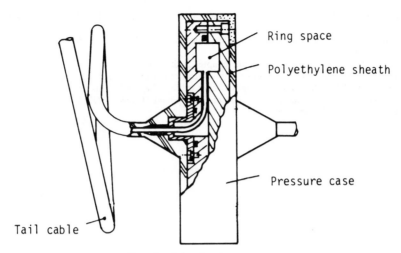

Fig. 13.   Joint chamber structure.

CONCLUSION

The development of an advanced FS-400M submarine optical repeater housing was reported. Fundamental design concepts, and key component design and characteristics were discussed as follows.

1) The housing was designed to accommodate a maximum of 3 systems (6 repeater circuits) and to be compatible with present cable ships. The key to realizing this housing was in locating the small joint chamber in the cable coupling.

2) A temperature difference of 3°C between inner housing and pressure housing was achieved by a heat-dispersive and shock-absorbing structure using a cushion of rubber mixed with Al powder and metal fins, which contributed to LD and system reliability.

3) An optical and electrical composite feedthrough using epoxy resin and solder was developed and exceeds performance specifications.

4) A caulking-type cable coupling using a multi-hold caulking sleve was developed which has a terminating tensile strength of 11 tF with high reliability.

In addition to the successful results in the first field trial, the FS-400M submarine optical repeater housing has performance and reliability sufficient to withstand both cable laying and recovery operations and the severe deep-sea environment.

ACKNOWLEDGMENT

The authors are grateful to Dr. E. Iwahashi, K. Fujisaki, and H. Fukinuki for their constant encouragement and thoughtful suggestions.

REFERENCES

[1]   R. Kaizu *et al.*, "CS-36M submarine coaxial cable repeater," *Rev. Electrical Communication Labs.*, vol. 22, no. 5–6, pp. 398–422, 1974.
[2]   K. Miyauchi *et al.*, "Consideration on the undersea optical transmission system," ICC, 1981.

[3] Y. Nomura *et al.*, "1.3 μm, 400 Mb/s undersea optical repeater sea trial," *Electron Lett.*, vol. 17, no. 23, 1981.

[4] M. Washio *et al.*, "400 mb/s submarine optical fiber system field trial," in *8th ECOC* (Cannes, France), 1982, pp. 472–477.

[5] H. Fukinuki *et al.*, "The FS-400M system," chapter 5, this book.

[6] S. Nishi *et al.*, "Optical fiber feedthrough in the pressure housing," chapter 24, this book.

[7] Y. Negishi *et al.*, "Deepsea submarine optical fiber cable," chapter 18, this book.

[8] Y. Niiro, "Optical fiber submarine cable system development at KDD," *IEEE J. Select. Areas Commun.*, vol. SAC-1, pp. 467–478, Apr. 1983.

23

Design and Test Results of an Optical Fiber Feedthrough for an Optical Submarine Repeater

YOSHIHIKO YAMAZAKI, YOSHIHIRO EJIRI, AND
KAHEI FURUSAWA

INTRODUCTION

Development of a long-haul optical fiber submarine cable system is now continuing, in various countries in view of establishing a transoceanic submarine cable system [1]–[7]. Realization of a highly reliable optical submarine repeater is indispensable for the practical use of an optical fiber submarine cable system. Development of a highly reliable repeater housing is very important, in this sense, to protect repeater circuits against high hydraulic pressure [8]–[9]. It is a very important subject for developing an optical submarine repeater housing to realize a highly reliable optical fiber feedthrough through which the optical fibers pass from the optical submarine cable into the repeater housing hermetically. This feedthrough is required to protect optical circuits without deteriorating mechanical and optical characteristics of optical fibers during the lifetime of the system. This chapter describes design, manufacture on trial, and evaluation results of a highly reliable optical fiber feedthrough for an optical submarine repeater.

DESIGN REQUIREMENTS

Optical fibers in the optical fiber submarine cable system must be protected chemically and mechanically against seawater and water pressure in the entire system, including particularly the optical submarine cable, the jointing part between cable and repeater, and the inside of the optical repeater. Therefore, seawater pressure should not be applied, during operation of the system, on a hermetic sealing part of optical fibers of the feedthrough. However, if cable or cable coupling has any fault such as shear, high hydraulic pressure corresponding to depth is applied on the feedthrough. The feedthrough protects repeater circuits from high hydraulic pressure, to have high reliability, and not to influence transmission characteristic during system operation but rather to function when faults occur. The requirements which the optical fiber feedthrough should have are summarized below.

1) During system operation:
   —optical fiber transmission characteristic should be stable for 25 years or more;
   —relative humidity rise within repeater housing should be 20 percent or less during 25 years.
2) In case of a system fault:
   —ingress of seawater into repeater housing must be prevented;
   —humidity rise inside repeater housing must be prevented;
   —optical fiber transmission characteristic should not change before and after repair;
   —deterioration in mechanical strength of optical fiber should not occur even after repair.

The above requirements can be classified into the two following items from the point of view of long-term reliability of the feedthrough:

1) achievement of airtightness during the life of the system;
2) guarantee of operating life of the fiber at the hermetic sealing part during the life of the system.

Investigations have been made regarding the two items mentioned above in order to clarify the basic concept for long-term reliability of the feedthrough.

STRUCTURE OF THE OPTICAL FIBER FEEDTHROUGH

Hermetic Sealing Structure of Optical Fibers

As an optical fiber hermetic sealing method, a glass seal, plastic seal, and metal seal are considered, but the method which gives little thermal, mechanical, and chemical damages to the optical fiber should be employed. Therefore, a method of sealing optical fiber with adhesive and a method of sealing by solder after applying metallic coating to optical fiber are considered as practical. The method using adhesive can be realized easily but is inferior in airtightness because permeation of resin cannot be avoided as compared with sealing by solder. Sealing by solder generates more heat in the metallic coating process or sealing process as compared with sealing by adhesive, but it is expected to realize higher reliability of feedthrough by applying metallic coating around the primary coating in order to prevent deterioration of optical fiber. Discussed here is a feedthrough structure where metallic coating is applied around the primary coating of the optical fiber and it is sealed by solder.

Design of Hermetic Sealing

1) Stress on Optical Fiber While Hydraulic Pressure Is Applied: Figure 1 shows the condition where a thin metal is coated around the primary coating of an optical fiber and moreover a solder layer is coated on such thin metal. Here, when a pressure P_0 is applied from the cross-sectional and circumferential direction of such optical fiber, a stress working on the optical fiber within such a structure is calculated.

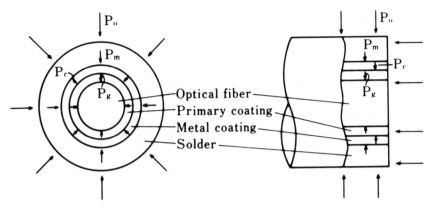

Fig. 1. Cross and longitudinal sections of solder coated fiber. Arrows show the directions of outer pressure P_0 and inner pressures P_m, P_c, and P_s.

When the outer radius of glass is represented by R_1, that of the primary coating of R_2, that of the metal coating by R_3, that of solder by R_4, and r_m is defined in the range of $R_3 \leqslant r_m \leqslant R_4$, internal stress P_{rm} of solder working on the r_m surface can be expressed as follows:

$$P_{rm} = \frac{2R_4^2/\left\{ E_4\left(R_4^2 - r_m^2 \right) \right\}}{(1-\nu_{rm})/E_{rm} + \left\{ \left(R_4^2 + r_m^2 \right)/\left(R_4^2 - r_m^2 \right) + \nu_4 \right\}/E_4} \cdot P_0 \tag{1}$$

where, E_4, ν_4 are Young's modulus and Poisson's ratio of solder, respectively, while E_{rm}, ν_{rm} are equivalent Young's modulus and equivalent Poisson's ratio up to the r_m surface, respectively. E_{rm} is expressed by the following equation:

$$E_{rm} = \left\{ E_1 R_1^2 + E_2\left(R_2^2 - R_1^2 \right) + E_3\left(R_3^2 - R_2^2 \right) + E_4\left(r_m^2 - R_3^2 \right) \right\}/r_m^2 \tag{2}$$

where, E_1, E_2, E_3 are Young's modulus of glass, primary coating, and metallic coating, respectively. When $r_m = R_3$ in (1) and (2), a stress P_m working on the surface of metallic coating can be obtained.

When r_c is defined within the range of $R_2 \leqslant r_c \leqslant R_3$, an internal stress P_{rc} working on the surface r_c is expressed as follows:

$$P_{rc} = \frac{2R_3^2/\left\{ E_3\left(R_3^2 - r_c^2 \right) \right\}}{(1-\nu_{rc})/E_{rc} + \left\{ \left(R_3^2 + r_c^2 \right)/\left(R_3^2 - r_c^2 \right) + \nu_3 \right\}/E_3} \cdot P_m \tag{3}$$

where, ν_3 is Poisson's ratio of metallic coating and E_{rc}, ν_{rc} are equivalent Young's modulus and Poisson's ratio up to the surface r_c, respectively. E_{rc} can be expressed by the following equation:

$$E_{rc} = \left\{ E_1 R_1^2 + E_2\left(R_2^2 - R_1^2 \right) + E_3\left(r_c^2 - R_2^2 \right) \right\}/r_c^2. \tag{4}$$

When $r_c = R_2$ in both (3) and (4), a stress P_c working on the surface of primary coating can be obtained.

Next, when r_g is defined in the range of $R_1 \leqslant r_g \leqslant R_2$, an internal stress P_{rg} working on the surface r_g is expressed as follows:

$$P_{rg} = \frac{2R_2^2 / \left\{ E_2 \left(R_2^2 - r_g^2 \right) \right\}}{(1 - \nu_{rg})/E_{rg} + \left\{ \left(R_2^2 + r_g^2 \right) / \left(R_2^2 - r_g^2 \right) + \nu_2 \right\} / E_2} \cdot P_c \tag{5}$$

where, ν_2 is Poisson's ratio of the primary coating and E_{rg}, ν_{rg} are equivalent Young's modulus and Poisson's ratio up to the surface r_g, respectively. E_{rg} is expressed as follows:

$$E_{rg} = \left\{ E_1 R_1^2 + E_2 \left(r_g^2 - R_1^2 \right) \right\} / r_g^2. \tag{6}$$

When $r_g = R_1$ in (5), (6), an internal stress P_g working on the glass surface of optical fiber can be obtained.

Where an intrinsic adhesive force between solder and metallic coating when no external pressure is applied is represented by F_m, such force between the metallic coating and primary coating by F_c, and the same force between the primary coating and optical fiber glass by F_g, the conditions for generating no leak of sea water or vapor between layers when a water pressure P_0 is applied is indicated below:

1) between solder and metallic coating

$$P_0 < P_m + F_m. \tag{7}$$

2) between metallic coating and primary coating

$$P_0 < P_c + F_c. \tag{8}$$

3) between primary coating and optical fiber glass

$$P_0 < P_g + F_g. \tag{9}$$

By modifying (7)–(9)

$$P_m/P_0 > 1 - F_m/P_0$$
$$P_c/P_0 > 1 - F_c/P_0$$
$$P_g/P_0 > 1 - F_g/P_0. \tag{10}$$

Equation (10) shows that leak of seawater or vapor can be reliably prevented without regard to an intrinsic adhesive force between layers when P_i/P_0 $(i = m, c, g)$ is larger than 1.

2) Calculation Results: Table I shows material properties used for calculation. From (1)–(6) using values indicated in Table I, an internal stress P_r $(R_1 \leqslant r \leqslant R_4)$ when an external pressure P_0 is applied is calculated in the form of P_r/P_0. The result is shown in Figs. 2–4.

Figure 2 shows results of calculations for a structure where copper is coated in the thickness of 2 μm around the primary coating in the thickness of 10 μm and moreover solder is coated in the thickness of 10 μm around such copper. According to the results, the minimum interlayer intrinsic adhesive force of 0.076 kg/mm^2 is required for external pressure of 800 kg/cm^2 in order to prevent leakage between the copper and primary coating, and there is no leakage between other layers without relation of interlayer intrinsic adhesive force.

TABLE I
MATERIAL PROPERTIES

Material	E (kg/mm^2)	v	α (/°C)
Silica glass	7300	0.17	0.4×10^{-6}
Polyimide	200	0.40	150×10^{-6}
Cu	12900	0.34	16×10^{-6}
Sn	5500	0.34	21×10^{-6}
Solder	3060	0.40	25×10^{-6}
Kovar	15000	0.35	5×10^{-6}

In the case of Fig. 3, the thickness of the solder is 20 μm, two times than that of Fig. 2. In Fig. 4, the thickness of copper is 1 μm half than that of Fig. 2. In any case, leakage does not occur even in the interlayer under these conditions.

Therefore, in the case where metallic coating is applied (about 1–2 μm thick) to an optical fiber on which the primary coating is carried out (less than 10 μm), and the solder is coated around a metal coating at least 20 μm thick, a self-sealing effect is generated inside by a tightening stress in accordance with an external force during system operation and the occurrence of fault. Accordingly, a feedthrough which ensures high airtightness for a long period of time can be realized.

3) Permeability of Primary Coating: Since the primary coating is executed with an organic material, permeation is inevitable. Amount of permeation Q can be expressed as follow according to the Fick-type diffusion equation

$$Q = (P \cdot \Delta p \cdot a \cdot t)/L \tag{11}$$

where, P is a coefficient of permeation of the primary coating, Δp is the difference of

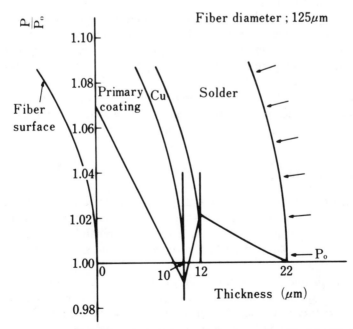

Fig. 2. Inner stress distribution where thickness of primary coating, Cu and solder is 10, 2, and 10 μm, respectively.

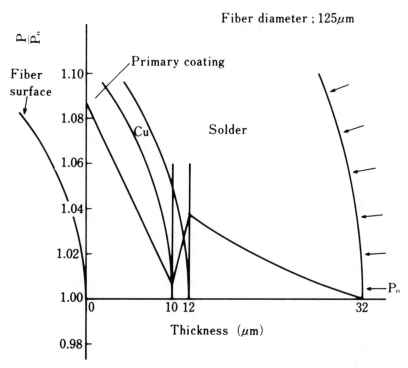

Fig. 3.   Inner stress distribution where thickness of primary coating, Cu and solder is 10, 2, and 20 μm, respectively.

vapor pressures, L is the length of the sealed area, a is the cross-sectional area of the primary coating, and t is the time.

Here, the thickness of primary coating is approximately 10 μm, the length of the sealing area is 5 mm, the feedthrough thus obtained is exposed to seawater for 25 years under the depth of 8,000 m, and the primary coating is applied with a material (silicone resin, etc.) having comparatively high permeability ($P = 10^4$ g·mil/m^2·24 h·atm). In this case, Q is 2.59×10^{-5} g, which causes a humidity rise of only 0.04 percent within the repeater housing. Actually, the feedthrough is exposed to seawater only for 1 to 2 months and therefore the amount of permeation can be neglected. The material of the primary coating gives little influence of the design of hermetic sealing part.

Design for Structural Strain

An optical fiber feedthrough suffers a variety of mechanical strains during or after fabrication [10]. This strain effects the transmission characteristic and mechanical strength of optical fiber and probably deteriorates a long-term reliability. Here, the structural design for suppressing strain is described in order to ensure a high reliability of optical fiber at the sealing part.

1) Strain by Temperature: An optical fiber, primary coating, and metallic coating serve in the composite condition as a metal-coated fiber. When the Young's modulus, cross-sectional area, and coefficient of thermal expansion of them are respectively represented by (E_1, A_1, α_1), (E_2, A_2, α_2) and (E_3, A_3, α_3), an equivalent coefficient αe of thermal expan-

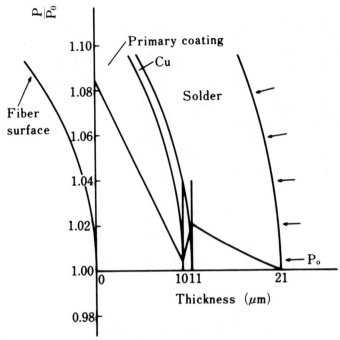

Fig. 4. Inner stress distribution where thickness of primary coating, Cu and solder is 10, 1, and 10 μm, respectively.

sion of metal-coated fiber is expressed as follows:

$$\alpha e = (E_1 A_1 \alpha_1 + E_2 A_2 \alpha_2 + E_3 A_3 \alpha_3)/(E_1 A_1 + E_2 A_2 + E_3 A_3). \tag{12}$$

Temperature Strain During Fabrication: A metal-coated fiber passes through a temperature rise process for sealing by solder and is then cooled. When solder is solidified in the cooling process, a metal-coated fiber is restricted to contract by solder and metal sleeve, resulting in residual strain ϵ_f after the cooling process. When a temperature difference between the solder solidified temperature and room temperature is represented by ΔT_1, ϵ_f can be expressed by the following equation:

$$\epsilon_f = \frac{\displaystyle\sum_{j=4}^{n} (\alpha_j - \alpha_e) E_j A_j}{\displaystyle\sum_{i=1}^{n} E_i A_i} \cdot \Delta T_1 \tag{13}$$

where, $(E_4, A_4, \alpha_4), \cdots, (E_n, A_n, \alpha_n)$ indicate the Young's modulus, cross-sectional area, and the coefficient of thermal expansion of the components (solder, metal sleeve, etc.) of feedthrough other than a metal-coated fiber.

Temperature Strain After Fabrication: A feedthrough which is fabricated generates internal strain due to temperature change ΔT_2. The condition just after fabrication is considered as the initial condition, strain ϵ_T due to temperature change ΔT_2 is expressed

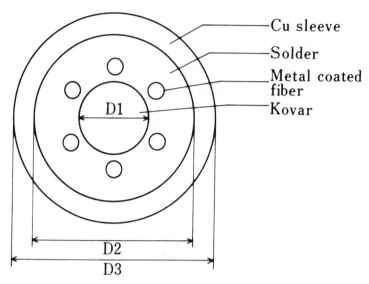

Fig. 5.   Hermetic sealing part cross section. Six fibers are sealed.

as follows:

$$\epsilon_T = \frac{\sum\limits_{i=1}^{3} E_i A_i \alpha_e + \sum\limits_{j=4}^{n} E_j A_j \alpha_j}{\sum\limits_{i=1}^{n} E_i A_i} \cdot \Delta T_2. \tag{14}$$

2) Calculation Results: It is effective to seal plural optical fibers within one hermetic sealing part for a reduction in size of the optical fiber feedthrough. Therefore, a cross-sectional structure of the sealing part is shown in Fig. 5. As a metal support provided at the center, Kovar having a small coefficient of thermal expansion is used, and as an outer sleeve, copper is used. In this case, a temperature strain during fabrication calculated by (13) is shown in Fig. 6.

Figure 6 shows that the larger the Kovar diameter is and the thinner the thickness of copper sleeve is, the smaller the temperature strain becomes.

Figure 7 shows an example of the calculation of strain from (14) when the temperature is lowered to $-10°C$ from $25°C$. This result shows the same as the tendency in strain generated during fabrication and such strain is about $1/4$ of that generated during fabrication.

Therefore, it is essential to suppress strain by enlarging the cross section of Kovar, by thinning the thickness of the copper sleeve, and by reducing the soldering area.

3) Strain by Pressure: When water pressure is applied on the feedthrough on a cable fault, ϵ_p applied on the fiber is expressed as follows when the water pressure is P:

$$\epsilon_p = \frac{\sum\limits_{i=1}^{n} A_i}{\sum\limits_{i=1}^{n} E_i A_i} \cdot P. \tag{15}$$

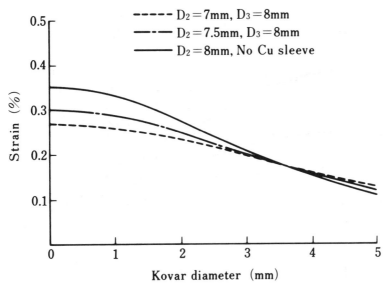

Fig. 6. Calculated curves of strain on manufacturing as a function of Kovar diameter.

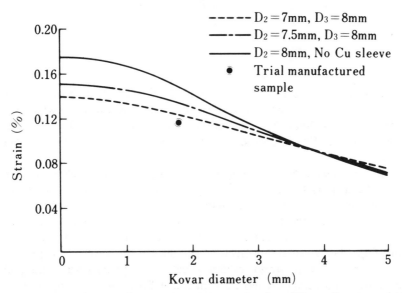

Fig. 7. Calculated curves of temperature strain as a function of Kovar diameter, and trial manufactured sample's strain is also shown.

4) Calculation Results: Figure 8 shows the calculation results of strain by pressure. The magnitude of strain is almost the same as that by the temperature generated during fabrication. According to this result, the strain becomes large when the copper sleeve is not used because solder is compressed to a large extent, but actually it is also true when the soldering area is wide even in the case when the copper sleeve is provided. Therefore, it is essential to thicken the Kovar and narrow the soldering area in order to keep the strain minimal.

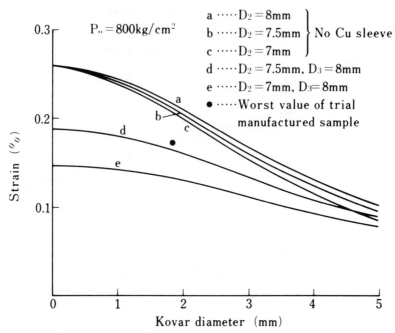

Fig. 8. Calculated curves of pressure strain as a function of Kovar diameter, and trial manufactured sample's strain is also shown.

Discussions

In a structure where a metal is coated around the primary coating of the optical fiber and solder is filled in the circumference of such a fiber, a feedthrough which sufficiently assures airtightness only with the solder in the thickness of about 20 μm can be realized and the solder coating of such a thickness can be easily realized with presoldering for the fiber.

It can also be understood that it is better to use a metal (Kovar, etc.) having a coefficient of thermal expansion which is similar to that of glass as the center support of the hermetic sealing part, to thicken the diameter, to arrange an optical fiber as closely as possible to such support, and to reduce the soldering area, for minimizing a temperature strain, and pressure strain to be applied on an optical fiber. Moreover, a proof-test level, for guaranteeing fiber strength for such strain during the system life period, can be determined, in order that the strain to be applied on an optical fiber of the sealing part can be calculated. Accordingly, a feedthrough which does not show any deterioration of strength for the system life period can be realized even when temperature change and pressure is applied on the feedthrough after fabrication, by executing the screening in such a level to the optical fiber of the sealing part before fabrication of feedthrough.

RESULTS OF MANUFACTURE ON TRIAL

Structure of Feedthrough Manufactured on Trial

Figure 9 shows structure of feedthrough manufactured on trial. In this case, a total of six optical fibers are sealed at a time. The solder is mounded for Kovar of the central support for the sealing.

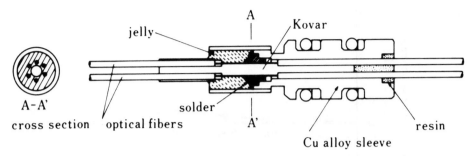

Fig. 9. Trial manufactured feedthrough structure.

TABLE II
LEAK-TEST RESULTS OF MANUFACTURED FEEDTHROUGH

Test	Result
He gas, 800 kg/cm^2	$< 3.0 \times 10^{-10}$ cc/s
Heat cycle	60°C $\leftrightarrow$ −10°C 5 cycles
He gas, 800 kg/cm^2	$< 5.0 \times 10^{-10}$ cc/s
Shock impression	50 G
He gas, 800 kg/cm^2	$< 1.0 \times 10^{-10}$ cc/s
Water, 800 kg/cm^2	more than a half year No leak

Test Results

Table II shows results of the air and water tightness test of a feedthrough manufactured on trial. As a result of leakage test under the He gas pressure up to 800 kg/cm^2, leakage was under the minimum detectable leak rate (10^{-10} cc/s). The airtightness was not deteriorated even after the heat cycle test was conducted within temperature range from

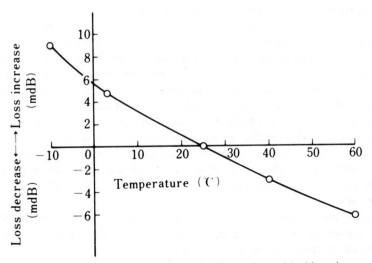

Fig. 10. Temperature-loss characteristics of manufactured feedthrough.

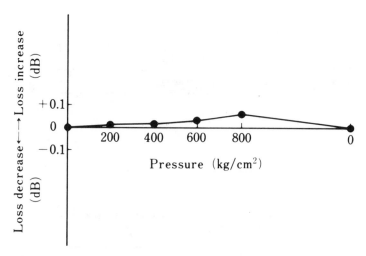

Fig. 11.   Pressure-loss characteristics of manufactured feedthrough.

60°C to −10°C and moreover the shock of 50 G in maximum was given. The watertightness was also confirmed for a long-term water pressure test of 6 months or longer.

Figure 10 shows change of loss in optical fiber for temperature change. The optical fiber showed change of loss as very small as 0.015 dB for temperature change from 60°C to −10°C.

Figure 11 shows change of loss in optical fiber in the case when a pressure is applied on the feedthrough. It was found that the maximum loss of 0.06 dB was generated when a water pressure of 800 kg/cm² was applied, but such a loss was returned to that before pressure was applied by reducing the pressure.

Discussions

It has been found that satisfactory airtightness and watertightness are maintained by the structure where plural optical fibers are sealed by soldering.

In the case of a structure manufactured on trial, the strain of fibers is calculated as shown in Figs. 7 and 8, but the strain by pressure is about 5 times than that of the strain by temperature. In comparison of loss between strains by pressure and temperature, strain by pressure shows a loss in the temperature range same as that for calculation, about 8 times than that of the strain by temperature. It means, considering that bending loss exponentially increases for bending radius, that a bending strain applied on fibers is also very little in order that the ratio of loss is not as large as that for the ratio of strain. It can also be understood from that an absolute value of loss is very small. As previously explained, this strain can be alleviated moreover by improvement in structure and pretension during fabrication. Therefore, it is obvious that an extremely reliable feedthrough can be realized from the point of view of ensuring the lifetime of the fiber at the sealing part.

As explained above, it was proven that a feedthrough manufactured on trial satisfied the basic characteristics, resulting in an outlook of realizing highly reliable optical feedthrough for an optical submarine repeater.

Summary

The authors of this chapter have proposed a structure where metallic coating is applied for the surface of optical fibers and such metal coating and external protection metal are hermetrically sealed by soldering, as an optical fiber hermetic sealing structure of feedthrough to be used in an optical submarine repeater housing. It is also theoretically shown that this feedthrough assures reliability of airtightness for a long period of time because the solder is filled at the circumference of individual optical fibers and thereby the solder filled shows the self-seal effect when pressure is applied and a high degree of airtightness is obtained.

Moreover, investigations about a strain to be applied on the hermetic sealing part proves that strain can be reduced by using a material having a coefficient of thermal expansion which is similar to that of glass such as Kovar as the support of fiber, which enlargens the cross section and reduces the cross section of soldered area.

According to the airtightness test of a feedthrough manufactured on trial, the amount of leakage for the He gas of 800 kg/cm^2 after the heat cycle (60°C − −10°C) and shock (50 G) was 10^{-10} cc/s or less.

In addition, an increment of loss for temperature change from 60°C to −10°C was 0.015 dB or less and an increment of loss of 0.06 dB or less was observed when a pressure of 800 kg/cm^2 was applied, however no residual loss could be found.

Realization of a feedthrough which is superior in airtightness and shows less strain provides a bright outlook for use in an optical submarine repeater.

Acknowledgment

The authors wish to thank Y. Niiro at the KDD Head Office for fruitful discussions. They also wish to thank Y. Iwamoto at KDD Laboratories for his encouragement.

References

[1] Y. Niiro, "Optical fiber submarine cable system development at KDD," *IEEE J. Selected Areas Commun.*, vol. SAC-1, no. 3, pp. 467–478, Apr. 1983.
[2] K. D. Fitchew, "Technology requirements for optical fiber submarine systems," this book, see ch. 2, p. 23.
[3] R. L. Williamson and M. Chown, "The NL1 submarine system," this book, see ch. 9, p. 119.
[4] P. K. Runge and P. R. Trischitta, "The SL undersea lightguide system," *IEEE J. Selected Areas Commun.*, vol. SAC-1, no. 3, pp. 459–466, Apr. 1983.
[5] G. Le Noane and M. Lenoir, "Submarine optical fiber cable development in France," ICC, June 1982.
[6] I. Yamashita, Y. Negishi, M. Nunokawa, and H. Wakabayashi, "The application of optical fibers in submarine cable systems," *Telecom. J.*, vol. 49-II, pp. 118–124, 1982.
[7] K. Miyauchi, Y. Nomura, and K. Fujisaki, "Consideration on the undersea optical transmission system," presented at ECOC, 1981.
[8] Y. Yamazaki, Y. Ejiri, K. Furusawa, and Y. Niiro, "Design and test results of an optical submarine repeater housing," Monograph Tech. Group Comm. Syst., IECE Japan, CS-81-60.
[9] K. Furusawa, Y. Ejiri, and Y. Yamazaki, "A study on repeater housing for use in optical fiber submarine cable systems," Monograph Tech. Group Commun. Syst., IECE Japan, CS-79-154.
[10] Y. Ejiri, Y. Yamazaki, and K. Furusawa, "A study on the structure of an optical fiber feedthrough for an optical submarine repeater," in *National Conv. Rec. IECE Japan*, 1982, p. 1880.

24
Optical-Fiber Feedthrough in Pressure Housing

SHIGENDO NISHI AND NOBUHARU TAKAHARA

INTRODUCTION

In order to put a submarine optical repeatered transmission system in practice, it is necessary to establish a repeater housing which protects the repeater circuits against the severe environment of the seabed. The most important components of a repeater housing are a thermally designed pressure housing, the cable coupling, and optical-fiber feedthrough. This chapter presents the design of an optical-fiber feedthrough and describes its optical and mechanical characteristics drawn from experiments.

REQUIREMENTS [1]

Figure 1 shows the current type of submarine optical repeater housing which can be laid on the seabed 8000 m deep. The function of the optical-fiber feedthrough is to provide an optical signal transmission path into the pressure housing. The fibers which connect the optical-fiber feedthrough and the cable coupling are protected from hydraulic pressure by a pressure proof pipe, so that the optical fibers in the optical-fiber feedthrough are not exposed to the hydraulic pressure. However, if the submarine optical cable is damaged and seawater intrudes into the pressure proof pipe of the cable, the optical-fiber feedthrough must withstand the full water pressure to protect the repeater circuits in the pressure housing. The highly reliable repeater circuits are required to function for a long time without maintenance, so they must be operated in low humidity conditions. Therefore, the optical-fiber feedthrough must be gas tight to prevent water vapor from intruding into the pressure housing. Additionally, low optical loss of the optical-fiber feedthrough is essential to allow for long repeater intervals. The requirements for the optical-fiber feedthrough are summarized in Table I.

DESIGN

Sealing Method Selection

The main technological subject of the optical-fiber feedthrough is the structure to seal the gap between the fibers and the hole for introducing them. The following sealing methods are considered: 1) resin seal; and 2) solder seal.

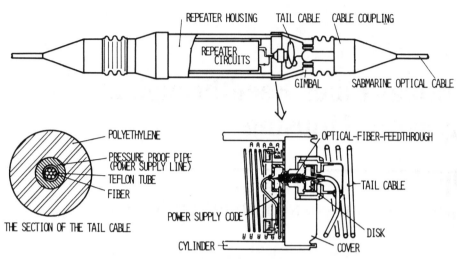

Fig. 1.   Current optical submarine repeater configuration.

TABLE I
REQUIREMENTS FOR THE OPTICAL-FIBER FEEDTHROUGH

Item	Content
Strength against Hydraulic Pressure	Above 800 Kg/cm²
Gastight	Not raising the relative humidity within the repeater housing by more than 20% over 25 years
Optical Loss	Under 0.1 dB for the temperature from −2 to 30°C
Life Time	25 years

1) Resin Seal [2], [3]: In this method, the fibers are fixed by epoxy resin. The adhesive strength of epoxy resin has proved to be enough to withstand water pressure up to 800 kg/cm², as shown in Fig. 2, and optical loss is low enough throughout the temperature range from −2 to 30°C.

The water vapor leakage rate $M(t)$ of the permeable part made of epoxy resin is given by the following equation:

$$M(t) = \text{Pe} \cdot A \cdot \Delta p / l \left(2(1-m)^{1/2} \cdot K(m)/\pi \right)^{1/2} \tag{1}$$

$$t = l^2 K(1-m)/(D \cdot K(m)) \tag{2}$$

where

D diffusion constant;
Pe   permeable constant of epoxy resin;
A section of permeable part;
l length of permeable part;
Δp difference of water-vapor pressure;
K elliptic function;
m parameter of elliptic function;
t time.

SAMPLE NO.	PARAMETER		PRESSURE (Kg/cm²)	RESULT
	θ(DEG.)	D(μm)		
1	5	128		
2				
3		150		
4				
5	10	128		
6			2000	NO WATER
7		150		LEAKAGE
8				
9	15	128		
10				
11		150		
12				

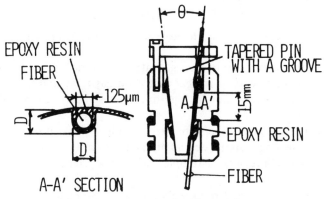

Fig. 2.   Resin seal strength against water pressure.

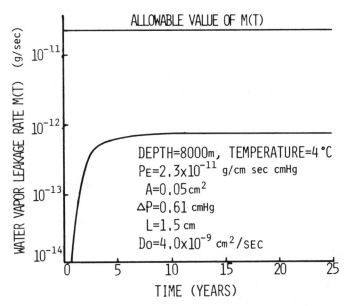

Fig. 3.   Water-vapor leakage rate of resin seal.

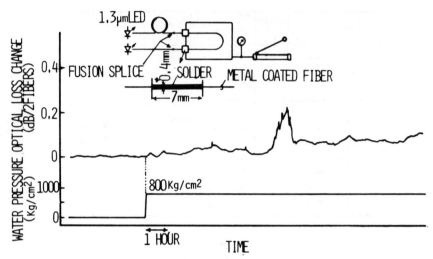

Fig. 4.   Optical loss charge for solder seal under high water pressure.

Figure 3 shows the water vapor leakage rate over time and the allowable value calculated from the requirements. It is clear that good gas-tight characteristics prevail. In this case, however, the concern lies in long-time adhesive-strength stability in a high humidity environment.

2) Solder Seal: In this method, fibers are fixed by solder. Here, it seems that there is no water leakage and no sealing material degradation. However, there is concern about solder creep under high water pressure. Examples of solder creep are shown in Figs. 4 and 5.

As mentioned above, solder and resin seals each have problems, but the problems are not of the same kind. Accordingly, we adopted joint use of resin and solder in the seal, so that the defects can compensate each other. In this case, the solder serves as a barrier against water vapor by remaining gas tight and so prevents degradation of the epoxy resin, and the epoxy resin supports the solder to avoid creep and withstand the high water pressure.

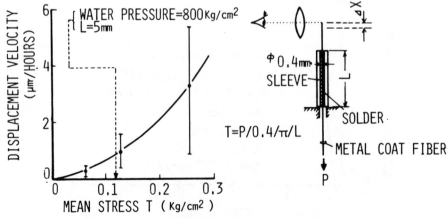

Fig. 5.   Creep deformation of solder seal caused by fiber tension.

(WATER-SIDE)   (REPEATER-SIDE)

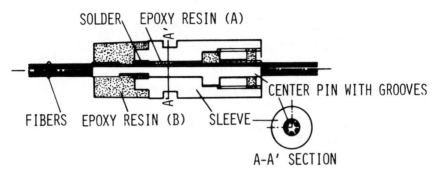

Fig. 6.   Structure of the optical-fiber feedthrough.

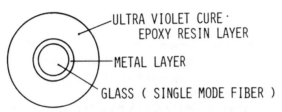

Fig. 7.   Cross section of a fiber.

Structure

The structure of the optical-fiber feedthrough can be seen in Fig. 6. Fibers are buried in the grooves of the center pin, which is inserted in the hole of the sleeve. The gap between the center pin and the sleeve is filled with solder, and the space inside the solder is filled with epoxy resin ((A) in Fig. 6). The outside surface of the solder is also covered with epoxy resin ((B) in Fig. 6). A section of the fiber is shown in Fig. 7. The metal coating is accomplished by dipping, and ultraviolet cured epoxy resin is coated over the metal

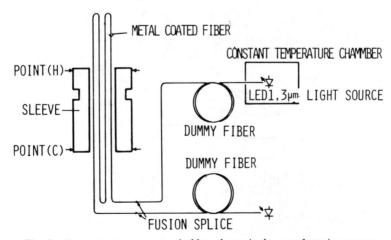

Fig. 8.   Apparatus to measure optical loss change in the manufacturing process.

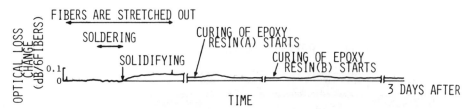

Fig. 9. Optical loss change in the manufacturing process.

coating during the fiber manufacturing process. The coat of ultraviolet cured epoxy resin is removed in the soldered area.

EXPERIMENTAL

Optical Characteristics

1) Optical Loss Increases in the Manufacturing Process: The optical loss change during the manufacturing process is measured with the apparatus shown in Fig. 8. The result is shown in Fig. 9. Maximum change of optical loss is 0.06 dB/6 fibers, and the final optical loss increase from the beginning is 0.03 dB/6 fibers.

In Fig. 8, the sleeve is heated at Point (H) by passing current for soldering, and cooled at Point (C) by water. The fibers are stretched out in order to avoid bending during soldering.

Figure 10 shows the loss change under various soldering conditions. From this experiment, the condition [1], i.e., the condition for Fig. 9, is found to be good.

2) Optical Loss Due to Temperature Change: The optical loss variation for temperature change can be examined in Fig. 11. As the temperature decreases, the optical loss increases. The rate of loss increase is steeper when the temperature is lower. The rate of variation is 0.06 dB/6 fibers/30°C.

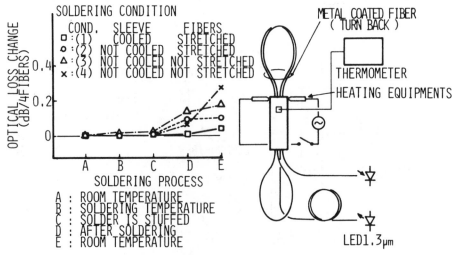

Fig. 10. Optical loss change under various soldering conditions.

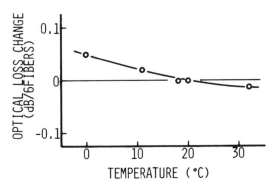

Fig. 11.   Optical loss versus temperature.

Mechanical Characteristics

1) Gas-Tight Test: High pressured helium gas was applied to the samples and the rate of the leaking gas was measured with a mass spectrometer.

In this test, the samples did not have epoxy resin (*A*) and (*B*) in order to shorten the test time because it takes a very long time for helium gas to permeate the epoxy resin. The result of the test is that in all samples ($N = 8$), the rate of leaking was below the sensitivity of the mass spectrometer. It is evident from this result, that there was no water vapor leakage caused by soldering imperfections.

2) Pressure Proof Test: The sample of the optical-fiber feedthrough was placed in the bulkhead of a pressure vessel, and high pressure water was forced into the vessel. Water leakage was checked under a water pressure of 800 kg/cm^2.

The result was that no samples ($N = 8$) showed water leakage, so they proved to have adequate strength against high water pressure.

3) Heat Cycle Test: While applying heat cycles to the sample of the optical-fiber feedthrough, the optical loss changes were observed. In the early cycles, as shown in Fig. 12, the optical loss pattern changed a little, but after that, a stable loss pattern was repeated.

4) Tensile Strength Test: Considering that tension may be applied to the fibers in the feedthrough during assembly, $T - T$ fiber joint, and so on, it is necessary to evaluate the tensile strength of the optical-fiber feedthrough.

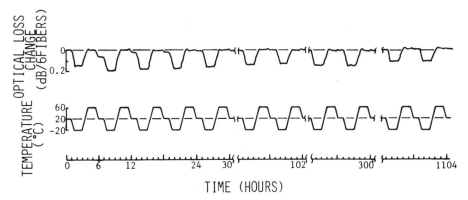

Fig. 12.   Optical loss change under heat cycle.

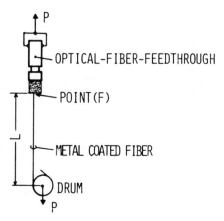

Fig. 13. Apparatus for tensile strength test.

The apparatus for the tensile strength test is shown in Fig. 13, and the results are shown in Fig. 14. In all samples, fiber failure occurred, and failure points are concentrated at Point (F), shown in Fig. 13.

Compared to the original strength, the metal coated fibers of the optical-fiber feedthrough become weaker, but they are still about three times stronger than in the case of the optical-fiber feedthrough with polymer-coated fibers.

CONSIDERATION

In the previous section, it became evident that the type of optical-fiber feedthrough having a structure using resin and solder jointly shows good enough characteristics to satisfy the requirements.

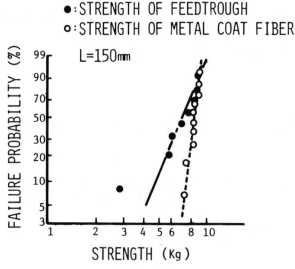

Fig. 14. Tensile strength of the optical-fiber feedthrough.

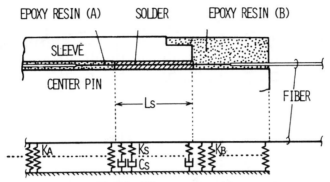

Fig. 15. Fiber support model.

In this section, we consider the following points concerning the results of the previously mentioned experiments

1) the mechanism of optical loss increase;
2) methods to minimize the optical loss increase during soldering; and
3) fiber strength decrease when it is assembled into the optical-fiber feedthrough.

The Mechanism of Optical Loss Increase

The optical loss increase of the optical-fiber feedthrough is caused by the bending of the fiber. The curvature of the bending can be presumed to become large when the center pin, in which fibers are buried, contracts with cold. In order to analyze the mechanism of optical loss increase, we assume that the support of the fiber in the optical-fiber feedthrough can be described by the model, as shown in Fig. 15. For temperatures from −2 to 30°C, fiber buckling does not happen because the axial compression force of the fiber is lower than the critical buckling force. However, if there is initial bending deformation, the fiber curvature becomes larger when the temperature becomes lower. As for the fiber inside the epoxy resin (A) or (B), there is no reason why initial deformation should occur in the manufacturing process. But, in the solder, initial fiber deformation can occur when the solder is solidified. If no counterplan is taken, there is a strong possibility that the fiber will be deformed by a nonuniform force applied to the fiber, as shown in Fig. 16.

In addition to this, the curvature of the fiber probably becomes larger due to the creep deformation of the solder, as shown in Fig. 17, and the optical loss increases as the curvature increases.

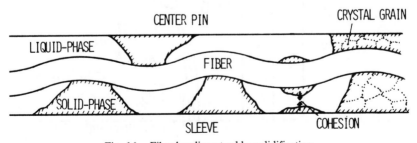

Fig. 16. Fiber bending at solder solidification.

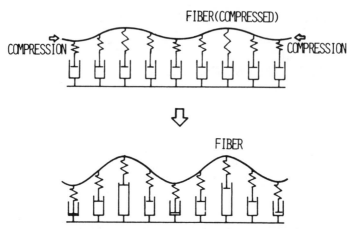

Fig. 17.   Optical loss increase caused by solder creep deformation.

Methods to Minimize the Optical Loss Increase During Soldering

As is mentioned in the previous section, we present the following two methods to minimize the optical loss increase in the process of soldering: 1) fiber stretching; and 2) sleeve cooling.

1) Fiber stretching wherein a small tension is applied, the fiber is forcibly prevented from bending. However, fiber bending can not be completely avoided by this method only.

2) Sleeve cooling is the method used for the following purposes:

a) to make the solidifying point move from the cooled side (repeater side) to the heated side (water side) and to prevent a void caused by volume shrinkage; and

b) to minimize the solder length aberration to below the critical value.

Figure 18 shows the temperature distribution calculated by the following cooling fin equation:

$$T = Te + (To - Te) \cdot \cos hm(1 - x)/\cos hml$$

$$m = (h \cdot Lp/(k \cdot A))^{1/2} \tag{3}$$

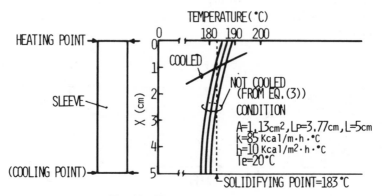

Fig. 18.   Temperature distribution.

where

    T the temperature at x,
    Te the room temperature,
    To the temperature at the heated point,
    A the section of the sleeve and pin,
    Lp the circumference of the sleeve section,
    h heat transfer coefficient,
    k heat conductivity.

It is evident that as the temperature gradient becomes larger, the solder length aberration becomes smaller for the heating point temperature error.

Decrease in Fiber Strength When it is Assembled into the Optical-Fiber Feedthrough

When manufacturing an optical-fiber feedthrough using fibers whose primary coat is a polymer such as silicone, it is necessary to remove the primary coat in the sealing area in order to accomplish a perfectly gas-tight seal. Therefore, small flaws appear on the glass fiber surface from the process of removing the primary coat, and the fiber is weakened.

However, in the case of using a metal-coated fiber, it is not necessary to remove the primary metal coat, and no such flaws appear. Although, the metal-coated fiber assembled into the optical-fiber feedthrough seems to be weaker than its original strength, as shown in Fig. 14. The reason for this fiber strength decrease is worth consideration.

Fiber failure concentrates at Point (F), as shown in Fig. 13. Probably, this is concerned with stress concentration. We assume that the fiber comes to the failure through the following stages, also shown in Fig. 19.

First Stage: The ultraviolet curved epoxy resin layer is cracked.
Second Stage: Almost all tension is applied on the glass section.
Third Stage: L Glass fails.

Therefore, the ratio of the tension when the fiber fails is given as follows:

$$R = T\mathrm{oft}/T\mathrm{fiber} = Ag \cdot Eg / (Ag \cdot Eg + Ae \cdot Ee) \qquad (4)$$

where

    Toft   strength of the fiber assembled into the optical-fiber feedthrough;
    T fiber original strength of the fiber;

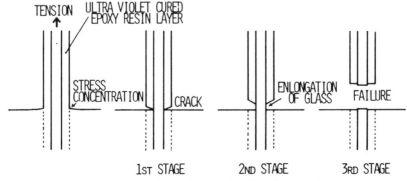

Fig. 19.   Fiber failure in the optical-fiber feedthrough.

Ag   cross section of the glass;
Eg   Young's modulus of the glass;
Ae   section of ultraviolet cured epoxy resin layer;
Ee   Young's modulus of ultraviolet cured epoxy resin.

For the values of $Eg = 7300$ kg/cm^2, $Ee = 150$ kg/cm^2, $Ag = 0.000123$ cm^2, $Ae = 0.00134$ cm^2

$$R \doteq 0.82.$$

This value almost coincides with the ratio of the mean strength calculated from the result of the strength test shown in Fig. 14. Therefore, it can be presumed that the strength of the glass fiber does not decrease by assembling it into the optical-fiber feedthrough.

CONCLUSION

To accomplish the structure for an optical-fiber feedthrough that satisfies the requirements induced by a severe undersea environment, we adopted a structure using resin and solder, so that their defects can compensate each other.

From experiments on the optical and mechanical characteristics, we obtained results that agreed with those expected.

ACKNOWLEDGMENT

The authors are grateful to Dr. E. Iwahasi, Dr. S. Shimada, K. Fujisaki, and H. Fukinuki for their encouragement and their effective suggestions, and the authors wish to thank I. Kitazawa and H. Okamura for their helpful advice.

REFERENCES

[1] S. Nishi, S. Nakamura, and H. Okamura, "A design of submarine optical repeater housing," *Monograph Tech. Group Commun. Syst. IECE Japan*, CS83-113.
[2] S. Nishi and Y. Kawakami, "A study on the undersea optical feedthrough using tapered pin with grooves," *Monograph Tech. Group Commun. Syst. IECE Japan*, CS81-59.
[3] Y. Kawakami, S. Nishi, and I. Kitazawa, "Studies on feedthrough applied to submarine optical repeater," *Monograph Tech. Group Commun. Syst. IECE Japan*, CS79-155.

25
Physical Design of the SL Branching Repeater

MICHAEL W. PERRY, GORDON A. REINOLD, AND
PAUL A. YEISLEY

INTRODUCTION

We have designed a branching repeater for the SL fiber-optic undersea cable system which will be used in the Transatlantic Cable #8 (TAT 8) to be placed into service in mid 1988. The branching repeater will connect a cable from the United States with branches going to the United Kingdom and France. It will be installed off the shores of the United Kingdom and France and is expected

1) to be capable of switching the high-voltage power so that the transatlantic circuits can be restored in the event of a cable failure in either branch;
2) to be capable of remote reconfiguration of the internal interconnections to allow for such contingencies as

   a) rerouting transatlantic traffic to one branch leg in case of failure of the other,
   b) loop-back to allow location of faults,
   c) switching to allow the 1 for 2 SL scheme of standby line, and
   d) easier testing during installation of the branching repeater and associated links; and

3) to provide regeneration for the three cable spans which enter the device.

Throughout this repeater we are using as much SL line repeater hardware as possible to minimize design effort. This concept has resulted in the block diagram which is shown as Fig. 1.

If we examine this figure, we find relays K1, K2, and K3 which provide the necessary high-voltage reconfiguration capability. These relays operate appropriately in a logic sequence controlled by the order in which the branch legs are powered.

Additionally, we find two power supply circuits and nine regenerators. These circuits are the same as those used in the repeater. Since 1.6 A is not sufficient to power all of these we have mounted six regenerators, and their related power supply and supervisory circuit on a separate chassis to be powered as if they were a line repeater. The remaining hardware is mounted on another chassis (insulated from the first) and is powered in series with the above.

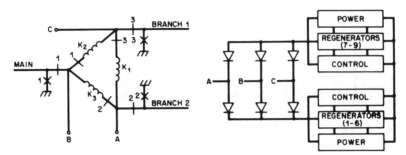

Fig. 1.   SL branching repeater block diagram.

Finally, we find the supervisory circuits which, although they use the same hardware as the repeater, will differ from them in the application of jumpers and unequipped positions as appropriate. The resultant physical package is shown in Fig. 2.

POWERING OF THE UNDERSEA SYSTEM

The scheme of powering an undersea cable system is to supply a well-controlled direct current to the conductor of the cable. Each repeater is connected in series to the conductor and develops the necessary biasing voltages through, in the case of analog systems, power separation filters, and in the case of the SL system, zener power diodes. To accomplish this, a dc voltage of up to 7.5 kV (the actual voltage is dependent on overall system length, earth potential, and the cable and repeater resistances) is impressed on the conductor at one shore station and up to −7.5 kV at the other.

In shorter systems it is possible to operate the system (by independently varying the terminal voltages, while maintaining system current) with any point in the system at

Fig. 2.   SL branching repeater network.

sea-ground potential. For transatlantic length systems, however, this becomes impossible because the necessary powering voltages become too high. We are, therefore, unable to run our system with the branching repeater at a zero voltage point. It, therefore, must have the capability of receiving the system high-voltage power.

As previously mentioned, it is necessary to be able to remotely switch this repeater so that powering can occur from either branch. It is also necessary that the relay which accomplishes this be capable of handling voltages of up to 7.5 kV.

Power Switching Scheme

If we examine Fig. 1 closely, we can find relay coils K1, K2, and K3. These coils are connected to the branch 1, branch 2, and main cable power conductors. For sake of illustration, we bring up the power on the main cable and branch 1 approximately simultaneously. Current starts to flow through the coil of relay 2 (K2). When the current reaches approximately 100 mA, the relay operates and the contacts labeled 2 change state. In this state the power stations on the main cable and branch 1 have taken control of the transatlantic powering, and branch 2 has been switched to a local sea ground at the branching repeater. Branch 2 can now be powered from its shore terminal. If we had powered up the main station and branch 2 first, they would have similarly taken control and branch 1 would have been connected to the sea ground.

With this arrangement then, we can power the transatlantic portion of the system through either branch depending solely on the order in which we turn on the power. Additionally, once the transatlantic powering link is established, the other branch can be powered independently of the other two. This becomes important when considering the polarity of the current flowing in the sea grounds if branching repeaters are cascaded (a consideration for potential future applications of this device).

Sea Electrodes

In the design of a sea ground, there are two basic considerations. First, it is necessary to assure ourself that over the design life of the system (25 years), the electrode will continue to maintain the proper resistance to ground and the necessary physical integrity. Second, it must not produce any by-products which are harmful to the cable system.

To achieve immunity from potential by-products there are two possible approaches. First we can remove the electrode to a location where the by-products do no damage. Second we can assure ourselves that the by-products do no damage to the cable materials. If we remove the electrode to a safe distance it becomes necessary to handle a fourth cable, thereby complicating the installation process. Additionally, the fourth cable must be protected from the same by-products. These complications led us to prefer a grounding system which would allow the ground to be near the branching repeater. It remained necessary to show that the by-products would do no damage to the cable system.

Because of our method of system powering, there are only two types of sea electrodes which can be used.[1] First the sea ground could be an anode, i.e., system current could be of a polarity as to leave from the anode to the ocean. Secondly it could be a cathode, i.e.,

---

[1] Some developers have produced systems in which the branch repeaters can operate using one half the system current. In this case, the branching repeater operates without a switch or a sea ground. We have chosen not to do this because it forces the repeaters in an unfailed branch to accommodate twice their rated current should the other branch fail, and the scheme is no longer usable in the case of cascaded branching repeaters.

system current enters the cathode from the ocean. As we will see, the effects of the system current are quite different at the two types of electrodes.

1) Anodes: At an anode in seawater an electrochemical reaction takes place which results in the formation of chlorine gas, oxygen, and hydrogen ions. This makes the immediate vicinity of an anode acidic and strongly oxidizing. In contact with such a solution, most polymers will be degraded and most metals will be corroded. These anode products will of course be dispersed and diluted in the ocean but over long times will be hazardous to the cable and repeater. Although it is possible to separate the anode to reduce the threat of such damage, this separation produces the difficulty of installation of a fourth cable. It is also necessary to attach the sea-ground cable to this anode and assure that it will remain attached to the cable with the proper resistance for the life of the system. Since most polymers are attacked by the reactions at the anode, and since those which are not attacked are not suitable for the manufacture of cable feedthroughs, it is apparent that anodic sea grounds are undesirable for long-term reliability.

2) Cathodes: The cathodic electrochemical reaction is one which produces hydrogen gas and alkalinity. These do not pose a significant threat to the materials which make up the system.

Although a variety of materials could have been chosen for use as cathodes in establishing sea grounds, our considerations were limited to the copper-beryllium and copper-nickel-tin spinodal alloys used in the SL repeater. Both of these alloys are highly corrosion resistant in seawater and, when made cathodic they are virtually immune to attack. Additionally, they are not subject to hydrogen embrittlement, as is steel when made cathodic, nor are they subject to "cathodic" corrosion as are aluminum or lead. We have chosen to use the copper-beryllium alloy, since it is the one with which we are most familiar from previous use.

Although the polyethylene and copper-beryllium are not affected by the increased alkalinity expected near this cathode, it still is best for the cathode design to provide for the maximum dispersion of the products from critical items. To do this we provide insulation of the sea-ground lead from the sea until it is connected internally to the ground. Additionally, since it is mechanically desirable for the electrode to encircle the main cable leg near the repeater, we provide an insulating layer on the inner diameter of the ground so that only the outer surface becomes cathodic. The hydrogen gas and alkalinity are thus constrained to be on the outside, where they can be washed away easily by ocean currents.

In addition, hydrogen gas [1] can cause increased loss in optical fibers. Because this cathode is on the outside of the cable, and is isolated from the cable by a layer of copper which would require any hydrogen which formed to diffuse through two barrier layers of copper before reaching the fibers, our calculations indicate that this effect will be negligible over the life of this system.

Finally, our concern was over the known mechanisms [2] of calcareous deposit formation on cathodes in seawater. It is known that at low current densities, tightly adhering calcareous deposits form on the surface of materials in seawater. However, it has also been observed that at higher current densities these materials have a tendency to become "frothy" and flake off as being formed. It was originally feared that these deposits would increase the resistance of this ground to the sea, thereby affecting our ability to power the branch. However, we have now operated several grounds on previous AT&T Bell Laboratories designed underwater systems and found that no such problems have occurred. We conclude that the mechanism of formation at high current densities has kept the cathode "clean." Further experiments are planned to substantiate this finding.

Branching Repeater Assembly

Pressure Vessel

Due to the need to accommodate two cable entries on the branching end of this repeater, and because of our desire to use the same cable couplings as used on the repeater, it was necessary to increase the diameter of the pressure vessel. This increased size allowed for greater space in which to perform the necessary fiber splicing, jointing, and splicing. It also provides the necessary extra volume to apply a nine regenerator scheme for fiber routing and switching.

The branching repeater has the same high reliability and lifetime requirements as the line repeaters. Environmental requirements are identical except for the pressure depth, which was set a 3000 fathoms.

Copper–beryllium (1.65-percent Be, balance Cu) was chosen as the material for the pressure vessel. It was chosen for several reasons.

1) We have had many years of experience with the proper foundry practice for the use of the beryllium–copper material.
2) Tooling was available to make Direct Chill (DC) billets of adequate sizes.
3) The housing was of a large size which had never been made before with the Cu–Ni–Sn alloy chosen for the repeaters.
4) Only a few of these housings would be made for each system installation, causing it to be uneconomical to produce special tooling to make them.

Mounting hardware (screws, washers etc.) are made from the same spinodal alloy (10-percent Ni, 8-percent Sn, balance Cu) as the repeater because of its superior stress corrosion characteristics.

Figure 3 shows an SL branching repeater. The cones and sea-ground details are made by conventional sand casting techniques. Covers and the sea ground are 100-percent machined from DC billets which have been forged. The pressure hull will be 100-percent machined from a forged and back extruded DC billet. Rings are ring forged from DC billets which have been upset and pierced. All forgings are made from DC billets which have a minimum upset of 30-percent and a final grain size of less than 1.0 mm after heat treatment.

Epoxy Insulation

This repeater must maintain good heat transfer from its interior to the ocean. It also must maintain voltage isolation between sea and chassis grounds. To do this, an insulating layer of epoxy is located on the inside diameter of the pressure hull. An aluminum cylinder is shrunk fit into this assembly, similarly to the SL repeater [3]. This sandwich provides the necessary thermal transfer and voltage isolation required for 25-year life.

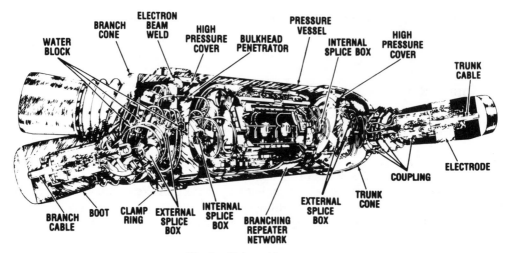

Fig. 3.   SL branching repeater.

Branching Repeater Network

The philosophy in the physical design of the branching repeater network is one of design commonality with the line repeater. This philosophy was chosen to minimize design and testing efforts through the use of proven building blocks whose fundamental life and environmental requirements are the same. The notable areas of difference between the repeater and the branching repeater are

a) accommodating a nine regenerator scheme, including the necessary signal switching to provide signal restoration and fault localization;
b) accommodating the power switching scheme; and
c) providing for serial powering of the internal units.

Description

The overall design of the branching repeater network has been broken down into three regenerator modules. The modules are common in that they are equipped with regenerators, whereas they differ in the other networks which comprise each of them.

One module operates functionally as a "half repeater." It consists of a regenerator unit, a power switching network assembly, a supervisory network assembly, and a dedicated power supply. It is electrically isolated from the other two modules.

The power switching network assembly consists of the high-voltage relays for "power switching."

The supervisory network assembly consists of the apparatus and hardware necessary to perform the supervisory functions of monitoring and control of various circuits.

The other two regenerator modules share a common supervisory network assembly and a common power supply. One module is equipped with the power supply and the other module is equipped with the supervisory network assembly.

The configuration (Fig. 4) is similar to that of the repeater, however, because the diameter was increased to accommodate two couplings, it's outside diameter is roughly 18 in. Comparably, the repeater's external diameter is roughly 13 in. Lengthwise, the branching repeater network is the same as the repeater unit.

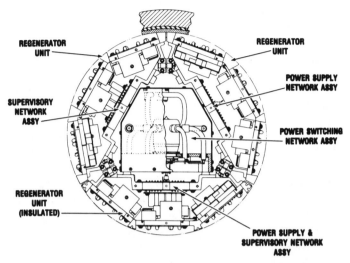

Fig. 4.   End view of the branching repeater network.

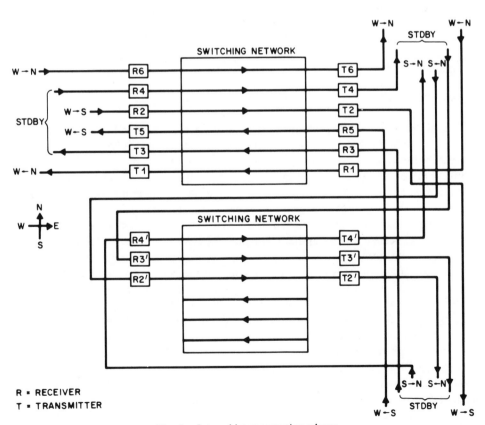

Fig. 5.   Internal interconnection scheme.

As shown, the design has been functionally partitioned into a) a basic transmission section which incorporates the regenerator units; b) a high voltage section which incorporates the power supply and supervisory network assemblies. As in the case of the repeater, the critical temperature sensitive items, such as the laser transmitter, are located such that their thermal impedance to the external ambient is minimized.

Nine-Regenerator Scheme

Our branching repeater scheme is designed under the assumption that each of the cables entering it should perform as a repeater span. This then, since in the most general case there will be two working and one standby pair in each cable, leads to the conclusion that nine regenerators are required.

1) Internal Switching Arrangement: It was our goal to pick a nominal interconnection scheme for this branching repeater which would allow as much flexibility as possible in its use. The scheme which appeared to give the most benefits is shown in Fig. 5. This scheme allows for remote loop-back of any fiber pair for fault localization. It also allows for the remote replacement of any failed span entering the branching repeater, and for the use of the regenerator switching networks used in the repeater.

2) Thermal Design: A thermal analysis has been made of the branching repeater. The analysis, which consisted of scaling the smaller diameter repeater to the expected diameter of the branching repeater, predicts that the thermal impedance will be roughly 0.1°C per Watt, from the regenerator chassis to the external ambient. By comparison, the repeater thermal impedance is 0.13°C per Watt. This results in an expected temperature rise on the regenerator chassis of approximately 7–8°C for 74-W dissipation (1.6 A and 46 V). The comparable temperature rise in the line repeater is approximately 4°C. Therefore, it is expected that (similarly to the repeater) the maximum device temperatures will stay well below critical limits.

Power Switching Relays

The capability of remotely switching system power in a repeater body has never been provided in undersea cable systems. As such, many different things had to be considered. Foremost in this regard was the selection of a mechanism for accomplishing power switching. For the SL system, a decision was made to use high-voltage relays.

1) Relay Requirement and Selection: Under normal operating conditions, the maximum system voltage in SL is not expected to exceed a magnitude of 7.5 kV. Likewise, the system current is expected to be 1.6 A. For all conceivable situations other than an open, the maximum expected voltage and current that the relays will be called upon to switch are 1.5 kV and 100 mA, respectively. In the case of an open, the relays might be called upon to switch surge voltages as high as 6 kV and surge currents as high as 200 A.

Based in part on these requirements, the relay initially chosen for this application was a four-pole, double-throw (4PDT) high-voltage vacuum relay. This particular relay was selected because of its prior usage in the power feed equipment of the SD, SF, and SG analog undersea cable systems. For these applications, the relays have an established reputation for high reliability.

Since only 2PDT relays are actually required, we are also evaluating 2PDT relays as candidates for power switching. Although both the 2PDT and the 4PDT relays had an established reputation in their previous applications, additional qualification tests had to be conducted for SL. This is in view of the more stringent environment and the higher reliability requirements.

2) Testing: Testing of the relays is taking place in three phases: a first phase to evaluate the mechanical and electrical performance of the 4PDT, a second phase involving further evaluations based on the first phase results, and a third phase involving evaluation of the 2PDT relays. Much of the first phase work has been completed. Work on the second and third phases is still in progress.

The testing was broken down into the following three categories:

1) destructive tests;
2) characterization and stress tests; and
3) long-term switching tests.

The purpose of the destructive tests was to establish design margins, with respect to the relays intended SL application, whereas the characterization and stress tests were designed to establish the in-service capability, integrity, and ruggedness of the relay. The long-term switching tests were designed to determine the switching behavior of the relay after long periods in the energized or unenergized position.

The results obtained thus far indicates that the 4PDT design is a suitable candidate for power switching int he branching repeater. However, in light of the fact that the additional poles are not needed, present testing is being restricted to qualifying it as a backup. At this stage, only preliminary testing has been performed to qualify the 2PDT relays. The initial results are promising, and inspection of the design suggests that because of greater clearances within the relay, it will have greater voltage-handling capabilities.

INSTALLATION TECHNIQUES

There are many ways to install such a branching repeater. We show one here as an illustration of the way one might be installed. Specific considerations might force the usage of other methods.

Since the branching repeater has three cable legs and weighs approximately three times that of a repeater, nonstandard shipboard handling and laying procedures must be used. For example, the laying procedure must allow for all three cable legs to be tested separately prior to joining into a system.

Reference to Fig. 6 shows that the branches can be laid separately and tested to and from the beach prior to buoying in a predetermined location. The trunk cable can be laid and tested through to its terminal (through the branching repeater) from the two legs aboard ship that are destined to be joined to their respective branches. After it has been determined that the trunk and branching repeater are satisfactory, the trunk can be buoyed off with the branching repeater and its three cable legs on the ship. The ship can then proceed to one branch, recover the buoy, and splice the branching repeater into that branch with a cable to cable splice. This leg can then be tested through to its terminal. After completing the testing, the ship can steam to the other branch buoy while paving out cable over the bow. At the buoy the branch cable can be brought on board and spliced into the branching repeater with a cable to cable splice. After satisfactory testing of this branch through the branching repeater to its terminal, the two branch legs can be paid out until there is tension on the trunk cable. With tension on the trunk the branching repeater can be placed on the bottom and the ship can proceed to the trunk buoy. On reaching the trunk buoy, the trunk is brought on board and the final cable-to-cable splice is made and slipped into the water.

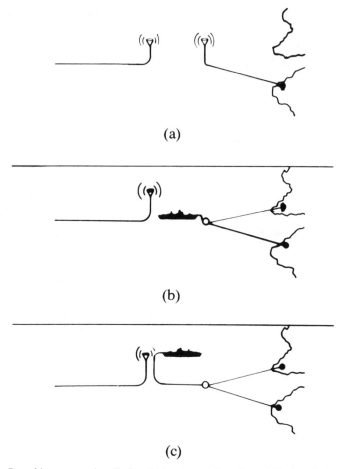

Fig. 6. Branching repeater installation. (a) Lay branch I, main could be installed as shown or laid continuously with branching repeater. (b) Lay branch II, join branch I to B.R, B.R. (c) Pick up end of main, join, test, lay final splice.

Conclusion

AT&T Bell Laboratories has designed a branching repeater for use in the SL ocean cable system. The existence of this branching repeater is made possible by the use of fiber cables which carry multiple transmission paths, and, therefore, can be routed to different destinations through switching networks in the repeater.

It is expected that the flexibility which results from the use of this hardware will result in an undersea networking capability which has been heretofore uneconomical to consider.

References

[1]   K. Mochizuki, Y. Namihira, and H. Yamamot, "Transmission loss increase in optical fibers due to hydrogen permeation," *Electron Lett.*, Sept. 1, 1983.
[2]   W. T. Hartt, C. H. Culberson, S. W. Smith, "Calcareous deposits on metal surfaces in sea water," in *Corrosion 83*, Paper no. 59.

Part VI
Undersea Electrical Components

Inside the repeater housing, electrical components receive, retime and regenerate the incoming optical signal. The high cost of deep sea repair of undersea systems place stringent reliability requirements on the electrical components inside the undersea repeaters. Five chapters in this part describe the integrated circuit technology used to design highly reliable undersea optoelectronic regenerations.

26

A Highly Integrated Regenerator for 295.6-Mbit/s Undersea Optical Transmission

DAVID G. ROSS, MEMBER, IEEE, ROBERT M. PASKI, MEMBER, IEEE, DAVID G. EHRENBERG, AND GLEN M. HOMSEY, MEMBER, IEEE

INTRODUCTION

Repeater circuitry for undersea systems must exhibit a unique combination of performance and reliability. The SL system, which in 1988 will constitute 85 percent of the first transoceanic application of digital lightwave technology [1], must operate at 295.6 Mbits/s over a design life of 25 years without adjustment. In this chapter, we discuss the design of the regenerators for the undersea repeater: the design goals, the special constraints imposed by the application, and the way they are reflected in the design methodology for each functional block. We also discuss the measured performance of a preproduction repeater, showing that the transmission quality meets our objectives.

DESIGN GOALS AND CONSTRAINTS

Reliability

Paramount among our concerns are those relating to reliability; these play a part in shaping every aspect of the regenerator design. Among them are: minimizing current consumption; providing the ability to individually test every regenerator function, including those embedded in integrated circuits; providing the ability to trim functions to their design center values; protecting the repeater against current surges induced by line faults; protecting the repeater against short-circuit failures in components; preventing loss of transmission in the event of a local clock failure; minimizing the number of active components; providing redundancy in components whose reliability has not yet been demonstrated; and providing the capability to monitor certain performance aspects to allow rapid localization of faults.

Performance

Consistent with reliability objectives, a prime goal is to achieve the best possible sensitivity, to minimize the number of repeaters. Dynamic range must be adequate to allow for manufacturing, and installation variations, and flexibility in repairing line faults.

Perhaps the most stringent performance requirements are those associated with timing jitter. Since a mistake in this area may require replacement of an entire system, every effort must be made to understand and control the mechanisms which govern jitter generation and accumulation [1].

OVERALL METHODOLOGY

Modular Construction

Our design strategy began with partitioning the regenerator into five major functional modules: the receiver, which converts the incoming light pulses to electrical pulses; the automatic gain control (AGC) amplifier, which adjusts the pulse amplitude and shape for best regeneration; the retiming circuit, which extracts a local clock from the data stream; the decision circuit, which uses the AGC and retiming outputs to reconstruct the data stream with precise amplitude and transition spacing; and the transmitter array, which consists of up to four transmitters, any one of which may be connected (through optical and electrical relays) to transmit the regenerated pulses on the output fiber. This partitioning allowed us to optimize each module separately and to then form a complete regenerator by interconnecting the modules with simple single-ended transmission lines and unfiltered power supply lines. A photograph and block diagram of the regenerator are given in Fig. 1.

Each module is constructed on a ceramic substrate using thin-film resistors, chip capacitors, and silicon integrated circuitry in a 28-pin leadless chip carrier. With the exception of the transmitter, each module uses a single custom IC. The modules are held and interconnected by a printed-wiring motherboard; this board provides common ground and power lines, stripline interconnects, and also carries a few leaded components for signal-path filtering.

The performance of the individual modules may be adjusted by laser-trimming thin-film resistors on the substrate or choosing ceramic chip capacitors. This allows us to improve reproducibility and operating margin, and in some cases, to tolerate wider manufacturing windows in the IC processing.

Powering

All the silicon circuitry is designed to operate from no more than 6.8 V above and below the local ground. Each module contains sufficient filtering on the hybrid integrated circuit (HIC) such that no failure in the module (not even a dead short) will cause the repeater to fail. This is done by ensuring that each functional subsection of the chip is separately powerable (this also aids in testing and trimming), by providing separate *RC* filtering on the power lines for each subsection (this also aids in suppression of spurious feedback) and finally by designing each subsection for minimum current drain (this also

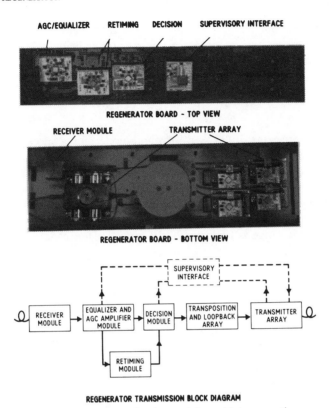

AGC/EQUALIZER   RETIMING   DECISION   SUPERVISORY INTERFACE

REGENERATOR BOARD - TOP VIEW

RECEIVER MODULE   TRANSMITTER ARRAY

REGENERATOR BOARD - BOTTOM VIEW

REGENERATOR TRANSMISSION BLOCK DIAGRAM

Fig. 1.   SL regenerator showing modules and interconnections.

aids in reducing line current). A detailed description of the powering arrangement is provided later in this paper.

Filtering

Consistent with our desire for simple interconnects, we made all signal-path filters and single-ended impedance-matched RLC bridged-T designs using on-HIC resistors and capacitors and motherboard-mounted inductors. The exception is the narrow-band filter in the retiming module where the surface-acoustic-wave (SAW) device represents the most reliable reproducible alternative at our baud frequency [2].

Monitoring

Critical parameters relating to the average light input, laser bias current, and regenerator error rate may be monitored in service from buffered outputs on the AGC, transmitter, and decision modules, respectively. Details on the use of these monitors in system maintenance are provided elsewhere [3].

DESIGN OF REGENERATOR MODULES

As we stated previously, the modular regenerator concept allowed us to develop the modules separately. In fact, the authors developed the modules of the electrical regenerator (the AGC, retiming, and decision modules) as well as the motherboard, while the receiver and the components of the transmitter array were developed by colleagues working to our interface specifications. In the discussion below, we will mention those specifications, where appropriate, but will refer the reader elsewhere for more detail on the modules.

Receiver

The receiver comprises an InGaAs p-i-n photodiode and a monolithic silicon trans-impedance preamplifier. Its hermetic package plugs into the regenerator motherboard. For any input optical power -34.2 dBm or greater, its output is at east 3 mV peak-to-peak, and its signal-to-noise ratio is good enough to allow regeneration with BER of 10^{-9} or better in the SL regenerator. The transmission passband characteristic is flat within ± 1 dB, with a high-frequency cutoff typically in excess of 250 MHz, allowing simple low-pass equalization. Dynamic range is typically 20-dB optical. The reader should see Snodgrass and Klinman [4] for further details.

AGC Module

We may treat the AGC module in two separate sections: the channel equalizer, which shapes the pulses, and the AGC amplifier, which adjusts their amplitude to a constant average value.

Channel Equalizer

The pulse shape we sought is of the Nyquist characteristic, with $\beta = 1$, which offers an excellent balance of sensitivity and jitter tolerance [5]. A close approximation to this characteristic is obtained by mating the electronics in the channel (transmitter, receiver, and AGC amplifier) with a two-section bridged-T low-pass equalizer. Each RLC section provides one pole and one zero. We may move the pole and zero loci by laser-trimming resistors, which yields a response adjustment of about 2 dB at the half-baud frequency. The resistors and capacitors in the filter reside on the HIC, and the inductors are lead mounted on the motherboard.

AGC Amplifier

Pulses leaving the receiver may have any amplitude from 3 mV to nearly 300 mV peak-to-peak. The wide range is due to manufacturing variations and changes with age in the transmitter, cable, and receiver. For optimum regeneration, we must present the decision and retiming modules with pulses of a uniform predictable level. Thus, we use an amplifier with AGC, which senses its average output level and adjusts its gain to hold that level constant. The design we used can vary its gain from about 150 to less than unity, holding the output level constant within 10 percent over a hundred fold variation in input level. The output level is set by laser trimming a thin-film resistor on the HIC, to a minimum of 300 mV peak-to-peak.

The AGC module also provides a monitor voltage proportional to the amplifier gain, yielding an indicator of the input signal power. The monitor is most sensitive at the critical low signal levels since it is proportional to the inverse of the incoming power.

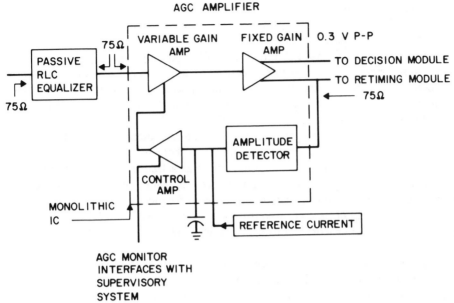

Fig. 2. AGC/equalizer module block diagram.

We incorporated all of the active circuitry into a silicon IC in a 28-pin chip carrier. The circuitry and the IC technology are described in detail elsewhere [6], [7]. A block diagram of the AGC circuitry is shown in Fig. 2. The signal enters a variable-gain amplifier and proceeds to a fixed-gain amplifier from which we take the two outputs. One of the outputs is sampled by a nonlinear device which charges a capacitor to a voltage proportional to the output pulse level. This is compared to a factory-trimmed reference to produce a control voltage for the variable-gain section, applied to the control amplifier, to complete the AGC loop.

Retiming Module

The pulses at the output of the AGC module contain negligible energy at the baud frequency, but it is possible, through a combination of linear filtering and nonlinear processing, to derive a high-quality local clock from them [5]. This is the function of the retiming module.

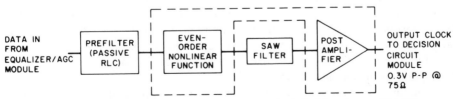

Fig. 3. Retiming module block diagram. Circuitry in dashed lines is monolithic IC.

A block diagram of the module is provided in Fig. 3. The circuitry comprises a passive prefilter, a nonlinear device, a narrow-band filter at the baud frequency, and a VHF amplifier; these are discussed below.

The prefilter is an RLC bridged-T high-pass filter, the purpose of which is to derive from the pulse stream a signal whose spectrum is approximately symmetric about the half-baud frequency. Such a signal has uniformly spaced zero crossings at the half-baud rate. The benefits of achieving precise spectral symmetry have been described in the literature [8], [9]. We have found that most of the benefit may be obtained by using a simple single-zero high pass to produce approximate symmetry close to the half baud. This filter is constructed with components similar to those in the channel filter.

The nonlinear device is an even function serving as a frequency doubler, producing at its output a signal having uniformly spaced zero crossings at the baud rate, with a spectrum approximately symmetrical about the baud frequency.

The narrow-band filter is a transversal SAW device with a Q of 800. The filter technology and the various considerations by which it was chosen are discussed elsewhere [10]–[12].

We configured the amplifier to provide more than 40-dB gain at the baud frequency to allow for insertion loss up to 15 dB in the SAW device.

As in the AGC module, the active circuitry is all included in a single integrated circuit. In fact, the high-speed amplifiers and nonlinear device on the IC used in the AGC module are quite suitable for the retiming application as well, so we have used the same chip in both modules—a significant economy.

Static phase of the output clock must be stable with temperature and time. The electronics and SAW device are designed to produce a projected stability within ± 0.25 rad in any condition [12]. To minimize the penalty due to expected drifts and to maximize tolerance to jitter, we must set the static timing phase at the time of manufacture to within ± 0.09 rad. The setting mechanism must be simple, absolutely stable, and adjustable over a full clock cycle. Custom-cut coaxial cable is the only technique which meets the requirements.

We place our most stringent manufacturing requirements on the phase jitter. We must control the jitter spectral density and especially the regenerator's phase transfer function [13] so that the overall system jitter will be within tolerable limits. The SL supervisory readback technique [3] imposes a further constraint on the minimum jitter transfer ratio at 26.4 kHz. These conditions are all satisfied by the typical retiming module.

Decision Module

The decision module's basic function is to reconstruct the data stream with precise amplitude and transition spacing using the pulse output of the AGC module and the local clock output of the retiming module. In designing this module, we had two goals: to minimize the margin penalty due to decision ambiguity [14], and to ensure that the data could be transmitted even in the event of a failure in the local clock. We also introduced circuitry which operates on the data to produce a separate output signal that is used to monitor the regenerator's error performance and for signaling on the parity channel [3]. A block diagram of the module is shown in Fig. 4.

The basic function is provided by a quantizer, which sets the amplitude of each pulse to a precise one or zero level; a delay flip-flop, which resets the transitions in response to the

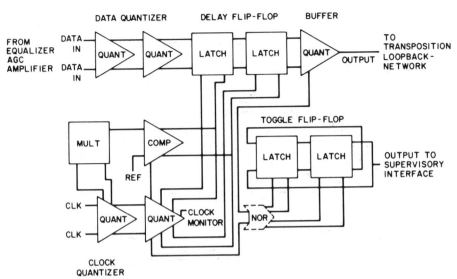

Fig. 4.   SL decision module block diagram.

clock circuitry; and an output buffer, which interfaces with the repeater's transposition and loopback network.

The clock circuitry receives the local clock from the retiming module and supplies it both to the failsafe circuit and, in quantized form, to the delay flip-flop.

The design of the logic circuitry allows us to achieve a very low ambiguity level of only 2 mV at 300 Mbits/s, yielding the most nearly ideal decision performance of any device we have seen.

The failsafe circuitry is a threshold detector which senses when the clock level has dropped below a preset level and disables the latches in the delay flip-flop, allowing the data to be passed to the output buffer with the pulses reshaped, but not retimed. The penalty for only reshaping is a small amount of additional jitter on the pulse transitions, which is tolerable in a single repeater span and very much preferred over a potential loss of transmission.

The error monitor and signaling circuitry comprises an NOR-gate NRZ-to-RZ converter and a toggle flip-flop modulo-2 divider, producing an output which allows the supervisory circuitry to detect variations from even parity in our 25-bit data blocks [3].

All of this active circuitry is included on a custom silicon integrated circuit, which is described in greater detail elsewhere [6].

Transmitter Array

The transmitter array consists of up to four transmitter modules [15], a network of electrical relays, and an optical relay [16]. The input of one of the transmitter modules is selected by the electrical relays to be connected to the loopback and transposition network. The output of that transmitter is selected by the optical relay and applied to the output fiber. All of this is done through the supervisory circuitry [3].

Each transmitter module contains an InGaAsP laser diode; a silicon integrated modulation circuit; an InGaAs p-i-n photodetector, which monitors the laser's backface

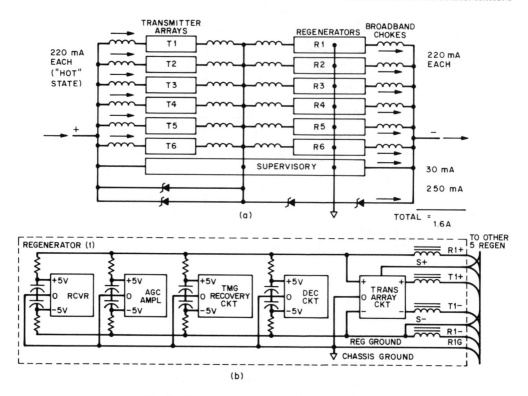

Fig. 5. Repeater and regenerator powering schemes.

output; and a silicon integrated circuit, which uses the backface monitor output in a feedback loop to hold the pulse output level constant. The bias circuit also provides a monitor output to the supervisory system; this output is an analog of the laser bias current, allowing an assessment of laser aging rate.

POWERING CONSIDERATIONS

Repeaters in an undersea system are powered by a dc current fed down the cable from the shore terminal. The local power supplies within each repeater are voltage sources derived from regulator diodes. The design of the SL repeater power supply reflects two important considerations. First, powering and supervision are done most simply and economically if all the regenerators are connected in parallel. Second, the laser bias currents vary widely at the beginning of life and also increase with time at various rates. The most flexibility in powering the lasers, with minimal impact on overall line current, is had by connecting the transmitters in series with their associated regenerators; this makes most of the regenerator current available for the laser. The powering plan for the SL regenerator is shown in Fig. 5(a). The efficient design of our integrated circuits allows each regenerator to operate on 220 mA; with 1.6-A line current, the regulating diodes have an ample 0.25-A regulation margin. The regenerator modules have sufficient resistive isolation in their supply lines [Fig. 5(b)] that a short circuit in any subsection will not exhaust the power supply's margin. Isolation between regenerators is done with series choke elements; no shunt capacitors are used since a short in such a device would cause repeater failure.

PERFORMANCE

A preproduction version of the SL repeater, comprising six regenerators, was tested with the following results:

		Worst	Best	Objective
Sensitivity	($\bar{P}$@10^{-9} BER)	-35.2	-36.1	-34.2 dBm
Overload	($\bar{P}$@10^{-9} BER)	-12.0 dBm	-9.0 dBm	-19 dBm
Jitter	rms	$1.5°$	$0.9°$	$3.0°$
	peak-peak	$15.9°$	$6.5°$	$30°$
	peak transfer gain	0.0 dB	0.0 dB	0.05 dB
Output Level		-2.0 dBm	0.0 dBm	-2.0 dBm

All tests were done with a $(2^{23} - 1)$ bit pseudorandom sequence.

Jitter accumulated through the six in cascade was 2.3° rms, as we would expect with no jitter peaking. The cascade would tolerate 178° peak-to-peak sinusoidal phase jitter at 200 kHz. Error-free operation through the cascade was sustained with line currents from the peak available current of 1.6 A down to 1.2 A. No evidence of interaction between the regenerators could be observed; that is, performance of the regenerators within the repeater environment, with the units operating simultaneously, was no different from that of the units operating independently on a laboratory bench.

Special conditions unique to undersea transmission, viz., superposition of a 25-Hz modulation on the power supply current and operation in the presence of high-voltage corona discharges in the repeater housing, also had no adverse effect on performance of the test regenerators.

CONCLUSION

The special requirements of undersea lightwave transmission, especially those of reliability, dictate a unique electrooptical regenerator design characterized by low part count and current consumption, fault tolerance, and monitoring capability. We have realized a design to meet the performance objectives for 295.6-Mbit/s transoceanic transmission using a modular approach. Each of the modules uses a single custom microwave integrated circuit (one module also uses a custom bias circuit), resulting in a simple robust construction. A preproduction version of the SL repeater, comprising six of these regenerators, has been tested and meets system performance goals. The ultimate test —the TAT-8 transatlantic system—is targeted for 1988.

ACKNOWLEDGMENT

We are indebted to our many colleagues at AT&T Bell Laboratories and AT&T Technologies for the provision of regenerator components, physical design of hardware, and assembly and testing of regenerators.

REFERENCES

[1] P. K. Runge and P. R. Trischitta, "The SL undersea lightwave system," this book, ch. 4, p. 51.
[2] T. R. Meeker and W. R. Grise, "Packaging and reliability of SAW filters," in *Proc. 1983 IEEE Ultrason. Symp.*, 1983.

[3] C. D. Anderson and D. L. Keller, "The SL supervisory system," this book, ch. 40, p. 567.

[4] M. L. Snodgrass and R. Klinman, "A high reliability, high sensitivity lightwave receiver for submarine cable applications," this book, ch. 33, p. 475.

[5] W. R. Bennett and J. R. Davey, *Data Transmissions*. New York: McGraw-Hill, 1965.

[6] D. G. Ross, R. M. Paski, D. G. Ehrenberg, S. F. Moyer, and W. H. Eckton, "Regenerator chip set for high speed digital transmission," in *Dig. IEEE Int. Conf. Solid-State Circuits*, 1984.

[7] ____, "Regenerator chip set for high speed digital transmission," to be published.

[8] E. Roza, "Analysis of phase-locked timing extraction circuits for pulse code modulation," *IEEE Trans. Commun.*, vol. COM-22, Sept. 1974.

[9] L. E. Franks and J. P. Bubrowski, "Statistical properties of timing jitter in a PAM timing recovery scheme," *IEEE Trans. Commun.*, vol. COM-22, July 1974.

[10] R. L. Rosenberg, D. G. Ross, and P. R. Trischitta, "SAW filter requirements for clock recovery in digital long haul optical fiber communications," in *Proc. IEEE Ultrason. Symp.*, Oct. 1981.

[11] R. L. Rosenberg, D. G. Ross, P. R. Trischitta, D. A. Fishman, and C. B. Armitage, "Optical fiber repeatered transmission systems utilizing SAW filters," *IEEE Trans. Sonics Ultrason.*, May 1983.

[12] R. L. Rosenberg, C. Chamzas, and D. A. Fishman, "Timing recovery with SAW transversal filters in the regenerators of undersea long haul fiber transmission systems," this book, ch. 28, p. 401.

[13] F. M. Gardner, *Phaselock Techniques*. New York: Wiley, 1966.

[14] P. Bylanski and D. G. W. Ingram, "Digital transmission systems," *Proc. IEE*, 1980.

[15] F. Bosch, G. M. Palmer, and C. B. Swan, "Compact 1.3 μm laser transmitter for the SL undersea lightwave system," this book, ch. 31, p. 445.

[16] S. Kaufman and R. L. Reynolds, "Optical switch for the SL submarine cable," this book, ch. 34, p. 487.

27

Design and Experimental Results of OS-280M Optical Submarine Repeater Circuits

HIROHARU WAKABAYASHI, KO-HICHI TATEKURA, MEMBER, IEEE, HIROAKI YANO, AND YASUHIKO NIIRO

INTRODUCTION

Optical fiber submarine cable systems are considered to be the international trunk line in the near future taking the place of existing coaxial cable systems because of advantages such as economy, large capacity, and digital transmission [1]–[3].

The OS-280M optical fiber submarine cable system, which provides a maximum of about 12 000 telephone channels, is now being developed. The optical submarine repeater is the key facility in a submarine system [4], [5]. In designing the repeater circuits, the following items are considered: 1) high reliability; 2) long repeater spacing; and 3) the supervisory system.

In this Chapter, design and experimental results of monolithic IC regenerating and supervisory circuits are described.

DESIGN PARAMETERS

Table I shows the design parameters of the OS-280M optical submarine repeater.

The information rate is 280 Mbit/s, which is a bundle of two CCITT 140 Mbit/s, and it corresponds to 3-, 780-, 640-kbit/s telephone channels based on the CCITT 2-Mbit/s primary multiplex.

The maximum system length is 8,000 km, which is capable of crossing the ocean.

An InGaAsP laser diode in the wavelength range of 1.30–1.32 μm and a 0.5-dB/km single-mode fiber are used to achieve a long repeater spacing of more than 50 km.

Table II shows a provisional loss budget of the OS-280M system. Nominal repeater gain is 33 dB, which includes the 8-dB margin. The margin is allocated to aging and some power penalties mainly due to the mode partition noise caused by unexpected degradation of the laser diode. A repair margin will not be allocated for the deep-water section.

TABLE I
DESIGN PARAMETERS OF THE OS-280M SYSTEM

Items	Design values
Transmission speed	295.6 Mbps
Transmission code	Scrambled Binary, NRZ
Wavelength	1.31 micro meter
Optical source	InGaAsP LD (DCPBH, VSB)
Optical detector	Ge-APD (100 µm∅)
Sensitivity (BER $10^{-''}$)	less than −37 dBm
Optical output power	more than 4 dBm
Jitter	less than 1 deg. (rms)
Maximum system length	8000 km
Feeding current	more than 1 A
Life time	25 years
Reliability	not more than 3 ship repairs in service life

Table III shows a provisional reliability budget for one regenerator circuit. In order to achieve a reliability such as no more than three ship repairs in 25 years, reliability as much as 21 fits/regenerator is required.

Although the reliability of the 1.3-μm laser diode has been improved rapidly, one cold standby laser seems to be necessary in order to put it into commercial use in the early stage. In the long haul system with as many as 150 repeaters, the required reliability is 1/30 of the nonredundant system.

DESIGN AND TEST RESULTS OF THE MONOLITHIC IC REGENERATOR

In the long haul optical submarine cable system, use of the monolithic IC for regenerator circuits is essential in order to achieve a high reliability and reduce the physical size and power consumption.

TABLE II
PROVISIONAL LOSS BUDGET OF THE OS-280M SYSTEM

repeater output power	more than −4 dBm
sensitivity (BER $10^{-''}$)	less than −37 dBm
nominal repeater gain	more than 33 dB
cable loss (including splice loss)	25 dB/repeater span
repeater aging	3 dB
cable laying effect	3 dB cable aging
others	2 dB
repeater spacing	more than 50 km

TABLE III
PROVISIONAL RELIABILITY BUDGET FOR THE REGENERATOR

Components	Reliability Objectives	Quantity	Total (Fits)
MIC	1 Fits	max 8	8
LD Module	30	2	1
APD Module	1	1	1
Optical SW	1	1	1
Optical SHT	1	1	1
Registors	0.01–0.02	150	2
Capacitor	0.02	100	2
SAW, MCF	0.5	2	1
Diodes	0.2	10	2
Others	—	—	2
Total	21 Fits/Regenerator		

When monolithic IC's are introduced to regenerator circuits, the following items should be considered.

1) The quantity and types of IC's to be used in one regenerator are as little as possible.
2) Power consumption and chip size are restricted for high reliability.
3) Function allocation of IC's.

Table IV shows the main function of monolithic IC's designed for OS-280M regenerator circuits. Each IC is fabricated by a Si-bipolar process which has abundant experience for high-speed computer use. The number of integrated active elements in one monolithic IC is 150–200 for MIC-1, -2, -3, and 4. Power consumption is less than 1 W and supply voltage is -6 V.

Figure 1 shows the block diagram of monolithic IC regenerator circuits.

TABLE IV
MONOLITHIC IC'S FOR THE OS-280M REGENERATOR

Names	Functions	
MIC-1	Equalizer Amp	Amplify, Equalizing AGC, DC level control
MIC-2	Decision	Decision Signal distribution
MIC-3	Timing	Clock extraction Limiting Signal-off detection
MIC-4	LD drive	LD drive, APC LD bias monitor
MIC-5-1	Supervisory	Command receive LD redundancy control Loopback control Monitoring
MIC-5-2	Supervisory	In-service error monitor
MIC-6	DC/DC conv.	Generation of APD bias

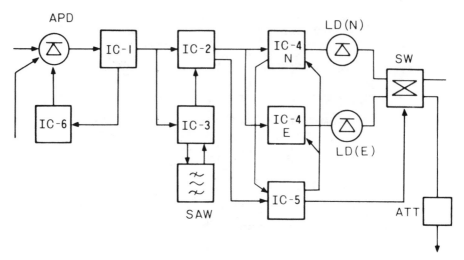

Fig. 1.   Block diagram of the monolithic IC regenerator.

Receiver Circuit

Ge-APD was selected for the optical detector because of the following reasons.

1) High sensitivity will be achieved in combination with the Si-bipolar front end.
2) It is widely used in other systems.
3) Wide dynamic range will be easily achieved.

The optimum avalanche gain of APD is about 12 dB in the case of standard APD characteristics. As the APD gain is adjusted to maintain the output pulse amplitude constant by controlling the biasing voltage, the pulse peak amplitude at the output of APD is 3.5 μA when the optical input power is from -38.5 to -26.5 dBm.

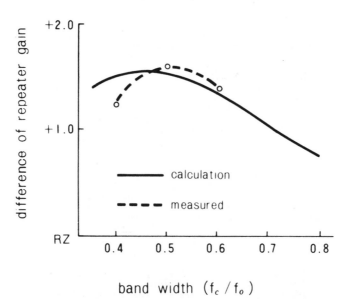

Fig. 2.   Difference of repeater gain between NRZ and RZ.

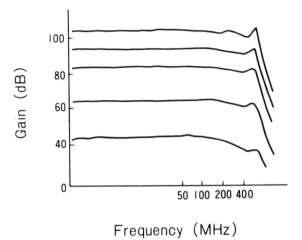

Fig. 3.   Gain-frequency characteristics of MIC-1.

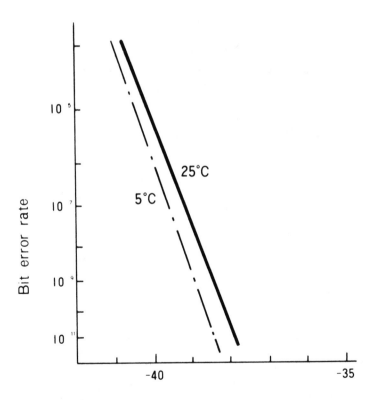

Fig. 4.   BER of the OS-280M regenerator.

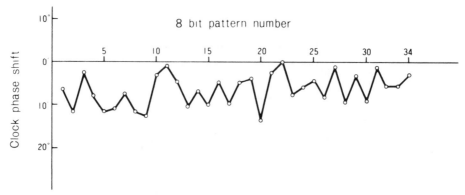

Fig. 5.   Clock phase shift to 8-bit repetitive patterns.

The necessary gain of the IC amplifier is about 100 dB in order to get the pulse peak voltage of 300 mV at the output of the amplifier. The required adjustable gain range of the AGC amplifier is +15 dB to achieve the wide optical dynamic range of more than 25 dB.

The NRZ format was adopted to reduce the required bandwidth in electronic circuits and to achieve longer repeater spacing. Figure 2 explains the theoretical and experimental difference of repeater gain between NRZ and RZ formats. By adopting the NRZ format, repeater gain increases by 1.5 dB at the optimum equalizing bandwidth.

Figure 3 shows the gain-frequency characteristics of MIC-1. Figure 4 shows the BER characteristics to the change of optical input power. The minimum average optical power at the BER 10^{-11} is about −38 dBm at 25°C, and it is improved about 0.5 dB at 5°C due to the reduction of APD dark current.

Timing Circuits

MIC-3 is a timing extraction circuit. A nonlinear circuit to generate the spectrum at clock frequency and a limiter amplifier are integrated in MIC-3. A SAW filter is used for reasons of sharp selectability and stability. The required Q is about 800.

Figure 5 shows the clock phase shift to the 34 8-bit repetitive patterns [6]. The peak-to-peak phase shift is 17°. The output jitter for the random input pattern will be estimated at about 0.5° rms from Fig. 5.

The signal off detection circuit is also equipped in MIC-3.

LD Driver Circuits

MIC-4 is an LD driver circuit. A high-speed current switch and an APC circuit are integrated in MIC-4. The maximum feeding pulse current to a laser diode is 30 mA to get the output peak power of 5 mW/facet.

DC-PBH and VSB lasers to be used in the OS-280M system have a low threshold current, as much as 20 mA. The maximum feeding biasing current is designed up to 60 mA, considering the degradation of laser diodes and the operation on the cable ship. APC adjusts the averaged optical output power constant against the change of temperature.

Figure 6 shows the example of the optical output pulse shape. The jitter at the pulse arising point is less than 100 ps and over/under shoot is small.

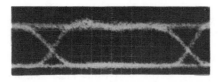

500ps/div

Fig. 6.   An example of the optical output pulse.

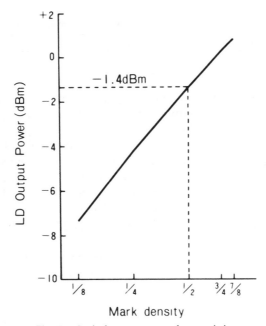

Fig. 7.   Optical output power characteristics.

Figure 7 shows the optical power at the pigtail of the laser diode to the change of mark density. The optical output power is −1.4 dBm when the mark density is 1/2. The extinction ratio will be estimated to be more than 15 dB from Fig. 7.

DESIGN AND EXPERIMENTAL RESULTS OF THE SUPERVISORY CIRCUIT

In the design of the supervisory circuit, the following items are considered:

1) control of the redundant laser diode without any service degradation;
2) accurate fault location within one repeater section for cable breaks and repeater faults;
3) in-service fault location for the intermittent bit errors;
4) level diagram monitoring for the system test.

Table V shows the supervisory items of the OS-280M system. The OS-280M system has two types of repeater supervisory systems, namely, in service and out of service.

TABLE V
SUPERVISORY ITEMS OF THE OS-280M SYSTEM

Items of Supervisory	In/Out service	Supervisory Work
Overall characteristics	IN	Parity, Frame synchronization, Signal off
Repeater characteristics	IN	Error monitoring
	OUT	1) error measurement by optical loopback 2) monitoring of in/output optical power
Laser diode	IN	Automatic switching
Switching	OUT	Station control

Redundant laser switching and error monitoring at each repeater are carried out by the in-service condition, and input/output power monitoring and optical loopback are operated by the out-of-service condition.

Figure 8 describes the block diagram of the supervisory circuit. MIC5-1 is designed to be used for out-of-service supervisory and MIC5-2 is for in-service supervisory. The supervisory circuit described in Fig. 8 is individually equipped to each regenerator circuit. Two supervisory monolithic IC's will be fabricated in the Si-bipolar process, and the circuits are the hybrid of low speed/high speed and analog/digital circuits. The integrated number of active elements in one monolithic IC is about 2000, and power consumption is less than 1 W.

Figure 9 summarizes the function of the OS-280M repeater supervisory system.

In this section, design considerations and test results of the experimental model are described.

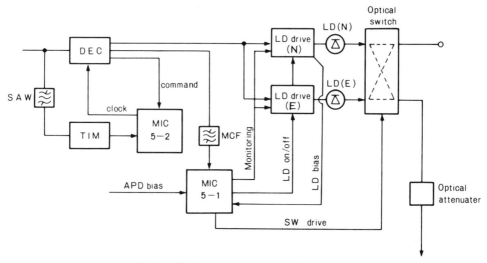

Fig. 8.   Block diagram of the supervisory circuit.

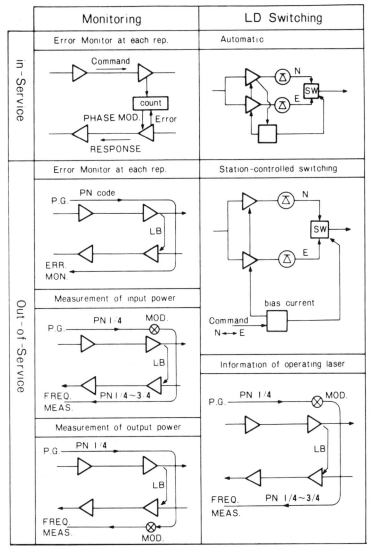

Fig. 9. Functions of the OS-280M repeater supervisory system.

Optical Loopback Circuit

In the OS-280M system, the optical loopback circuit is equipped for out-of-service fault location in which the optical output signal is loopbacked to the APD of the backward regenerator in the same subsystem. Adoption of the optical loopback circuit has some advantages as follows:

1) Accurate fault location is possible because the loopback is made between both ends of the regenerator.
2) As each regenerator has no electrical interface for loopback, it is possible to insulate all regenerator circuits accommodated in the same repeater housing. It makes it possible to reduce the system feeding current.

TABLE VI
CHARACTERISTICS OF THE OPTICAL SWITCH AND ATTENUATOR

Items	Optical SW	Optical ATT
Insertion Loss (25°C, at 1.3 μm)	$m = 1.1$ dB $\sigma = 0.2$ dB	23.5–36 dB
Loss change (5–40°C)	$m = 0.1$ dB	$m = 0.2$ dB $\sigma = 0.1$ dB
Cross talk	$m = 24$ dB $\sigma = 2$ dB	$\geqslant 45$ dB
Operating Power	1.0 V, 75 mA	—
Operating Speed	$m = 350$ μsec $\sigma = \partial\mathcal{J}$ μsec	—

3) Interface between each regenerator in the same subsystem is very simple.
4) It is possible to monitor the optical output power of the regenerator by use of APD biasing monitoring at the same time.

The optical loopback circuit consists of the optical switch, the optical attenuator, and the dual input APD module as shown in Figs. 1 and 8. In the figures, the optical switch is a 2×2 matrix type magnetooptics switch using YIG crystal, and it has the latching function of an electrical magnet.

The operation mode of the optical switch is reversely controlled, parallel or cross, by 1-ms pulse current with an amplitude of 100 mA. In the parallel mode, connection between LD(N) and the main fiber, and that between LD(N) and the loopback fiber are completed. In the cross mode, LD(N) and LD(E) are connected to the loopback fiber and the main fiber, respectively. In the normal condition, since either of two lasers operates, the optical signal does not appear at the loopback fiber. In the case that LD(E) is operating, the optical switch is controlled from the parallel to the cross for optical loopback, and in the case that LD(E) is operating, it is controlled from the cross to the parallel.

The optical signal is loopbacked to one port of the dual input APD module through the optical attenuator, which adjusts the loopback power level to be nominal.

Tables VI and VII show the characteristics of the experimental optical switch, the optical attenuator, and the dual input APD module. The insertion loss of the optical switch is less than 1.5 dB including the loss change due to temperature change over a range from 5 to 40°C. The drive current is so small that the switch is directly controlled by a monolithic IC. The quantum efficiency of two input ports of the APD module are almost the same and more than 70 percent.

TABLE VII
CHARACTERISTICS OF THE DUAL INPUT APD MODULE

Items	Port 1	Port 2
Quantum Efficiency (25°C, at 1.3 μm)	$m = 77.6\%$	$m = 77\%$
Change of Efficiency (5–35°C)	$m = 1\%$	$m = 1\%$
Dark Current	$m = 106$ nA	

sample number 30

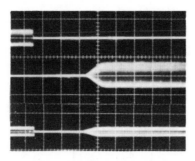

```
top;      output of LD(N)
middle;   output of LD(E)
bottom;   output of optical SW
holiz.;   200 us/div
```
Fig. 10. Optical output signals in laser switching.

Laser Redundancy Circuit

In the OS-280M system, one cold standby laser diode is equipped for redundancy. Two methods for laser switching are prepared: one of them is in-service automatic switching, and the other is out-of-service station controlled switching.

For the in-service automatic switching, biasing current to $LD(N)$ is continuously monitored. When the biasing current exceeds the predetermined threshold level, $LD(E)$ starts its operation instead of $LD(N)$ and the optical switch is controlled from the parallel to the cross.

The threshold level to decide the laser switching is designed to have the same temperature coefficient as that of the laser diode to prevent the malfunction due to temperature change. The threshold level is fixed at the value, up to which the laser diode degradations like pulsation, and broadening of the spectral width is statistically approved to be too small to have an effect on the transmission characteristics. The laser diode is preventatively changed over before the transmission degradation occurs.

As the in-service automatic laser switching system does not need the in-service laser monitoring at the station and in-service laser control from the station, control circuits in the repeater and the system maintenance at the station will be very simple.

A simple out-of-service station controlled switching is also equipped for the system test. It is controlled by the out-of-service 50-ms command signal from the shore end station.

Figure 10 shows the optical outputs of $LD(N)$, $LD(E)$, and the optical switch in the case that $LD(N)$ is switched to $LD(E)$.

Breaks due to laser switching are less than 1 ms.

In-Service Error Monitoring Circuit

When intermittent errors occur with a frequency and duration which do not meet the overall system objectives, fault location by the out-of-service loopback system will need the long time service break because the loopback circuit is open until the next errors occur.

The in-service error monitoring system is equipped which measures the bit errors at each repeater simultaneously.

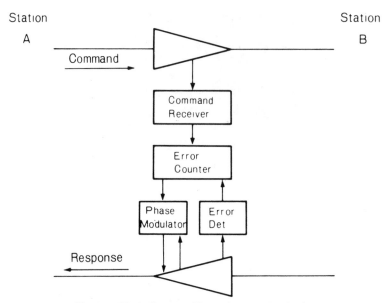

Fig. 11.   Block diagram of in-service error monitoring.

Figure 11 shows the block diagram of the in-service error monitoring system.

Every repeater starts and stops the error counting simultaneously according to the in-service command signals using the method of parity bit violation from Station A [7]. The readout information from the error counter is transmitted to Station A by the method of low-speed clock phase modulation on the transmission signal from Station B to A. Readout is sequentially made according to the in-service command to the individual repeater. It is possible to carry out the same operation by the command from Station B.

APD Bias Monitoring Circuit

The information of the input and output power of each repeater is useful for the system laying, commissioning, repairing, and fault location because it makes the system level diagram clear.

In the optical receiver circuit, biasing voltage of APD is proportional to the input optical power because the biasing voltage is controlled to maintain the detected pulse amplitude constant.

In the OS-280M system, the APD biasing voltage is converted into 300–700-kHz sinusoidal signal through a V/F converter. The 295.6-Mbit/s p-n signal from the shore end station is modulated by the above sinusoidal signal using the method of out-of-service mark density modulation. Optical output power is also measured by using APD biasing monitoring and optical loopback operation at the same time, as shown in Fig. 9.

Command Signal and Receiver Circuits

The OS-280M system has two types of command signals, namely, the out-of-service command signal and the in-service command signal. Table VIII shows the aspects of two command signals. Each signal has its own property: out-of-service command has a large

TABLE VIII
COMMAND SIGNALS OF THE OS-280M SYSTEM

	out-of-service	in-service
baseband format	same as shown in Fig. 12	
modulation	mark density modulation PN signal	periodical parity violation on signal
modulation frequency	10–17 MHz individual to repeater	26 kHz common to repeaters
property	1) high SNR 2) large immunity for bit errors	no effect on service signal transmission quality

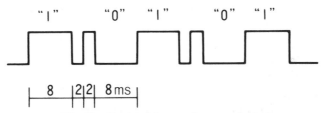

Fig. 12. Baseband format of command signals.

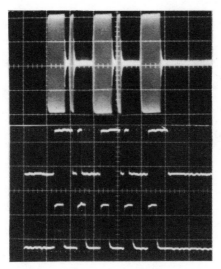

```
top    ; output of MCF
middle ; output of detecter
bottom ; generated timing
holiz. ; 10ms/div
```

Fig. 13. Operation of the receiver circuit.

immunity for bit errors, and in-service command does not have an effect on the transmission signal.

Two types of command signals have the same baseband format, asynchronous pulse-width modulation, as shown in Fig. 12. Receiver circuits consist of a rectifier, timing generator, and shift register. In order to prevent the circuits from malfunction due to the accidental disparity on transmission signals and bit errors, a timer circuit which limits the receiver operation time to less than 1 s is also equipped. Fig. 13 shows the operation of the receiver circuits.

Conclusion

Design and experimental results of the OS-280M repeater circuits are described. Experimental monolithic IC regenerator circuits were manufactured and evaluated. To the end of putting the long haul OS-280M system into commercial use, refining of the circuits and reliability tests of repeater components are being conducted.

Acknowledgment

The authors wish to thank Y. Ishikawa, Director of the Department of Submarine Cable Engineering, for his encouragement, and the NEC Corporation, Fujitsu Ltd. and the Nippon Telegraph and Telephone Public Corporation for their cooperation.

References

[1] Y. Niiro, "Optical fiber submarine cable system development at KDD," *IEEE J. Select. Areas Commun.*, vol. SAC-1, no. 3, Apr. 1983.
[2] R. L. Williamson and M. Chown, "The NL1 submarine system," this book, ch. 9, p. 119.
[3] P. K. Runge and P. R. Trischitta, "The SL undersea lightguide system," *IEEE J. Select. Areas Commun.*, vol. SAC-1, no. 3, Apr. 1983.
[4] Y. Niiro, H. Wakabayashi, and H. Tokiwa, "Design and an experimental result of the optical submarine repeater circuit," in *Proc. 7th ECOC*, 1981, paper 15.1.
[5] K. Amano *et al.*, "50 km, 2 repeaters sea trial of optical fiber submarine cable system," *Electron. Lett.*, vol. 18, no. 22, Oct. 1982.
[6] C. J. Byrne *et al.*, "Systematic jitter in a chain of digital regenerators," *Bell Syst. Tech. J.*, vol. 42, Nov. 1963.
[7] CCITT contribution, "COM XVIII-NO. 59-E," Aug. 1977.

28

Timing Recovery with SAW Transversal Filters in the Regenerators of Undersea Long-Haul Fiber Transmission Systems

ROBERT L. ROSENBERG, SENIOR MEMBER, IEEE, CHRISTODOULOS CHAMZAS, MEMBER, IEEE, AND DANIEL A. FISHMAN

INTRODUCTION

As with other aspects of transoceanic fiber transmission systems, ongoing development has fostered a growing sophistication in the regenerator timing-recovery loop. A broad-brush study of repeatered systems with surface-acoustic-wave (SAW) retiming filters, reported about two years ago [1], has already become dated. Progress since that time on three important topics is summarized here. Two of the topics came to the fore when a firm decision was made in favor of transversal SAW filters for the transatlantic cable. Such filters are notorious for the difficulty of eliminating passband ripple. The effects of ripple on timing jitter and alignment jitter have been characterized analytically [2], with results that are somewhat alarming under worst-case conditions. Fortunately, the worst-case conditions are substantially softened [3] by natural variations in the passband shapes of manufactured filters. The jitter situation is described in the section "Passband Ripple Effects on Jitter". Another notable consequence of using the transversal type of filter is that its effective "ringing" time is much longer than the ringing time of a minimum-phase (resonator) filter with the same 3-dB fractional bandwidth ("$1/Q$"). The reasons and the system implications are discussed in the section "Transversal Filter 'Ringing' Behavior".

Finally, the moderate temperature variability of SAW filter characteristics on quartz has proved to be more useful than the absolute temperature stability that has been pursued for many years. The temperature variation of filter phase can be employed to compensate much of the phase-temperature variation of the timing-loop electronics [4]. As a result, the thermal offset of the decision point from its optimum location in the eye diagram can be well controlled over the required temperature range. The compensation scheme is described in the section "Thermal Effects in SAW Retiming" where we also assess the cost, an increase in static frequency detuning. Results are summarized in the section "Summary".

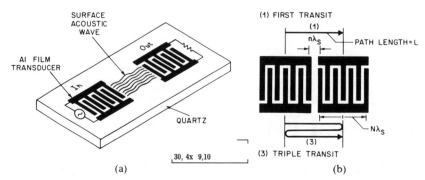

Fig. 1. (a) Basic structure of a transversal filter. The electrical source and load are shown schematically to indicate standard connections to the bus-bar pairs of the transducers. (b) Illustration of triple-transit reflections, which interfere coherently with simultaneously received first transits. For narrow-band retiming filters, the number of periods N_T per transducer is often several hundred. λ_s is the wavelength at the "synchronous frequency" where λ_s matches the transducer period. A quarter-wave change in the transducer separation $n\lambda_s$ can convert a dip to a peak at passband center [2]. Comparable phase changes over the propagation path length L may occur through random variations of λ_s caused by fabrication errors.

To establish terminology and simplify later discussion, we begin with a basic picture of a SAW transversal filter in the next section.

A BASIC SAW TRANSVERSAL FILTER

The simplest form of the SAW transversal filter is shown in Fig. 1(a) [5]. Piezoelectric transducers, which convert an RF electrode voltage difference into propagating SAW's or vice versa, are arranged as transmitting and receiving end-fire arrays along the propagation axis. Each transducer is a pair of interdigitated metallic-film combs deposited on a piezoelectric substrate (quartz in the case of narrow-band timing-recovery filters). A voltage applied at the bus bars of the transmitter combs creates strong fields of alternating sign between successive pairs of fingers to generate localized alternating strains (acoustic standing waves equivalent to a pair of running waves) in the substrate. The inverse process occurs at the receiver.

Contrary to the case of minimum-phase (resonator) filters, the amplitude and phase characteristics of a transversal filter are nearly independent of each other. The transmission phase of a narrow-band transversal filter is linear in frequency to a good approximation since the phase is determined primarily by constant-velocity acoustic propagation over a path of many wavelengths between transmitter and receiver.

In scattering terms, the simple transducers shown are 3-ports [6], [7], with an electrical port at the bus bars and two acoustic ports at the ends of each array. As with any narrow-band filter, the passband shape is sensitive to the terminations at the electrical ports. It is characteristic of these 3-ports that if the impedance match at the electrical port is improved to raise the transduction efficiency (thereby reducing the insertion loss of the filter), there is an attendant increase in the reflection of incoming acoustic waves at the acoustic ports [6], [7]. In a filter, the result is multiple acoustic reflections between transducers. The largest of these is the "triple-transit" reflection shown in Fig. 1(b). The triple transit is a delayed version of the directly transmitted signal and interferes with it coherently to produce a passband ripple (Fig. 2) that is difficult to suppress. Although

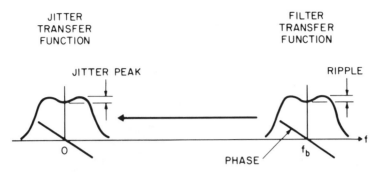

Fig. 2.   Illustration of the simple relationship between a ripply filter transfer function and the associated jitter transfer function when the filter transfer function has symmetric amplitude and antisymmetric phase about the baud frequency. In that case, the ripple translates into a jitter peak.

there is a voluminous literature dealing with ripple reduction, in hardware the ripple never goes away entirely. The jitter studies were undertaken to determine the consequences of ripple effects in a long chain of regenerators and to learn how much ripple might be acceptable in a transoceanic system.

Passband Ripple Effects on Jitter

The first question that arises is whether ripple on the filter transmission passband can cause jitter peaking in the associated regenerator transfer function for phase jitter [1], [8]. The answer turns out to be simple in the case of a filter with symmetric ripple and linear phase. The phase-jitter transfer function is then effectively the same as the filter transfer function [9] translated to baseband and normalized to 1 at dc, as shown in Fig. 2. The ripple becomes the jitter peak. Once the presence of a jitter peak is established, there is a real risk that timing jitter accumulation along a chain of regenerators can run away exponentially with the number of regenerators [8]. The remaining question is "Where along the chain does runaway growth take over?" The first studies [2] examined this question by applying the Chapman model [9] of cascaded regenerators to compute timing jitter accumulation and alignment jitter when all filters have the same symmetric ripple magnitude and peak frequencies. The filter model used for calculation was based on the crossed-field circuit model [2], [6] of transducers like those in Fig. 1, and included triple transits to describe the ripple in a natural way. The transmitting and receiving transducers were assumed to be identical. Sample results of the computations are shown in Figs. 3–5.

Figure 3 shows computed spectra of both systematic and random timing jitter for chains of 50 and 200 regenerators. All the filters in the chain are assumed to be identical, with a ripple of 0.2 dB and a Q of about 800. The vertical scale is normalized to 0 dB at low frequencies for a single regenerator, and the horizontal scale is normalized to the 3-dB point of the jitter filter response in the absence of multiple reflections. The systematic-jitter spectra show typical cancellation effects resulting from a coherent sum of jitter transfer functions: deep minima and a depression of the corner region. In addition, a spectral peak emerges in the region of the jitter peak. The effect of the peak on jitter accumulation can be appreciated by observing that both scales are plotted logarithmically and that the squared rms jitter is proportional to the numerical area under each spectral curve. The random-jitter spectra lack the phase cancellations that tend to pull down the corner in

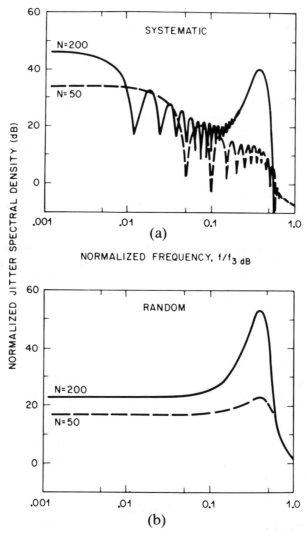

Fig. 3. Spectra of (a) systematic and (b) random jitter accumulated through 50 and 200 cascaded regenerators with identical jitter peaks of 0.2 dB and Q's of 800. The vertical scale has been normalized to 0 dB at low frequencies for the jitter spectral density after one regenerator. The horizontal scale is normalized to the 3-dB point of the ripple-free model [2].

systematic-jitter spectra, so that the jitter peak takes full effect. As a result, integrated random jitter tends to accumulate along a chain more rapidly than integrated systematic jitter in the presence of jitter peaking. This effect may be seen in Fig. 4, which shows the rms timing jitter accumulation at the end of a chain of N regenerators using identical filters. For comparison, we also show the rms jitter accumulation for nonpeaking filters with the same $Q(800)$. These curves follow the familiar fractional power laws [9]–[11], with random jitter the more slowly accumulating. For each curve, the normalization parameter σ_1 is the rms value of the indicated type of jitter emerging from the first regenerator. The peaking results shown here are quite analogous to those found with phase-locked-loop retiming [8], [12] where jitter peaking is usually present to some degree.

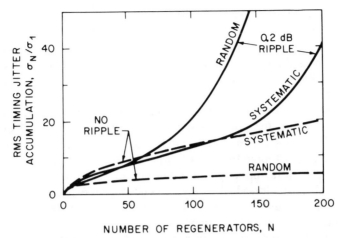

Fig. 4. rms timing jitter accumulation as a function of N, the number of identical cascaded regenerators in a chain. Both the systematic and random jitter components are shown for a peak of 0.2 dB in the jitter transfer function. Curves are also shown for the case of no jitter peaking.

In the absence of jitter peaking, it is known that alignment jitter does not accumulate [2], [3], [9], [11]. With peaking present, however, the rms alignment jitter will in fact grow along the chain, as shown in Fig. 5 [2], [3]. The rate of growth is faster for random jitter than for systematic, just as in the case of timing jitter accumulation. Alignment jitter in the last regenerator of a chain must be carefully controlled to avoid excessive decision errors.

In assessing the implications of Figs. 4 and 5, one must keep in mind that peak-to-peak jitter is some characteristic multiple of rms jitter. Multiples of 6–13 have been seen in various systems. If the multiple is 10, for example, and if σ_1 is $\frac{1}{2}°$ for random jitter, Fig. 5

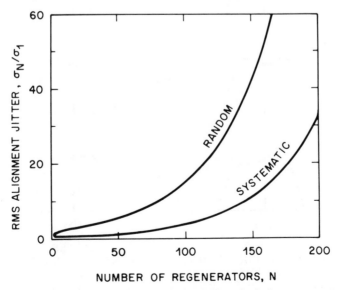

Fig. 5. Illustration of exponential growth of alignment jitter in the last regenerator of a chain when the regenerators have identical jitter peaks of 0.2 dB.

yields a peak-to-peak random alignment jitter of about 210° in the 150th regenerator. Its decision circuit would then be operating with random alignment swings of ±105° on top of the other static and dynamic misalignments. In most cases, the result would be a very high error rate. Under similar conditions, the full peak-to-peak accumulated jitter implied by the curves of Fig. 4 would have to be accommodated in the elastic store of the receiving terminal.

The above results appear open to question because the use of a transfer function implies a linear relation between the jitter coming into a regenerator and the jitter leaving it. Linearity is readily justified when jitter accumulation is small or when it is concentrated near dc, as in the case of ripple-free filters. However, when jitter accumulates away from dc, as in the case of jitter peaking, the input-output relation may become effectively nonlinear and thereby introduce deviations from the transfer-function predictions. For the small amounts of peaking under consideration here, estimates [3] indicate that nonlinear effects will be quite unimportant.

The next practical question is how to choose an acceptable degree of filter passband ripple. The answer depends strongly on the random variations in a population of manufactured filters. A suitable analytical formalism has been developed to deal with

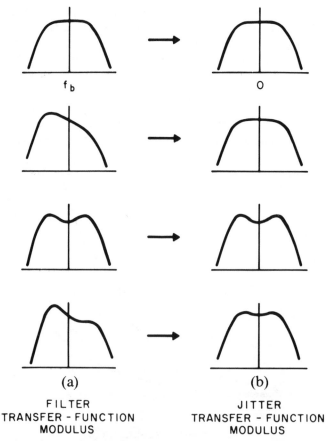

Fig. 6. (a) Examples of filter passband shapes made possible by shifting the phase of triple-transit interference with the first transit. (b) Corresponding jitter transfer moduli, constructed as described in the Appendix. The jitter modulus narrows and rolls off faster as the baud shifts toward either edge of the passband in (a).

distributions of filter passband shapes and center-frequency mistuning [3]. The shapes include asymmetric as well as symmetric passbands, with or without ripple, as shown in Fig. 6(a). The jitter transfer function [3] corresponding to each case is shown in Fig. 6(b). The connection between the complex filter transfer function and the complex jitter transfer function is discussed in the Appendix where the two functions are shown to be effectively related through simple symmetry operations.

To simulate a population of manufactured filters, a whole family of passband shapes, including all but the first in Fig. 6(a), has been represented analytically through a filter model described in [2]. Passband shapes were varied through a single random variable, the effective phase shift between the triple-transit and first-transit signals. If that phase were stepped incrementally through a change of 360°, the passband shapes would run the gamut from a central peak to strong asymmetry to a central dip to reversed asymmetry and back to a central peak. Somewhat more than half the full range, centered on a central dip (largest jitter peak), appears ample for approximate simulation of a prototype filter population for TAT-8. With reference to Fig. 1, the interference phase variation was described by a random variation of the transducer separation parameter $n = 20 + \epsilon$, with ϵ uniformly distributed between limits $(0.1, 0.25)$, equivalent to $(0.1-0.4)$ for symmetry reasons. For purposes of comparison to Fig. 4, where all filters had identical symmetric dips yielding jitter peaks of 0.2 dB, the symmetrically rippled member of the population has been arbitrarily adjusted to a peak of 0.2 dB. The magnitude of individual transducer reflections is thereby determined in the model. The analytical model is too simple to describe flat-topped band shapes, so that ripple effects in the manufactured population are overstated by the model.

The statistical effects on jitter accumulation are illustrated in Fig. 7. The solid lines represent a fully stochastic computation [3] of rms systematic and random jitter accumulation. The surrounding bands extend one standard deviation on either side. In the

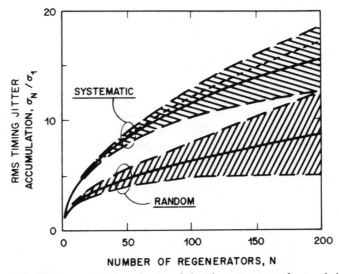

Fig. 7. Timing jitter accumulation along a chain of regenerators whose retiming filters exhibit a distribution of passband shapes. The distribution has been constructed as a "natural" generalization of the single passband shape used to obtain Fig. 4. The solid curves correspond to jitter computed from a fully stochastic computation [3]. Surrounding bands display the regions spanned by ± 1 standard deviation in each case. For the systematic case, a dashed line shows an approximation described in the text.

systematic case, the dashed line corresponds to a chain of identical regenerators whose filter transfer function is an average [3] taken over the filter population. The corresponding curve for random jitter accumulation is too close to its solid curve to be shown.

Figure 7 contains several striking results. First, the exponential jitter accumulation produced by a chain of peaking regenerators has gone away. The reason is that in the distributed population, the part of the population with jitter peaks is counterbalanced by another part that rolls off rather rapidly. The average jitter transfer function in the example population is in fact almost maximally flat. Second, the systematic and random accumulation curves have approximately recovered the power-law behavior of nonpeaking regenerator chains, with the systematic jitter accumulating faster than the random jitter. Third, the use of an average filter characteristic in the traditional unvarying-transfer-function computation has yielded a reasonable representation of a chain of regenerators with a distributed filter population.

We conclude that filter passband ripple may be either damaging or innocuous in a long-haul system. The effects of ripple ultimately depend on the magnitude of the associated jitter peaks and on the character and range of variation of passband shapes. Note that the circuits terminating the filters will play an important role since the passband shape of a narrow-band filter is sensitive to the filter terminations.

Transversal Filter "Ringing" Behavior

A timing-recovery filter must "ring" long enough to maintain a useful timing wave at the decision circuit during the longest string of zeros or NRZ ones expected over system life. Maximum string length was previously estimated to be 62 unit intervals in a 25-year transoceanic system with a generous safety margin [1]. During the transitionless string, filter deexcitation has to be compensated through the dynamic range of the postfilter amplifier. Filter deexcitation was described [1] by a power-decay factor of $\exp(-2\pi/Q)$ per unit interval, appropriate to a simple resonator with an inverse fractional bandwidth of Q. That decay law led to a mild lower-bound constraint on filter Q implied by the ringing requirement. Since the ringing mechanism in transversal filters is quite different from that in resonator filters, the transversal-filter decay law has been examined, with results described here.

A resonator filter is a minimum-phase infinite-impulse-response device. It loses a fixed fraction of the trapped energy in every transit across the resonant cavity, but the balance remains trapped for later transits. The 3-dB bandwidth is a direct measure of the fractional loss per transit. On the other hand, a transversal (or delay-line) filter (Fig. 1) is a nonminimum-phase finite-impulse-response device with transmission amplitude and phase nearly independent of each other. The filter is essentially a single-pass device that does not ring in the resonator sense. Energy is stored in the filter only during the time needed for the acoustic wave to travel the length of the transmitting and receiving transducer structures. The 3-dB bandwidth is also controlled by propagation: the transmission is ideally a maximum at the synchronous frequency, where the acoustic wavelength matches the period of the transducer finger pattern, and rolls off at other frequencies as the mismatch between wavelength and finger period increases. In evaluating the filter "ringing" behavior, we shall limit consideration to the synchronous case.

Figure 8 illustrates the coherent buildup of acoustic waves traveling toward the exit face of the transmitter nearest to the receiver. For simplicity, both transducers are assumed to

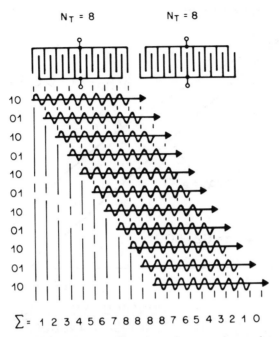

Fig. 8. Illustration of the coherent buildup of a surface acoustic wave by a superposition of wavetrains generated by code transitions in successive unit intervals. Startup and shutdown are characterized by linear signal tapers.

have N_T finger periods. We also assume for the moment that the transmitter is excited by a message transition in every unit interval, i.e., that the incoming binary code is $\cdots 101010 \cdots$, so that after the prefilter and squarer of an NRZ timing loop, there is a driving pulse in each time slot. The corresponding transitions are shown by the digrams [13] in the left-hand column. The voltage applied to the transmitter bus bars generates an acoustic wave train of the same length as the transducer at each instant. Although generation is continuous in time, we show the instantaneously generated wave trains only, at instants separated by multiples of the unit interval. Acoustic waves generated in successive cycles cause a progressive buildup in acoustic amplitude; at the transmitter output face, the amplitude is N_T times that produced by a single period of the transducer. Wave sums are indicated at the bottom for $N_T = 8$. In CW operation, a fully enhanced wave uniformly spans the entire receiver, whose N_T periods provide another multiplication by N_T in the received signal. Figure 8 also illustrates the linear buildup and decay of the acoustic wave when excitation begins and ceases. The linear taper in a transversal filter is analogous to the exponential taper in a resonator filter and is similarly responsible for the ringing. For a random binary message, acoustic generation occurs, on the average, during one out of two cycles, so that on average, the transmitter wave-sum is half as large as that shown. Fluctuations about the average tend to be insignificant for typical retiming filters with hundreds of finger periods in the transducers. Note that the transmitted wave still spans the entire receiver in the presence of continuing traffic, so that there is no additional factor-of-2 loss of amplitude in the receiver.

The above idealized model has been used to obtain the minimum filter transmission when m consecutive transitionless time slots are imbedded in a long wave train whose mean transition probability outside the m intervals is $1/2$. Results are shown in the upper

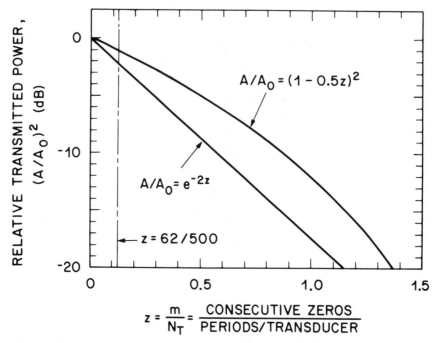

Fig. 9. Minimum transmitted timing-wave power through transversal and simple resonator filters having the same "Q" when normal traffic (transition probability $=1/2$) is interrupted by a string of m transitionless time slots. The transversal-filter curve is derived from the simplified modeling indicated in Fig. 8.

curve in Fig. 9 where transmission is normalized to that for $m = 0$ and where m is normalized to N_T on the abscissa. For comparison, we also show the corresponding transmission (lower curve) for a simple resonator filter having the same Q value as the transversal filter. Q has been approximated here by an expression derived from a familiar model of a SAW transducer [2], [6]: triple-transit effects in the filter have been omitted:

$$Q = \pi N_T/2. \tag{1}$$

Since Q scales with N_T, larger values of the abscissa correspond to lower Q's.

For a transversal filter on quartz with a Q of 800, N_T will be roughly 500. If the message contains one string of 62 zeros or ones, the filter transmission will dip by 1.1 dB, about half as many decibels as the transmission dip of a simple resonator filter of the same Q. The factor of 2 in decibels applies for all values of z below roughly 0.2. As in [1], we can relate the minimum transmitted power during a long transitionless string to the dynamic range D of the postfilter amplifier needed to restore the timing wave. With D in decibels, we can combine (1) and the decay law shown in Fig. 9 to obtain

$$D \simeq -40 \log \left[1 - \frac{m\pi}{4Q} \right]. \tag{2}$$

With a maximum transitionless string length of 62 time slots during system life, transversal-filter retiming with a Q of 800 should require only 1.1 dB of dynamic range in the postfilter amplifier.

Thermal Effects in SAW Retiming

To maintain a low bit-error-ratio (BER) over a specified temperature range (e.g., 0–30°C), the decision cross har in the eye diagram must be stable against temperature variations. Thermal stability of the eye diagram involves the temperature behavior of three distinct circuit segments: i) the circuit branch that carries the conditioned pulse train to the decision point, ii) the timing-loop electronics, and iii) the timing-recovery filter [1]. For simplicity, we shall start by assuming that part i) is constant over the temperature range of interest. The burden of stability then falls on the complete timing loop, parts ii) and iii). In an NRZ system, part ii) typically includes a prefilter, a squaring circuit, a high-gain limiting postfilter amplifier, and a static phase adjuster. The combination tends to result in a considerable temperature variation of transmission phase, which must be compensated as well as possible [4] through the thermal characteristics of the filter transmission phase.

The thermal behavior of the filter is controlled primarily by the orientation of the quartz substrate [14] and secondarily by the metal-film transducers [15]. Filter transmission phase versus temperature is illustrated in Fig. 10 for a typical high-stability orientation of the quartz. The phase is a parabolic function of the temperature, with the apex, at "turnover temperature" T_t, adjustable from well below 0°C to well above 70°C. The parabola curvature, expressed by the product cfs_t (symbols defined in Fig. 10 caption), may depend somewhat on both T_t and the metal transducers [15], [16]. Because T_t is controllable in manufacture, and because the phase variations of the filter and the electronics are of comparable range, it is possible to achieve good thermal stabilization of

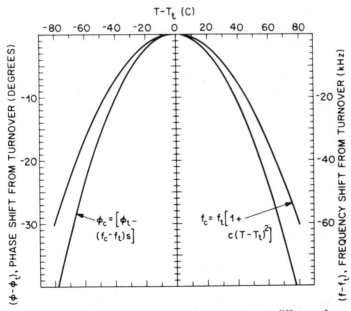

Fig. 10. Dependence of filter center frequency f_c on temperature difference from turnover temperature $T - T_t$ for a filter with a parabola constant $c = -32$ ppb/(°C)^2. A corresponding phase-temperature parabola is shown for a filter with linear phase slope $s = d\phi/df = -0.7$ deg/kHz. $\phi_c =$ transmission phase at f_c, ϕ_t and f_t are phase and frequency at the turnover temperature, and the baud frequency f_b has been substituted for f_t with sufficient accuracy for the figure.

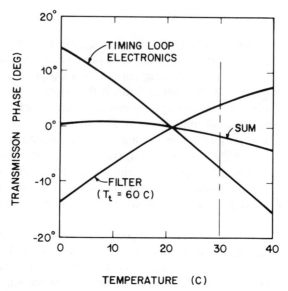

Fig. 11. Example of thermal stabilization of timing-loop electronics by utilizing the SAW filter phase-temperature characteristic. The filter is assumed to behave as in Fig. 10, with T_t placed at 60°C. The phase origin is arbitrary.

the complete timing loop over various temperature ranges of practical interest. This concept was introduced in a terrestrial system design described in [4].

Figure 11 shows an example more typical of present undersea regenerators employing monolithic integrated circuits. The phase characteristics of the electronics shows a pronounced curvature at the lower temperatures. Since the filter is in cascade with the electronics, phase stabilization only involves a sum of phase shifts. As the figure shows, with the filter turnover temperature in Fig. 10 placed at 60°C, the total timing-loop phase variation is less than 2° between 0 and 30°C. Notice that if the circuit carrying the pulse train to the decision circuit, segment i) above, should exhibit significant thermal phase variations, the timing loop could be adjusted for a similar thermal phase variation to maintain the correct decision phase.

The stabilization scheme described above has an unavoidable consequence. For a filter with a linear phase-frequency characteristic (e.g., a narrow-band transversal filter on quartz), one can show that the temperature variation of phase is tracked by a temperature variation of passband center frequency (see Fig. 10). In other words, as temperature moves away from T_t, the whole passband shifts downward in frequency. For a given operating temperature range (T_L, T_U) (lower and upper limits), the frequency shift across the range is minimized by placing T_t at $\bar{T} = (T_L + T_U)/2$. However, if T_t is shifted away from $\bar{T}$ to stabilize the timing-loop phase, then the frequency shift Δf across (T_L, T_U) necessarily increases. Since this source of static detuning must be controlled along with other sources, there is a limit to how far T_t can be pushed from the middle of the temperature range. Figure 12(a) illustrates how the total frequency range Δf across 0–30°C depends on T_t when the curvature c of the frequency parabola is -32 ppb/(°C)^2, a value appropriate to ST-X quartz. For the purpose of this illustration, it is assumed that c is independent of T_t, a commonly satisfied condition. When $T_t > T_U$, the curve is a straight line

$$f_c(T_U) - f_c(T_L) = -2cf_t(T_U - T_L)(T_t - \bar{T}) \tag{3}$$

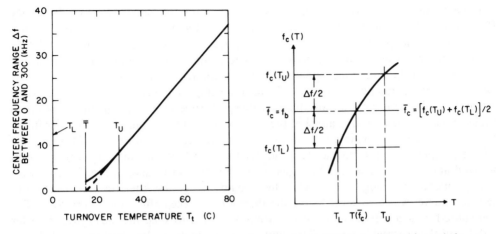

Fig. 12. (a) Center frequency range swept between 0 and 30°C by a filter with parabola constant $c = -32$ ppb/(°C), as a function of turnover temperature T_t. $f_c(T_U) - f_c(T_L)$ reverses sign at $T_t = \overline{T}$. $\phi_c(T_U) - \phi_c(T_L)$ behaves similarly. (b) Example of frequency-trimming target $f_c = f_b$ to minimize thermal frequency deviations from f_b between T_L and T_U.

where $\overline{T}$ is the center of the operating range. With T_t between T_U and $\overline{T}$, the curve is a parabola

$$f_c(T_t) - f_c(T_L) = -cf_t(T_t - T_L)^2. \qquad (4)$$

When $T_t < \overline{T}$, (3) and (4) apply if T_L and T_U are interchanged.

Figure 12(b) shows the way to minimize thermal frequency variations when $T_t \neq \overline{T}$: the filter is frequency-trimmed to make the filter center frequency equal to the baud when centered in Δf. Frequency swings will then be $\pm \Delta f/2$.

In addition to thermal detuning, other sources of static frequency detuning are trimming errors in filter manufacture, aging of the filter center frequency, and errors

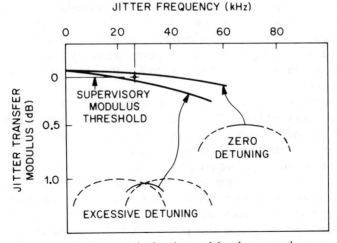

Fig. 13. Illustration of a jitter-transfer-function modulus that meets the mean supervisory requirement at the SL subcarrier frequency (upper curve) and a modulus that falls below the mean supervisory requirement. As shown by sketches, both curves derive from filter passbands of identical shape, with excessive static detuning in the lower case. (See the Appendix.) Curvatures are exaggerated for clarity. The vertical scale is illustrative only.

(including aging drift) in the terminal master clock. Only the aging and clock frequency offsets have a direct impact on static phase offsets, via the phase slope $d\phi/df$ of the filter transmission. The phase effects of thermal frequency detuning are compensated as described above, while frequency errors in manufacture translate into phase errors that are corrected by the timing-loop phase adjuster during regenerator assembly.

All frequency detuning mechanisms do have an impact, however, on the supervisory system devised for the TAT-8 transatlantic cable [17]. Supervisory information is returned to a terminal via the optical channel in the form of tone bursts of an FM subcarrier in the 15–30-kHz range. To maintain a useful subcarrier signal-to-noise (S/N) ratio, the average regenerator must have a jitter transfer function that rolls off very little between the baud and the subcarrier sideband, as shown by the upper curve in Fig. 13. Excessive static frequency detuning of the SAW filters could change the filter transfer function to the lower curve, which rolls off below the threshold of acceptable supervisory S/N. The total static frequency detuning must be carefully limited to keep the mean jitter transfer modulus in the acceptable region. The limit tends to tighten as the filter passband shapes become more varied.

Summary

The jitter effects of retiming-filter passband ripple in long regenerator chains have been examined within the transfer-function approximation. Approximations made along the way are believed to be conservative in the sense that stricter modeling would yield results less harmful to system performance. For a chain of identical regenerators, symmetric ripple has been found capable of causing a) exponential accumulation of timing jitter along the chain, b) random timing jitter accumulation faster than systematic, and c) exponential growth of alignment jitter along the chain, with random jitter again outpacing systematic. However, for a chain of regenerators with a distribution of filter transfer functions, constructed in a realistic manner from the symmetrically rippled case, all the above effects can disappear. When the average filter in the distribution is ripple free, jitter behavior effectively reverts to the ripple-free case. This conclusion is based on ripple not in excess of 0.2 dB. For larger ripple, the stochastic variance could become large enough to allow a significant probability of undesirable effects in a randomly assembled chain. Under common manufacturing limitations, it appears necessary to characterize a population of filters to determine whether a system built with them will behave well or badly with respect to jitter.

The ringing behavior of SAW transversal retiming filters has been derived from a highly simplified model. Although the model omits many features of real filters, those features are not expected to alter the ringing behavior appreciably. A SAW transversal filter has been found to drop its output level only half as much in decibels as a simple resonator filter of the same Q after excitation is disrupted by the longest transitionless string expected in 25 years of random traffic. As a result, with random traffic the timing loop can utilize a postfilter amplifier with half the dynamic range previously estimated [1].

Recent developments have been described concerning thermal stabilization of the decision phase in a regenerator. The filter phase-temperature characteristic can be used to stabilize the phase-temperature characteristic of the electronics [4] to within a few degrees in undersea systems. Across the operating temperature range, a penalty in the form of increased thermal shifting of the filter passband has been identified and evaluated. The importance of these shifts will depend on the specific parameters of timing and decision circuits. Implications for center-frequency filter trimming have also been noted.

A simple transformation is described in the Appendix for converting a filter transfer function into the corresponding jitter transfer function. The general features of the latter can be quickly visualized after a little practice.

APPENDIX: RELATION OF JITTER TRANSFER FUNCTION TO FILTER TRANSFER FUNCTION

As shown in [3], the complex *jitter* transfer function $W(f)$ can be expressed as follows in terms of the complex *filter* transfer function $H(f)$. Let

$$H(f) = A(f)e^{j\phi(f)}. \tag{5}$$

Define a low-pass equivalent transfer function by

$$H_L(f) = H(f_b + f)/H(f_b) \tag{6}$$

where f_b is the baud frequency and f is the deviation from the baud, i.e., a baseband frequency. Then the jitter transfer function can be written

$$W(f) = [H_L(f) + H_L^*(-f)]/2 \tag{7}$$

where * means complex conjugate. A similar result, given by Mengali and Pirani [18], was used in [2] to treat the case of detuned resonator filters. We use the symmetries implicit in (7) to obtain a simple construction for $W(f)$.

For that purpose, it is convenient to introduce new amplitude and phase symbols

$$B(f) = A(f_b + f)/A(f_b) = B_e(f) + B_o(f) \tag{8}$$

$$\psi(f) = \phi(f_b + f) - \phi(f_b) = \psi_e(f) + \psi_o(f) \tag{9}$$

where the subscripts "e" and "o" identify the even and odd parts of the functions. Note that $\psi(f)$ is the phase of the filter transfer function measured relative to its phase at the baud. With these relations, we may write

$$W(f) = e^{j\psi_o}(B_e \cos\psi_e + jB_o \sin\psi_e)$$
$$\simeq B_e e^{j\psi_o} \tag{10}$$

where the last approximation is valid mainly because ψ is effectively linear in frequency across the passband of simple transversal filters, so that $\psi_e \simeq 0$.

According to (10), the effective jitter transfer function is formed from the even part of B,

$$B_e(f) = \frac{A(f_b + f) + A(f_b - f)}{2A(f_b)}, \tag{11}$$

and the odd part of ψ

$$\psi_o = \tfrac{1}{2}[\phi(f_b + f) - \phi(f_b)] - \tfrac{1}{2}[\phi(f_b - f) - \phi(f_b)] \tag{12}$$

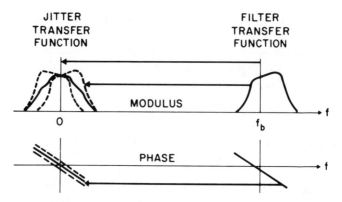

Fig. 14. Transformations showing how a jitter transfer function is constructed from a filter transfer function when the passband phase is sufficiently linear. The jitter transfer function is given by the solid curves on the left.

both defined relative to the baud. These relations have a simple geometrical significance illustrated in Fig. 14: apart from normalization, the jitter transfer function is formed by a) translating the filter transfer function to baseband, with the baud moving to the origin, b) averaging the shifted amplitude response with its mirror reflection in the vertical axis, in accord with (11), and c) averaging the shifted phase response with its transform obtained by inversion through the origin, in accord with (12). Because of construction b), a filter passband with a linear amplitude slope across the baud will have a jitter transfer function that is flat near $f = 0$. Construction b) also helps to explain why two-section-Butterworth and simple resonator filters show no jitter peaking with detuning until the detuning exceeds the 3-dB bandwidth [1]. At smaller values of detuning, there is no central dip in the average of the filter transfer function with its reflection through the baud.

ACKNOWLEDGMENT

This work has benefited from interactions with many colleagues. Special thanks are due to T. R. Meeker, W. R. Grisé, and W.-S. Tsay; to G. M. Homsey, R. M. Paski, and P. R. Trischitta; and to C. B. Armitage for early information on the feasibility of thermal phase stabilization of the timing loops.

REFERENCES

[1] R. L. Rosenberg, D. G. Ross, P. R. Trischitta, D. A. Fishman, and C. B. Armitage, "Optical fiber repeatered transmission systems utilizing SAW filters," in *1982 Ultrason. Symp. Proc.*, IEEE cat. no. 82CH1823-4, pp. 238–246; see also *IEEE Trans. Sonics Ultrason.*, vol. SU-30, pp. 119–126, 1983.
[2] D. A. Fishman, R. L. Rosenberg, and C. Chamzas, "Analysis of jitter-peaking effects in digital long-haul transmission systems using SAW filter retiming," *IEEE Trans. Commun.*, to be published.
[3] C. Chamzas, "Accumulation of jitter: A stochastic model," *AT&T Tech. J.*, vol. 64, pp. 43–76, 1985.
[4] C. B. Armitage, "SAW filter retiming in the AT&T 432 Mb/s lightwave regenerator," *Proc. 10th ECOC*, pp. 102–103, 1984.
[5] R. M. White and R. W. Voltmer, "Direct piezoelectric coupling to surface elastic waves," *Appl. Phys. Lett.*, vol. 7, pp. 314–316, 1965.
[6] W. R. Smith, H. M. Gerard, J. H. Collins, T. M. Reeder, and H. J. Shaw, "Analysis of interdigital surface wave transducers by use of an equivalent circuit model," *IEEE Trans. Microwave Theory Tech.*, vol. MTT-17, pp. 856–864, 1969.

[7] R. L. Rosenberg, "Wave-scattering properties of interdigital SAW transducers," *IEEE Trans. Sonics Ultrason.*, vol. SU-28, pp. 26–41, 1981.

[8] E. Roza, "Analysis of phase locked timing extraction circuits for pulse code modulation," *IEEE Trans. Commun.*, vol. COM-22, pp. 1236–1249, 1974.

[9] C. J. Byrne, B. J. Karafin, and D. B. Robinson, Jr., "Systematic jitter in a chain of digital regenerators," *Bell Syst. Tech. J.*, vol. 42, pp. 2679–2714, 1963.

[10] O. E. de Lange, "The timing of high-speed regenerative repeaters," *Bell Syst. Tech. J.*, vol. 37, pp. 1455–1486, 1958.

[11] H. E. Rowe, "Timing in a long chain of regenerative binary repeaters," *Bell Syst. Tech. J.*, vol. 37, pp. 1543–1598, 1958.

[12] J. Wu and E. L. Varma, "Analysis of jitter accumulation in a chain of digital regenerators," in *Globecom '82 Conf. Rec.*, vol. 2, 1982, pp. 653–657.

[13] W. R. Bennett, "Statistics of regenerative digital transmission," *Bell Syst. Tech. J.*, vol. 37, pp. 1501–1542, 1968.

[14] J. F. Dias, H. E. Karrer, J. A. Kusters, J. H. Matsinger, and M. B. Schulz, "The temperature coefficient of delay time for X-propagating acoustic surface waves on rotated Y-cuts of alpha quartz," *IEEE Trans. Sonics Ultrason.*, vol. SU-22, pp. 46–60, 1975.

[15] J. Minowa, N. Nakagawa, K. Okuno, Y. Kobayaski, and M. Morimoto, "400 MHz SAW timing filter for optical fiber transmission systems," in *1978 Ultrasonics Symp. Proc.*, IEEE cat. no. 78CH1344-1SU, pp. 490–493.

[16] Y. Shimizu and Y. Yamamoto, "SAW propagation characteristics of complete cut of quartz and new cuts with zero temperature coefficient of delay," in *IEEE 1980 Ultrasonics Symp. Proc.*, IEEE cat. no. 80CH1602-2, vol. 1, pp. 420–423.

[17] C. D. Anderson and D. L. Keller, "The SL supervisory system," this book, ch. 40, p. 567.

[18] U. Mengali and G. Pirani, "Jitter accumulation in PAM systems," *IEEE Trans. Commun.*, vol. COM-28, pp. 1172–1183, 1980.

29

An Undersea Fiber-Optic Regenerator Using an Integral-Substrate Package and Flip-Chip SAW Mounting

PAUL A. DAWSON AND S. PAUL ROGERSON

INTRODUCTION

The overriding need for reliability in undersea transmission systems has led, in the case of optical fiber systems, to a preference for thick-film hybrid circuits mounted in hermetically sealed containers. The same constraint has led to a general choice of surface–acoustic wave devices (SAWF's) as a timing recovery filters in undersea repeaters. These SAW's are conventionally packaged in TO-8 or similar cans, which are best suited to mounting on printed circuit board.

We describe here an electrical undersea regenerator module, operating at a line rate of 324 Mbaud with 7B8B line code, in which quartz SAWF's are fixed directly to the thick-film substrate together with the associated electronic components. This forms the base of an integral-substrate package (ISP), the sealing of which provides the necessary hermeticity. The SAWF is bonded without the use of organic compounds, which also improves reliability.

THE INTEGRAL SUBSTRATE PACKAGE

The reliability of hybrid circuits can be greatly influenced by the method of encapsulation used. One packaging technique which has proved to be very reliable is the ISP [1]. This form of package does not use any organic compounds as no adhesive is needed to attach the substrate to the can.

The ISP (Fig. 1) is formed by using the thick-film substrate as the base of the package. An insulating layer is deposited around the perimeter of the board over which a 60/40 lead–tin solder layer is laid. A KOVAR ring is then soldered to the prepared substrate. The lid is laser welded to the ring to form the hermetic seal. The overall dimension is our

Fig. 1.   Photograph of ISP.

ISP is 2×3 in. These were custom made because the largest commercially available ISP is at present 2×2 in. However, the custom made ISP's are cheaper. To ensure that a good hermetic seal could be obtained, sealing tests using the helium back pressure method [2] were carried out on a sample of the ISP's. The leak rate of the ISP's was initially found to be in the range of $2*10^{-9}$ mbar $1/s$ to $1*10^{-8}$ mbar $1/s$. The initial measurement also includes the effect of loss of helium from the small voids in the surface of the package [3]. This effect quickly becomes negligible. The leak rate was therefore remeasured after a period of three days and was found to be in the order of $6*10^{-10}$ mbar $1/s$. These results indicate that the hermeticity of the ISP is extremely good, and compare well with other published data [4].

FLIP-CHIP SAWF MOUNTING

Mounting Method

SAWF's devices have conventionally been mounted in TO-8 or similar enclosures. Other packages have been investigated [5], [6]. However, these mainly concentrate on the mounting of the SAWF inside the package rather than the mounting of the package on to different substrates. Any SAWF mounting arrangement has to satisfy a number of requirements. For undersea applications these requirements can be ranked, as follows:

i)   compatibility with thick-film technology;
ii) hermetic sealing for low aging rates;
iii) avoidance of organic compounds;
iv) stress free and shock proof mounting;
v)   built-in ground plane;
vi) simple construction.

The mounting method described here was developed to meet these requirements.

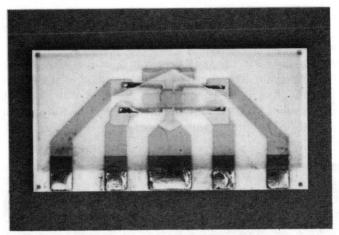

Fig. 2.   Flip-chip SAW mounted on test substrate.

The SAW transducers are metallized with aluminum on to a quartz substrate. The bonding regions of the SAW are formed by covering the aluminum with a thin barrier layer of titanium over which a layer of gold is deposited. The use of titanium prevents any danger of Au–Al intermetallic formation in the bonding region. The wafer is then cut to form the desired substrate shape. Metal tapes [7] are attached to the gold bonding area using thermocompression bonding. The arrangement is turned upside down and placed onto the ISP. The metal tapes are then attached to the thick-film conductors. A photograph of the flip-chip SAW mounted onto a thick-film test substrate is shown in Fig. 2.

This mounting method satisfies criteria i)–vi) above in the following ways:

i) The method follows the use of surface mounting commonly used in thick-film circuits and increasingly in PCB construction.

ii) Hermetic sealing occurs when the lid of the ISP is welded into position. Many workers during the past decade have investigated the effect of packaging on the aging rates of SAWF's. Parker [8], Latham and Saunders [9], Shreve et al. [10], and Minowa et al., [11] have all shown that a low aging rate demands a hermetic package. Their work has also demonstrated that the use of organic compounds can adversely affect the aging of the SAWF.

iii) By specially shaping the SAWF substrate, edge reflections have been reduced without the use of acoustic damping material. The SAWF response can be degraded by reflections of the surface wave from the substrate edge. This effect is commonly known as "edge reflection." This problem is normally cured by coating the edges of the SAWF substrate with an acoustic absorber, e.g., epoxy or polyimide. This technique is not suitable for undersea systems because it uses organic compounds. The passband ripple of a SAWF without acoustic absorbers may be unacceptable in this application. The frequency of the passband ripple, f_r, is related to the length of the substrate by the relationship,

$$f_r \, \text{Hz} = \frac{v \, \text{ms}}{d \, \text{m}}$$

where d is the shortest SAW propagation path from one transducer to the SAWF substrate apex, and $v = 3158\text{m/s}$ which is the SAW velocity on quartz. From the substrate dimensions f_r is predicted to be 343 KHz.

iv), v) The bonding tapes serve as both the electrical and physical connection between the SAWF and the substrate. The mounting method uses specially prepared gold tapes that have been well annealed. Large bonding areas have been provided on the SAWF and mounting substrate so that three or more bonds can be made for each tape connection. Tape bonding achieves a low profile which, in conjunction with the earth plane on the

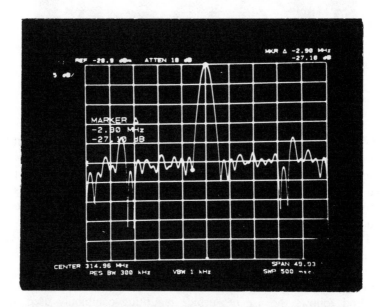

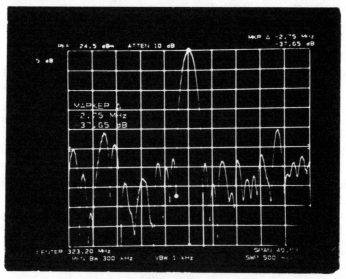

Fig. 3.   (a) Amplitude response of SAW filter mounted in TO-8 can. (b) Amplitude response of flip-chip SAW filter. (N.B. Both SAW filters measured unbalanced.)

ISP, reduces electro-magnetic (EM) breakthrough between the SAWF transducers. It also helps to reduce stress transmission from the hybrid substrate to the quartz. The stress due to thermal effects have been reduced by matching the coefficient of thermal expansion of the substrate to that of the SAWF [12], [13]. This method contrasts very favorably with the use of epoxy or polyimide to secure the SAWF inside the package. Neither of these compounds can be used because of the need to achieve low aging rates—see i). The only alternative would be to use the signal bond wires for the securing mechanism: however, as the bondwires are not designed as restraining items they would be liable to break if subjected to any movement. The success of this method in reducing EM breakthrough compares well with the alternatives: a) the evaporation of metal onto the SAWF substrate to form an earthed strip between the transducers, and b) an earthed metal plate perpendicular to the substrate between the transducers.

vi) This method uses the fewest operations between the fabrication of the SAWF and its mounting onto the ISP. The SAWF substrate is transparent, allowing visual inspection of all the tape connections. The mounting method needs to be simple, as greater complexity reduces the probable yield. Furthermore, undersea systems must have a very low failure rate, typically two faults in 25 years thus the cost of quality assurance (QA) effort is correspondingly high. Making the mounting operation as simple as possible reduces the time required for QA. For example, the absence of organic compounds removes the need to monitor the outgassing from these materials which can occur after curing.

Performance of Flip-Chip SAW

Figure 3 shows that flip-chip mounting improves the stopband rejection of the SAW response by 10 dB. This improvement is due to the elimination of EM breakthrough between the transducers. The frequency of the ripple introduced into the amplitude response is approximately 333 KHz, which is in close agreement with the theoretical value of 343 KHz. The amplitude of the ripple is approximately ± 0.045 dB. By comparison the ripple of a SAWF having simular electrical design but constructed on a rhombic substrate, has an amplitude of $= \pm 0.08$ dB.

The attachment of gold tapes to the transducer titanium/gold bond pads gave pull strengths in excess of 50 grams and failure of these devices was due to lead breakage. In contrast, although gold leads could be bonded directly to the aluminum bonding pads on the SAWF chip, the pull strength of the leads was only a few grams and failure occurred by the aluminum lifting from the substrate. The flip chip SAWF's have also been subjected to shock and vibration tests, details of which are given in Table I.

TABLE I
MECHANICAL TESTS

Vibration:	10—150 Hz at $1g_n$ for 30 minutes in each of 3 orthogonal planes.
Bump:	667 bumps of $40g_n$, 6ms duration, half sine pulse in each of six directions.
Shock:	$500g_n$.

Test based on BS2011 Part Ea and Fd
No failures of any of the tested devices was observed.

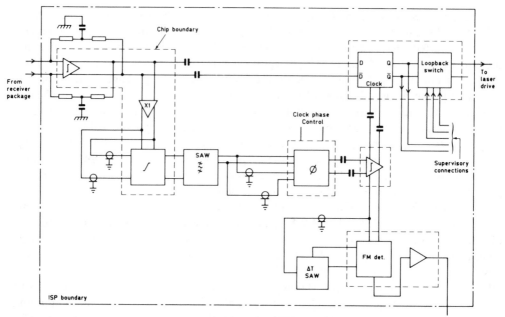

Fig. 4.   Block schematic of ISP regenerator.

Electrical Aspects

Block Schematic

A block schematic of the regenerator is shown in Fig. 4. All the chips are realized on an uncommitted logic array (ULA) in ECL-40 technology [14]. The main amplification path consists of two limiting amplifiers. These are a modification of a design by Wörner, and together give a gain of 40 dB over a 470-MHz bandwidth [15].

The SAWF is a quartz-substrate transversal filter with a typical loss between 50-Ω terminations of 25 dB, and an effective bandwidth Q of 150. The loss is compensated by a combination of high retiming gain, and the use of an exclusive-OR. This can give a higher output than other nonlinear processors, as it produces a pulse of half-unit interval length on every transition in the data.

Following the SAWF is the timing epoch adjustment, described more fully in the following subsection. The retiming amplifier is identical with that in the main gain path.

The master–slave decision bistable has a maximum toggle rate of 700 MHz. It shares a chip with a loopback gate which is necessary for supervisory circuits. Also included on this module is a chip for FM demodulation which forms part of the supervisory circuit. Thus the regenerator functions are realized on four chips: the use of a 28-pin chip carrier will reduce this to three because two of the chips are identical but different parts are bonded out. The power consumption of the regenerator function chips is 1.3 W.

Timing Epoch Adjustment

Timing phase is conventionally adjusted by trimming a piece of coaxial cable in the retiming path. This is an unsatisfactory method because tests must be performed after

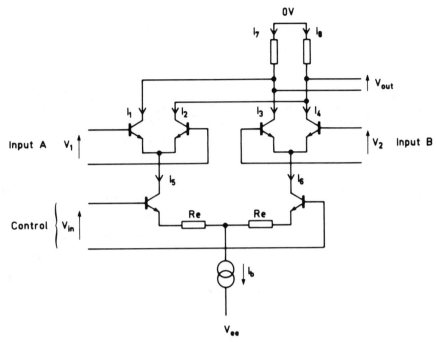

Fig. 5. Timing phase adjustment circuit.

each adjustment of the cable; and this is time consuming and therefore expensive in production.

A circuit diagram of the timing adjustment circuit is shown in Fig. 5.[1] It works by adding variable proportions of two signals at the clock frequency which differ in phase, nominally by 120°. The emitter degradation resistors improve the linearity of the phase/voltage characteristic and reduce the sensitivity to control voltage, making adjustment easier. Not only is the timing epoch easier to optimize with this circuit, but it can be set up more accurately. This will decrease the performance margin allowance for this adjustment.

It is known that a closed-form expression for the large-signal transadmittance of a differential amplifier with emitter resistance cannot be found. Therefore a closed-form expression for the phase-shifter characteristic cannot be found. However, the sensitivity $(d\theta/dV_{in})$ for a given value of V can be predicted. If $R_e = 0$ it can be shown (Appendix) that

$$\frac{d\theta}{dV_{in}}\bigg|_{V_{in} = 0} = \frac{q}{2kT} \tan(\phi/2)$$

where

θ relative phase angle of output,
q electronic charge,
k Boltzmann's constant,
T absolute temperature,
ϕ input phase difference.

[1] This circuit was conceived in collaboration witlh Dr. D. W. Faulkner.

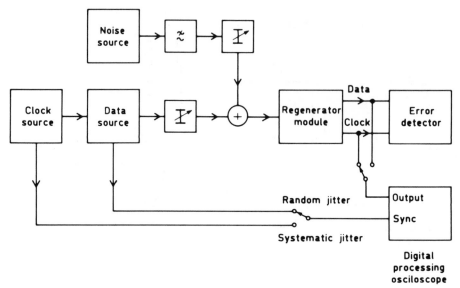

Fig. 6.   BER and jitter measurement.

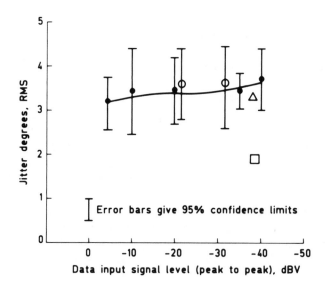

Fig. 7.   Data eye jitter of ISP regenerator.

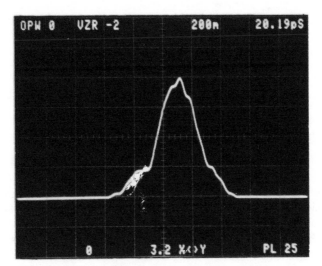

Fig. 8.   Regenerated data eye jitter histogram.

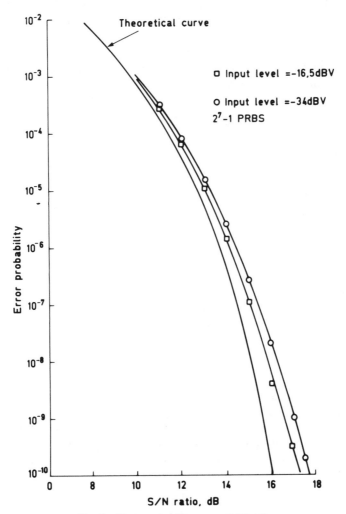

Fig. 9.   Error probability versus S/N ratio.

Now if $R_e \neq 0$, the transadmittance of the lower pair is reduced by a factor

$$1 + (qI_b R_e / 2kT)$$

when $V_{in} = 0$, and it can be supposed that $d\theta/dV$ is reduced in the same proportion. The output amplitude variation is given by the expression

$$\Delta V_{out} = \cos(\phi/2).$$

It is seen that even for $\phi = 120$, the amplitude variation is only 6 dB, which is easily compensated by the following limiting amplifier.

Performance

1) Measurement Procedure: The jitter and BER measurement setup is shown in Fig. 6. The jitter was measured on a digital processing oscilloscope [16]. This enables the jitter of the regenerated data as well as the recovered clock to be measured. Since the regenerator was designed to handle 7B8B code, a $2^7 - 1$ PRBS was used for most measurements, although some were repeated with $2^{15} - 1$.

2) Results: A plot of rms jitter of regenerated data against input level using a $2^7 - 1$ PRBS is shown at Fig. 7. Also shown are i) data jitter using $2^{15} - 1$ pattern for two input levels only; ii) values of both systematic and random recovered clock jitter at one input level only (-38 dBV). Fig. 8 shows a data jitter histogram for a $2^7 - 1$ pattern and -38-dBV input.

Total clock jitter was 3.3°rms, the random jitter component being 1.9°. BER was plotted against signal-to-noise ratio for input levels of -16.5 and -34 dBV, and the results are shown at Fig. 9. The measured characteristic of the phase-shifter circuit is shown at Fig. 10.

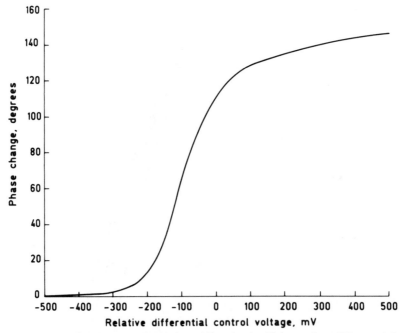

Fig. 10.   Measured characteristics of phase shifter circuit with input phase difference (ϕ) $\simeq$ 120.

DISCUSSION

Phase Shifter Circuit

According to the theory, the circuit should have a maximum slope of 0.2 deg/mV. About twice this value was observed. The theory is probably too elementary at these frequencies: for example, no account is taken of high-frequency behavior of the transistors. It is known that the circuit is operating close to its high frequency breakpoint. The bandwidths of the two amplifiers which make up the circuit will vary with their emitter currents so, therefore, will their phase responses. This will cause the phase/voltage response of the circuit to differ from that predicted by the theory given above. Nevertheless, the performance of the circuit is satisfactory.

Jitter Performance

The low values of jitter of the regenerated data, which is below 3.5° rms over the whole desired range of input levels, is attributable to three main factors. The first is the use of balanced circuitry throughout, which minimizes the effect of crosstalk. The second is careful attention during chip layout to matters of balance and supply and earth rail paths. The third is the good amplitude–phase conversion characteristic of the limiting amplifiers used to provide gain throughout.

The near equality of the clock and data jitter values shows that the loopback gate introduces no appreciable jitter. Finally it can be seen that the use of a $2^{15}-1$ PRBS increases jitter by only 15 percent or so.

BER Performance

The impairment at $P_e = 10^{-9}$ is approximately 0.8 dB with an input of -16.5 dBV. This worsens by 0.6 dB when the input is reduced to -34 dBV. Since the error rate at low S/N ratios closely approaches the theoretical, it is reasonable to assume that this impairment is predominantly due to amplifier noise. Assuming this is Gaussian, it can be estimated that the module has a margin of 6 dB over 10^{-9} error rate for -16.5-dBV input, and 4.2 dB for -34-dBV input. That is to say, a change in margin of 1.8 dB results from a 17.5-dB change of input level. This relative invariance with input level is thought to result from the use of limiting amplifiers. At low input levels these will behave like linear amplifiers, but as the input level rises they must give distortion and signal/noise intermodulation products which slightly impair regeneration. Unfortunately, the theory of their operation in this field has so far proved intractable.

ISP Mechanical Performance

The ISP regenerator was subjected to a range of vibration and shock [17] tests which are summerized below.

Vibration: 5–2000 Hz at one octave per minute, $\lg_n$, constant acceleration
Transient: 100 at $17g_n$, $50 + $ve pulses
$50 - $ve pulses
Shock: Maximum of 10 shocks at $315g_n$ per shock.

Throughout these tests the ISP regenerator performance was monitored using Bit Error Rate Test (BERT) equipment. No electrical errors due to the ISP performance were recorded. During all these tests no discernable mechinical damage or suspect responses were observed.

Conclusion

A 324-Mbit/s electrical regenerator module intended for undersea optical-fiber systems has been described. It has a sensitivity of 20 mV peak to peak, at which the total jitter is 3.3° rms, a low figure achieved by good attention to balance at chip and circuit board level. Timing phase adjustment is by means of a single select-on-test resistor, which enables the timing to be optimized quicker and more accurately than by conventional means.

The regenerator functions are realized using four chips in the proven ECL-40 technology. This number reduces to three if a 28-pin chip carrier is used. The power consumption of these chips totals 1.3 W.

The module incorporates a novel method ("flip-chip") of mounting SAWF which is mechanically robust without the use of organic compounds, and which improves SAWF performance by the virtual elimination of EM breakthrough.

Appendix

Referring to the circuit of Fig. 5, let the inputs to the upper pairs be

$$V_1 = V \sin \omega t$$

and

$$V_2 = V \sin (\omega t + \phi)$$

and let V be small so that these pairs operate linearly. Then

$$I_5 = A \cdot I_b$$

where

$$A = 1 / [1 + \exp(qV_{in}/kT)]$$

and

$$I_6 = I_b - I_5.$$

Hence if $R_e = 0$ the differential output current is given by

$$I_0 = I_7 - I_8$$
$$= I_b [A \sin \omega t + (1 - A) \sin (\omega t + \phi)]$$
$$= \frac{qV}{kT} \cdot I_b \{ [A + (1 - A) \cos \phi] \sin \omega t + (1 - A) \sin \phi \cos \omega t \}.$$

Hence

$$\frac{|I_0|}{I_b} = \frac{qV}{kT}\sqrt{\left[(A + B\cos\phi)^2 + (B\sin\phi)^2\right]} \tag{A1}$$

$$\theta = I_0 = \arctan\left[\frac{B\sin\phi}{A + B\cos\phi}\right] \tag{A2}$$

where $B = 1 - A$. Hence

$$\frac{d\theta}{dV_{\text{in}}} = \frac{q}{2kT} \cdot \frac{AB\sin\phi}{A^2 + B^2 + 2AB\cos\phi}$$

$$\left.\frac{d\theta}{dV_{\text{in}}}\right|_{V_{\text{in}}=0} = \frac{q}{2kT}\tan(\phi/2). \tag{A3}$$

The output amplitude variation may be found from (A1). If $R_e \neq 0$,

$$V_{\text{in}} = \frac{kT}{q}\ln\left(\frac{I_5}{I_b - I_5}\right) + (2I_5 - I_b)R_e$$

$$\frac{dV_{\text{in}}}{dI_5} = \frac{4kT}{qI_b} + 2R_e$$

$$\left.\frac{dI_5}{dV_{\text{in}}}\right|_{V_{\text{in}}=0} = \left.\frac{dI_5}{dV_{\text{in}}}\right|_{I_5 = I_b/2} = \frac{I_b}{(4kT/q) + 2I_bR_e}$$

i.e., if $R_e \neq 0$, sensitivity is reduced by a factor $(1 + qI_bR_e/2kT)$.

ACKNOWLEDGMENT

The authors would like to thank the staff at the Microelectronics Division and the VLSI Technology Division at BTRL for fabrication of the integrated circuits and for the fabrication and assembly of the ISP. In particular, thanks are due to P. D. Walmsley and B. M. MacDonald for collaboration in the development of the "flip-chip" SAW mounting; The Welding Institute for the attachment of the ISP lid to its frame; and A. W. Koszykowski of General Mitronics Ltd. for fabrication of the ISP frames and lids. Acknowledgment is also due to P. R. Harris, S. C. Fenning, and M. K. Compton for their assistance during this project. Acknowledgment is made to the Director of British Telecommunications Research Laboratories for permission to publish this chapter.

REFERENCES

[1] J. Sargent and N. Silverstein, "Integrated substrate packages," *Proc. Internepcon*, 1979.
[2] R. P. Merrett, "Methods and limitations of assessing the hermeticity of semiconductor component encapsulation," *Brit. Telecom Technol. J.*, vol. 2, no. 3, pp. 79–89, July 1984.
[3] I. G. Whyte, BTRL Internal private communication.
[4] G. Simpson, "Laser-welding the large MIC, A new approach to hermetic sealing," *Microwave J.*, Nov. 1984.
[5] C. A. Erikson, Jr., and D. R. LeSiege, "Vertical inline SAW package," *IEEE Ultrason. Symp. Proc.*, 1983.

[6]   J. Minowa, T. Morikawa, and H. Abe, "Stability of SAW filters with narrow pass-band," *Elec. Commun. Lab. Tech. J.* (Japan), vol. 27, no. 2, pp 429–439 (1978).

[7]   B. M. MacDonald, S. P. Rogerson, and P. D. Walmsley, "Flip chip assembly of surface acoustic wave devices," *IEEE Ultrason. Sysmp. Proc.*, vol. 1, pp. 36–39, 1984.

[8]   T. E. Parker, "Analysis of aging data on SAW oscillators," *Proc. 34th Ann. Freq. Control Symp.*, USAERADCOM, Ft. Monmouth, NJ, May 1980.

[9]   J. I. Latham and D. R. Saunders, "Aging and mounting developments for SAW resonators," *IEEE Ultrason. Symp. Proc.*, 1978.

[10] W. R. Shreve, J. A. Kusters, and C. A. Adams, "Fabrication of SAW resonators for improved long term aging," *IEEE Ultrason. Symp. Proc.*, 1978.

[11] J. Minowa, K. Sawamoto, and K. Tanaka, "The aging characteristic of surface acoustic wave filters," *Trans. IECE Japan*, vol. J62-a, no. 1, pp. 103–4, 1979.

[12] R. B. Stokes and M. J. Delaney, "Aging mechanisms in SAW oscillators," *IEEE Ultrason. Symp. Proc.*, 1983.

[13] S. J. Dolochycki, E. J. Staples, J. Wise, and J. S. Schoenwald, "Hybrid SAW oscillator fabrication and packaging," Proc. 33rd Ann. Freq. Contr. Symp., 1979.

[14] D. Baker, "High reliability transistors for submarine systems," ESSDERC, Inst. Physics Conf. Series, no. 4, 1977.

[15] D. W. Faulkner, "A wide-band limiting amplifier for optical fibre repeaters," *IEEE J. Solid-State Circuits*, vol. SC-18, pp. 333–340, June 1983.

[16] P. Cochrane, I. W. Barley, and J. J. O'Reilly, "A direct jitter measurement technique," *Electron. Lett.*, vol. 15, no. 24, pp. 774–776, Nov. 1979.

[17] R. Bowers, "Regenerator card—Vibration and shock tests," Brit. Telecom. M&C Internal Rep., 1984.

30
Ultra-High Reliability Ultra-High Speed Silicon Integrated Circuits for Undersea Optical Communications Systems

LEWIS E. MILLER, SENIOR MEMBER, IEEE

INTRODUCTION

A line of silicon integrated circuits have been designed for use in the regenerator that will be a crucial part of the new optical undersea system known as SL [1]. These circuits, microwave junction isolated monolithic (MJIM) and microwave complementary bipolar integrated circuit (MCBIC) have been described in previous publications [2], [3].

They are of a medium scale of integration (MSI) complexity which are realized with microwave transistors as the active components. MJIM circuits employ only n-p-n transistors, whereas the MCBIC circuits have complementary n-p-n and p-n-p transistors with extrapolated unity gain frequency (f_t) in the range of 4 and 2.5 GHz, respectively.

These circuits are fabricated using an all ion-implanted technology, are silicon nitride passivated, and use a single-level metal formed from a composite layer of titanium, titanium nitride, platinum, and gold. This technology is identical to that used in a broad class of discrete and integrated devices designed and manufactured by AT&T [4].

These UHF IC's for SL actually represent the sixth generation of semiconductor devices developed for undersea applications, the SF [5] and SG [6] systems as well as AT&T Bell Laboratories designed military systems.

Over the two decades that these developments spanned, a methodology for assuring reliability of components for undersea applications has been devised. This methodology had its roots in the procedures adopted for the electron tubes used in the earliest undersea cables realized with electron tube repeaters [7].

However, the early electron tube methodology has been altered and refined significantly over this period. The methodology as applied to discrete semiconductors has been described elsewhere [8] and has evolved because of the following factors.

First, semiconductors are becoming more designable. This is derived from their well-behaved adherence to thoroughly understood and documented physical principles. This permits a greater reliance on analysis and less need for elaborate empirical procedures adopted in the early vacuum tube repeater developments.

Also, experience with development and manufacture of the semiconductors for the five previous generations of undersea devices has evolved a battery of proven qualification and screening procedures which can identify potentially short-lived devices on lots of devices and censoring such potential short-lived devices from a population being considered for repeater manufacture.

Finally, these UHF IC's share a common technology with the discrete silicon transistor that has a proven record of reliability in the SG Undersea systems as well as the discrete microwave transistor that has also established a record of high reliability in the T4M terrestrial high-speed (274-Mbit) digital coaxial system [9].

Thus the IC's for SL represent an extrapolation of a proven technology and methodology to a structure providing an improvement in performance for use in lightwave systems.

COMPARISON OF MJIM AND MCBIC TO PROTOTYPE DISCRETES

As noted in the previous section, the MJIM and MCBIC circuits that will be used in SL are derivatives from two older proven product lines and share a common technology. These are the SG transistor and the microwave transistor, which is the active element in these UHF IC's.

These components not only share a common technology, their development overlapped in time and, because of this, a synergy developed which benefited each, in turn.

For example, the discrete microwave transistor was introduced into manufacture in 1973 and was used in terrestrial high bit rate coaxial installations. These devices were not hermetically packaged but in beam-lead form, were applied to thin-film microwave circuits, and coated with room-temperature vulcanizing (RTV) silicone rubber [10].

There was no screening procedure or burn-in used on these devices or circuits. In spite of the presence of a small number of early failures embedded in the statistics, they have compiled a record of 20 FIT's in a total of 4,000,000,000 device hours of service by mid-1979.

During the planning period in the early 1970's prior to availability of SG transistors, life acceleration tests performed on the discrete microwave transistor were used to evaluate whether its technology was capable of meeting SG reliability goals as shown in Fig. 1.

This life-test acceleration curve evaluated the microwave transistor against the usual transistor life-test end points of stability of breakdown voltages, forward resistances, and leakage currents as well as the more stringent requirements of stability of dc current gain. The devices exhibited no changes in breakdown voltages and forward resistances and a decrease in leakage currents accompanied by an increase in current gain. The failure mode was a general drift upwards in current gain beyond the 2.5-percent limit, a change characteristic of that resulting from annealing of surface states. The activation energy of approximately 1.5 eV displayed by Fig. 1 is also typical of that observed when surface states are annealed.

Subsequently, when prototype MJIM circuits were fabricated, their long-term reliability capability was compared not only to the discrete microwave transistor but the low-frequency discrete transistor ($f_t \sim 300$ MHz) which is the active element in all of the general purpose digital and analog bipolar integrated circuits manufactured by AT&T. These VHF IC's are currently being manufactured by AT&T by the hundreds of millions per year and are ubiquitous in virtually all of AT&T electronic apparatus. Life perfor-

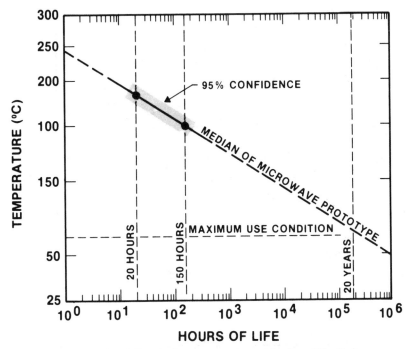

Fig. 1. Capability of microwave transistor technology, SG criteria.

mance of these circuits have been carefully monitored and reproducibly demonstrate FIT rates of better than 10 without the need for burn-in [11].

The technology for these circuits has been refined over the 20-year period of their development and manufacture and represents the core technology from which the technology for the discrete microwave transistor, the SG transistor, and the IC's was derived.

A comparison of the results of life-test acceleration curves as shown in Figs. 2 and 3 will be discussed in some detail because they provide a self-consistent summary of the long-term life capability of the MJIM/MCBIC technology.

These figures follow the usual custom of plotting junction temperature in degrees centigrade (in a scale which is linear in 1/K) versus log time for several relevant devices [12]. The shaded area in Fig. 2 represents the range within which the median life of the VHF prototype transistor falls. It is plotted for reference since, as was noted, it is the basic active element used in all the bipolar logic and linear silicon integrated circuits manufactured by AT&T and because the technology used in its manufacture has been refined for these as well as the discrete microwave transistor depicted in Fig. 2 by the experimental points high-lighted by the solid and dashed curves. These curves show that in spite of the fact that narrower lines and spaces and thinner metallization and dielectrics are used in the microwave transistor, the same long-term life will be realized.

If one extrapolates to use condition, one predicts extremely low failure rate. And, in fact, the failure rate in the high-speed AT&T digital coaxial installation, as has been noted, has by now verified this prediction.

The data of Fig. 2 is repeated on Fig. 3. Superimposed on Fig. 3 are some results with prototype MJIM circuits. The points arranged vertically were obtained by using short

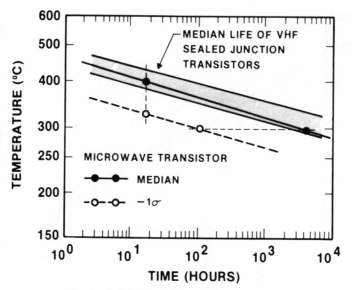

Fig. 2.   Reliability of discrete silicon transistors.

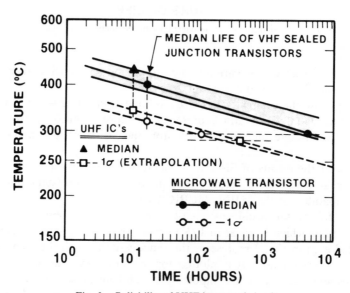

Fig. 3.   Reliability of UHF integrated circuits.

time (10-h) intervals and increasing stress in increments until the failure distribution represented by the sigma and the median plotted on the figure were obtained. The solid square is derived from a constant stress test where a small number of failures were used to extrapolate to the 16th percentile. If one were to extrapolate further, one would find that the median of the MJIM circuits would also fall within the shaded area represented by the VHF discrete (as well as the microwave discrete).

The dashed curve representing the sigma of the MJIM failure distribution is parallel to the median line of the VHF and microwave discretes. It also is very close to the sigma of the microwave discrete failure distribution.

This self-consistency of behavior of the prototype MJIM circuits with the discrete microwave prototype as well as the VHF discrete from which they were all derived is very persuasive evidence that the basic technology used in these devices provides devices of similar long-term life capability and that it is extremely repeatable. (The data was generated over a long period of time by different individuals and in different structures made in several manufacturing locations.)

Moreover, as is shown in Fig. 1 and the discussion associated with it, the SG transistor is a part of this body of data on the reliability capability of the basic technology. Data accumulated on the SG transistor during the production lot qualification procedures to be discussed in the next sections adds significantly to the confidence of this data base.

QUALIFICATION PROCEDURES

As was noted in an earlier section, accelerated life tests conducted on the discrete microwave prototype transistors were used as a basis for deciding that SG reliability goals could be met.

When production of the SG transistor was initiated, stress acceleration tests similar to those in Figs. 1, 2, and 3 were used to verify that each lot of devices measured up to expectations. For this type of evaluation, a lot is generally defined by the wafers in a

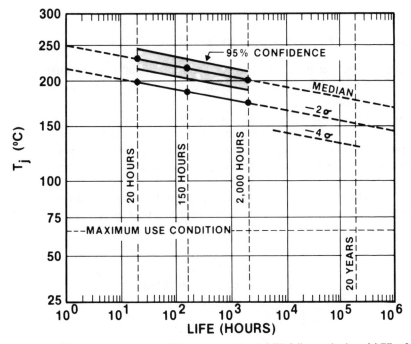

Fig. 4.   Qualification tests on the SG transistor. $P_t = 1.5$-W failure criterion $\Delta hFE \leqslant 2.5$ percent.

diffusion run, the common denominator which establishes the quality of passivation. This might range from five wafers in development to 50 wafers in high-level production. Typically, in production of devices for undersea applications, wafer lot sizes range from 5 to 20 wafers.

Figure 4 represents a compilation of that lot-by-lot qualification procedure which was used on the production runs for the SG transistors used in TAT6 and TAT7.

Three samples from each lot were stress-tested by dissipating 1.5 W in each device while maintaining constant heatsink temperatures for 20, 150, and 2000 h, respectively. The heatsink temperature was raised an increment after completion of each time interval and the failure distribution at each time of stress interval using an end point of $\Delta h_{FE} \leqslant 2.5$ percent was determined.

The shaded area in Fig. 4 represented a growing data base within which 95 percent of all the SG transistor production sample medians fell. The criterion for acceptance of the lot for continued processing was that the sample median fall within the shaded area and that the failure distribution also fall within normal limits as typified by the solid curves plotted on Fig. 4. The significance of the dashed -4σ line indicates that one should expect less than 1 device failure in a 330 repeater system in 20 years if no early failures are contained in the population.

As in the case of Fig. 1, the failure mode displayed by these devices is a drift upwards in current gain beyond the 2.5-percent limit. There were no failures due to breakdown voltages, forward resistances, or leakage currents. Since these devices received a gain stabilization bake, the activation energy of the acceleration curve increased from 1.5 to 2 eV because the more shallow surface states were annealed by the gain stabilization treatment.

Plan for Reliability Assurance of TAT8 SL SIC's

The methodology which has evolved for assuring reliability of semiconductor devices for undersea applications, which has been described in detail in [8], involves three important procedures.

The first is to use a technology that offers an intrinsic long life so that the failures due to wear-out mechanisms (the rising portion of the familiar bathtub curve) shown in Fig. 5 are not present during normal life; and to use lot-by-lot qualification techniques as demonstrated in Fig. 4 to assure that wear-out mechanisms extend beyond the planned useful life of the system.

The second, which is increasingly more important with the realization of the former, is to devise techniques which censor atypical devices: those which fall in the decreasing

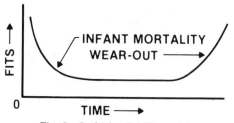

Fig. 5.   Bathtub reliability model.

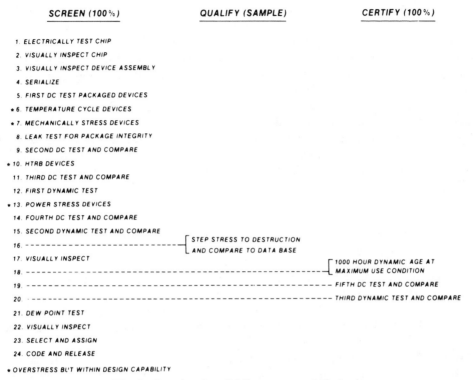

SCREEN (100%)   QUALIFY (SAMPLE)   CERTIFY (100%)

1. ELECTRICALLY TEST CHIP
2. VISUALLY INSPECT CHIP
3. VISUALLY INSPECT DEVICE ASSEMBLY
4. SERIALIZE
5. FIRST DC TEST PACKAGED DEVICES
∗ 6. TEMPERATURE CYCLE DEVICES
∗ 7. MECHANICALLY STRESS DEVICES
8. LEAK TEST FOR PACKAGE INTEGRITY
9. SECOND DC TEST AND COMPARE
∗ 10. HTRB DEVICES
11. THIRD DC TEST AND COMPARE
12. FIRST DYNAMIC TEST
∗ 13. POWER STRESS DEVICES
14. FOURTH DC TEST AND COMPARE
15. SECOND DYNAMIC TEST AND COMPARE
16. - [STEP STRESS TO DESTRUCTION
                                              [AND COMPARE TO DATA BASE
17. VISUALLY INSPECT
18. - [1000 HOUR DYNAMIC AGE AT
                                                                        [MAXIMUM USE CONDITION
19. - FIFTH DC TEST AND COMPARE
20. - THIRD DYNAMIC TEST AND COMPARE
21. DEW POINT TEST
22. VISUALLY INSPECT
23. SELECT AND ASSIGN
24. CODE AND RELEASE

∗ OVERSTRESS BUT WITHIN DESIGN CAPABILITY

Fig. 6.   Procedure for reliability assurance of SL circuits.

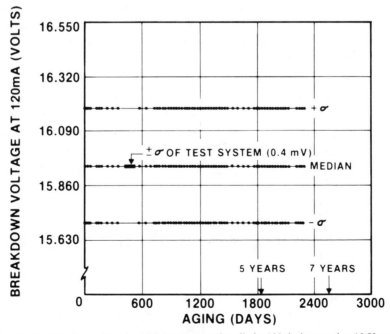

Fig. 7.   Certification aging of an SG surge protection diode. 100 devices aged at 13-V reverse bias.

region of the curve of Fig. 5. This process has been characterized as censoring, screening, or, more recently, purging devices exhibiting infant mortality.

Finally, a burn-in procedure which is intended to verify (certify) that the previously described two procedures have been effective is used. This certification procedure is usually an extended (typically 1000 h) operational life test at or modestly above the maximum use condition of the device.

Figure 6 contains a concise summary of the steps included in the 3 phases of reliability assurance which is being used for SL devices.

The methodology summarized in the preceding paragraphs and in Fig. 6 was used on the prototype MJIM circuits which were prepared for the SL system field trial thereby demonstrating for a short period the efficacy of the methodology when applied to UHF IC's [13].

Results Obtained Using Undersea Component Reliability Assurance Methodology

Certification procedures for the discrete transistors used in SF and SG systems were dominated by the need to forecast long-term stability of current gain (h_{FE}) to assure that these analog systems did not drift out of alignment [14]. Therefore, precision long-term aging equipment was established in the manufacturing line to measure [15], and hopefully, to permit extrapolation of current gain [16].

In spite of the fact that this life-test measurement system was designed and functioned at a precision unprecedented in device life-test art, the stability of the devices was such that one could not distinguish parameter aging trends as shown by Figs. 7 and 8.

Data from certification aging such as described above, as well as records of in-service performance, have been carefully recorded and periodically updated [17]. These data are summarized in Fig. 9 which depicts 20 years of experience with undersea semiconductor device reliability. It includes the cumulative number of device life hours in certification aging, as well as hours in service.

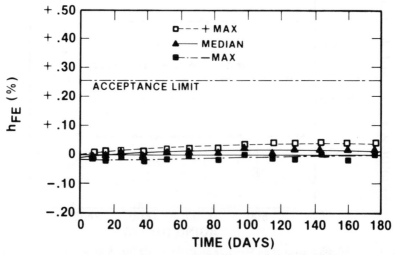

Fig. 8.   SG transistor certification. 100 devices aged at 1.5 W.

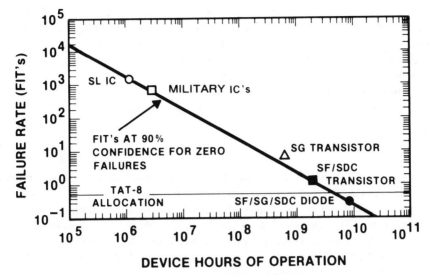

Fig. 9. Experience with undersea devices.

Figure 9 plots failure rate in FIT's versus cumulative device hours of operation. The solid line represents a theoretical curve that represents the best FIT rate that one can demonstrate with a 90-percent confidence level as a function of device life-test hours when one has no failures.

Also shown on Fig. 9 are the SL IC allocation of 0.5 FIT and data representing five different kinds of components; silicon diodes, discrete germanium UHF power transistors, discrete silicon UHF power transistors, silicon VHF IC's, and silicon UHF IC's.

The components which have accumulated the most life hours fall in the lower right portion of Fig. 9, while those with the least are in the upper left.

The diodes which are used in the SF, SG, and SDC systems are all of the same design and have accumulated the most life hours with no failures. These components have demonstrated a FIT rate of approximately 0.3, thus surpassing the target allocation for SL.

The SF germanium transistor, which was first placed on certification aging in 1965 and entered service in 1968, has accumulated a total of 2×10^9 device hours and no failures have been experienced. These results establish a FIT rate of 1.2 at a 90-percent confidence level. The SF system FIT rate goal of 5.0 was demonstrated to a 90-percent confidence level in 1980, 12 years after the SF system entered service.

Similarly, the first SG transistor was placed on the LTA racks in 1973 and entered service in 1976. With the one failure that occurred during system laying, 7.2 FIT's have been demonstrated in 5.4×10^8 device hours of aging.

The results discussed above emphasize the tyranny of numbers associated with attempting to demonstrate, via laboratory or factory tests, the very low FIT rates considered necessary for undersea system reliability.

While the data just summarized demonstrates the effectiveness of the qualification, screening, and certification procedures for discrete semiconductors, similar procedures have been applied to integrated circuits used in a new undersea system designed and manufactured by AT&T for the U.S. Government.

The results of screening these devices, as well as certification age, as of mid-August 1984, are also plotted on Fig. 9. These IC's, which represent the first significant use of ICs in undersea repeaters, have demonstrated a FIT rate of approximately 864, a figure which is approximately equal to that demonstrated by the SF and SG transistors (individually) at the time when the systems were placed in service.

Finally, the data point in the upper left portion of Fig. 9 represents the accumulated experience of 1500 FIT's with prototype SL UHF IC's in August 1984.

Conclusion

This paper has reviewed the methodology which has been developed and refined for providing assurance of reliable operation of semiconductor components in undersea applications as well as the application of this methodology to increasingly more sophisticated components. The empirical data supports the thesis that the methodology is universally applicable. Its most recent application has been to the prototype silicon UHF IC's used in the AT&T SL sea trial, and it is now being applied to the IC's which will be deployed in the SL undersea lightwave system.

References

[1] P. K. Runge and P. R. Trischitta, *The SL Undersea Lightwave System*, chapter 4, this book.
[2] D. G. Ross, "Integrated circuits for high speed digital transmission on optical fiber," presented at High Speed Digital Technol. Conf., San Diego, CA, Jan. 1981.
[3] D. G. Ross, R. M. Paski, D. G. Ehrenberg, W. H. Eckton, Jr., and S. F. Moyer, "A regenerator chip set for high speed digital transmission," presented at ISSCC, San Francisco, CA, Feb. 1984.
[4] R. S. Payne, R. J. Scavuzzo, K. H. Olson, J. M. Nacci, and R. A. Moline, "Fully ion-implanted bipolar transistors," *IEEE Trans. Electron Devices*, vol. ED-21, Apr. 1974.
[5] S. T. Brewer, "S.F. submarine cable system: Foreward," *Bell Syst. J.*, vol. 49, pp. 601–604, May–June 1970.
[6] S. T. Brewer, R. L. Easton, H. Soulier, and S. A. Taylor, "Requirements and performance of the SG undersea cable system," vol. 57, pp. 2319–2354, Sept. 1978.
[7] J. D. McNally, G. H. Nelson, E. A. Veazie, and M. F. Holmes, "Electron tubes for transatlantic cable systems," *Bell Syst. Tech. J.*, vol. 36, pp. 163–188, Jan. 1957.
[8] L. E. Miller, "Reliability of semiconductor devices for submarine cable systems," *Proc. IEEE*, vol. 62, pp. 230–244, Feb. 1974.
[9] J. M. Sipress, "T4M: A new superhighway for metropolitan communications, *Bell Labs. Rec.*, vol. 53, no. 9, pp. 352–359, Oct. 1975.
[10] D. S. Peck, "Reliability of beam-lead sealed junction devices, in *Proc. 1969 Ann. Reliability Symp.* (Chicago, IL), 1969, pp. 191–201.
[11] D. Stewart Peck and C. H. Zierdt, Jr., "The reliability of semiconductor devices in the Bell System," *Proc. IEEE*, vol. 62, pp. 185–211, Feb. 1974.
[12] G. A. Dodson and B. T. Howard, "High stress aging to failure of semiconductor device," in *Proc. 7th Nat. Symp. Reliability Durability Control* (Philadelphia, PA), 1961, pp. 262–272.
[13] P. K. Runge, "Deep sea trial of an undersea lightwave system," *Tech. Dig. Opt. Fibre Commun Conf.* (New Orleans, LA)., Feb. 1983, MD2 paper 8.
[14] A. J. Wahl, W. McMahon, N. G. Lesh, and W. J. Thompson, "SF system: Transistors, diodes and components," *Bell Syst. Tech. Dig.*, vol. 49, pp. 683–698, May–June 1970.
[15] R. L. Odenweller, Jr., "An aging and testing facility for high reliability transistors and diodes," *Western Elec. Eng.*, vol. 11, pp. 20–29, July 1967.
[16] I. G. Abrahamson *et al.*, "Statistical methods for studying aging and for selecting semiconductor devices," in *Trans. 23rd Ann. Tech. Conf. Amer. Soc. for Quality Control* (Los Angeles, CA), May 5–7, 1969, pp. 533–540.
[17] A. J. Wahl, "Ten years of power aging of the same group of submarine cable semiconductor devices," *Bell Syst. Tech. J.*, vol. 56, pp. 987–1005, July–Aug. 1977.

Part VII
Undersea Optical Devices

Undersea optical devices, in particular, the laser transmitters, the optical receivers and any laser sparing component must be designed so that a high performance optical transmission system is achieved. The transmission performance requirements placed on these new technology devices are examined separately in five chapters of this part.

31
Compact 1.3-μm Laser Transmitter for the SL Undersea Lightwave System

FRIDOLIN BOSCH, G. M. PALMER, CHARLES D. SALLADA, AND
C. BURKE SWAN, SENIOR MEMBER, IEEE

INTRODUCTION

The transmitter converts the regenerated electrical signal in a repeater [1] to optical pulses by modulating an injection laser. The transmitter circuitry has three main functions: 1) to generate modulation current pulses with subnanosecond rise and fall times; 2) to maintain the laser light output within specifications throughout the life of the system; and 3) to provide a safe environment for the laser by protecting it from any damaging transients.

A stable fraction of the laser light is coupled to a single-mode fiber for transmission to the next repeater and also to a monitor photodiode in the feedback loop controlling the laser bias.

As is evident, a variety of technologies are part of the transmitter design. Since reliability is of paramount importance for undersea systems, we use proven high-reliability technology wherever possible, specifically in the circuit design and in much of the mechanical design. The introduction of new lightwave technologies involving lasers, photodiodes, and fibers have been carefully planned to facilitate the qualification of the transmitter design and certification of the manufactured produce for undersea use.

Proven technology implies the existence of extensive processing, performance, and reliability records which were gained in similar system applications. Silicon bipolar IC technology combined with thin-film hybrid integrated circuit technology on alumina substrates meet these criteria and were chosen for the transmitter circuit realization.

Certification procedures for the IC's, the laser, and the photodiode monitor impose mutually exclusive temperature requirements. Our approach is to mount these components in separate, hermetic packages which permit individually optimized certification testing for each component prior to assembly in the transmitter.

A new reliability assurance technique, called "purging," is used for lasers and photodiodes [2]. It is based on a combination of stresses, both thermal and electronic, with the aim to eliminate all weak devices and to obtain in short intervals a responsible estimate of device reliability for the remaining robust population.

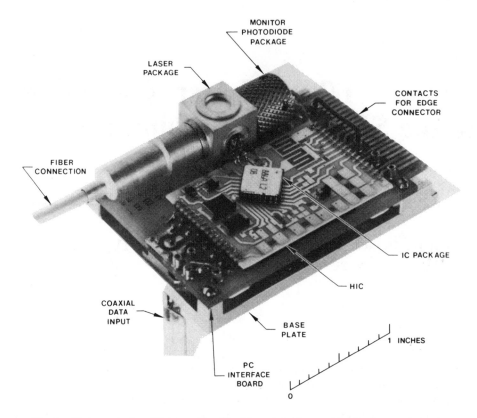

Fig. 1. Photograph of an SL transmitter. The integrated driver and feedback control circuits are in a single IC package. The edge connector contacts are used only for testing during manufacture. The data input and the dc power leads are on the bottom of the transmitter housing.

TRANSMITTER LAYOUT

The transmitter is divided into three hermetically packaged building blocks: the laser package with fiber and monitor light output, the photodiode package, and the IC package. As shown in Fig. 1, the monitor photodiode is attached to the laser package which is mounted firmly to a base plate through a stud which effectively heatsinks the laser. The base plate also supports a printed circuit board providing the many external connections to the hybrid integrated driver circuit. The driver circuit, which consists of an alumina thin-film substrate, mounted above the PC board with compliant electrical connections, carries the hermetic IC package and the discrete chip capacitors. The heart of the hybrid driver is the IC package containing two integrated circuit chips: a high-speed modulation chip and a low-speed feedback control chip. A very low thermal resistance path is provided from the back of the alumina substrate, directly under the IC package, to the base plate. The IC's and the laser are, therefore, well heatsinked since the base plate is in good thermal contact with the repeater housing.

By these measures, the maximum rise of the laser junction temperature and the average IC chip temperature above the base-plate temperature is limited to approximately 12°C and 7°C, respectively. This occurs when the laser reaches its maximum end-of-life bias current which also entails the maximum IC dissipation.

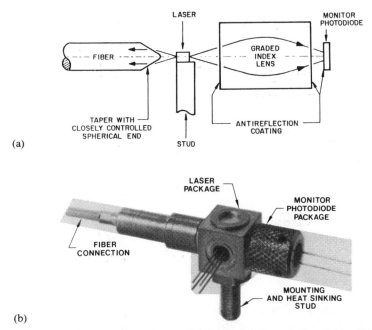

Fig. 2. SL laser package configuration. (a) Schematic showing coupling of laser light to a tapered single-mode fiber with a hemispherical tip. The light from the back facet is coupled through a graded-index lens window in the laser package to the separately encapsulated p-i-n photodiode monitor. (b) Photograph showing the monitor photodiode affixed to the laser package. Two pairs of parallel leads assure a low inductance connection to the electrically insulated laser chip.

The transmitter has a fiber connection and four electrical connections in the repeater. The electrical connections, a semirigid coaxial input for data, two power supply connections, and a laser bias monitor go through the base plate. An edge connector on the PC board permits a convenient, simultaneous connection to all test points for automated functional and life testing. This connector is not used in system operation.

CHARACTERISTICS OF PACKAGED LASER–FIBER–PHOTODIODE COMBINATION

An index-guided laser inherently provides the light beam stability necessary for long-term stable coupling into a single-mode fiber. The InGaAsP BH-laser [3] selected for the SL system, was the first high-performance 1.3-μm index-guided laser with high reliability potential. In addition, this laser has high output (> 5 mW) with fundamental mode stability, low threshold current (typically $\lesssim 25$ mA), small modulation current (~ 30 mA), an almost symmetrical beam pattern allowing high fiber coupling efficiencies (~ 4 dB), and flat wide-band modulation response [4] (up to 2 GHz) adequate for direct digital gigabit per second modulation [5], [6].

A tapered fiber with a well-controlled spherical tip provides an efficient laser to fiber coupling [7], [8] (see Fig. 2(a)). In addition to simplicity, the lensed fiber end assures low reflections back into the laser. This effectively minimizes one source of optical feedback. Small reflections from other sources are still present, but the BH-laser exhibits a low reflection sensitivity and a corresponding low system performance penalty [9] from the resulting modified laser behavior [10], [11].

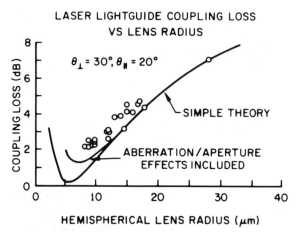

Fig. 3.   Laser–fiber coupling performance compared with theory for hemispherical lenses of various radii (from R. T. Ku [7]).

The laser backface light is coupled to a photodiode through a graded-index lens. The active photodiode area is large compared to the defocused light spot image to guarantee long-term laser light monitoring stability. The InGaAs monitor photodiode is of planar structure and is especially designed for high reliability [12]. It has the same structure as the receiver photodiode [13], but has a larger active area. Light reflections from the monitor arrangement back into the laser are kept low by antireflection coatings on the graded-index lens and the photodiode and by a dileberate defocusing of the photodiode [14]. These packaging design considerations resulted in a compact, stable, and robust package combination for the laser and photodiode, as shown in Fig. 2(b).

Theoretical and experimental laser-to-fiber coupling results as a function of lens radius are in good agreement (Fig. 3). The stable laser–fiber and laser–monitor coupling is

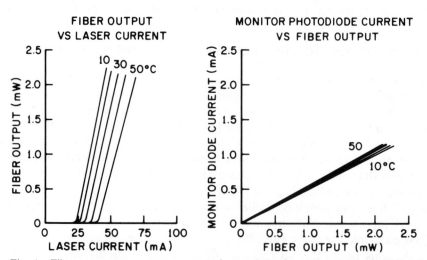

Fig. 4.   Fiber output power versus current characteristics for a typical 1.3-μm InGaAsP BH-laser used for SL. The monitor current versus fiber output tracking characteristics show both the temperature dependence and the power dependence of the fiber coupling.

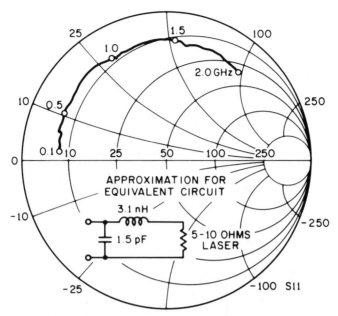

Fig. 5. Drive impedance versus frequency for the SL laser package. The laser chip can be represented by a 5–10-Ω resistor. The package inductance of 3.1 nH gives a series reactance < 10 Ω up to 500 MHz.

reflected in a good tracking characteristic between fiber and monitor output, as seen in Fig. 4.

A well-controlled electrical RF drive impedance for the laser package is also important in order to make use of the laser modulation potential. The laser is electrically insulated from the package, which eliminates dc-potential restrictions. A hermetic header with four insulated leads connects the laser to the outside. One lead pair is connected to each laser terminal to keep the inductance low. The package input impedance up to 2 GHz and its first-order equivalent circuit are shown in Fig. 5. The laser chip itself can be well approximated by a 5–10-Ω resistor. The inductive series reactance remains below 10 Ω up to about 0.5 GHz, giving a more than adequate frequency range for 300-Mbit/s operation. The lead shunt capacitance plays a role only at much higher frequencies causing in conjunction with the inductance, a parallel resonance between 2 and 3 GHz.

Automatic Light Feedback Control Strategy

An automatic light control that adjusts the bias current level is needed to accommodate both the threshold current increase with aging and the inherent temperature sensitivity [15]–[20]. Five different light control approaches are listed in Table I in order of increasing circuit complexity and required photodiode speed. The single-loop averaging light feedback control, proportional to the average data level has been selected. The required circuit is simple and attractive from the circuit reliability point of view. Although it accommodates only small slope changes of the laser light–current characteristic, this is not a limitation with high quality BH-lasers. Aging experiments show that a substantial part of the laser population has acceptable small slope changes and these can be selected by screening.

TABLE I
Comparison of Automatic Light Feedback Control Options

CONTROLLED QUANTITY	FEEDBACK LOOPS	ADJUSTED PARAMETERS	DISADVANTAGES	ADVANTAGES	COMPLEXITY FOR INTEGRATED FEEDBACK CIRCUIT
AVERAGE LIGHT (INDEPENDENT OF DATA)	ONE	BIAS CURRENT	DUTY CYCLE SENSITIVE, SMALL LASER SLOPE AGING REQUIRED	LOW OFF-STATE BELOW THRESHOLD	LOW
AVERAGE LIGHT (PROPORTIONAL TO AVERAGE DATA LEVEL)	ONE	BIAS CURRENT	SMALL LASER SLOPE AGING REQUIRED	DUTY CYCLE INSENSITIVE, LOW OFF-STATE BELOW THRESHOLD	
AVERAGE LIGHT AND OFF-STATE SLOPE (dL/dI) (CLOSE TO THRESHOLD)	TWO	MODULATION AND BIAS CURRENT	LASER AT THRESHOLD MOST SENSITIVE TO LIGHT REFLECTIONS	DUTY CYCLE INSENSITIVE, SIGNIFICANT LASER SLOPE CHANGE ACCEPTABLE	MEDIUM
ON-STATE AND OFF-STATE LIGHT	"	"	OFF-STATE CRITICAL FOR TURN-ON DELAY	"	HIGH
TIMING AND ON-STATE OR AVERAGE LIGHT	"	"	VERY HIGH-SPEED MONITOR AND CIRCUIT REQUIRED	"	HIGH

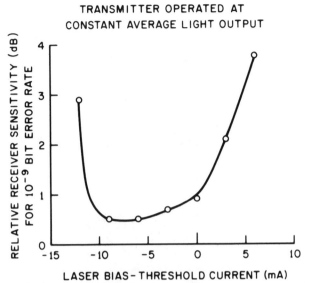

Fig. 6. Relative transmitter power required for a 10^{-9} bit error rate as a function of the laser bias. The transmitter was operated with constant average light output. In system use the receiver is operated with higher received power and with an error rate many orders of magnitude lower than the 10^{-9} used to facilitate this measurement.

For a single-loop light control, a stable feedback signal derived directly from the laser light is essential. The backface monitor arrangement was designed for highly stable coupling to the laser. This assures that the quality of the light pulse (e.g., extinction ratio and timing) is unaffected by any change in the laser–fiber coupling.

An alternative light source stabilization scheme, which was also considered, involves monitoring the fiber output with a suitable optical tap. With this approach, however, fiber coupling variations with aging (when using a fiber-tap monitoring scheme in combination with a single-loop feedback) would lead either to a loss in extinction ratio or excessive turn-on delay, depending on whether the coupling decreases or increases. A dual-loop feedback would then be required and this was rejected because of the extra complexity without compensating reliability benefits.

The laser bias in the transmitter is set a few milliamps below threshold. The digital modulation current is adjusted to give about 5-mW ON light per laser facet. If light slope changes occur for a laser under average light feedback control, the bias setting with regard to threshold will also change. The influence on this bias change on the system performance was evaluated with the following experiment: laser light slope changes leading to bias changes were simulated by varying the modulation current of a laser under feedback control. The relative required receiver power for 10^{-9} bit error rate as a function of bias was determined and is given in Fig. 6 for a typical laser. The system performance changes only slightly for a bias between 0 and 10 mA below threshold. Below this range there is a hard limit because of light pulse degradation through pattern dependent turn-on delay [21]. One finds a soft limit for bias levels above threshold due to a gradual reduction in extinction ratio.

The high-performance bias range is, however, wide enough to build reliable transmitters with single-loop feedback using screened and purged BH-lasers.

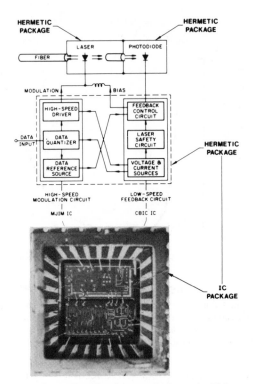

Fig. 7. Schematic of transmitter electronics. The photograph shows the high-speed driver IC and the feedback control IC which are mounted together in a hybrid integrated circuit package.

INTEGRATED TRANSMITTER CIRCUITS

Reliability and manufacturability considerations mandate that the transmitter circuit be highly integrated. As outlined in the transmitter schematic of Fig. 7, the circuit is divided into two sections with different functional requirements: 1) a high-speed modulation section with subnanosecond switching times for currents up to 60 mA, and 2) a low-speed automatic light feedback control circuit with a response time in the millisecond range. The feedback circuit provides up to 200 mA of bias current, which allows a large margin for end-of-life BH-laser bias requirements.

The medium-current high-speed section and the high-current low-speed section were designed in two different, proven silicon bipolar technologies. The first was realized with microwave junction isolated monolithic (MJIM) n-p-n technology [22] containing transistors with 4-GHz unity gain frequency f_T. Complementary bipolar integrated circuit (CBIC) technology [23], [24] with vertical n-p-n and p-n-p transistors having f_T's > 400 MHz was used for the second.

The entire active transmitter circuit consists, therefore, of two silicon IC's which are mounted together in one hermetic carrier with 28 contacts (see Fig. 7). The carrier is mounted to an alumina substrate which provides good heatsinking.

A relatively low transmitter supply voltage, 6.8–7.2 V, was adopted in order to keep the power dissipation small. The dissipation of the MJIM IC, adjusted for 30-mA typical modulation current, is approximately 500 mW and the CBIC IC dissipation with 25-mA typical bias current is about 200 mW.

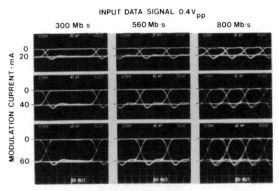

Fig. 8. Eye diagrams for the laser modulation current provided by the high-speed MJIM integrated circuit. Data are shown for currents of 20, 40, and 60 mA and for modulation rates of 300, 560, and 800 Mbit/s. The BH laser used required typically < 30-mA modulation current and the SL system bit rate is 295.6 Mbit/s. The eye diagram is the superposition of all possible word patterns using a pseudorandom word generator.

MODULATION IC AND FEEDBACK IC FUNCTIONS

The modulation IC contains a data quantizer, a highspeed driver and a data reference source, as schematically shown in Fig. 7.

The input signal is requantized to make the transmitter performance insensitive to input amplitude variations. As a result, practically no performance change is seen for an input signal range from 0.3–0.6 V. The requantized signal is applied to the high-speed driver which provides the laser modulation current.

The MJIM high-speed modulation performance is shown in Fig. 8. The 20-, 40-, and 60-mA current eye diagrams are given for 300, 560, and 800 Mbit/s. The rise and fall times increase slightly with current levels. For 60 mA, the rise time is still ≤ 0.4 ns and the fall time ≤ 0.6 ns (with correction for the 0.35-ns scope rise time).

The MJIM circuit has also been used successfully in high-speed LED transmitters delivering 120-mA modulation current with ∼1-ns rise and fall times.

The data reference source provides a reference signal for automatic light control which is proportional to the data duty cycle.

The modulation current and data reference signal are insensitive to temperature and supply voltage variations. They are controlled through stabilized currents injected from the CBIC circuit.

The CBIC feedback control circuit receives as its error correction signal the difference between the monitor photodiode current and the data reference signal from the MJIM circuit. The overall feedback response time is a few milliseconds.

A safety circuit makes sure that the feedback circuit becomes operational only after proper voltages have been established everywhere. It guarantees that the laser reaches its final operating point without experiencing damaging transients. It blocks any laser bias if the supply voltage remains or drops below approximately 6 V.

The safety circuit performance and the feedback control transient have been tested by square wave modulating the supply voltage. Figure 9 shows three cases, where the supply voltage is stepped from 0 to 8, from 5.8 to 8 and from 6.2 to 8 V. In the first two cases the 6-V safety switching point is crossed and the light switches completely on and off. In the third case the light stays on all the time and the supply voltage steps cause only very

MODULATION FREQUENCY: 8.3 Hz

SAFETY SWITCHING VOLTAGE ~ 6 VOLTS

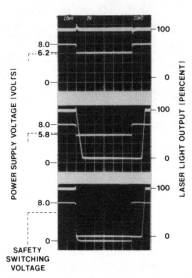

Fig. 9. Demonstration of the operation of the laser safety circuit using a square-wave modulation on the supply voltage. This circuit protects the laser by assuring that it cannot be powered when there is insufficient voltage for proper transmitter operation. In the figure the modulation test frequency is 8.3 Hz and the safety circuit switches at 6 V.

minor light ripples. The feedback controlled laers turn-on occurs essentially without overshoot and free of transient oscillations.

Two monitor terminals permit monitoring the average laser light and the bias current while the transmitter is in operation. The monitoring of the bias current is used in the SL system to give an indication of the laser health.

Transmitter Performance

The transmitter performance is shown in Fig. 10 with NRZ light eye diagrams and light pulse sequences at 295.6 Mbit/s using a $2^{15} - 1$ digit pseudorandom pattern. Switching times of < 0.75 ns are achieved, and there is little amplitude and timing pattern dependence. The light turn-on time from below threshold is actually much shorter, < 0.2 ns, than the turn-off time due to the gain-switching mechanism governing the laser dynamics [25], [26]. The light eye diagrams of Fig. 10 resemble closely, as expected, the inverted current eye diagrams shown in Fig. 8. Due to the automatic feedback control, the eye diagram is virtually constant from 10° to 50°C.

The transmitter can accommodate a 30-percent supply voltage surge, which is part of the system specifications. The circuits are designed so that the laser bias and modulation currents remain regulated during such a surge, and the transmitter performance is thus not degraded.

The laser wavelength is limited to the 1290–1330-nm range with an rms spectral width of 2 nm. A CW optical spectrum at the 2-mW fiber level and a pulsed spectrum at the 1-mW average level are shown in Fig. 11. In this example, the rms spectral width is 1.29-nm CW and 1.55-nm pulsed.

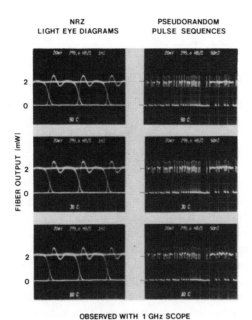

Fig. 10. NRZ light eye diagrams and pulse sequences for the SL transmitter at 295.6 Mbit/s using a 2^{15}-1 digit pseudorandom word and a 1-GHz bandwidth oscilloscope. The automatic feedback control assures that the eye diagram is not degraded over the 10–50°C temperature range.

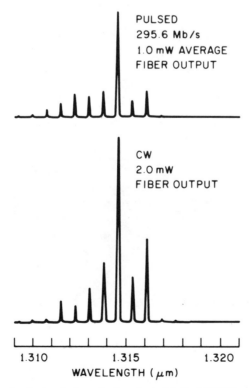

Fig. 11. CW and pulsed optical spectra for a typical InGaAsP BH-laser used in SL.

Each transmitter is subjected to both a functional test and a stringent certification test before it is committed to the system.

TRANSMITTER RELIABILITY

The system reliability requirements for a transatlantic installation lead to a failure rate allocation of 70 FIT's for each transmitter when used without redundancy and with a switchable standby line [27].

At the present time, projections of ongoing reliability testing indicate laser functional lifetimes $\geqslant 10^7$ h, with a corresponding failure rate of $\leqslant 100$ FIT's. This is still about 100 times larger than failure rates allocated for other individual components.

To provide adequate reliability margin for TAT-8, the system design uses two levels of redundancy [28]–[30]. The first is achieved in the regenerator by providing two transmitters per line, one powered and one unpowered standby unit. In addition, the system is designed with a complete switchable standby line as protection for the two active lines in each direction. This permits transatlantic system reliability objectives to be met while allowing approximately 700 FIT's per laser [27].

The transmitter array module has been designed to accommodate four transmitters per regenerator, allowing as many as three redundant transmitters. Mechanically, the trans-

SCHEMATIC

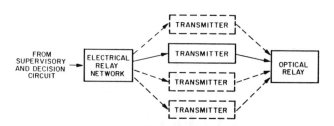

MECHANICAL TRANSMITTER ARRANGEMENT

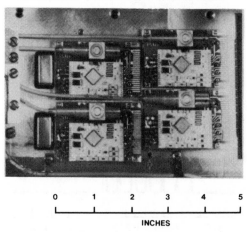

Fig. 12.   Schematic and photograph of an array of four SL transmitters which allows the use of three standby units. The current plan is to use only two transmitters per line in the typical TAT-8 repeater with one active and one standby unit.

mitter layout permits a compact mounting of a group of four in a 4.5″×3.5″ area, as shown in Fig. 12. As indicated schematically, an electrical relay switches electrical signals and power connections from one transmitter to another, and an optical relay [31] switches the optical signals from the various transmitters onto a single output fiber in response to supervisory control signals.

In manufacture, the transmitter reliability is assured by careful certification of all components used in combination with elaborate characterization and burn-in testing of each transmitter.

Environmental Tests

In addition to the long-term stability requirements for the transmitter, it must also withstand the shock, vibration and temperature cycling conditions experienced in repeater assembly and bake-out, storage and transportation, and cable laying and recovery operations.

The SL prototype transmitters were subjected to shock tests up to 800 g along three mutually perpendicular axis without harm. This is eight times the specification level of 100 g.

The transmitters passed also a 3-g vibration test from 10–2000 Hz without any measurable change. This is three times the specified acceleration level.

Temperature cycling was performed with both powered and unpowered transmitters. The unpowered cycling was intended to insure the ability of the transmitter to withstand storage and transportation in adverse conditions. This was performed over the $-20°$ to $+60°$C temperature range. The powered cycling approximated more closely the manufacturing and service temperatures. The powered transmitters were cycled over the $0°$ to $+50°$C temperature range without degrading their performance.

Summary

A high-performance compact laser transmitter for undersea application has been designed and put into manufacture. The transmitter circuit is integrated in proven silicon bipolar technology.

Transmitter redundancy has been designed into the system to permit system reliability appropriate for long-haul undersea use to be obtained with currently available high-reliability lasers.

To facilitate design qualification and product certification of the different technologies involved, the laser, monitor photodiode, and IC's are put into separate hermetic packages. This permits optimized testing and certification prior to their commitment to a transmitter.

A special feature of the transmitter is the access to multiple test points through an edge board connector for automated tuning and testing.

Acknowledgment

We acknowledge the contributions of many individuals in different groups at AT&T Bell Laboratories.

REFERENCES

[1]   D. G. Ross *et al.*, "A highly integrated regenerator for 280 Mbit/s undersea optical transmission," this book, ch. 26, p. 377.

[2]   E. I. Gordon, F. R. Nash, and R. L. Hartman, "Purging: A reliability assurance technique for new technology semiconductor devices," *IEEE Electron Device Lett.*, vol. ED-4, pp. 465–468, Dec. 1983.

[3]   M. Hiraro, A. Doi, S. Tsuji, M. Nakamura, and K. Aiki, "Fabrication and characterization of narrow stripe InGaAsP/InP buried heterostructure lasers," *J. Appl. Phys.*, vol. 51, pp. 4539–4540, Aug. 1980.

[4]   R. S. Tucker and I. P. Kaminow, "High frequency characteristics of directly modulated InGaAsP ridge waveguide and buried heterostructure lasers," *IEEE J. Lightwave Technol.*, vol. LT-2, pp. 385–393, 1984.

[5]   W. Albrecht, C. Baak, G. Elze, B. Enning, G. Heydt, L. Ihlenburg, G. Walf, and G. Wenke, "Optical digital high speed transmission: General considerations and experimental results," *IEEE J. Quantum Electron.*, vol. QE-18, pp. 1547–1559, Oct. 1982.

[6]   K. Hagimoto, N. Ohta, and K. Nakagawa, "4 Gibt/s direct modulation of 1.3 μm InGaAsP/InP semiconductor lasers," *Electron. Lett.*, vol. 18, no. 8, pp. 796–798, Sept. 2, 1982.

[7]   R. T. Ku and W. H. Dufft, "Hemispherical microlens coupling of semiconductor lasers to single-mode fiber," in *Proc. OFC'82* (Phoenix, AZ), Apr. 13–15, 1982, paper THDD7, pp. 60–62.

[8]   R. T. Ku, "Progress in efficient/reliable semiconductor laser-to-single-mode-fiber coupler development," in *Proc. OFC'84* (New Orleans, LA), Jan. 23–25, 1984, paper MB3, pp. 4–6.

[9]   V. J Mazurczyk, "Sensitivity of single mode buried heterostructure lasers to reflected power at 274 Mbit/s," *Electron. Lett.*, vol. 17, no. 3, pp. 143–144, Feb. 5, 1981.

[10] R. Lang, and K. Kobayshi, "External optical feedback effects on semiconductor injection laser properties," *IEEE J. Quantum Electron.*, vol. QE-16, pp. 347–355, Mar. 1980.

[11] K. Y. Liou, F. Bosch, and C. B. Swan, "Comparison of optical feedback effects for laser diodes coupled to multimode and single-mode fibers," in *Proc. OFC* (New Orleans), Feb. 28–Mar. 2, 1983, pp. 68–71.

[12] R. H. Saul, F. S. Chen, and P. W. Shumate, "Reliability of planar InGaAs photodiodes," in *Proc. OFC'84* (New Orleans, LA), Jan. 23–25, 1984, paper TUH2, pp. 50–51.

[13] L. Snodgrass and R. Klinman "A high reliability, high sensitivity lightwave receiver for submarine cable applications," this book, ch. 33, p. 475.

[14] G. Arnold, "Influence of optical feedback on the noise behavior of injection lasers," in *Proc. 7th European Conf. Optical Commun.* (Copenhagen, Denmark), 1981, paper 10.4–14.

[15] P. W. Shumate, F. S. Chen, and P. W. Dornam, "GaAlAs laser transmitters for lightwave transmission systems," *Bell Syst. Tech. J.*, vol. 57-6, pp. 1823–1836, July–Aug. 1978.

[16] D. W. Smith and T. G. Hodgkinson, "Laser level control for high bit rate optical fiber systems," in *Proc. 13th Circuits Syst. Int. Symp.* (Houston, TX), Apr. 1980, pp. 926–930.

[17] J. Gruber, P. Marten, R. Petschacher, and P. Russer, "Electronic circuits for high bit rate digital fiber optic communication systems," *IEEE Trans. Commun.*, vol. COM-26, pp. 1088–1098, July 1978.

[18] F. S. Chen, "Simultaneous feedback control of bias and modulation currents for injection lasers," *Electron. Lett.*, vol. 16, no. 1, pp. 7–8, January 3, 1980.

[19] R. E. Epworth, "Subsystems for high speed optical links," in *Proc. Second European Conf. Optical Commun.* (Paris, France), 1976, p. 377–382.

[20] S. R. Salter, D. R. Smith, R. P. Webb, and B. R. White, "Laser automatic level control circuits for optical communications systems," in *Proc. 3rd European Conf. Optical Commun.*, (Munich), 1977, pp. 209–210.

[21] S. M. Sze, *Physics of Semiconductor Devices*, 2nd ed. New York: Wiley, 1981, Section 12.5.3: *Turn-On Delay and Modulation Frequency*, pp. 735–737.

[22] W. Kruppa and F. D. Waldhauer, "A UHF monolithic operational amplifier," in *Proc. 1978 IEEE ISCC*, (San Francisco, CA), Feb. 15–17, 1978, paper WPM 7.2, pp. 74–75, and 267.

[23] P. C. Davis and S. F. Moyer, "Ion implanted, compatible, complementary p-n-p's for high slew rate operational amplifier," in *IEDM Tech. Dig.* (Washington, DC), Dec. 4–6, 1972, pp. 18–20.

[24] D. W. Aull, D. A. Spires, P. C. Davis and S. F. Moyer, "A high voltage IC for a transformerless trunk and subscriber line interface," *IEEE J. Solid-State Circuits*, vol. SC-16, pp. 261–266, Aug. 1981.

[25] P. M. Boers, M. T. Vlaardingerbroek, and M. Danielson, "Dynamic behavior of semiconductor lasers," *Electron. Lett.*, vol. 11, no. 10, pp. 206–208, May 15, 1975.

[26] C. Lin, "Microwave frequency intensity modulation and gain switching in semiconductor injection lasers," in *Proc. SPIE EAST*, 1984.

[27] R. L. Easton, private communication.

[28] C. D. Anderson, R. F. Gleason, P. T. Hutchison, and P. K. Runge, "An undersea communication system using fiberguide cables," *Proc. IEEE*, vol. 68, no. 10, pp. 1299–1303, Oct. 1980.

[29] P. K. Runge and P. R. Trischitta, "The SL undersea light guide system," *IEEE J. Select. Areas Commun.*, vol. SAC-1, pp. 459–466, Apr. 1983.

[30] D. K. Paul, K. H. Greene, and G. A. Koeff, "Undersea fiber-optic cable communications system of the future: Operational, reliability, and system considerations," in *OFC'84*, (New Orleans, LA), Jan. 23–25, 1984, paper TUN18, pp. 84–85.

[31] S. Kaufman and R. L. Reynolds, "Optical switch for the SL submarine cable," pp. 975–979.

32

System Penalty Effects Caused by Spectral Variations and Chromatic Dispersion in Single-Mode Fiber-Optic Systems

PETER J. ANSLOW, JEFFREY G. FARRINGTON,
ISOBEL J. GODDARD, AND W. R. THROSSELL

INTRODUCTION

The difficulty in obtaining access to cables and repeaters after their deposition on the seabed makes it necessary to design and manufacture submarine communication equipment to the highest standards. A detailed study has to be made of all factors affecting performance which may change with operating conditions or during service life. The spectral characteristics of the lasers used as optical sources are of particular importance in this respect. Ideally, the laser would operate monochromatically at the wavelength for which the single-mode fiber has zero chromatic dispersion, but although various expedients are available to ensure single-mode operation in the laboratory none is at present applicable for unattended long-term use in environments with a wide range of temperatures. Multimode operation, with a spectral envelope extending over a few nanometers, must therefore be expected. Nominally similar lasers may differ considerably from each other in spectral width and its dependence on output and temperature. Moreover, because of production tolerances, some degree of mismatch between the central wavelength of the laser and the zero-dispersion wavelength will usually be present.

The effects of these factors have been examined by several authors. Marcuse [1] and Kapron [2] considered the limits on system bandwidth for a single spectral line of finite width and fixed wavelength. The length–bandwidth product is very high even when the source wavelength differs by tens of nanometers from the zero-dispersion wavelength, and it does not constitute a limitation on the systems which concern us here. The width of individual laser lines is small enough for us to treat them as monochromatic sources at bit rates up to at least 1 GHz.

With a multimode spectrum and a dispersive fiber there will be some degree of pulse broadening which may necessitate an increase in receiver bandwidth to avoid inter-symbol interference. This in turn will lead to an increase in noise level [3]. More important, however, are the variations in the spectrum. Multimode spectra are invariably subject to change during modulation and, even if not modulated, to a continuous interchange of energy between modes. The effect of this partition has been analyzed by Yamamoto *et al.* [4] and by Ogawa [5], [6].

The former authors adapt Personick's treatment [7] of system performance in the presence of various noise sources by summing the contribution from each spectral line, with appropriate delays imposed by chromatic dispersion, and calculating the variance due to partition between modes. The major conclusion is that the noise contribution from mode partition effects is essentially a function of a delay parameter s' defined as the product of the chromatic dispersion (ns per nm), the width of the spectral envelope (nm), and the bit rate (Gbit/s). The sensitivity to pulse shape is quite small, so a single curve of signal-to-noise ratio against s' is approximately valid over a wide range of bit rates and delays. It is estimated that a Gaussian distribution for this noise would lead in the absence of any other noise source to an error ratio of 10^{-9} when s' is in the region of 0.28. Ogawa's treatment of the problem is simpler but leads to a rather similar result. Put in the same terms, it predicts an error of 10^{-9} when the delay parameter (as defined above) $= 0.25$ and there are no other sources present. To take an example, this would correspond to a data rate of 1 Gbit/s, an envelope width (FWHM) of 2.5 nm, and a chromatic dispersion of 1 ns/nm. Both results are "worst cases" in that they are based on the assumption of the highest possible level of mode partition. That is, at any instant all the laser output is in only one of the available modes. For less severe mode partition, Ogawa applies a correction factor k to the delay parameter, where k $(0 < k < 1)$ is a measure of the fluctuations in the instantaneous powers in each mode.

These results can be extended, on the basis of fairly simple assumptions, to the estimation of "spectral penalties" in a working system, with noise contributions from the receiver. "Spectral penalty" is defined as the increase in received power, over that sufficient in the absence of dispersion, which is required for the maintenance of a set error ratio. It is not very dependent on receiver performances for receivers of the same type, but a distinction needs to be made between those using avalanche photodiodes at optimum gain and those using p-i-n diodes where the shot-noise contribution is small: The penalty is up to twice as great in the former case, depending on the excess noise factor of the APD. [8]. An exact comparison between the theoretical predictions for different systems is therefore not easy, but both authors find the penalty to be negligible for $s' < 0.1$. It then increases sharply with s', becoming infinite when s' attains the limiting value for the error ratio.

The measurements reported in this paper for a 320-Mbit/s system using p-i-n photodiode receivers reveal, for many lasers, a somewhat more severe penalty at low s' values than would be expected. In seeking an explanation for this, we have examined the contribution of spectral vagaries other than inter-mode partition and have also re-examined the relationship between partition effects and dispersion-induced noise. We consider in particular the influence of regular changes in wavelength associated with modulation. Because of the range of these effects, even for lasers of the same type, there is a need for a test facility which will measure dispersion penalties quickly and easily while exploring as far as possible the full range of environmental conditions which may be encountered. Equipment for this purpose is described and comparisons are made between results and the theoretical expectations.

Types of Spectral Variation and Their Effects

It is convenient to classify the temporal changes in a laser spectrum under three headings:

1) partition between the lasing modes;
2) changes in the wavelength of individual modes;
3) changes in the mean wavelength due to modulation.

These are considered separately below.

Partition Between Modes

Partition of power between the modes of a multimode laser has been examined by many workers [9]–[11]. It is easily revealed as variations in the output from a spectrometer which is set to isolate one mode. The magnitude of the variation is dependent on temperature and derive conditions, commonly having its greatest value when there are two major spectral lines with the same (time-averaged) intensities [9]. The variance of the intensity for strong lines may then approach one third of the mean-square value, indicating an equal probability for all instantaneous intensities from 0 to 100 percent of the laser output. The most extreme form of partition, where the instantaneous power is always 100 percent in one of the available modes (corresponding to $k = 1$ in Ogawa's formulation [10]) does not appear to occur in practice.

In considering the effects of this phenomenon in the presence of chromatic dispersion, it is necessary to consider how it is affected by digital modulation. One procedure is to associate each "on" pulse with a specific intensity for each mode which is constant during the pulse, the sum of these intensities being fixed by the total output of the laser. In other words, the partition process is taken to be linked to the modulation, and the power only changes from pulse to pulse. We know, however, that partition occurs even without modulation, with a radio-frequency spectrum extending over hundreds of megahertz (Fig. 1). We may in fact envisage a situation where partition is independent of modulation. The difference is illustrated in Fig. 2, which shows the RF spectra for a single spectral line, under modulation, for the two models. Whereas the noise spectrum for the modulation-

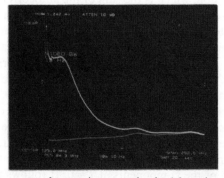

Fig. 1. Frequency spectrum of output in one mode of a 1.3-μm laser, biased 25 mA above threshold, not modulated. The lower trace shows the noise on the total output (all modes) of the laser, scaled to the same mean power.

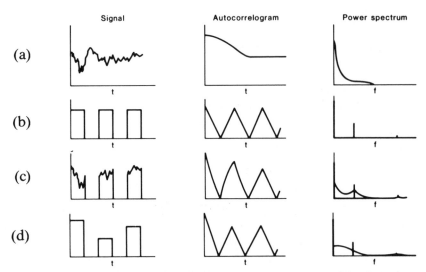

Fig. 2. Partition spectra models. (a) RF spectrum for one mode, unmodulated. (b) Square wave. (c) (a) Modulated by (b). (d) Modulation-linked partition.

linked model is a sinc^2 function, the other model shows a different frequency content, its exact form depending upon the spectrum of the unmodulated noise. In the analysis which follows we use this second model, while recognizing that the behavior of a real laser will be somewhere between the two extremes.

The Effect of Dispersion: Suppose that we have an unmodulated laser with partition of power between $2n+1$ modes, the instantaneous fractional powers in each being

$$a_{-n}, a_{-n+1}, a_{-n+2}, \cdots a_0, a_1, a_2, \cdots a_n.$$

Each mode differs from its neighbors by $\Delta\lambda$ in wavelength, and the fiber length and dispersion are such that the mode a_i is retarded by a time equal to $i\tau$ referred to the central model. The effect of these delays is to produce noise in the output signal. We obtain the power spectrum of this noise, in the usual way, by taking the Fourier transform of the autocorrelation function, which is the sum of the autocorrelation functions for each mode, together with cross-correlation terms. For white noise, each correlogram consists of a δ-function at the origin together with a uniform value corresponding to the mean-square signal. In general, the noise will not be white, but it is reasonable to assume that the correlograms will be alike in shape, differing only by scale factors related to the average intensities of each mode. The cross-correlograms are displaced from the time origin because of the differential delays and contain negative "spikes" corresponding to the negative correlation between mode intensities. For a symmetrical optical spectrum we can pair the contributions of a_i and a_{-i}. The autocorrelogram for the complete signal is then symmetrical (Fig. 3). The constant total power condition implies that the sum of the variances is equal and opposite to the sum of the covariances so the sum of the negative spikes is equal to the sum of the components at $t=0$. It follows that the noise spectrum has a value of zero at the origin and is periodic, with minima at separations of $1/\tau$. For most cases of practical interest, $\tau \ll$ the bit period, so the first part of the spectrum is the most important.

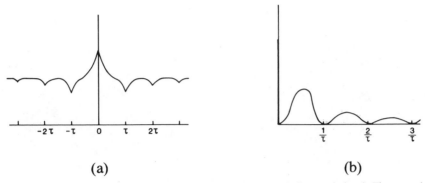

Fig. 3. Noise induced by dispersion. (a) Autocorrelogram of dispersed signal. The central peak is due to the sum of the autocorrelation contributions from each mode. The negative peaks are due to the sums of cross-correlation terms. (b) Spectrum (single-sided). It is the partition spectrum multiplied by a periodic function.

The noise spectrum, then, consists of the noise spectrum of the partition process $F(\omega)$, multiplied by the sum

$$\left\{ \sum b_i + \sum c_{i\,i+1} \cos \omega\tau + \sum c_{i\,i+2} \cos 2\omega\tau \cdots \right\} \tag{1}$$

where b_i and c_{ij} are variances and covariances, respectively.

It is convenient to normalize $F(\omega)$ with respect to the mean signal so that, for each mode in isolation, we have

$$\frac{\text{mean square noise}}{\text{signal}^2} = b_i \int_0^\infty F(\omega)\, d\omega.$$

We can relate the variance and covariance terms to the average values $\bar{a}_i$ by the following argument. If we assume that partition is a random process but weighted according to the average power in each mode, we see that c_{ij}, the covariance between the ith and jth modes, will be equal to $-k^2 \bar{a}_i \bar{a}_j$, where k is a measure of the magnitude of partition fluctuations. The covariance between the ith mode and all other modes will be $-k^2 \bar{a}_i (1 - \bar{a}_i)$. It follows that $b_i = k^2 \bar{a}_i (1 - \bar{a}_i)$. The multiplier can then be expressed as

$$k^2 \left\{ \sum \bar{a}_i (1 - \bar{a}_i) - \sum \bar{a}_i \bar{a}_{i+1} \cos \omega\tau - \sum \bar{a}_i \bar{a}_{i+2} \cos 2\omega\tau \cdots \right\}. \tag{2}$$

For small $\omega\tau$, we use the approximation

$$\cos n\omega\tau \approx 1 - \frac{n^2 \omega^2 \tau^2}{2}$$

and obtain

$$G(\omega) = F(\omega) \cdot \frac{k^2 \omega^2 \tau^2}{2} \left\{ \sum \bar{a}_i \bar{a}_{i+1} + 4 \sum \bar{a}_i \bar{a}_{i+2} + 9 \sum \bar{a}_i \bar{a}_{i+3} \cdots \right\}$$

$$= F(\omega) k^2 \omega^2 \tau^2 A. \tag{3}$$

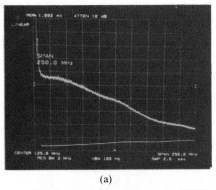
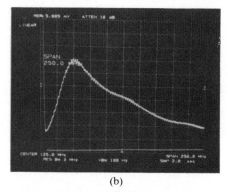

(a)   (b)

Fig. 4. (a) RF spectrum of intensity variations for one mode of an unmodulated laser. (b) RF spectrum of total laser output after transmission through fiber with a dispersion of 200 ps per laser mode. The mean signal is 16 dB (optical) higher than that in (a).

The value of A is sensitive to the shape and width of the spectral envelope. Some values are

0.12   for two modes only

0.6   for a Gaussian envelope $\bar{a}_i = 1/\sigma\sqrt{2\pi}\,e^{-i^2/2\sigma^2}$ and $\sigma = 1$

2.0   for a Gaussian envelope with $\sigma = 2$.

$G(\omega)$, the spectrum of the noise induced by chromatic dispersion, then, is the product of the mode partition spectrum $F(\omega)$ and a factor $A\cdot(k\omega\tau)^2$ where A lies between 0 (single mode) and 2 or more (for multimode operation). The factor is zero at $\omega = 0$. We see that the signal-to-noise ratio is much higher than that for a single mode in isolation (Fig. 4) and is very large at low frequency.

An upper bound for the dispersion noise can be calculated if we assume that the partition spectrum is flat up to an abrupt bandwidth limit ω_0. $F(\omega)$ is then $1/\omega_0$, ($\omega \leqslant \omega_0$) and the total chromatic noise is, using (3)

$$\int_0^\infty G(\omega)\,d\omega = \frac{k^2}{\omega_0}\int_0^\infty A\omega^2\tau^2\,d\omega$$

$$= \frac{Ak^2\omega_0^2\tau^2}{3}. \tag{4}$$

We may use noise levels obtained in this way to estimate error ratios and spectral penalties in a complete system.

It is well known that, subject to certain assumptions [12], an error ratio of 1 in 10^{-9} in a digital system corresponds to a signal-to-noise ratio Q of 12, with the decision threshold set at 50 percent of the mean "1" level. It is easily shown, on the same assumptions, that the spectral penalty α in a system with other noise sources is approximately $-5\log(1 - (144/Q_s^2))$ where Q_s is the signal-to-noise ratio due to dispersion alone. An exact expression, taking account of the difference between "1 as 0" and "0 as 1" error rates, does not lead to significantly different results except at very low penalties.

Substitution from (4) gives

$$\alpha = -5\log\left(1 - 48A\beta^2\omega^2\tau^2\right).\tag{5}$$

In the more general case, where the partition spectrum is not flat, we can write

$$\alpha = -5\log\left(1 - D\tau^2\right)\tag{6}$$

where D depends on the bandwidth and on k^2A. This is a relation which can be tested experimentally by taking measurements of α over a range of dispersions. (Section on "Results and Discussion" of this chapter.)

Line Shifts

For simplicity we consider the effect of changes in the wavelength of a single spectral line, no partition noise being present.

Suppose that the output power P of the laser is made to vary sinusoidally with time and that the wavelength changes correspondingly. We can then write

$$P = P_0(1 + \sin\theta), \qquad \theta = \omega t\tag{7}$$

and

$$\lambda = \lambda_0(1 + g\sin\theta)\tag{8}$$

where $g \ll 1$. The parameter g is, in general, frequency dependent, having its highest values at low and very high frequencies [13]. At the end of a length of dispersive fiber the relative lag for the wavelength λ compared with λ_0 is

$$\pm \tau g\lambda_0 \sin\omega t\tag{9}$$

with t defined by (7) above.

The sign depends on whether the fiber dispersion is negative or positive at the laser wavelength. The relation between θ at t', the time measured at the receiver, and t is then

$$t' = t \pm \tau g\lambda_0 \sin\theta.$$

The power received at the time t' is

$$P_0(1 + \sin\theta)\frac{dt}{d\theta}\bigg/\frac{dt'}{d\theta} = \frac{P_0(1 + \sin\omega t)}{1 \pm \omega\tau g\lambda_0\cos\omega t}.\tag{10}$$

It is evident that if $\omega\tau g\lambda_0$ approached unity we should have the possibility of a concentration of power into a small fraction of the modulation cycle. However, the values of the parameters which are observed in practice at high modulation frequencies are sufficiently small to preclude this possibility in most systems. Some indications of the effect have been observed at a dispersion of 1200 ps·nm [14].

Envelope Shifts

The effect of change in the drive current on a multimode laser spectrum is not only to change the wavelength of each line but also to shift the envelope of the lines to a different wavelength. The change in the center wavelength is usually greater than the line shift, and changes of a few nanometers may occur within the first few nanoseconds after turn-on.

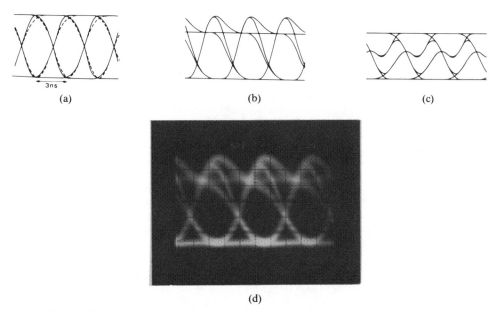

Fig. 5. Simulated eye diagrams showing the effect of transmission through dispersive fiber.
(a) Sources with no wavelength variations. The transmitted pulses are full-width NRZ and the
simulated receiver has a three-pole Bessel filter characteristic. The solid line is for a single-line
spectrum; the dotted line is for a multi-line spectrum 8 nm wide. Dispersion is 0.24 ns per nm.
(b) Source wavelength increasing by 6 nm after 1 ns in each pulse. (c) As (b) above but with
dispersion of opposite sign. (d) Observed eye diagram with "sensitive" laser (cf. Fig. 8(e)).

Their effect on the system will depend on the rate at which they occur and does not lend itself readily to analysis. It can, however, be simulated by computation, with results which appear to accord well with observation. For this purpose, a rectangular pulse was assumed to undergo a change in mean wavelength at a certain time after the start of the pulse. The exit pulse, modified as a result of the differential delay for the two wavelengths, was then submitted to a simulated receiver with appropriate frequency response. By superimposing different pulse patterns, it was possible to simulate an eye diagram (Fig. 5). There is an evident correspondence between the observed eye and the simulated eye for negative dispersion and a jump to a higher wavelength. The third simulated eye is due to positive dispersion. For every laser so far evaluated the fiber dispersion has been negative at the laser wavelength.

It should perhaps be stressed that this behavior, though far from unique in the lasers we have examined, is not typical. It illustrates the difference between a "normal" laser, where the envelope shift during modulation is of the order of 1 nm, and an abnormal one which should be excluded from system use.

THE AUTOMATED TEST FACILITY

The equipment described below is a test facility for lasers which provides

a) automatic plotting of error rate as a function of received optical power;
b) automatic plotting, as a function of laser temperature, of the received power required to maintain a specified bit error ratio (BER).

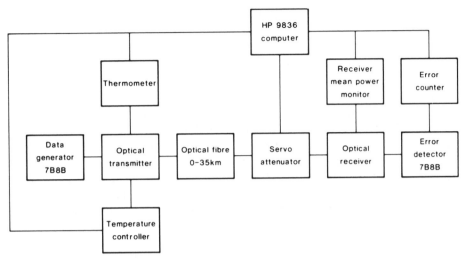

Fig. 6. Schematic of Fiber Optic Automatic System Test.

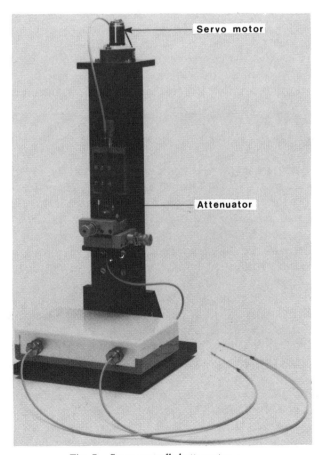

Fig. 7. Servo-controlled attenuator.

By making measurements with and without a dispersive fiber it is possible to obtain quantitative results for laser behavior which are independent of the performance of the optical receiver.

The arrangement is shown schematically in Fig. 6. It uses a standard Error Test Set (Anritsu MS 65A) which is interfaced to a Hewlett-Packard 9836 or 9816 computer and plotter for control and output purposes. The attenuator (Fig. 7) consists of two precisely aligned single-mode fibers, immersed in index-matched liquid, with a longitudinal separation which can be varied by a servo-operated lead-screw to provide attenuations up to 40 optical dB. For the plotting of the error ratio, the attenuator is made to provide successive decrements in received power by instruction from the processor. For the maintenance of a set error ratio, the attenuator changes in response to the output of the test set. A BER above the set ratio brings about a decrease in attenuation and vice versa. In ordinary operation a "set" BER of 10^{-6} results in recorded values between 3×10^{-6} and 3×10^{-7}. Any excursion outside these limits is signaled in the print-out.

The optical transmitter and receiver are described elsewhere in this book [15]. The receiver used in the test facility had less-than-optimum sensitivity (-34 dBm) because of a fiber-alignment fault, subsequently curved. For the measurement of spectral penalties we used special fiber [16] with a zero-dispersion wavelength of 1644 nm and a high dispersion at 1300 nm (22 ps·nm^{-1} km^{-1}). A relatively short length of fiber (5.5 km) can then provide low attenuation and approximately the same dispersion ~120 ps·nm^{-1} for all lasers operating in the region of 1300 nm. Longer lengths were used for investigations into the effect of higher dispersion.

A temperature cycle at the error ratio of 1 in 10^{6} can be completed in 5 min and is sufficient to reveal the temperatures at which the error is greatest. Detailed measurements of received power at lower error ratios can then be made at these temperatures.

THE MEASUREMENT OF SPECTRAL VARIATION

Of the three types of spectral variation discussed previously, two partition fluctuations and envelope-shifts are significant for their effects on system performance. Both have the effect of changing the mean wavelength of the spectrum, and we may therefore use a measurement of this change for characterization purposes. As far as partition effects are concerned, the correspondence is not complete because the shape of the spectral envelope is an additional variable, but we may expect the variance of the mean wavelength to be a reasonably satisfactory indication both of partition noise and of the noise it produces with chromatic dispersion.

For measurements of the change, it is convenient to use a narrow-band interference filter with a peak transmission at the longest laser wavelength which will be encountered. Such filters are available with an approximately linear dependence of transmission on wavelength extending over some 15 nm, sufficient to contain a complete laser spectrum and all but the most extreme variations. Moreover, the whole filter characteristic can be shifted over a range of 50 nm by altering the angle of incidence, with very little change in the transmission profile, so that a single filter is suitable for measurements on all lasers with wavelengths in the region of 1300 nm. The filter characteristic and instrument layout are shown in Fig. 8, together with some results which illustrate the different varieties of spectral change which may have been encountered. The spectral penalties for the same lasers are recorded in the next section.

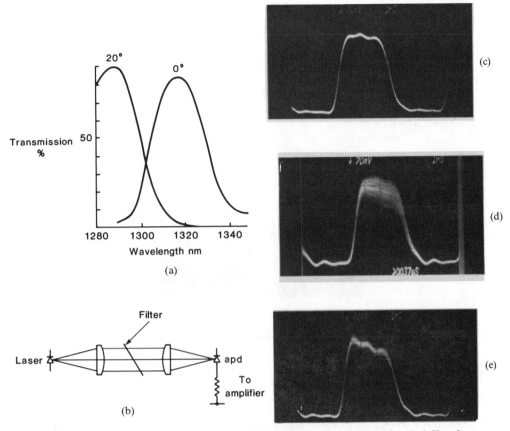

Fig. 8. Measurement of spectral shifts. (a) Transmission characteristic of optical filter for two angles of incidence. (b) Measurement system. (c) Laser with stable spectrum. (d) Laser A. Strong partition. (e) Laser B. Wavelength change during each pulse.

RESULTS AND DISCUSSION

The lasers tested in these investigations were buried heterostructure GaAsInP devices of the type described elsewhere [17] operating at wavelengths in the range 1290–1330 nm. The spectrum was usually multimode with two or three prominent lines at intervals of 1 nm, but single-mode operation was not uncommon under some conditions. Single-mode fiber tails were coupled to the lasers via fusion-formed lens ends, with a transfer efficiency of about 30 percent. The connection to the attenuators and dispersive fiber was made with the aid of a micropositioning jig, with index matching to eliminate reflection from the tail end.

Figure 9 shows a comparison between received power requirements with and without dispersion for a "good" laser. The spectral penalty at a BER of 10^{-6} does not exceed 0.2 dB and, at lower BER's, is still less than 0.5 dB. The spectral changes accompanying modulation are likewise small.

Figure 10, on the other hand, shows results for two lasers with significant penalties and illustrates the need for a test facility which explores all temperatures in the working range. These penalties were due predominantly to mode partition (laser A) and a regular,

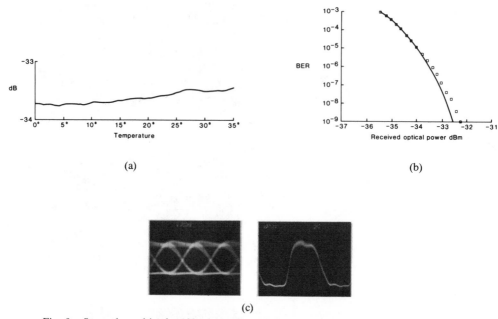

(a)

(b)

(c)

Fig. 9. Spectral penalties for 120 ps/nm dispersion (best lasers). (a) Received power (arbitrary scale) for 10^{-6} BER as a function of temperature. (The plotted curve for zero dispersion agrees within 0.1 dB at all temperatures indicating that the spectral penalty for this laser is very small.) (b) BER versus received power for the same laser. The solid line corresponds to Gaussian receiver noise, with no contribution from the signal. (c) Eye diagram and spectral deviation measurement for the same laser.

modulation-induced change of mean wavelength (laser B). The eye diagram for laser B is shown and may be compared with the simulation shown in Fig. 5.

In order to examine further the contribution of partition effects on the results we can compare observed spectral penalties at different dispersions with theoretical predictions, Fig. 11 shows penalties for two lasers plotted against our (6) for selected values of the parameters D. It will be seen that a certain value of D gives fair agreement with experiment, although the penalties at low dispersions are somewhat higher than predicted. If all the partition fluctuations fell within the system bandwidth, with a uniform frequency distribution, we could apply (5) to estimate D, obtaining $D = 48k^2A\omega^2$.

For a noise bandwidth of 250 MHx and $D = 20$ (corresponding to the upper curve on Fig. 11) we obtain $k^2A = 0.17$. It is not an easy matter to compare realistically this estimate of k^2A with experimental values, for the following reasons:

a) There is evidence, for our lasers at least, that k is not the same for all laser modes. Whereas k values in the region of 0.5 are typical for the central modes, much lower values—less than 0.2—are observed for the weaker modes at the edges of the spectral envelope. Similar findings have been reported elsewhere [10]. Since these modes make a major contribution to the factor A, it follows that the effective values of both k and A will be smaller than a single measurement of k on a strong mode would lead us to expect.

b) Neither the assumption that the partition process occurs independently of modulation, nor the contrary one that it occurs only at each change in laser output, is strictly valid. The first assumption is refuted by the observation that estimated k factors depend upon modulation frequencies [10], the second by the fact that partition is observed in the

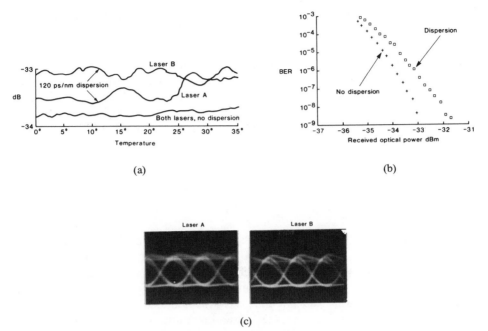

Fig. 10. Spectral penalties for 120 ps/nm dispersion (shifting lasers). (a) Received power for 10^{-6} BER as function of temperature. (b) BER versus received power for laser A at temperature of maximum penalty. Results for laser B are very similar. (c) Eye diagrams for received signals.

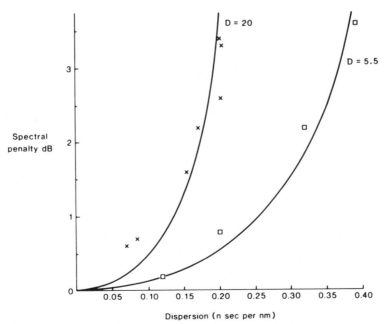

Fig. 11. Spectral penalties versus dispersion. Experimental points for two lasers were obtained with a BER of 10^{-9} and various lengths of dispersive fiber. The curves are plots of the function $-5\log(1 - D\tau^2)$ with D chosen for best fit.

absence of modulation. Any single value of k is therefore open to objection, particularly in an NRZ system of modulation such as we use.

c) Partition noise is highly sensitive to reflections, even very weak ones, from discontinuities in the fiber. With some tens of kilometer lengths of fiber, well-spliced, serious discontinuities will not be present, but in making measurements with shorter lengths it is important to eliminate reflection effects. If this is not done the observed partition noise will usually be considerably higher than that for reflection-free conditions.

It is preferable, therefore, to regard D as a semi-empirical constant which depends upon the laser and the bit-rate.

Conclusion

By using an automated test facility which allows a wide range of working conditions to be explored, we have measured the spectral penalties arising from the use of lasers in a monomode fiber-optic system with some chromatic dispersion. For a specified set of modulation conditions the penalties are commonly quite sensitive to temperature, and the highest penalty could easily remain undetected without such a facility. We confirm previous findings that partition between laser modes is a major contributor to the observed penalty. However, we also find that regular changes of mean wavelength, brought about by modulation within each bit period, are for some lasers more significant. The penalty increases sharply with dispersion and can be represented by the formula: penalty (dB) $= -5\log(1 - D\tau^2)$ where D is a constant and τ the relative delay between adjacent modes.

References

[1]   D. Marcuse, "Pulse distortion in single-mode fibers," *Appl. Opt.*, vol. 19, no. 10, 1653–1660, 1980.
[2]   F. P. Kapron, "Dispersion slope parameters for monomode fiber bandwidth," in *Tech. Dig. Conf. on Optical Fiber Communication* (New Orleans, LA), Jan. 23–25, 1984.
[3]   N. K. Cheung, "Dispersion penalties for 432 Mbit/s single-mode fiber transmission systems in the 1.3 μm wavelength region," in *Proc. 9th Eur. Conf. on Optical Communication* (Geneva), 1983.
[4]   S. Yamamoto, H. Sakaguchi, and N. Seki, "Repeater spacing of 280 Mbit/s single mode fiber optic transmission system using 1.55 μm laser diode source," *IEEE J. Quantum Electron.*, vol. QE-18, no. 2, pp. 264–273, 1982.
[5]   K. Ogawa, "Analysis of mode partition noise in laser transmission systems," *IEEE J. Quantum Electron.*, vol. QE-18, no. 5, pp. 849–855, 1982.
[6]   K. Ogawa, "Considerations for single mode fiber systems," *Bell Syst. Tech. J.*, vol. 61, no. 8, pp. 1919–1931, 1982.
[7]   S. Personick, "Receiver design for digital fiber optic communication systems," *Bell Syst. Tech. J.*, vol. 52, pp. 843–886, 1973.
[8]   Y. Okano, N. Nakayawa, and I. Ito, "Laser mode partition noise evaluation for optical fiber transmission," *IEEE Trans. Communications*, vol. COM-28, no. 2, pp. 238–243, 1980.
[9]   K. Petermann and T. Weidel, "Semiconductor laser noise in an interferometer system," *IEEE J. Quantum Electron.*, vol. QE-17, no. 7, pp. 1251–1256, 1981.
[10] K. Ogawa and R. S. Vodhanel, "Measurement of mode partition noise of laser diodes," *IEEE J. Quantum Electron.*, vol. QE-18, no. 7, pp. 1090–1093, 1982.
[11] R. Schimpe, "Intensity noise associated with the lasing mode of a (GaAl)As diode laser," *IEEE J. Quantum Electron.*, vol. QE-19, no. 6, pp. 895–897, 1983.
[12] R. G. Smith and S. D. Personick, *Semiconductor Devices for Optical Communication*, H. Kressel, Ed. New York: Springer, 1980, p. 133.

[13] S. Kobayashi, G. Yamamoto, M. Ito, and I. Kimura, "Direct frequency modulation in AlGaAs semiconductors lasers," *IEEE J. Quantum Electron.*, vol. QE-18, no. 4, pp. 582–595, 1982.

[14] D. A. Frisch and J. O. Henning, "The effect of laser chips on optical systems," *Electron. Lett.*, vol. 20, 15, pp. 631–633, 1984.

[15] G. A. Heath and M. Chown, "The UK-Belgium no. 5 optical fiber submarine system," this book, ch. 10, p. 129.

[16] B. J. Ainslie, K. Beales, D. M. Cooper, C. R. Day, and J. D. Rush, "Monomode fibre with ultra low loss and minimum dispersion at 1.55 μm," *Electron. Lett.*, vol. 18, pp. 842–4, 1982.

[17] S. E. Turley, G. D. Henshall, P. D. Greene, V. P. Knight, D. M. Moule, and S. A. Wheeler, "Properties of inverted rib-waveguide lasers operating at 1.3 μm wavelength," *Electron. Lett.*, vol. 17, no. 23, pp. 868–870, 1981.

33
A High Reliability, High Sensitivity Lightwave Receiver for the SL Undersea Lightwave System

MICHAEL L. SNODGRASS AND RICHARD KLINMAN, MEMBER, IEEE

INTRODUCTION

Reliability is crucial to any undersea communications system since deep water repairs are very costly, both in lost revenue and repair costs. A requirement for high reliability of the receiver as an isolated component is obvious since the reliability budget for a system would typically allocate negligible probability for a system failure due to the failure of any individual receiver during the nominal 25-year life of the system. A requirement for high sensitivity in the receiver is also relevant to reliability: improving the sensitivity will increase the allowed span length, thereby reducing the number of repeaters and thus improving total system reliability (and also probably reducing initial system cost). Unfortunately, the requirements of reliability and sensitivity often cause conflicts in design since many of the usual procedures for improving the reliability of a receiver result in reduced sensitivity and vice versa.

In this chapter we will discuss the choices which we made to optimize both reliability and sensitivity in a lightwave receiver designed to be used in the SL undersea lightwave system. We will then discuss the performance of prototype receivers manufactured for use in laboratory and undersea experiments.

DESIGN CHOICES

Overall Design Process

The receiver's design has been chosen to provide the best possible assurance of reliability while meeting critical system needs for performance such as sensitivity, dynamic range, immunity to electrical noise, and low power consumption. All design tradeoffs were made with reliability as the primary criterion. Components having reliability proven by extensive field experience in similar applications were used where possible. Components involving new technologies and the complete receiver are being qualified for

475

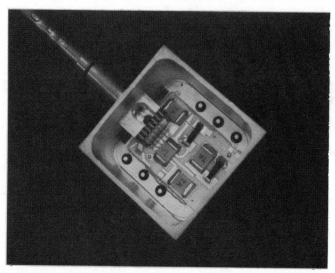

Fig. 1.   Photograph of receiver prior to final bake-out and lid seal. Visible are the optical chip carrier (OCC) which contains both active devices, the termination of the optical fiber pigtail, and the passive electrical components.

undersea use by a rigorous laboratory program designed to test for potential failure modes. The complete receiver is also being qualified through a series of field installations. The program to develop this receiver, which has been under way for more than four years, involves many iterations of design and testing. The manufacturing processes have been designed so that every component as well as each complete receiver will be 100 percent tested in order to certify reliability. This includes purging procedures demonstrated to remove weak devices, plus an extended burn-in.

An example of our prototype receiver is illustrated in Fig. 1. The following paragraphs consider why each of the components was chosen and how they were integrated into this module.

Photodetector

Properties of the optical fiber dictate operation in the wavelength band from 1.2 to 1.6 μm, with the SL system being optimized for 1.3 μm. This effectively limited our choice of photodetectors to those fabricated in either germanium or the InGaAsP alloy family.

Although an avalanche photodiode (APD) has the potential to provide a receiver with higher sensitivity than that possible with a p-i-n photodiode, the APD has a number of drawbacks which weigh against its use. First, an APD needs a significantly higher bias voltage, which implies either the use of much higher voltages at the cable's shore terminals or the use of dc–dc converters in each repeater. Higher voltages at the shore terminals greatly increase the design requirements for insulation in the cable and in each repeater, while dc–dc converters increase the overall parts count with an attendant reduction in reliability. Second, these higher voltages tend to accelerate some failure modes. Third, since gain and noise of the APD are dependent upon temperature, either the bias voltage must have temperature-compensated regulation or the APD must be temperature controlled (e.g., by a thermoelectric cooler). Neither approach has a simple and reliable implementation. Fourth, in order to achieve good dynamic range, a receiver utilizing an APD also requires feedback control of the APD's bias, which leads to a more complicated and less reliable circuit. Production Ge APD's have problems with dark current, passivation, and excess noise. The experimental Ge APD receivers which we tried had only a modest sensitivity advantage over competing p-i-n receivers. This sensitivity advantage was judged insufficient to outweigh all the drawbacks. At present, only laboratory models of InGaAsP APD's are available, and no investigations of their reliability have begun. Therefore, the InGaAsP APD was not considered for the SL receiver. However, InGaAsP APD's offer significant improvements in sensitivity at higher bit rates [1]. They will be seriously considered for future submarine systems once they are routinely produced and more is known about their reliability.

We have chosen to use a back-illuminated planar p-i-n InGaAs photodiode since such a device has the following desirable features compared to devices which are front illuminated, have a mesa structure, or are fabricated from germanium:

1) reproducible fabrication (no critical etching steps);

2) surface topology allows easy passivation with SiN, which seals the junction and stabilizes surface states;

3) low stable dark current;

4) ability to purge weak devices at high temperatures without degrading the junction or the SiN passivation;

5) physical structure has low capacitance, which is important for good sensitivity;

6) the back face of the diode is the InP substrate, which can serve as the "window" in a hermetically sealed package containing the diode; and

7) the back face is easily antireflection coated for improved efficiency in optoelectronic conservation.

More details on this photodiode and its qualification for undersea cable applications may be found elsewhere [2]–[5].

Amplifier

When a photodetector with unity gain is employed, many critical properties of the receiver are governed by characteristics of the amplifier, which converts the detected photocurrent into a higher level signal. The SL system requires a receiver with the conflicting properties of high sensitivity, large dynamic range, and broad bandwidth. We have already discussed the need for high sensitivity. The need for large dynamic range is due to allowance for: 1) variation in fiber loss among the six fibers in the SL cable; and 2) variation in launched power arising from initial differences in transmitters or from laser aging and/or switching to redundant lasers. The need for broad bandwidth is due to the SL system's use of NRZ data at 296 Mbit/s. Furthermore, the band shape should be insensitive to variations in either the amplifier or any requisite equalizer due to aging or temperature (stable response after installation) or due to manufacturing tolerance (simplifies regenerator assembly).

When a p-i-n photodiode is used, the sensitivity of the receiver is limited by the total equivalent noise current at the input of the amplifier. The total equivalent noise current is determined by integrating the spectral noise density, weighted by the band shape required by the signal. The spectral noise density as a function of frequency can be expressed as

$$d\langle i_N^2(f)\rangle/df = [4kT/R] + \left[S_1 + (2\pi Cf)^2 S_2\right] + S_3 \tag{1}$$

where k = Plank's constant, T = absolute temperature, f = frequency, R = either load resistance for an integrating (shunt impedance) amplifier or feedback resistance for a transimpedance amplifier, $S_1(f)$ = equivalent shunt noise of first transistor, $S_2(f)$ = equivalent series noise of first transistor, C = capacitance shuntin S_2, and $S_3(f)$ = noise due to other sources in the amplifier [6]–[9]. Note that we have assumed that noise from the photodiode is negligible. By making R very large, its noise contribution can be made negligible compared to S_1 and S_2. By making the gain of the stage containing the first transistor large enough, the noise due to other sources can also be made negligible. In an idealized amplifier, the noise is therefore dominated by the input transistor alone.

Thus, in simple theory, the same sensitivity can be achieved by a receiver using an amplifier with an integrating front end and by a receiver using a transimpedance amplifier. In practice, very large values of R result in limited bandwidth and dynamic range. The receiver with the integrating front end can achieve the required bandwidth by inserting a broad-band equalizer at some later stage in the amplification process. However, the integrating amplifier can meet the requirements for dynamic range only through special coding of the digital data stream, which forces the baud rate to be higher than the bit rate. Because of the overhead associated with the coding, the required bandwidth increases somewhat, which allows more noise to be passed for a given amount of signal. The coding also causes the required bandwidth to be shifted to higher frequencies. Since the transistor's noise per unit bandwidth increases with frequency, this also means more noise for a given amount of signal. Hence, at the higher baud rate, the sensitivity is lower. Due to feedback in the transimpedance amplifier, the dynamic range can be quite large, and the value of the effective capacitance which governs the bandwidth can be much smaller than that in an integrating amplifier. However, this capacitance is still finite, so there is an upper limit on the value of the feedback resistor, which is set by the bandwidth requirement (assuming no equalization). The thermal noise due to this finite value of resistance degrades the receiver's sensitivity from the ideal. In practice, sensitivities for transimpedance and integrating receivers at comparable bit rates are always nearly equal. Since the transimpedance circuit is less complex (i.e., no equalizer) than the integrating one, and since the system designers preferred simple NRZ data, the transimpedance amplifier was chosen.

Bandwidth and sensitivity are both improved by low capacitance at the amplifier's input. This fact, coupled with the design goal of a low parts count in order to improve reliability, dictates the use of a monolithic integrated circuit (IC). There are three candidate semiconductor technologies: Si BJT, fine-line Si MOSFET, and fine-line GaAs FET. In theory, an amplifier with a GaAs FET for its first stage should produce the most sensitive receiver for operation at rates around 300–400 Mbit/s. A modest advantage was confirmed for discrete transistor GaAs FET circuits, which we fabricated and operated at these bit rates. However, at the time this receiver was designed, there were no demonstrations of acceptable low noise IC amplifiers in either GaAs FET or Si MOSFET technologies. Only the Si BJT had a mature technology capable of producing an IC with reproducible and stable performance and with reliability confirmed by years of field experience in comparable applications. Furthermore, since the sensitivity of a BJT receiver can be optimized for a given bit rate by adjusting the designed collector current [7] and by adjusting the physical structure of the device (within the constraints of reliability), it was possible to fabricate Si BJT IC amplifiers which achieved sensitivities very close to those provided by discrete GaAs FET amplifiers.

These amplifiers have a circuit similar to that used in previous designs [10]. In addition to changing from a hybrid circuit with discrete transistors to a monolithic IC, improve-

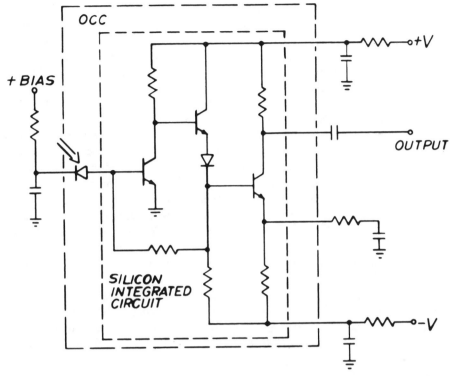

Fig. 2. Schematic of circuit. The dashed lines indicate the extent of the Si IC and of the OCC. The remaining components are thin-film resistors and chip capacitors.

ments have been added to increase the dynamic range still further, to raise the signal amplitude at the output, to stabilize the operating conditions, and to provide points for wafer probing, which will give a reasonable indication of final receiver performance (wafer probing circuits which can operate at frequencies of several hundred megahertz are very difficult to implement). A schematic diagram for the circuit of the complete receiver is shown in Fig. 2.

Overall, the best choice for the present system was judged to be the Si BJT IC. Development of GaAs FET IC amplifiers is proceeding rapidly, and improvements in performance and reliability could make them very attractive for use in the future undersea systems. Still further in the future is the possibility of a long wavelength photodetector integrated on the same substrate as a FET amplifier fabricated from an InGaAsP alloy [11]. Such alloys are potentially better than GaAs for FET amplifiers for use in receivers, and they have already been shown to produce good photodetectors.

Semiconductor Packaging

There are basically three options for packaging the semiconductor devices (photodetector and amplifier) before assembly into the overall package for the receiver: 1) bare (i.e., unencapsulated) devices; 2) separately encapsulated devices; and 3) both devices encapsulated in a single package. The first and third options allow for the minimum penalty to sensitivity due to parasitic capacitances since a direct chip-to-chip wire bond is possible. The second and third options allow for the most protection for the devices prior to completion of the receiver's overall assembly. In order to certify semiconductor components for undersea use they must undergo a high temperature bake-out and hermetic seal in a benign atmosphere, followed by a high temperature high bias purge to screen out weak devices [3]–[5], [12]–[14]. Because the first option relies upon the overall receiver package for its hermetic seal, the temperatures which can be used for baking-out and purging are limited by the optical and passive electrical components contained in the receiver. All these factors favor the use of the third option. It should be noted that the third option makes it difficult to test the performance of each semiconductor device separately, which means that the yield of good devices in the combined package is proportional to the product of the yields inherent in the first and second options where each device can be separately tested. This penalty has been reduced in our design because we have developed tests for the waver stage of fabrication for both devices which accurately predict packaged performance. Further reduction in this penalty was attained when we developed tests that determine if the amplifier is functional once it is mounted in the package, but before the photodiode has been mounted and connected to the amplifier's input. In any case, for a submarine cable system the impact of the third option upon sensitivity and reliability more than outweighs the cost of this penalty in manufacturing yield.

This combined packaging scheme was implemented by using a standard leadless ceramic chip carrier which has been modified by placing a hole in its base. The photodiode is eutectically bonded over the hole in a manner such that the diode's transparent substrate can be used to form part of the carrier's overall hermetic seal. Due

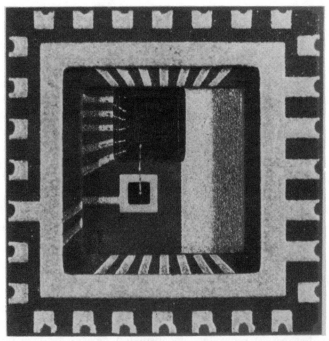

Fig. 3. Photograph of the OC prior to bake-out and lid seal. The large chip is the Si IC and the small chip is the back-illuminated photodiode, which is sealed over the hole through which the optical signal enters.

to the photodiode's back-illuminated structure, the hole permits light from the receiver's optical fiber pigtail to reach the active area of the photodiode. This package, which we have dubbed the optical chip carrier (OCC), is shown in Fig. 3 as it appears before bake-out and seal, with both semiconductor chips in place and wire bonds formed. In order to help assure reliability, the OCC is assembled using solder but no fluxes or epoxies, and is sealed in an inert atmosphere. The completed OCC can be fully tested before and after purging and burn-in in order to provide reliable devices which have good performance when assembled into a complete receiver.

At high bit rates an optical receiver's input noise tends to be proportional to the total front-end capacitance (see (1)). The OCC provided an improvement in receiver sensitivity of about 0.8 dB at 296 Mbits/s and about 1.1 dB at 432 Mbits/s compared to separately packaged devices when a BJT IC amplifier was used. Reduction of parasitic capacitance in this case decreased the total front-end capacitance by about 15 percent. We estimate that the improvement in sensitivity would be even more dramatic when comparing the OCC to separately packaged devices when a FET IC is used since parasitic packaging capacitance could easily be a larger fraction of the total front-end capacitance.

Passive Electrical Components

Electrical biasing for the OCC, including short-circuit protection for other repeater modules, filtering against noise and interference, and coupling to the next regenerator module (an AGC amplifier), is accomplished by a ceramic thin-film IC to which ceramic and solid tantalum chip capacitors have been soldered. The thin-film resistors on the ceramic substrate avoid the need for solder connections, can be laser trimmed to value, and have very small parasitic impedances. The chip capacitors have thermal expansion coefficients compatible with the substrate and have no leads with their attendant inductance and reliability risk. Use of these components provides for optimum high frequency performance, minimum size, maximum reproducibility, and maximum reliability. They have been qualified for submarine cable use on the basis of extensive field experience in similar high reliability systems.

Optical Components

The receiver's optical fiber pigtail is a section of multimode graded-index fiber, which is designed to act as a "photon bucket," allowing a low loss splice to be made with the incoming single-mode fiber from the cable. Extensive system tests have proven that no penalty due to modal noise in the multimode pigtail occurs in our receiver. In order to assure reliability, a number of precautions have been taken against the possibility of a receiver failure due either to a break developing in the fiber or to a degradation of the alignment between the fiber and the photodiode. Immediately after the process of drawing the fiber, a conformal coating is applied to the fiber. This coating seals the surface against contamination by water vapor or other substances which might tend to accelerate the growth of cracks and protects the glass against scratches. Then the fiber is proof tested before being accepted for use in the receiver. The optical fiber is further protected by a jacket containing a stranded high strength fiber reinforcement sandwiched between layers of plastic. This jacketing, plus procedures to assure compliance with minimum bend radius and maximum applied stress limits, reduces the probability of immediate or latent failures due to handling during assembly, testing, or installation. The fiber passes through the wall of the receiver via a hermetic seal formed by a proprietary glass-to-metal seal, which has minimal stress and does not leave any portion of the fiber external to the enclosure exposed to possible surface contamination. Inside the enclosure, the fiber is actively aligned to the photodiode and locked into place with a stable low stress glass-to-metal joint. In order to avoid inducing stress which might cause cracks in the fiber, and in order to prevent shifts in alignment, all the bulk materials mechanically coupled to the optical fiber have thermal expansion coefficients which are matched to that of the fiber. No epoxies are used in this or any other operation during the assembly of the receiver.

Overall Enclosure

The outer metal package is passivated and hermetic and utilizes only proven plating and sealing techniques. No metal susceptible to corrosion is left exposed. No metal susceptible to whisker growth is used. All chemically reactive compounds are thoroughly removed from the package by washing with carefully selected solvents, and this is followed by a bake-out of the package at a temperature which does not compromise the reliability of the passive electrical or optical components. Components inside are protected from degradation by filling the enclosure with an inert gas which has an extremely low moisture content.

The metal package aids in shielding the low noise circuitry of the receiver from external interference. Electrical connections to the outside are made via terminals sealed with glass, which has a compatible thermal expansion coefficient. Wire-bond and ribbon-bond interconnects inside the package are 100-percent pull tested at levels of force which will detect weak bonds, but which will not damage good bonds.

Screening Procedures

In addition to all the screening procedures designed to certify the reliability of each component and of partial subassemblies, we have developed a sequence of tests which will verify the reliability of each complete receiver before it is placed into service. Every receiver is subjected to thermal, mechanical, electrical, and optical stresses at levels which are significantly above those to be encountered in actual use, but which are significantly below those which will cause damage in receivers without defects. After applying these stresses, each receiver is burned-in for an extended period of time while monitoring a number of its characteristics. In any of these characteristics exhibits a drift which, when extrapolated 25 years, is in excess of that which can be tolerated for proper functionality, then the receiver will be rejected.

Results

Reliability

Table I presents the projected failure rates in FIT's (failures in 10^9 device hours) for each of the components in this receiver. Of course, these failure rates apply to components which have been certified by thorough screening. Values for these failure rates were derived from our program to qualify our design for undersea use. The first four entries in the table are firmly based on extensive field and laboratory experience with similar

TABLE I
RECEIVER RELIABILITY BUDGET

Component	Projected Failure Rates (FIT's)
Hermetically sealed package	0.10
Thin-film integrated circuit	0.15
Chip capacitors (6)	0.30
Bipolar Si integrated circuit	0.50
Planar p-i-n photodiode	1.00
Optical chip carrier	0.15
Optical fiber pigtail	0.15
Optical alignment assembly	0.10
Total for complete receiver	2.45

components in high reliability systems, including previous submarine cable systems. The second four entries involve items which represent relatively new technologies. These entries are conservative limits based both on testing the inherent reliability of the components under a comprehensive range of applied stresses and on limited (but rapidly growing) field experience with lightwave systems. The projected failure rate for the complete receiver is based on the assumption that all the components have simple exponential failure rates. While not exact, this is not a bad assumption. Our ongoing qualification program is designed to verify the failure rate of the complete receiver.

An extended period of time and an iterative program of design and experiment are necessary to enable a subsystem using new technologies to grow to a mature state of performance and reliability. This receiver design has evolved from one which we provided for an early deep-sea trial [15], [16]. The program to qualify the reliability of this receiver and each of its components has been under way for more than four years and will continue in order to provide a higher degree of confidence in these projected failure rates. One of the most important results of this program has been the identification of the magnitudes of stresses and combinations of stresses which will purge the population of components, removing those components which are weak without impairing the inherent reliability of good components.

This receiver is physically robust, being able to withstand thermal, mechanical, electrical, and optical stresses well in excess of any to be encountered in actual use. Repeated rapid thermal cycling between -40 and $+90°C$ causes no degradations. Mechanical shocks of up to 5000 g can be survived. External pressures up to at least 5 atm do not impair operation. No damage is induced by electrostatic discharges up to at least 3000 V. The receiver easily recovers from optical inputs of up to at least 0 dBm.

Electrooptical Performance

The upper curve in Fig. 4 shows the effective transimpedance of typical receiver versus the modulation frequency of an optical source. Units on the left-hand vertical scale are $20 \times \log$ (transimpedance) where the transimpedance is expressed in ohms. The response is flat over the bandwidth required by the 296-Mbit/s NRZ signal with an upper 3-dB cutoff frequency of about 300 MHz. The response agrees quite well with the predictions of a computer model of the circuit.

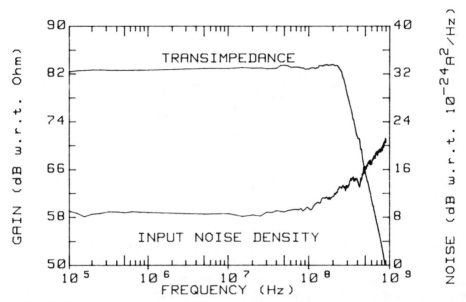

Fig. 4. Receiver's transimpedance (upper curve, left-hand scale) and equivalent input current noise (lower curve, right-hand scale) versus modulation frequency of an optical source. Response and noise of the measuring system have been normalized out.

The lower curve in Fig. 4 presents the input noise spectrum of the same receiver derived from the measured transimpedance and measured output noise spectrum. Units on the right-hand vertical scale are $10 \times \log$ (noise) where the equivalent current noise density is expressed in $10^{-24} A^2/Hz$. The noise spectrum agrees quite well with our computer model, which is based on a much more detailed version of (1). The portion of the spectrum which is independent of frequency is about 40 percent due to the feedback resistor and about 59 percent due to the input transistor. The portion of the spectrum which increases as the second power of frequency is nearly 100 percent due to the input transistor. When this noise spectral density is integrated over the useful bandwidth, the receiver's sensitivity is found to be limited about 22 percent due to the feedback resistor and about 77 percent due to the input transistor.

If this receiver were placed in an otherwise ideal system and if the optical detection efficiency at 1.3 μm were 100 percent, the sensitivity (minimum average optical power required for BER $=10^{-9}$) limited only by receiver noise would typically be -37.8 dBm at 296 Mbits/s. (Receivers made with twice the feedback resistance and half the bandwidth had 0.8 dB better sensitivity at this bit rate, but the bandwidth proved to be too marginal for conservative system design.) The typical penalty for detection efficiency is 1.1 dB, including internal quantum efficiency of the photodiode, surface reflection losses at the diode and at the end of the fiber aligned to it, misalignment of the fiber and diode, and a single-mode to multimode connector loss. The typical penalty to account for nonideality of the lightwave transmitter, linear channel, timing recovery, and decision circuit of our test regenerators is around 1.1 dB. A cumulative distribution of sensitivities, measured in the presence of these penalties, at 296 Mbits/s for 41 of our prototype receivers is shown in Fig. 5.

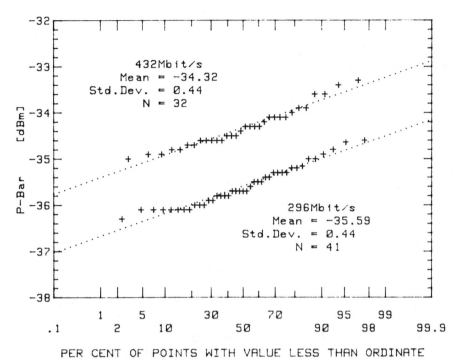

Fig. 5.   Cumulative distribution of sensitivities for receivers at 296 and 432 Mbit/s. Vertical scale is average optical power (at 1.3 μm) incident at the connector on the pigtail needed to attain BER $< 10^{-9}$ in a typical regenerator.

Versions of this receiver are used in systems operating with NRZ data from 140 to 440 Mbits/s. Also shown in Fig. 5 is the distribution of sensitivities for 32 prototypes at 432 Mbits/s for a version with slightly broader bandwidth. As can be seen from the two curves, the sensitivities are quite uniform, with standard deviation of about 0.4 dB, mostly due to the spread in efficiencies. The sensitivities at 296 Mbits/s (mean = -35.6 dBm, best = -36.3 dBm) and at 432 Mbits/s (mean = -34.2 dBm, best = -35.0 dBm) are very good considering the choice of p-i-n photodiode and bipolar Si amplifier, both designed to consider reliability rather than utmost sensitivity. The maximum average optical power which can be tolerated while maintaining BER $< 10^{-9}$ is typically in excess of -14 dBm.

This receiver is sensitive over the wavlength range from 1.0 to 1.6 μm. For example, one of these prototypes was used in a demonstration at 1.55 μm, which at the time set the record (119 km, 420 Mbits/s) for the product of distance and bit rate [17].

Conclusion

The receiver which we have described is now beginning to be manufactured for use in the longest portion of the TAT 8 transatlantic undersea lightwave cable system. The sensitivity, dynamic range, and projected reliability satisfy the overall system goals of long repeater spans and low risk of failure. The primary features of this receiver are an InGaAs planar p-i-n photodiode and a Si bipolar IC, both encapsulated in a special chip carrier. This optical chip carrier is crucial to certifying the reliability of the active devices without compromising the reliability of the passive components.

The next generation of undersea lightwave receivers will probably employ either the combination of an InGaAsP APD and a GaAs FET monolithic IC amplifier or a completely monolithically integrated detector/amplifier combination fabricated in InGaAsP alloys. Both of these configurations should have sensitivity better than our present design, but considerable work will be required to prove the necessary reliability.

ACKNOWLEDGMENT

The development of this receiver has been aided by the contributions of many of our colleagues at AT&T Bell Laboratories. In particular, we would like to acknowledge the assistance of R. M. Paski in the design of the amplifier.

REFERENCES

[1] J. C. Campbell, A. G. Dentai, W. S. Holden, and B. L. Kasper, "High-speed operation of InP/InGaAsP/InGaAs avalanche photodiodes," presented at the Conf. Opt. Fiber Commun., New Orleans, LA, Jan. 23–25, 1984, paper WA3.
[2] T.-P. Lee, C. A. Burrus, and A. G. Dentai, "InGaAs/InP p-i-n photodiodes for lightwave communications at the 0.95–1.65 μm wavelength," *IEEE J. Quantum Electron.*, vol. QE-17, pp. 232–238, 1981.
[3] R. H. Saul and F. S. Chen, "Reliability assurance for devices with a sudden-failure characteristic," *IEEE Trans. Electron Devices*, vol. ED-4, pp. 467–468, Dec. 1983.
[4] R. H. Saul, F. S. Chen, and P. W. Shumate, "Reliability of planar InGaAs photodiodes," presented at Conf. Opt. Fiber Commun., New Orleans, LA, Jan. 23–25, 1984, paper TUH2.
[5] ____, "Reliability of InGaAs photodiodes for SL applications," *Bell Syst. Tech. J.*, to be published.
[6] J. E. Goell, "Input amplifiers for optical PCM receivers," *Bell Syst. Tech. J.*, vol. 53, pp. 1771–1793, 1974.
[7] R. G. Smith and S. D. Personick, "Receiver design for optical fiber communication systems," in *Semiconductor Devices for Optical Communication*, H. Kressel Ed. Berlin: Springer-Verlag, 1980.
[8] S. D. Personick, *Optical Fiber Transmission Systems*. New York: Plenum, 1981.
[9] S. Moustakas and J. L. Hullett, "Noise modelling for broadband amplifier design," *Proc. Inst. Elec. Eng.*, vol. 128, pt. G, pp. 67–76, 1981.
[10] R. G. Smith, C. A. Brackett, and H. W. Reinbold, "Atlanta fiber experiment: Optical detector package," *Bell Syst. Tech. J.*, vol. 57, pp. 1809–1822, 1978.
[11] R. E. Nahory and R. F. Leheny, "In $_{0.53}$Ga$_{0.47}$As field effect transistors (FETs) and PIN-FETs," *Proc. SPIE (High Speed Photodetectors)* vol. 272, pp. 32–35, 1981.
[12] L. E. Miller, "Reliability of semiconductor devices for submarinecable systems," *Proc. IEEE*, vol. 62, pp. 230–244, 1974.
[13] A. J. Wahl, "Ten years of power aging of the same group of undersea cable semiconductor devices," *Bell. Syst. Tech. J.*, vol. 56, pp. 987–1005, 1977.
[14] W. M. Fox, W. H. Yocum, P. R. Munk, and E. F. Sartori, "SG undersea cable systems: Semiconductor devices and passive components," *Bell Syst. Tech. J.*, vol. 57, pp. 2405–2434, 1978.
[15] M. M. Boenke, R. E. Wagner, and D. J. Will, "Transmission experiments through 100 km and 84 km of single-mode fibre at 274 Mbits/s and 420 Mbits/s," *Electron. Lett.*, vol. 18, pp. 897–898, 1982.
[16] P. K. Runge, "Deep-sea trial of an undersea lightwave system," presented at the 6th Topical Meeting Opt. Fiber Commun., New Orleans, LA, Feb. 28–Mar. 2, 1983, paper MD2.
[17] W. T. Tsang *et al.*, "119 km, 420 Mb/s transmission with a 1.55 μm single-frequency laser," presented at the 6th Topical Meeting Opt. Fiber Commun., New Orleans, LA, Feb. 28–Mar. 2, 1983, paper PD9.

34
An Optical Switch for the SL Undersea Lightwave System

STANLEY KAUFMAN, ROBERT L. REYNOLDS, AND GEORGE C. LOEFFLER

INTRODUCTION

The SL undersea lightwave system [1] is a single-mode lightwave system operating at 1300 nm. For each active transmitter, one or more dormant transmitters will be provided. In the event that an active transmitter becomes marginal in performance, a spare transmitter can be activated. The optical path may be established by either a passive device or a mechanical switch.

A passive polarization independent device suffers an inherent minimum 3-dB bifurcation loss. The primary disadvantages of a mechanical switch are the moving parts which could cause reliability problems. This chapter describes a low-loss highly reliable 4×1 electromechanical optical switch which can provide for up to three spares or a total of four transmitters.

Previous work on multimode mechanical optical-fiber switches [2]–[4] was carefully considered for this single-mode application. The switch adopted for the proposed SL Undersea Lightwave System is a moving fiber switch [5] that was first made using preferentially etched silicon chips [6] previously developed for fiber-optic array splicing [7].

The chips have been subsequently modified for closer tolerances, larger groove spacing, and for fewer and deeper etched grooves to enhance the performance of the switch.

RELIABILITY CONSIDERATIONS

The switch must meet a reliability specification of 2 FIT's as a passive device (2 failures per 10^9 device hours) with a device life of 25 years. This equates to 1 failure per 2,300 devices over a 25-year period. Also, the probability of not performing the switch-over function in 25 years must be less than 0.001. A switch-over failure is defined as a failure to obey a switching command, even when that command is issued repeatedly.

Obviously for a new device with this low a FIT rate, one cannot rely only on statistics because of the limited number of samples and/or the limited time for testing. The

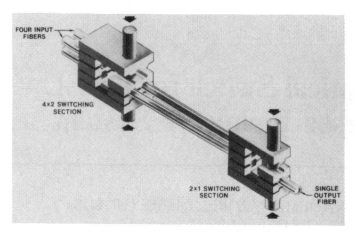

Fig. 1.   Elements of a 4×1 optical switch.

philosophy adopted is to engineer out all defects by design. This approach is enhanced by an ongoing program of subjecting, not only final models, but also where necessary subassemblies, to severe environmental testing.

Test results indicate that there are no adverse interactions between the matching fluid (silicone oil) in the switch and other materials in the device that come in contact with the fluid. There is no adhesion of the silicon chips due to long engagement under pressure in the fluid. The fiber seal, which is described in more detail later, has been designed, fabricated, tested and proved to be hermetic relative to the silicone oil.

An accelerated room temperature wear test of early design models gave no indication of significant wear in the mechanically moving parts of the relay. Some 14 relays, which were exercised for some 2.2 million operations total and cycled from $-20°C$ to $+50°C$, showed insignificant change in optical performance. Ongoing tests to determine the effect of long-term dormancy on actuation failure have shown no adverse effect to date.

PRINCIPLE OF OPERATION OF THE SWITCH

The 4×1 optical switch consists of two cascaded grooved silicon-chip fiber switches, a 4×2 and a 2×1 s, as shown in Fig. 1. Each switch consists of a movable two-high stack of (positive) grooved silicon chips with one row of fibers and a fixed three-high stack with two rows of fibers. Outside containment of these stacks is provided by silicon-chip assemblies (negative) with inverted grooves (tracks) which mate with those of the stacks and hence provide for alignment of the fibers.

SILICON-CHIP REQUIREMENTS

Of the three types of silicon chips used, the "positive chips" and "positive center chips" have precise parallel grooves etched into the top and bottom surfaces. The negative chips are etched on only one surface to generate two parallel tracks which seat into the two bottom grooves of the positive chips. The top sides of the positive chips have five grooves: one matches one of the grooves on the bottom surface for top and bottom mask alignment and four are used for fiber placement. The fifth groove also aids in device assembly by breaking up chip symmetry. The positive center chip has eight grooves (four on each side) to match the four-fiber grooves of the positive chip.

The tolerances on groove and track widths and on chip thickness may vary by several micrometers for chip-to-chip variation, but are held to 0.5 μm for groove width variations and 0.25 μm for thickness variations within a given chip. Center-to-center groove distances are held to 0.5-μm tolerances. These tight tolerances provide precise alignment between positive and negative chips and precise fiber placement.

FIBER REQUIREMENTS

Variations in fiber-cladding diameter and concentricity of the fiber core must be held to tight tolerances to insure precise alignment of the cores of abutting fibers.

Consider the fiber-groove geometry of Fig. 2. The variation in the distance from center of fiber-to-silicon chip surface due to two abutting fibers of different diameters is given by $\Delta g = \Delta d\sqrt{3}/2$. If the core is eccentric to the cladding by an amount "e" then the misalignment of the core due to diameter variations and eccentricity is $2e + \Delta d\sqrt{3}/2$; the factor of 2 is to account for two eccentric abutting fibers. If in addition there is a variation in the groove width "W", then the total misalignment ϕ which can result is

$$\phi = \frac{\Delta d\sqrt{3}}{2} + \frac{\Delta W\sqrt{2}}{2} + 2e. \tag{1}$$

Assuming a matched fluid, the transmission as a function of fiber offset is given by Marcuse [8] as $T = e^{-(\phi/W)^2}$, where W is the waist of the beam and is approximately

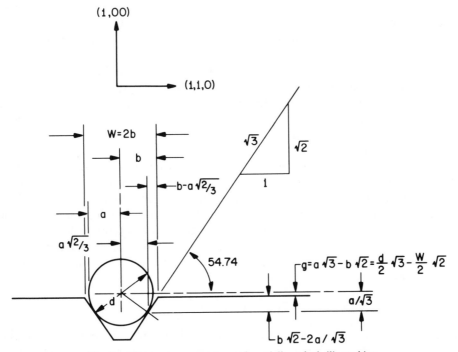

Fig. 2.   Fiber-groove geometry preferentially etched silicon chips.

TABLE I

Microns		
Δd	ΔW	e
0.75	0.75	0.31
0.75	1.0	0.22
1.0	0.75	0.20
1.0	1.0	0.11

$0.55d_c$, and where $d_c = 8.3\ \mu$m, the fiber core diameter. The loss in decibels due to offset is

$$L_\phi = 10\log_{10}\frac{1}{T} = 14.4\left(\frac{\phi}{d_c}\right)^2. \tag{2}$$

The maximum insertion loss for a 4×1 switch is 1.8 dB or approximately 0.9 dB for each of the two switching sections. In addition to offset loss, one might expect smaller contributions from angle variations of abutting fibers, from fiber separation and from pitch variations of mating grooves between positive and negative chips. With this in mind, we shall limit L_ϕ to 0.7 dB which results in a maximum allowed offset of

$$\phi = 8.3\sqrt{0.7/14.4} = 1.8\ \mu\text{m}. \tag{3}$$

Hence,

$$\frac{\Delta d\sqrt{3}}{2} + \frac{\Delta W\sqrt{2}}{2} + 2e = 1.8. \tag{4}$$

This relationship in terms of the eccentricity is $e = 0.9 - 0.433\Delta d - 0.354\Delta W$. If we let Δd, ΔW take on values of 0.75 μm and 1.0 μm, then the allowed eccentricity is presented in Table I above.

It is not reasonable to expect that the fiber core eccentricity can be held below 0.2 μm, therefore, variations on both fiber diameter and chip groove width should be held below 0.75 μm. It is extremely important to restrict chip groove and fiber variations to a minimum since insertion loss in decibels for offset goes as offset squared. It is our goal to restrict these variations to 0.5 μm.

Variations in fiber diameter are made up of two parts: the contraction and expansion of the fiber diameter along its length; and, the ellipticity (out of roundness) of the fiber. A functional test is used which accounts for the sum of concentricity ($2e$) plus ellipticity. This test consists of breaking the fiber and butting the two halves at the break in a V-groove, fixing one half, and allowing the other half to rotate. A light source is placed onto the end of the fixed half and detected at the end of the rotating half. Variations in light power are detected as the fiber is rotated 360°. The maximum variation is a measure of the offset due to concentricity and ellipticity. If we restrict these variations to 0.1 dB, then the resulting offset is

$$8.3\sqrt{0.1/14.4} = 0.7\ \mu\text{m}. \tag{5}$$

This substantially meets our goal if we allocate 0.2 μm to ellipticity and 0.5 μm to concentricity ($2e$). If, in addition, we restrict diameter variations to 0.5 μm over a 25-m

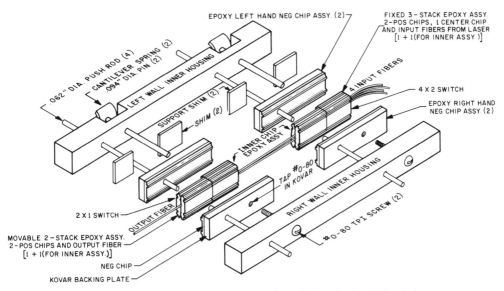

Fig. 3. Silicon chip-fiber assemblies schematic for a 4×1 optical switch.

length, which is more than sufficient to build a 4×1 switch, then our goal of fiber variations is met.

SILICON CHIP-FIBER STACK

A schematic of the silicon chip-fiber assemblies (or stacks) for the 4×1 optical switch is shown in Fig. 3. The movable two-stack consists of two positive chips with one row of fibers. The fixed three-stack consists of a positive chip, a row of fibers, a positive center chip, a second row of fibers, and a second positive chip. The inner chip assembly consists of a fixed three-stack and a movable two-stack. For stability of the stacks there are always fibers epoxied in the outer grooves of the positive chips. Note that the two light guide fibers in the inner assembly are horizontal in the fixed three-stack (two rows of fiber) and vertical in the movable two-stack (one row of fibers).

In order to reduce losses to a minimum, the positive chips are broken in half along their length. The halves are mated at this break in the fixed and movable stack assemblies. The mating ends of the positive chip stack assemblies that abut are lapped to produce a planar surface of fiber, silicon, and epoxy.

DEVICE ASSEMBLY

Each of the fixed three-stacks is sandwiched between two negative chip assemblies, right hand and left hand, as seen in Fig. 3. The two tracks on each negative chip mate with grooves in the positive chips. For structural integrity, the negative chip is epoxied to a Kovar backup plate. Kovar was chosen for its close coefficient of thermal expansion match to silicon which minimizes distortion during epoxy curing.

A pin secured by a cantilever spring bears against the lefthand negative chip assembly and in turn against the fixed three-stack, the right-hand negative chip assembly and the

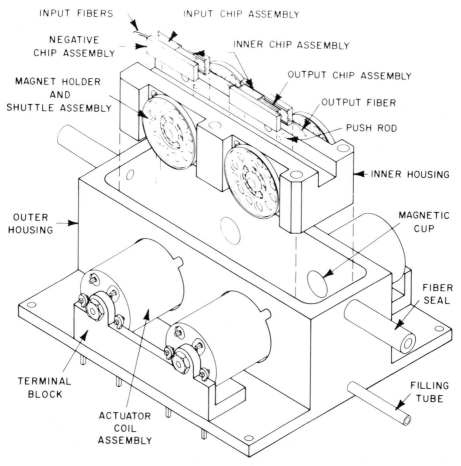

Fig. 4.   A 1×4 optical switch.

right-hand stainless-steel wall of the inner housing which is a ground surface. The left-hand negative chip assembly is further supported by a small shim which together with the adjacent shim, is epoxied during final assembly, to the ends of the plate and the housing wall. Longitudinal motion during assembly is prevented by the number 0 torque screw which holds the right-hand negative chip assembly to the inner housing wall. The right-hand negative chip assembly (toward the viewer in Fig. 4) rests on the ground bottom surface of the inner housing. An inner cover (not shown) rests flat on top of the inner housing.

As sketched in Fig. 4, the movable two-stacks are moved by push rods. These push rods and the corresponding reamed holes in the walls of the inner housing are held to close tolerances. Oversize holes are provided in the negative chip assemblies for push rod clearance. The push rods lie between the movable chip stacks and the inner surfaces of the magnet holders which are actuated by coils.

The actuators, consisting of magnet holder assemblies, coils, and magnetic cups, were designed at AT&T Bell Labs, Columbus, Ohio. A 4×1 switch has two sets of push rods and magnetic holder assemblies. A magnet holder assembly consists of two magnet holders screwed and epoxied fast to a shaft. The shaft and the push rods, which act as

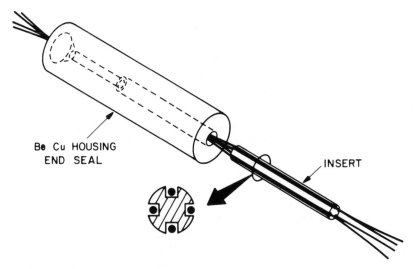

Fig. 5. An SL optical relay fiber to metal epoxy seals four-fiber input.

pistons, are fabricated from aluminum bronze. They move in a cylinder of stainless steel (the inner housing) and are somewhat lubricated by the silicone matching fluid.

Epoxy Seal for Glass Fiber

The coated glass fibers are inserted through cylindrical holes at the end seals of the outer housing (Fig. 4). Slotted beryllium–copper inserts are placed in the hole with a fiber in each slot. The assembly operation is shown schematically in Fig. 5. Epoxy is inserted into the larger diameter section of the hole and by piston action the insert forces epoxy through both the smaller section of the hole and back through the slots thus sealing the fibers between the insert and the housing end seal. Alternatively, the inserts may be coated with epoxy by use of a special tool and then placed in the end seal hole with the fibers in place.

Seals of this type were made by one of the authors and thermally cycled at least five times from $-60°C$ to $100°C$. A helium leak test was then performed. The leak rate in each of the five samples was less than 10^{-9} cm^3/s of helium. Further work recently was completed involving specimens containing matching fluid (silicone oil). All specimens were held nine days at $60°C$ and tested. The specimens were then thermally shocked at $-78°C$ and again held at $60°C$ for 19 days. Chromatograms substantiated the claim that the proposed fiber seal is sufficiently tight to silicone oil to serve the intended purpose. The switch will be in a dry nitrogen environment under sea. Therefore, it was not necessary to provide a seal to defend against diffusion of water and, hence, no testing for this was performed.

Outer Cover and Matching Fluid

After the fiber seal is made a flexible cover is sealed into the outer housing. Silicone matching fluid is then vacuum degassed and introduced into the outer housing through a tubulation, and then the tubulation is sealed off. At a wavelength of 1.30 μm the silicone fluid has an index of refraction of 1.39 which sufficiently closely matches the fiber index of 1.46. During temperature fluctuations, the flexible outer cover expands and contracts due to volume changes of the fluid. This cover motion is necessary to track the fluid motion in order to insure that no gas bubbles can form in the fluid which might degrade the performance of the switch. The flexible cover was analyzed by one of the authors using a large deflection finite element program [9]. The stress levels in the flexible cover due to temperature excursions and vacuum filling of the fluid were well below the yield point of the cover material. Even if yielding occurs, the condition is certainly not catastrophic, as the cover would simply conform to the contracted fluid surfaces.

Results

4×1 switches have been fabricated with insertion losses less than 1.8 dB (specification requirement). In addition to 4×1 switches, 2×1 switches have been manufactured utilizing only one of the two switching sections (see Fig. 2). The loss specifications for a 2×1 switch is 0.9-dB maximum. Fig. 6 depicts insertion-loss histograms for thirteen 2×1 switches and twenty seven 4×1 switches at room temperature with a light source wavelength of 1.31 μm. The average insertion loss for the 2×1 switches (26 ports) is 0.27 dB. The average insertion loss for the 4×1 switches (108 ports) is 0.92 dB. Spectral-loss dependence of a typical 4×1 switch has been measured by AT&T Technologies, Clark, NJ, and is shown in Fig. 7. The maximum insertion-loss specification must be met over the wavelength range from 1.29 to 1.33 μm. For this switch three of the four

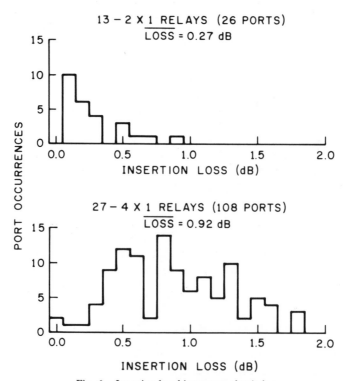

Fig. 6.   Insertion-loss histograms of switches.

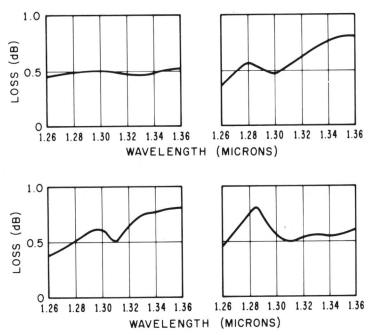

Fig. 7.   Spectral loss—atypical 4×1 optical switch (four ports).

ports show a variation of approximately 0.25 dB from 1.29 to 1.33 μm while the fourth port (upper left hand in Fig. 7) shows only a 0.05-dB variation. Loss variations due to temperature changes from 0°C to 30°C (the specification operation range) show a similar range for insertion loss.

SUMMARY AND CONCLUSION

One cannot rely on statistics alone to insure the reliability requirements of the proposed SL optical switch. The philosophy adopted is to engineer out all defects by design and testing. The design phase has substantially been completed. The ongoing testing program of subjecting subassemblies and completed switches to severe environmental conditions is well under way.

The reliability of the switch has yet to be proven. It is our opinion that the switch will meet all of the requirements set forth in the specifications, with the present design; and that further reliability testing will substantiate this opinion with perhaps only minor design modifications.

ACKNOWLEDGMENT

We would like to state, for the record, that the design of the optical switch was truly a team effort, and we would like to thank all who contributed to its success. The authors are especially indebted to F. Zwickel, AT&T Technology Systems, whose fabrication and design ingenuity made a lasting imprint on the Optical Switch.

REFERENCES

[1] P. K. Runge and P. R. Trischitta, "SL undersea lightwave system," this book, ch. 4, p. 510.

[2] P. G. Hale and R. Kompfner, *Electron Lett.*, vol. 12, no. 15, p. 388, July 22, 1976.

[3] R. B. Kummer, S. C. Mettler, and C. M. Miller, "A mechanically operated four-way optical switch," in *Proc. 6th European Conf. Opt. Commun.* (Amsterdam, The Netherlands), Sept. 1979, paper 6.4.

[4] H. Yamamoto and H. Ogiwara, *Appl. Opt.*, vol. 17, no. 22, p. 3675, Nov. 15, 1978.

[5] W. C. Young and L. Curtis, "Cascaded multimode switches for single mode and multimode optical fibers," *Electron. Lett.*, vol. 17, no. 16, Aug. 6, 1981.

[6] C. M. Schroeder, "Accurate silicon spacer chips for an optical-fiber cable connector," *Bell Syst. Tech. J.*, vol. 57, no. 1, pp. 91–97, Jan. 1978.

[7] C. M. Miller, "Fiber-optic array splicing with etched silicon chips," *Bell Syst. Tech. J.*, vol. 57, no. 1, pp. 75–90, Jan. 1978.

[8] D. Marcuse, "Loss analysis of single-mode fiber splices, *Bell Syst. Tech. J.*, vol. 56, no. 5, May–June 1977.

[9] Marc Analysis Research Corporation, "Marc general purpose finite element program," Palo Alto, CA.

35

LD Redundancy System Using Polarization Components for a Submarine Optical Transmission System

SHUNSUKE TSUTSUMI, YASUTAKA ICHIHASHI,
MASATOYO SUMIDA, AND HARUO KANO

INTRODUCTION

Laser diode (LD) standby redundant optical repeaters are necessary for achieving highly reliable transoceanic submarine transmission systems. This is because threshold current increases little by little over time. LD reliability and aging characteristics have been studied extensively over a long period of time, however, the results are not yet completely satisfactory.

Generally, an LD redundant system consists of a few LD's, and an optical switch or coupler which directs optical signals from an operating LD into a single-mode fiber transmission line. An important factor in the development of the LD redundant systems is the realization of high reliability in the optical switch or coupler.

A single-mode fiber-optic switch [1] and a YIG optical switch are proposed as optical switches for the LD redundant systems. A single-mode fiber-optic switch uses a movable array of precision etched silicon V-groove chips. However, there is an intrinsic problem in reliability because it contains movable parts. In order to overcome this problem, it is preferable to use an optical coupler without movable parts in the LD redundant system. However, in addition to insertion loss, conventional optical couplers have branching loss which increases with the number of input–output ports. For example, branching loss is 3 dB for a four-port (two input and two output ports) coupler. To lessen additional loss, a four-port coupler having a polarization filter is proposed in this chapter. Application of the polarization coupler to a redundant system having a standby LD is effective [2]. This is because of the linearly polarized emission which is inherent in Fabry–Perot type LD's. The use of a polarization maintaining optical fiber is essential to the polarization coupler. In this work, a polarization maintaining optical fiber named PANDA (polarization-maintaining and absorption-reducing fiber) is used [3], [4]. However, new technical problems

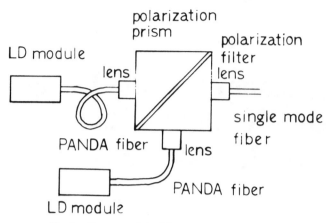

Fig. 1.   Structure of the LD standby redundant system.

have arisen in the fabrication of polarization sensitive components as well as in the fabrication of the LD module and lens circuit of the optical coupler. It is especially important to study the angular misalignment between the x axis of the PANDA fiber and the electric field vector in the TE mode of the operating LD. It is also necessary to study the angular misalignment between the x axis of the PANDA fiber and the p- or s-polarization plane of the polarization coupler.

This chapter describes the new laser redundant system, the design and characteristics of the LD module employing the PANDA fiber, the polarization coupler, and a new method in which the misaligned angle can be measured.

DESIGN OF POLARIZATION COMPONENT

The structure of the LD standby redundant system is shown in Fig. 1. This system makes use of laser intrinsic polarization and consists of two LD modules and a polarization coupler combined by the PANDA fiber. The characteristics of the PANDA fiber used [5] are shown in Table I.

Design of the LD Module Using the PANDA Fiber

A block diagram of the LD module using the PANDA fiber is shown in Fig. 2. The confocal lens circuit is used in this LD module. This confocal lens circuit for coupling the LD to the single-mode fiber in an LD module has been reported [6]. The far-field patterns

TABLE I
CHARACTERISTICS OF THE PANDA FIBER
USED IN THIS WORK

Fiber diameter	125 μm
Spot size	5.1 μm
Δ	0.29 percent
Birefringence	3.4×10^{-4}
Crosstalk	less than -40 dB
Optical loss	1.5 dB/km

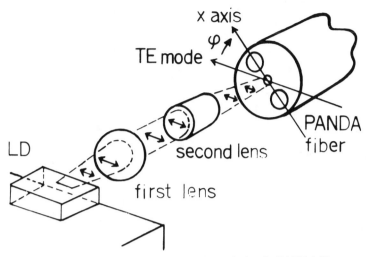

Fig. 2. Block diagram of the LD module employing the PANDA fiber.

of the PANDA fiber are shown in Fig. 3. When the x-polarized light is launched into a 2-m PANDA fiber, the far-field pattern of the PANDA fiber output is measured for both the x and y directions. This measurement is repeated for the y-polarized light. All spot sizes are 5.1 μm. Accordingly, the far-field pattern of the PANDA fiber is not affected by polarization. Coupling efficiency in this module is the same as that for a single-mode fiber module. The lasing light output from the double heterostructure (DH) LD is almost

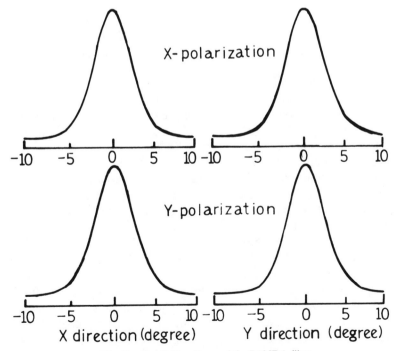

Fig. 3. Far-field patterns of the PANDA fiber.

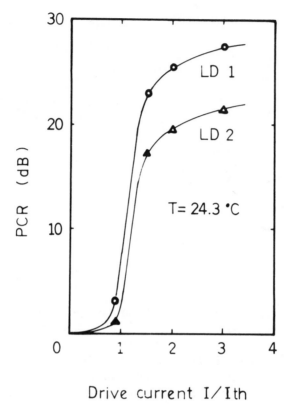

Drive current I / I th

Fig. 4. Typical polarization constituent ratio (PCR) of 1.3-μm InGaAsP BH LD with only the first lens.

completely polarized. The optical electric field vector for lasing light is parallel to the heterointerface plane. That is, the cavity mode which dominates lasing in the DH structure is the TE mode.

Figure 4 shows a typical polarization constituent ratio (the power ratio of the TE mode to the TM mode) of the 1.3-μm InGaAsP BH LD with the first lens (as shown in Fig. 2). This ratio is shown as a function of the driving current. The polarization constituent ratio (PCR) was more than 15 dB at a dc current $I > 1.2 I_{th}$. In the LD standby redundant system using the polarization filter, the polarized light from the LD is transported into the polarization coupler by means of a PANDA fiber to prevent cutoff of TE mode power of the LD by the polarization filter. In this LD module, it is important that the direction of the electric field vector of the TE mode coincide with the principal axis of the PANDA fiber. Figure 5 shows the measured PCR of PANDA fiber output power as a function of the angular misalignment between the TE mode from the LD and the principal axis of the PANDA fiber (in the x direction, shown in Fig. 2). The LD used was a DC-PBH LD with a threshold current of 23.8 mA at room temperature (24°C). The drive current of this LD was 39.4 mA for dc operation and the pulse and bias currents were 17.0 and 22.4 mA (445.8-Mbit/s p-n signal, direct modulation), respectively. As the angular misalignment φ increases, the PCR decreases. The PCR was more than 20 dB at $\varphi < 5°$. That is, at $\varphi = 5°$ the optical power cutoff due to the polarization filter is 1 percent of the LD output power.

In the case of angular misalignment, the PCR for this LD module is dependent on the length (L), misalignment angle (φ), wavelength (λ), and birefringence (B) of the

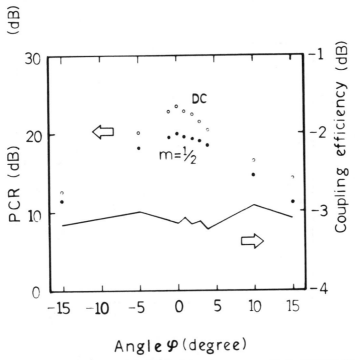

Fig. 5. Measured PCR of the LD module output power with the PANDA fiber.

PANDA fiber. The PCR (R) of LD module output power is represented as follows [7]:

$$R = 10\log\frac{1+P}{1-P} \tag{1}$$

$$P = \sqrt{1 - \sin^2 2\varphi \sin^2 \frac{2\pi \cdot B \cdot L}{\lambda}} . \tag{2}$$

Figure 6 shows both the measured PCR values and the calculated PCR values obtained by using (1) at $\varphi = 5°$, $\lambda = 1.302$ μm, and $B = 3.4 \times 10^{-4}$. The measured values agree well with the calculated values. As angular misalignment increases, the difference between the minimum and the maximum values of the PCR increases. At $\varphi = 0°$, the PCR is constant at the maximum value and is not dependent on the length or birefringence of the PANDA fiber. Moreover, it is not dependent on the wavelength of the LD. Fig. 7 shows the calculated values obtained from (1) as functions of the wavelength of the LD at $\varphi = 5°$. For instance, when the PANDA fiber length is 2 m and the wavelength of the LD is varied from 1.3015 to 1.3035 μm, the PCR changes from 21.5 to 19.2 dB. However, the maximum difference in the PCR is determined by only the angular misalignment. Consequently, the angular misalignment is determined to be less than $5°$ in order that the PCR of the LD module using the PANDA fiber be more than 20 dB.

Next, the method for measuring φ is discussed. In the coupling lens circuit shown in Fig. 2, TE and TM mode power have been transformed into the x axis mode and y axis mode in the PANDA fiber, respectively. The output power model at 445.8-Mbit/s direct

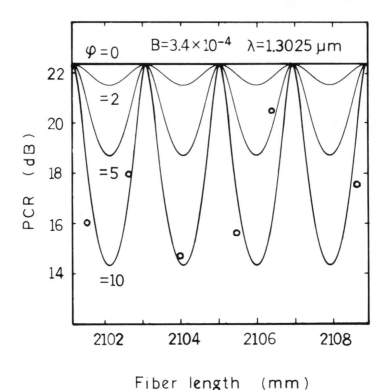

Fig. 6. Fiber length and birefringence dependence of the PCR of the LD module output power using the PANDA fiber.

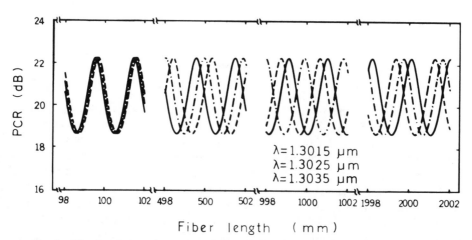

Fig. 7. Wavelength dependence of the PCR of the LD module output power.

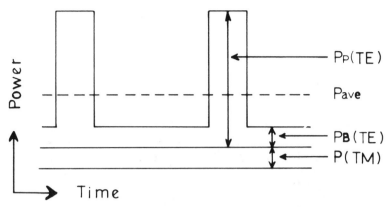

Fig. 8.   Output power model at 445.8-Mbit/s direct modulation.

modulation is shown in Fig. 8. The PCR for PANDA fiber output power is represented as follows:

$$R = 10\log \frac{K \cdot m \cdot \cos^2\varphi \cdot P_P(\text{TE}) + (1 - Km)\cos^2\varphi \cdot P_B(\text{TE}) + \sin^2\varphi \cdot P(\text{TM})}{K \cdot m \cdot \sin^2\varphi \cdot P_P(\text{TE}) + (1 - Km)\sin^2\varphi \cdot P_B(\text{TE}) + \cos^2\varphi \cdot P(\text{TM})} \tag{3}$$

$$\doteq 10\log \frac{Km\cos^2\varphi \dfrac{P_P(\text{TE})}{P(\text{TM})} + 1}{Km\sin^2\varphi \dfrac{P_P(\text{TE})}{P(\text{TM})} + 1} \qquad \left(\because \ \frac{P_P(\text{TE})}{P(\text{TM})} \gg \frac{P_B(\text{TE})}{P(\text{TM})} \right). \tag{4}$$

Here, the TM mode is not lasing. $P_P(\text{TE})$, $P_B(\text{TE})$, and $P(\text{TM})$ are TE mode power at

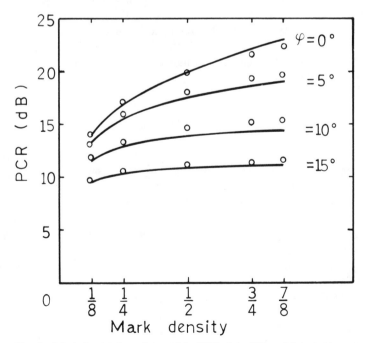

Fig. 9.   Mark density dependence of the PCR of the LD module output power.

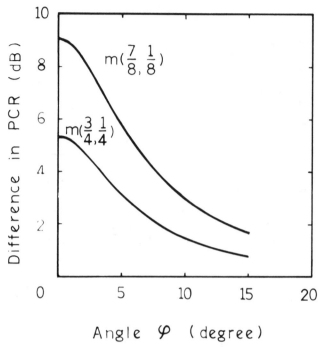

Fig. 10. Angular misalignment dependence of the difference in the PCR.

mark pulse, TE mode power at space pulse, and TM mode power, respectively. The notations k and m are the duty of the pulse and mark density, respectively. Figure 9 shows the measured PCR for a mark density of $\frac{7}{8}-\frac{1}{8}$ and the calculated values derived using (4), under the condition that the PCR is 22.3 dB at $m = \frac{1}{2}$, $k = \frac{1}{2}$, $\varphi = 0°$, and $\lambda = 1.302 \ \mu$m. There is little difference between the calculated values and the measured values. As φ increases, the dependence of PCR on the mark density decreases. Consequently, as shown in Fig. 10, φ can be estimated from the difference in PCR between either $m = \frac{7}{8}$ and $\frac{1}{8}$ or $m = \frac{3}{4}$ and $\frac{1}{4}$. For example, when the difference is nearly 6 dB between $\frac{7}{8}$ and $\frac{1}{8}$, φ is 5°.

Design of a Polarization Coupler

Polarization filters composed of dielectric layers exhibit little crosstalk between the p- and s-polarization waves. Figure 11 shows the polarization coupler having a polarization filter. The two PANDA fibers are arranged in the filter-to-fiber polarization coupler such that their polarization directions coincide with the p- and s-polarization directions of the polarization prism. Accordingly, both light beams from the two PANDA fibers are coupled into an output single-mode fiber. GRIN lenses were used as collimating or converging lenses.

The incident angular dependence of the optical loss and crosstalk in the polarization prism used in this study is shown in Figs. 12 and 13, respectively. The measurement setup is shown in Fig. 14. The optical source was a BH LD with a wavelength of 1.303 μm for dc operation at room temperature. The loss was measured at position A in the p-polarization wave and at position B in the s-polarization wave. The crosstalk was measured at the opposite positions as those for loss measurement. At $\theta = 45°$, the optical loss of p- and

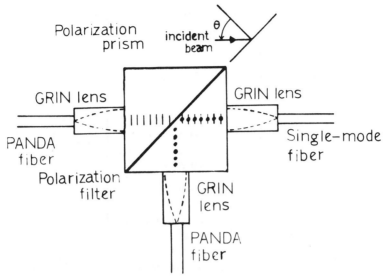

Fig. 11.   Polarization coupler.

s-polarization power was 0.35 and 0.34 dB, respectively, and the crosstalk of the p- and s-polarization power was 29 and 36 dB, respectively. As the incident angular θ is either larger or smaller than 45°, crosstalk and loss increase. The increase in crosstalk in the case of $\theta > 45°$ was larger than that when $\theta < 45°$. This is because the degree of the dielectric layer thickness due to variations in incident angular misalignment was different between the cases of $\theta > 45°$ an $\theta < 45°$. At $\theta = 46.3°$, the crosstalk of the p-polarization wave was less than 20 dB. In the case of a GRIN lens with a focal length of 1.92 mm, which was used as a collimating lens, an incident angular misalignment of 1.3° from 45° ($\theta = 46.3°$) is converted to a fiber-offset misalignment of 44 μm. Accordingly, it is easy to assemble an optical coupler having a crosstalk of more than 20 dB. When the PANDA fiber with a spot size of 5 μm and a single-mode fiber with a spot size of 5 μm have been assembled in this lens circuit, the coupling loss increases only 0.1 dB at a fiber offset of

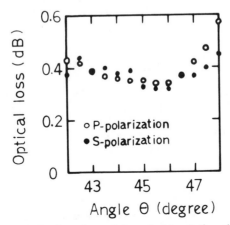

Fig. 12.   Incident angular dependence of the optical loss in the polarization prism.

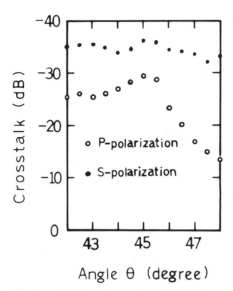

Fig. 13.   Incident angular dependence of crosstalk in the polarization prism.

1.5 μm. Consequently, this optical coupler is very stable from the standpoint of temperature dependence and highly reliable over a long period of time.

Next, angular misalignment between the x axis of the PANDA fiber and either the s- or p-polarization plane of the polarization prism is discussed in a similar manner to that for the LD module and the PANDA fiber. If the s- or p-polarization plane of the polarization prism does not coincide with the direction of the TE mode electric field vector of the LD light transmitted through the PANDA fiber, coupling loss appears in this optical coupler due to reflection and scattering on the surface of the fiber, lens, and polarization filter. This is in addition to cutoff loss due to the polarization filter, which arises from the angular misalignment γ between the x axis of the PANDA fiber and the s- and p-polarization plane of the polarization prism. At $\gamma = 5°$, the cutoff loss may be less than 0.1 dB.

The requirements of an LD standby redundant system using these polarization components for use in submarine optical repeaters are summarized in Table II.

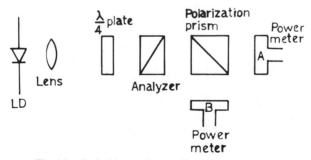

Fig. 14.   Optical loss and crosstalk measurement setup.

TABLE II
REQUIREMENTS AND DESIGNS OF THE LD STANDBY REDUNDANT
SYSTEM USING POLARIZATION COMPONENTS

LD module	
Coupling loss	less than 5 dB (0–40°C)
PCR	more than 20 dB (0–40°C)
Angular misalignment	less than 5.0°
PANDA fiber	
Fiber diameter	125 ± 3 μm
Spot size	5.0 ± 0.5 μm ($\lambda = 1.3$ μm)
Optical loss	10 dB/km ($\lambda = 1.3$ μm)
Bending radius	more than 25 mm
Twisting turns	less than 1 turn/m
Splice loss	less than 0.2 dB
Crosstalk	less than -20 dB
Polarization coupler	
Coupling loss	less than 1.7 dB ($\lambda = 1.3$ μm)
Crosstalk	less than -20 dB
Incident angular	
misalignment	less than 1.3°
Angular misalignment	less than 5.0°
Temperature dependence	less than 0.5 dB

CHARACTERISTICS OF AN LD REDUNDANT SYSTEM

A Polarization LD Module

Figure 15 shows the typical temperature dependence of the PCR for the LD having only the first lens at output powers of 3 and 5 mW. The PCR was more than 20 dB at 5–50°C. Figure 16 shows the typical temperature dependence of the PCR for the LD module employing the PANDA fiber, when the LD was modulated at 445.8-Mbit/s return-to-zero signals. The temperature dependence of the PCR was less than 1 dB. Moreover, the angular misalignment of this LD module estimated from the result in Fig. 10 was $\varphi = 5°$.

An Optical Coupler

Figure 17 shows the coupling loss of the polarization coupler. The average coupling loss was 1.4 dB. A typical breakdown of the coupling loss shows a reflection loss at the lens surface of 0.3 dB, a reflection loss at the fiber endface of 0.2 dB, a reflection and scattering loss in the polarization prism of 0.3 dB, and a misalignment and lens aberration loss of 0.6 dB. Coupling loss in the optical coupler due to temperature variations is less than 0.5 dB as shown in Fig. 18. The average fusion splice loss was 0.34 dB when the optical coupler was used in the LD module. An overall view of this polarization coupler is shown in Fig. 19.

If the LD redundant system is installed in the repeater housing, the PANDA fiber between the LD module and the optical coupler must be bent and twisted. As shown in Table II, the requirements of this LD redundant system concerning bending and twisting are a minimum bending radius of 25 mm and a turn 1 m when twisting. Figure 20 shows the measured optical output power variation of this LD redundant system when a

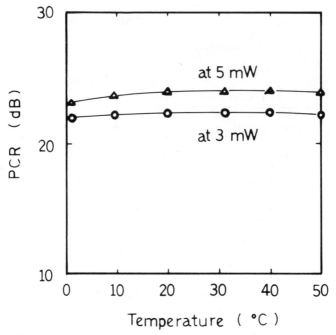

Fig. 15.   Typical temperature dependence of the PCR of the LD having only the first lens.

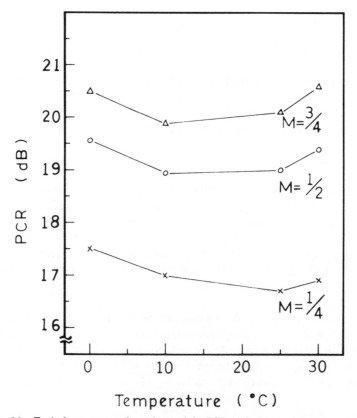

Fig. 16.   Typical temperature dependence of the PCR of the LD module output power.

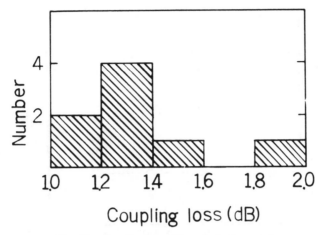

Fig. 17. Coupling loss of the polarization coupler.

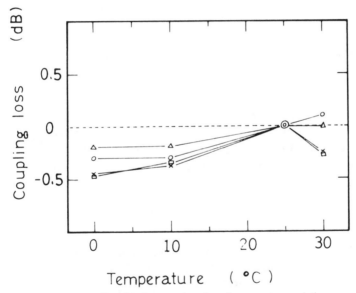

Fig. 18. Coupling loss as a function of temperature variation.

Fig. 19. Overall view of the polarization coupler.

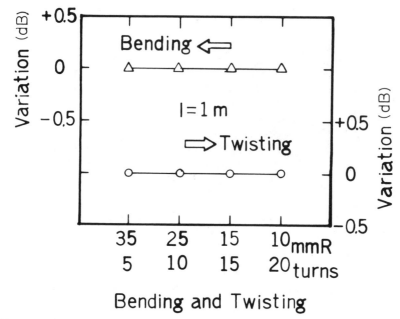

Fig. 20. Optical output power variation of the coupler due to PANDA fiber bending and twisting.

PANDA fiber of 1 m between the LD module and the optical coupler is bent or twisted. Optical output power variation was not observed at bending radii of 10–35 mm nor at twisting from 5–20 turns. Consequently, there were no variations in crosstalk and loss in the PANDA fiber connecting the LD module and the optical coupler due to bending or twisting.

LD Redundant System

The confocal lens circuit was adopted in the LD module employing the PANDA fiber. The first lens was a ball lens with a refractive index of 1.78 and a focal length of 0.46 mm. The second lens was a 0.2 pitch GRIN lens having a diameter of 1.8 mm and a focal length of 2.1 mm. The beam spot size of the PANDA fiber used was 4.2 μm. Coupling loss was 4.0 dB for both LD modules. The PCR of these LD modules was 24.6 and 20.2 dB, respectively. The two PANDA fibers were arranged in the fiber-to-fiber polarization coupler in such a way that their polarization directions coincided with the p- and s-polarization directions of the polarization prism as shown in Fig. 11. The prism was $10 \times 10 \times 10$ mm and contained a polarization filter. Crosstalk in the prism was 29 and 36 dB for p- and s-polarization, respectively. Two GRIN lenses having a diameter of 1.8 mm and pitches of 0.23 and 0.20 were used as the collimated and converging lenses, respectively. Coupling loss was 1.3 and 1.5 dB for s- and p-polarization, respectively. Consequently, a total loss of 5.5 dB was achieved for the optical circuit from the LD to the single-mode fiber output.

Conclusion

An LD redundant system consisting of two LD modules and a polarization optical coupler connected by a PANDA fiber has been fabricated. This system utilizes laser intrinsic polarization and has no movable parts. Total loss is 5.5 dB, which is approximately the same as that for conventional laser sparing schemes with movable parts. The temperature dependence of the LD's PCR, which has been the key weak point in this type of LD standby redundant system, was negligible in the temperature range of 5–50°C. This is because the PCR of the LD having only the first lens was more than 20 dB. The angular misalignment between the x axis of the PANDA fiber and the electric field vector in the TE mode of the operating LD has been studied experimentally and theoretically. This misalignment angle was determined to be less than 5° in order that the PCR of the LD module using the PANDA fiber was more than 20 dB. Moreover, coupling loss in these LD modules was equal to that for conventional LD modules using confocal lens circuits. Error in the newly proposed misalignment angle measuring method was less than 1°. The average coupling loss and the temperature dependence of the polarization coupler were 1.4 dB and less than 0.5 dB at 0–30°C, respectively. Finally, it has been concluded from these results that this LD redundant system can realize high reliability and high stability when applied to submarine optical transmission systems.

Acknowledgment

The authors wish to thank E. Iwahashi, S. Shimada, K. Fujisaki, and H. Fukinuki for their encouragement. The authors also wish to thank T. Ito for his fruitful discussion. They also wish to express their appreciation to H. Noda and Y. Sasaki for supplying the PANDA fiber.

References

[1] C. D. Anderson, R. F. Gleason, P. T. Hutchison, and P. K. Runge, "An undersea communication system using fiber guide cables," *Proc. IEEE*, vol. 68, no. 10, pp. 1290–1303, 1980.

[2] R. Kishimoto, "A consideration of an optical coupler for optical submarine transmission laser redundancy system," *IEEE Trans. Commun.*, vol. COM-31, no. 2, pp. 232–244, 1983.

[3] Y. Sasaki, K. Okamoto, T. Hosaka, and N. Shibata, "Polarization-maintaining and absorption reducing fibers," in *Proc. 5th OFC*, 1982, pp. 54–56.

[4] Y. Sasaki, T. Hosaka, K. Takada, and J. Noda, "8km-long polarization maintaining fiber with highly stable polarization state," *Electron. Lett.*, vol. 19, no. 19, pp. 792–794, 1983.

[5] Y. Sasaki, T. Hosaka, and J. Noda, "Polarization maintaining optical fibers used for a laser diode redundancy system in a submarine optical repeater," to be published.

[6] M. Saruwatari and T. Sugie, "Efficient laser diode to single-mode fiber coupling using a combination of two lenses in confocal condition," *IEEE J. Quantum Electron.*, vol. QE-17, no. 6, pp. 1021–1027, 1981.

[7] K. Okamoto, Y. Sasaki, T. Miya, M. Kawachi, and T. Edahiro, "Polarization characteristics in long length V.A.D. single-mode fiber," *Electron. Lett.*, vol. 16, no. 20, pp. 768–769, 1980.

Part VIII
Assuring the Reliability of Undersea Electrical and Optical Devices

Assuring the reliability of undersea electrical and optical devices is the most important and basic issue in the design of undersea lightwave communication systems circuit technology is used to minimize the number of components inside the repeaters. Rigid certification programs are developed to inspect, burn in and tests each and every component committed to undersea use. In addition, redundancy of critical components is employed to further enhance the system reliability. This part highlights the effects made worldwide assuring the reliability of undersea systems.

36

A Very High Reliability Fast Bipolar IC Technology for Use in Undersea Optical Fiber Links

JEAN YVES FOURRIER AND JEAN PIERRE PESTIE

INTRODUCTION

In the previous generation of long-haul analog undersea transmission links, discrete bipolar transistors were used. These transistors were built especially and manufactured to meet such requirements as wide-band, low noise or distortion, and high reliability.

The advent of digital transmission with high bit rates has spurred the need for new devices like integrated circuits (IC's). In 1975, it was decided to develop an in-house IC capability. This activity has led to the current DIFOX I technology. This technology, once originally used for terrestrial transmission equipment, has been upgraded to meet the stringent requirements of undersea systems. The changes were needed to provide wider quality/reliability margins.

The purpose of this chapter is to describe the basic features of this technology with the enhancements made for submarine application, to present some life-test results and to expose the basic methodology which will be applied to weed out potentially weak devices. The selection, qualification, and surveillance procedures will also be briefly reviewed, since at the moment the expertise is just being built. However, as will be seen, the philosophy behind these routines owes much to what has been successfully implemented on the submarine transistors during the last 15 years.

THE BASIC DIFOX I TECHNOLOGY

Scope of the Technology

The DIFOX I technology permits the manufacture of fast ECL bipolar IC's. Because of its balanced structure, ECL gives definite advantages to circuit designers in terms of device speed, symmetry, and complementary outputs. Some kind of flexibility is also offered by ECL for analog operation.

TABLE 1
SOME CHARACTERISTICS OF THE INTEGRATED TRANSISTORS

Emitter-base junction (depth/sheet resistivity)	0.25 μm, 30 Ω/□
Collector-base junction (depth/sheet resistivity)	0.60 μm, 750 Ω/□
Emitter + Contact width	2 μm
Emitter pitch	12 μm

In counterpart for its high speed, ECL has a propensity for high-power consumption and needs a process with a certain degree of complexity.

Consequently, the DIFOX I technology has been optimized keeping in mind such important goals as to provide:

- operation of the fastest IC's up to 1 GHz;
- operation of slow low-power IC's with less than 1 mA/gate;
- the highest degree of quality and reliability.

Description of the Technology

As with any IC technology, DIFOX I combines active elements with passive ones through the means of interconnect layers. The quality level of any IC technology cannot be better than its simpler elements. Therefore, it is of paramount importance to choose a simple element with a sound structure.

1) The Transistor Cell: The transistor cell adopted works at microwave frequencies up to 3 GHz. It is derived from the input transistor (LB01) used in the mid-seventies in CIT ALCATEL 60-MHz analog terrestrial transmission system. Such transistors have accumulated over 1.3×10^9 devices-hours in service without a failure. The emitters are washed and doped with arsenic. The main features are listed in Table I.

2) The Collector of Transistor: An extra step is needed in IC technology to contact the collector of the transistor on top of the structure. This step has been included prior to the transistor manufacture in order to introduce minimum changes in transistor performance. The collector series resistance is minimized through the use of an implanted buried layer. The transistor is fabricated on a high-quality SiH_4 grown layer doped with phosphorous (1.4 μm/0.8 Ω·cm). The substrate is p-type, oriented $\langle 111 \rangle$ with a resistivity of (5 ± 1) Ω·cm.

3) Isolation Scheme: Isolation must be provided between contiguous elements. A classical approach uses a reverse biased junction. Although advantageous in terms of process simplicity, such a solution is detrimental to the speed of operation and to the density of integration. Therefore, an oxide isolation process has been preferred. However, much consideration has been brought to the process to keep the surface as flat as possible and with smooth transisitions. The field oxide is fully recessed with a thickness of 1.5 μm.

4) Resistors: Resistors with highly reproducible values are desirable. In addition, they must not provide leakage path to the substrate or to other elements over the full operating range.

Two kinds of p-type resistors have been defined:

- a p^+-implanted resistor with a low sheet resistivity 80 Ω/□; and
- a p-diffused resistor (along with the transistor base) with a higher sheet resistivity.

5) Metallization System: The choice of a rugged and stable metallization system is a must for high-reliability IC's operating at high current densities and temperatures.

Although widely used in commercial parts, an aluminum-based metallization system is found to be inadequate for telecom applications. A metallization system using gold as a conductive layer must be used. Since gold cannot be used alone, various systems may be adopted. The LB01 transistor used a composite layer of molybdenum and gold. Such a metallization is well suited for IC's having more than one metal interconnect level. Reliability data pertaining to this system indicate an activation energy of 0.60 eV for electromigration. This value is perfectly compatible with the long operating life of the systems.

THE ENHANCED DIFOX I TECHNOLOGY FOR SUBMARINE APPLICATION

Although perfectly suited for use in terrestrial equipment, the DIFOX I technology has been enhanced for undersea application. The modifications affect only the layout ground rules and the metallization system.

Layout Ground Rules

The layout specialist follows the pertinent set of ground rules to transform the logic schematic or electrical diagram into geometrical patterns. Layout ground rules must be consistent with device physics as well as device manufacturing capability to guarantee the highest level of yield while permitting functionality of the part over the whole operating range.

In spite of redundancy schemes to enhance overall system reliability, the occurrence of a failure in an undersea system must primarily be considered at the device level. The layout ground rules enable one to achieve this goal by providing a very efficient way to buy reliability at the expense of a few tradeoffs:

- chip yield versus device size;
- device size versus chip size.

Several life-test experiments conducted on DIFOX I have shown that electromigration was the major cause of failure. Electromigration occurs mainly in spots where the current density exceeds a certain threshold value depending on local temperature. It is well described by an Arrhenius type law

$$MTFI = MTFO \exp(-Ea/kT) \tag{1}$$

where $Ea = 0.6$ eV typically for thin Mo Au films and MTF stands for the mean time between failures.

Once started, electromigration is a regenerative phenomenon which leads to device failure.

In order to improve resistance to electromigration, several known solutions exist consisting either in reducing the current density or lowering the temperature or, preferably, both. To lower the operating temperature one can either reduce the power consumption and/or improve the thermal resistance between the junction and the ambient. A combination of both approaches is the best compromise.

The DIFOX I layout rules have therefore been amended (see Table II). In addition, although the silicon surface is mainly flat on the IC, crossovers or via openings must be made outside regions having a disturbed topography.

TABLE II
CURRENT DENSITY

	Maximum current-density in 10^5 Acm^{-2}	
	Transistor	Other line
Terrestrial DIFOX I	3	3
Submarine DIFOX I	1.5	1.25

IC Packaging

The package not only provides a convenient protection of the chips, but also is an effective means of evacuating the dissipated heat. Its structure must be compatible with current manufacturing and assembly techniques and enable operation at high frequency.

Temperature has a huge influence in activating drift and failure mechanisms. One can appreciate the magnitude of the acceleration factor by looking at Table III referenced to an average temperature of 30°C.

Standard T0-86 14-lead flat-pack packages with a kovar bottom plate have a thermal resistance of 110 K/W (junction/air). The mere addition of a stud makes it possible to approximately divide this value by two with an efficient heatsink.

To avoid excessive acceleration factors, it is a good practice to limit the junction temperature to 45°C when the IC is dissipating. Under this condition it implies that packages with reduced thermal resistances must be used for chip power dissipation above 400 mW.

Metallization System Tradeoffs

For IC's operating at high frequency (above 400 Mb/s and in the case of complex IC's having 100–150 transistors) it is advantageous to use two levels of metal for the interconnect.

This leads to a smaller chip size, shorter electrical path, ease of layout, and lower power consumption. On the other hand, for IC's to be operated in undersea equipment at lower speed (280 MB/s), the prime consideration must be reliability.

Hence to achieve a low failure rate ($\lambda \leqslant 10^{-9}$) a simpler process is recommended. In effect, a simpler process guarantees:

- fewer steps;
- higher yield; and
- lower number of failure mechanisms.

Of course, the overall yield will be a function of the IC chip size. A compromise as to the use of one or two levels of metal will exist depending on chip complexity. No firm answer can be given at this point and each circuit must be analyzed on an individual basis.

TABLE III
TEMPERATURE ACCELERATION FACTOR

Temperature (°C)	30	45	60	75	90	105	120	135
Acceleration factor for $Ea = 0.6$ eV	1	2.96	7.92	19.5	44.6	95.4	192.6	369.4
Acceleration factor for $Ea = 0.75$ eV	1	3.87	13.3	41	115.2	298.1	717.7	1620

The following considerations have to be pondered.

1) All the strategic IC's such as those found on the signal path must have only one level of metal.

2) On an IC with one metal level, the power consumption will be higher, therefore the temperature will increase and accelerate the drift mechanisms. In addition, fewer candidates will be available for selection.

3) On an IC with two levels of metal, new failure mechanisms can occur, such as, second metal step coverage, possible shorts between first and second metal on crossovers due to a defect or pinhole in the isolation dielectric, and marginal contacts in the vias (contact between metal 1 and 2). In addition to the presence of a second level of metal, the associated dielectric further complicates the visual examination of each die under the microscope.

In the DIFOX technology, 12 extra steps are needed to add the second level of metal.

1) IC's with Single-Level Metal: For use in submarine systems the first metal thickness has been increased from 4500 Å in the original DIFOX I technology to a minimum of 9000 Å. Wider emitter metal fingers are also used on critical transistors.

2) IC's with Double-Level Metal: It has been experimentally demonstrated that the yield of an IC having double-level metal is highly dependent on how well the second metal layer is optimized.

The second metal layer is used mainly to dispatch the main power lines VEE or VCC and adds extra convenience for the interconnect. The second metal must eventually pass over the first metal without any breakage or width reduction. Therefore, the first metal thickness cannot be too high as good reliability practice would indicate for less electromigration. 5000-Å sputtered gold has been found adequate for the first level. The second metal layer is then sputtered consisting of 1000-Å Mo and 5000-Å Au. After patterning, the second gold is electroplated to a minimum thickness of 25 000 Å. In Figs. 1 and 2, one can observe the application of this technique to an IC used in the 560-Mb/s digital transmission system on coaxial cable. The dielectric is a high-quality CVD oxide with an average thickness of 4000 Å.

3) Improved Material for the Metallization System: Although Mo–Au is presently being used in DIFOX I, owing to its basically good properties and ease of processing, Ti–Pt–Au is the choice for the new technology. Dubbed DIFOX I A, it is currently under development, specifically for IC's having a single level of metal. In effect, its resistance to electromigration is higher ($Ea \geqslant 0.75$ eV) and moisture does not alter its integrity, which

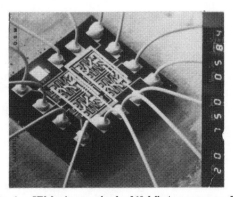

Fig. 1.   SEM micrograph of a 560-Mb/s regenerator IC.

Fig. 2. SEM close-view of the double-level metallization system.

is not the case for molybdenum. However, these advantages can be obtained only at the expense of a more acute etching process to pattern the platinum layer. In addition, the definition of a valid process sequence for circuits having two levels of metal is still under study, the main difficulty being to improve the etching selectivity. This is why, at least in the first generation of submarine IC's, Mo–Au will be used. It is not yet certain that Mo–Au will not be used for IC's having two levels of metal.

GENESIS OF AN IC FOR SUBMARINE APPLICATION

The design and manufacture cycle of an IC to be used in undersea equipment obeys stringent rules in order to make certain that it will fulfill its mission during its operational life (25 years).

Flow chart I in Fig. 3 indicates the major steps. Once in production, wafer fab, testing, assembly, burn in, and aging must provide, on a routine basis, practical and economical ways to guarantee a high level of quality and reliability.

Before entering into production, the IC must pass several important milestones. Flow chart II (Fig. 4) gives an overview of the major operations.

Design and Layout

After review and approval of the initial specifications, the circuit is designed using modern CAD tools. The circuit is then laid out. During this phase, close coupling exists between the circuit designer and the layout specialist. A final verification is automatically made using Electrical and Design Rules Check programs (ERC and DRC) in order to eliminate costly errors and delays. Then mask plates are fabricated. Only master masks are used in production (contact printing).

Fabrication

At all stages during the manufacturing cycle (wafer fab + assembly), the operations are carefully monitored. Surveillance procedures are enforced in order to avoid catastrophic process fluctuations, mishandling, or incorrect fabrication techniques. Foremen and

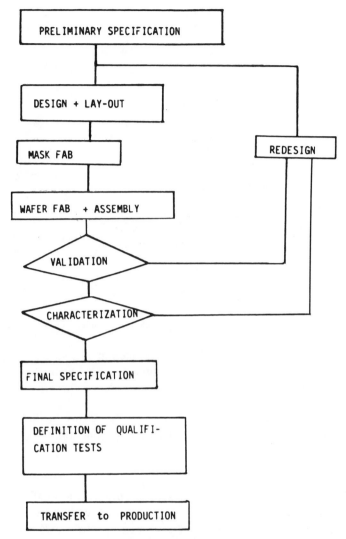

Fig. 3.   Flow chart I: Genesis of an IC for submarine operation.

supervisors have a key role in sensitizing their production personnel to the respect of preset standards.

Validation

Once manufactured, each new type of IC must be validated. At this stage the IC must demonstrate its capability to perform as specified in the full operating range (temperature, voltage, current, frequency) with wide enough margins so as to guarantee a secure behavior. If it fails, corrective actions must be taken (redesign).

Characterization

After validation, the IC will enter the characterization program. This program consists of tests of short duration (<1000 h) but with the application of high-amplitude stresses in order to quickly reveal and identify most likely degradation or drift mechanisms.

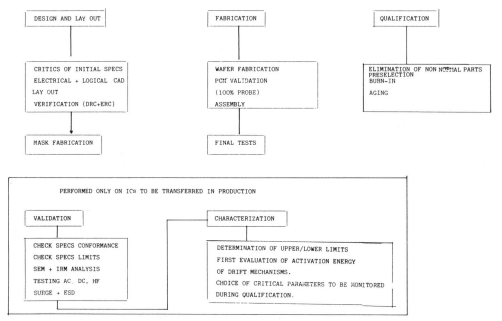

Fig. 4.   Flow chart II: Basic phases in the certification program of a submarine IC.

If necessary, corrective actions can still be undertaken at this stage. The characterization tests enable the definition of the tests to be applied during the qualification program,
and determine the sequencing, nature, level, and duration of these tests as well as the
sensitive parameters to be monitored.

Upon completion of the characterization program, final specifications and relevant
documents are written before the IC is transferred into production.

Qualification

Contrary to the validation and characterization programs, the qualification program
(QP) is run on each lot of IC's to be used in undersea equipment. As successfully
implemented with submarine transistors, one lot will be one wafer. This ensures the
highest degree of homogeneity between IC's belonging to the same lot (see Fig. 5, Flow
chart III).

The QP is conceived in such a way to avoid damage of the normal parts (i.e., the ones
following the normal law used in statistics), but with enough severity to weed out
marginal or potentially weak parts. Therefore, the level of the pertinent life tests are
tailored to each IC, taking into account their particular features or operating conditions
within an undersea repeater. When completed, the QP leads to the editing of a summary
test file. Each IC will be later individually approved or rejected under the control of a
commission. This commission is composed of several representatives of CIT and of the
PTT administration.

Surveillance

The surveillance procedure establishes practices and methods of control to be applied
during the whole manufacturing cycle, so as to make certain that a batch of wafers or
parts has been processed according to specified conditions. This ensures reproducible

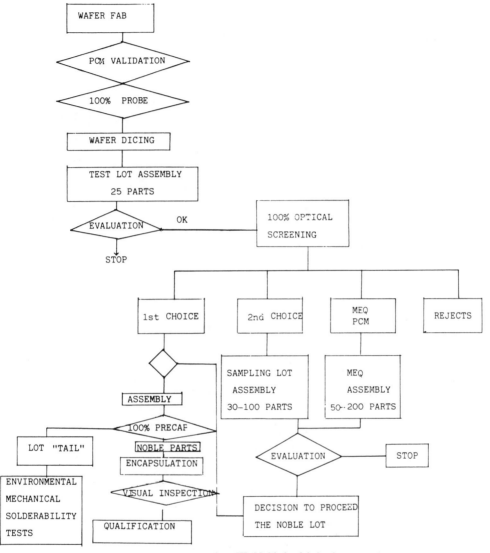

Fig. 5.   Flow chart III: Noble lot fabrication.

characteristics. Major operating parameters and results are carefully logged and are easy to retrieve. In the long run, when the fab has accumulated enough experience, these data can be further analyzed to provide a guideline for a subsequent optimization program (see Table IV).

LIFE EVALUATION

Basic Principles

Once fabricated in a well-controlled environment using a sound process, the intrinsic built-in Quality/Reliability should lead to a life-expectation several orders of magnitude in excess of the 25-year warranty period. Granted that random failure modes are avoided

TABLE IV
Major Parameters Monitored During Wafer, Lot, and Part Selections of Submarine IC's

TEST LOT EVALUATION	SAMPLE LOT EVALUATION	NOBLE LOT QUALIFICATION
NONOPTICALLY SCREENED GOOD ELECTRICAL PARTS	2° CHOICE OPTICAL, GOOD ELECTRICAL PARTS	FIRST CHOICE PARTS
VALIDATION OF FUNCTIONAL POTENTIALITY	VALIDATION OF WAFER QUALITY AND SOLIDITY	FINAL SELECTION OF IMMERGEABLE ICs
VERIFICATION OF DC+AC CHARACTERISTICS PARAMETERS CENTERING AND DEVIATION AT T = Tmax, Vmin, and Vmax	CHARACTERIZATION OF THE LOT AT HIGH STRESS LEVEL MECHANICAL TESTS ENVIRONMENTAL TESTS ELECTRICAL TESTS AT MAX TEMPERATURE, FREQUENCY DETERMINATION OF WAFER SOLIDITY ON PCM	†PRESELECTION ACCELERATION (30000g; Y_1, axis) VRT 5 cycles (−55, +150°C) LEAK TEST FINE LEAK(10^{-8} atm cm³/s) LEAK TEST GROSS LEAK INITIAL TEST STORAGE 168H, 200°C INTERMEDIATE TEST (1) †BURN IN 125°C, 168 H, V MAXIMUM INTERMEDIATE TEST (2) †AGING 125°C, V NOMINAL MEASUREMENTS at 250 H 500 H 1000 H 2000 H† PARAMETERS DISTRIBUTION ANALYSIS OF DRIFTS EDIT SUMMARY FILE †(only on critical IC's)

TEST CONDITIONS FOR NOBLE LOT QUALIFICATION

ALL STEPS : ROOM TEMPERATURE DC AND PARAMETRIC
INTERMEDIATE TEST 2 : DC + PARAMETRIC AT A HIGH TEMPERATURE
AND AFTER LAST: TO BE DETERMINED DURING THE CERTIFICATION PROGRAM
AGING STEP: AC AT ROOM TEMP

through the enforcement of stringent surveillance and qualification procedures, the occurrence of a defect should become a deterministic event.

This is all the more true, when the characterization program has been well executed. Hence, the occurrence of a defect can be mathematically predicted with a very high confidence level, provided the physics of degradation mechanisms is known. Referring to the structure of the IC, the drift mechanisms will take place in the integrated elements (transistors, resistors) or in the interconnect. As compared to discrete transistors, the IC structure adds a few more mechanisms. But if the whole design manufacturing and qualification cycle is well controlled, the transistor cell itself can be made the most sensitive, i.e., the weakest link in the IC, thus providing an effective means of monitoring any unwanted drift or incipient failure. Of course, a variation in one or several internal elements of the IC may not necessarily lead to a failure mode since the IC design can induce mutual compensation. In other words, it means that a nonobservable drift from the external pins of the IC may not always mean that some elements in the IC have not already started to drift. It is the purpose of the characterization program, and of the qualification program, to give 100-percent confidence in this correlation.

Present Results

Figure 6 has plotted the standard representation of the lifetime as a function of temperature. Depending on the criterion chosen for assessing the lifetime, different results can be obtained.

For instance, curve 1, which represents the drift mechanism of current gain of a LB01 type transistor, which has basically the same structure as the integrated transistor,

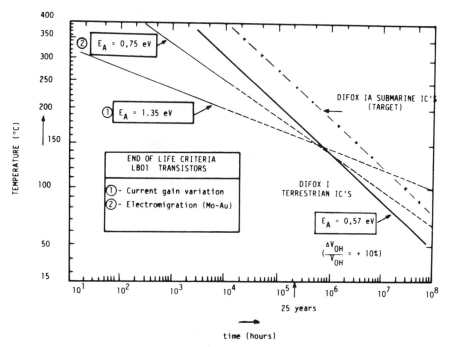

Fig. 6.   Arrhenius plots for terrestrial wide-band transistors (60-MHz system) LB01 comparison with IC's. (DIFOX I: Experimental data; DIFOX IA: target data.)

indicates a life in excess of 10^8 h at 100°C. This stems from the high activation energy ($Ea = 1.35$ eV).

For the same transistor, curve 2, indicates a life of 10^8 h at a lower temperature (65°C). This is true since the activation energy is lower ($Ea = 0.75$ eV) for electromigration. Referring to curve 3, drawn for DIFOX I IC's used in terrestrian application, one obtains a life of 10^8 h at still a lower temperature (50°C), once again because of a reduced

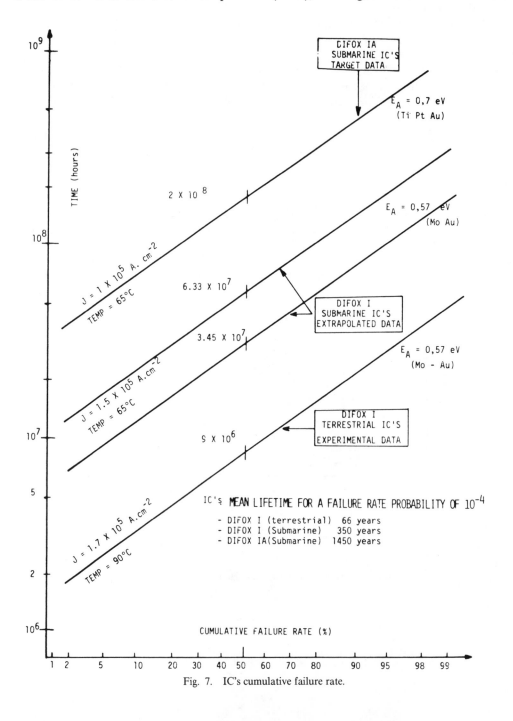

Fig. 7. IC's cumulative failure rate.

activation energy ($Ea = 0.57$ eV). The chosen criterion in this case $\Delta V_0 H / V_0 H = 10$ percent being valid for this particular digital IC. For other kinds of IC's, the characterization program implemented on each IC will provide information on the criterion to be adopted.

Figure 7 gives a representation of the cumulative failure rate which can be anticipated for submarine IC's. Right now, life in excess of 350 years with a failure rate probability of 10^{-4} can well be envisioned for the first generation of submarine IC's.

Lifetime Enhancement

Due to improvements in the technology both in the transistor structure and the metallization system (reduced current density use of TiPt Au in lieu of Mo–Au) the future technology dubbed DIFOX 1A, now currently in its development phase, will result in even higher lifetimes. For a failure rate probability of 10^{-4}, the anticipated lifetime should exceed 1450 years, thus making it possible to have a very good security margin. As progress is made, the confidence level will build up and it is likely that fiber optics submarine links beyond TAT 8 may well resort to simpler redundancy schemes, thus diminishing the costs of the repeaters. Silicon IC's have yet to prove that they can be as dependable as their discrete predecessors, the submarine transistors. However the basic properties of silicon make it a very realistic goal.

CONCLUSION

The manufacture of ultra-reliable IC's for use in undersea optical fiber submarine transmission systems represents a tremendous challenge in terms of device design, manufacturing, and selection.

In the case of discrete transistors, it was easy to address externally through simple electrical measurements all potential failure mechanisms by careful monitoring of even minute parameter variations. This approach is no longer possible with IC's, since their complexity prohibits individual testing of internal transistors or elements. Therefore, Quality/Reliability must be more than ever built in within the device.

Backed by its wide experience accumulated during 15 years with the manufacture of silicon submarine transistors, CIT-Alcatel has strived to define and achieve:

- a dependable IC technology: DIFOX 1,
- an effective manufacturing process, and
- efficient selection methodology and procedures.

As on going efforts and results tend to demonstrate, this approach is the right direction. Of course, at this early stage in the development of a new breed of components, no one can either guarantee or demonstrate that the goal of low 10^{-9} failure rates has really been reached. Still, more work remains to be done, to prove the validity of this endeavor.

Nevertheless, we are confident that through patient work, use of dedicated means, and through motivated and talented people, the goal of manufacturing ultrareliable IC's for use in undersea optical fiber links can be reached.

37

Integrated Optical Submarine Repeater Circuits Using a High Reliability Process Technology

MAMORU OHARA, MEMBER, IEEE, TSUTOMU KAMOTO, AND SHUICHI KANAMORI

INTRODUCTION

High-speed optical submarine transmission systems are expected to play an important role in domestic and international communications [1]. To realize such systems, a small sized repeater with high reliability and low power dissipation is essential. Thus we have developed integrated repeater circuits, such as an equalizing amplifier IC, timing amplifier IC, decision circuit IC, laser diode (LD) driver IC and supervisory IC's; relying on the progress in analog integrated circuit technology [2]–[4].

These IC's were fabricated using a Si shallow-junction bipolar process with 7 GHz f_T. To achieve high reliability in IC's, a two-level metallization system was developed. Cu-doped Al was used for the first layer with a current density under 1.5×10^5 A/cm², and a Ti-Pt-Au metal system was incorporated for the second layer. As a diffusion barrier, TiN films were inserted under the first layer and between the two layers. This resulted in high thermal stability at 550°C.

This chapter describes circuit configuration, performance, and the high reliability process of optical submarine repeater IC's.

REPEATER CIRCUIT CONFIGURATION

A block diagram of repeater circuits is shown in Fig. 1. The repeater IC's consist of six kinds of chips: an equalizing amplifier, a timing amplifier, a decision circuit, two LD drivers and two types of supervisory circuits. The repeater circuits were divided into seven IC's, for the following reasons: The first reason was to ensure amplifier stability. Since the equalizing amplifier and timing amplifier have high gain (50–60 dB) at high frequency (a few hundred MHz), these circuits have instability factors, such as parasitic oscillation caused by crosstalk due to input from large amplitude signals. Therefore, these analog circuits should be separated from the decision and LD driver circuits. The second reason

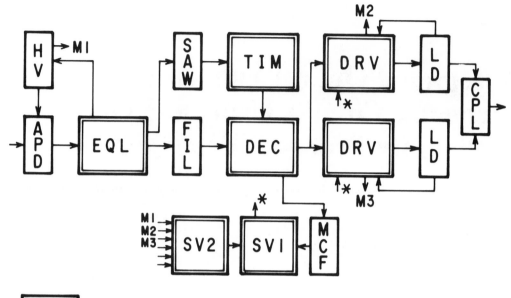

EQL : Equalizing amplifier   HV : High voltage circuit
DEC : Decision circuit   SAW : Surface acoustic wave filter
TIM : Timing circuit   FIL : Equalizing filter
DRV : LD driver   LD : Laser diode module
SVI,2 : Supervisory circuits   MCF : Monolithic crystal filter
APD : Avalanche photo diode   CPL : Optical coupler

Fig. 1.   Optical repeater block diagram

was due to the power dissipation limit per chip. In order to realize very high reliability of less than several FIT's (which is required of IC's used in submarine transmission systems) transistor junction temperature should be less than 65°C. So, the power dissipation per chip is limited to 1 W. The supply voltage of these IC's is −6 V. Design targets for these IC's are listed in Table I.

Repeater IC Configuration and Experimental Results

Equalizing Amplifier IC

The equalizing amplifier IC consists of a preamplifier, an automatic gain control (AGC) amplifier, a post amplifier, and control circuits which generate a gain control voltage and input-offset compensation voltage as shown in Fig. 2. A microphotograph of this IC is shown in Fig. 3. This IC is made up of about 200 elements.

The preamplifier consists of a single-ended shunt-series feedback amplifier with low noise characteristics.

TABLE I
IC DESIGN TARGETS

Items			Target value	Remarks
Equalizing amplifier	Transimpedance	Max	> 91 dB	s_{21} > 57 dB
		Min	< 71 dB	s_{21} < 37 dB
	Output amplitude		0.3 V	
	3 dB down bandwidth		> 310 MHz	after filtering
	Input current noise density		10 pA/$\sqrt{\text{Hz}}$	
	Power dissipation		0.6 W	
Timing amplifier	Output amplitude		0.8 V	
	Voltage gain		> 55 dB	
	Dynamic range		> 25 dB	
	Rise and fall time		0.3 nsec	20 ~ 80%
	Phase deviation		< 10°	
	Power dissipation		0.7 W	
Decision circuit	Output amplitude		0.8 V	
	Sensitivity		< 4%	
	Rise and fall time		0.3 nsec	20 ~ 80%
	Duty ratio		50%	
	Power dissipation		0.8 W	
LD driver	Pulse current		0 ~ 30 mA	
	Pre bias current		0 ~ 70 mA	
	Duty ratio		50%	
	Power dissipation	Operating	0.6 W	
		Waiting	0.3 W	

In the equalizing amplifier, constant output amplitude is achieved by controlling the avalanche photodiode (APD) multiplication factor and the AGC amplifier gain for the received optical power fluctuations. Multiplication factor is controlled through the high-voltage circuit. The variable gain range required for electrical gain control is more than 20 dB. The AGC amplifier configuration, in which a peaking technique is applied to obtain wide-band characteristics, is shown in Fig. 4 [3]. This circuit consists of two

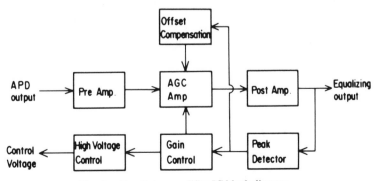

Fig. 2. Equalizing amplifier IC block diagram

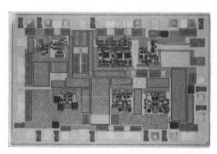

Fig. 3.   Microphotograph for equalizing amplifier IC (1.8 mm × 2.8 mm).

differential-pairs $Q1$-$Q2$ and $Q3$-$Q4$. As the load resistance R_L is common for these pairs, the total gain is given as a sum of the two differential-pair gains. The gain control is effected by changing the current ratio in the differential pair $Q5$-$Q6$, with the control voltage V_C. The variable gain range, obtained by a one-stage AGC amplifier is about 15 dB. In order to obtain the required variable gain range, a two-stage differential amplifier was utilized.

The post amplifier, which consists of an emitter peaking differential amplifier with 6-dB constant gain and 1-GHz bandwidth, produces an output amplitude of 300 mVp-p, and drives the decision and timing IC's.

The control circuits generate a gain control voltage and input-offset compensation voltage from the detected peak voltage of the equalizing amplifier output.

The equalizing amplifier experimental frequency response is shown in Fig. 5. Maximum gain (s_{21}), variable gain range, and bandwidth are 65 dB, 30 dB, and 300 MHz, respectively. These characteristics fully satisfy the design target. Computer simulation shows that 65-dB gain (s_{21}) is equivalent to 100-dB transimpedance, and the bandwidth is also equivalent to about 450 MHz when it is driven by an APD with high impedance. An output eye-diagram for this IC is shown in Fig. 6. Regarding noise characteristics, an equivalent input current noise density under 10 pA/$\sqrt{\text{Hz}}$ was obtained.

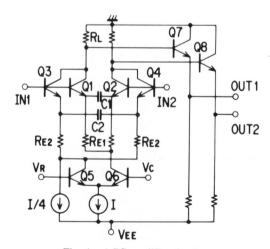

Fig. 4.   AGC amplifier circuit.

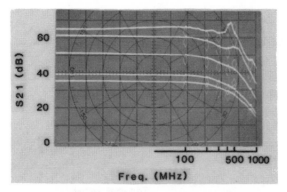

Fig. 5. Frequency response for equalizing amplifier IC.

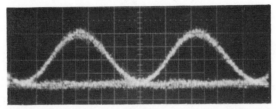

Fig. 6. Output waveform of equalizing amplifier IC at −30-dB optical input power (*V*: 100 mV/div, *H*: 500 ps/div).

The maximum gain-frequency response has peaking characteristics at about 500 MHz as shown in Fig. 5. This is caused by crosstalk from the output signal to the input. Computer simulation of this mechanism is shown in Fig. 7. When a parasitic capacitance of 0.005 pF exists between input and output, similar peaking characteristics as those measured can be simulated. Therefore, the reduction of crosstalk is important for preventing instabilities in high-gain and high-frequency circuits.

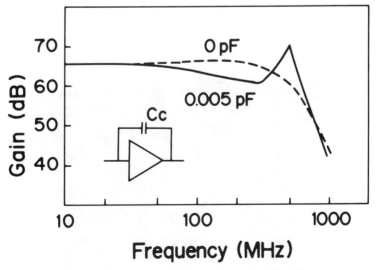

Fig. 7. Simulated crosstalk in equalizing amplifier IC.

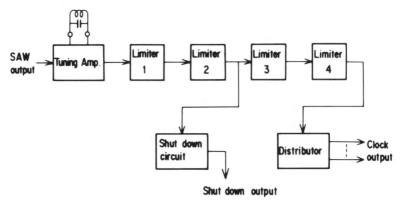

Fig. 8.   Timing amplifier IC block diagram.

Timing Amplifier IC

A timing amplifier IC block diagram is shown in Fig. 8. The equalizing amplifier IC output is inputted to the surface acoustic wave (SAW) filter, and the timing signal is extracted. Assuming that the equalizing amplifier has an output amplitude of 300 mVp-p, the output level of SAW filter ranges from −15 dBm to −40 dBm including insertion loss. Therefore, in the timing amplifier IC, about 50-dB gain was required to obtain a clock output amplitude of 800 mVp-p. This IC's maximum gain is designed at 55 dB to obtain a phase deviation of less than 10°.

A tuning amplifier, four limiter stages, and a timing clock distributor make up this IC. The tuning amplifier is utilized to improve jitter characteristics. Since each limiter has a high input impedance, several picofarad integrated capacitors can be used as interstage coupling capacitors. This makes dc biasing easier in each limiter.

Differential amplifiers are used in this IC, because the gain varies little with supply voltage, and phase deviation characteristics can be improved by using a bandwidth widening technique, such as emitter peaking or a cascade configuration.

The timing amplifier IC also includes a shut-down circuit. This circuit detects the existence of input signals from the transmission line by comparing a decision reference level and the output peak voltage in the middle stage limiter. When no signal is present,

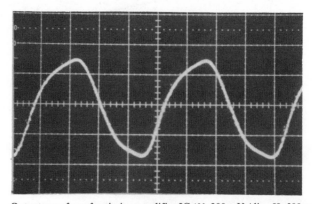

Fig. 9.   Output waveform for timing amplifier IC (*V*: 200 mV/div, *H*: 500 ps/div).

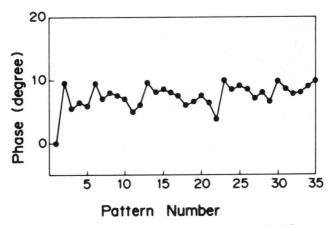

Fig. 10.   Jitter characteristics for timing amplifier IC.

the shut-down circuit fixes the decision IC output to prevent free-running due to random noise.

An output waveform for this IC is shown in Fig. 9. 60 dB is obtained at a tuned frequency of 445 MHz. The phase deviation is less than 10° as shown in Fig. 10.

Decision IC

A decision IC block diagram is shown in Fig. 11. This IC consists of a slice amplifier and a delayed flip-flop circuit. The slice amplifier gain is designed at 20 dB to obtain decision sensitivity under 4 percent for the equalizing amplifier. This IC's output is emitter coupled logic (ECL) compatible and the output emitter follower can drive a 50-Ω coaxial cable.

The slice amplifier's measured gain and bandwidth are 17 dB and 500 MHz, respectively. The output waveform of this IC is shown in Fig. 12. A 300-ps rise and fall time is obtained. The sensitivity is about 6 mV, as shown in Fig. 13.

LD Driver IC

An LD driver IC block diagram is shown in Fig. 14. The LD driver, consisting of differential circuits, switches 30-mA maximum current at a 300–500-ps rise and fall time. This IC has an automatic power control (APC) function. The APC circuit controls the duplicated LD pre-bias current, and exchanges a degradated LD for a new one automatically thereby supervising the optical power detected by APD. This circuit is designed to control up to a 70-mA pre-bias current. Since this IC includes transistors operating on

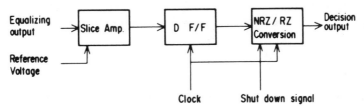

Fig. 11.   Decision IC block diagram.

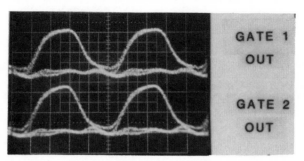

Fig. 12. Output waveform for decision IC through a 7.9-dB attenuator (V: 100 mV/div, H: 500 ps/div).

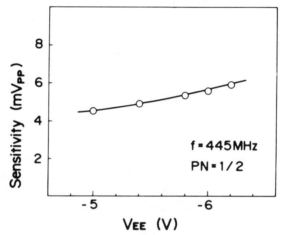

Fig. 13. Sensitivity for decision IC.

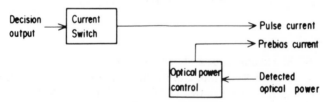

Fig. 14. LD driver IC block diagram.

high current, thermal concentration takes place in the IC. In order to lower the junction temperature to under 65°C, a pattern layout is designed to keep distance between the differential pair transistors.

A 500-ps rise and fall time is obtained for the optical output waveform, as shown in Fig. 15.

Supervisory IC's

Supervisory circuits monitor the APD bias voltage and LD pre-bias current and transform the monitored analog voltage to a frequency using the V-F convertor. This frequency modulates the mark density of the loop back pulse stream and the monitored

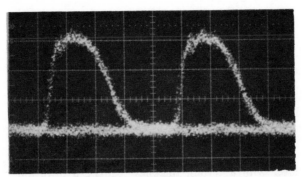

Fig. 15. Optical output waveform of LD module at -3.5-dBm output power (H: 500 ps/div).

information is transmitted to a shore terminal through the loop back line. Moreover, these circuits can exchange the degraded LD module for a new one according to a command from the shore terminal. These supervisory circuits consist of two IC's, a high speed loop back gate IC (SV1) and a low speed controlling and monitoring circuit IC (SV2).

Device Technology

A schematic transistor cross section is shown in Fig. 16. SiO_2 isolation, two-level metallization with polysilicon contact, and TiN diffusion barrier layers are used. The keys in achieving high reliability are multilevel metallization for high-temperature and high current density, a bonding technique free from metallurgical instability, and a protective coating to prevent mechanical damage and foreign metallic particles.

High reliability metallization in transistors for conventional submarine coaxial cable repeaters have been developed [5] by using a Au system such as the Au/Pt/Ti beam lead metal system. In high-speed bipolar IC's, interconnect resistance should be reduced by multilevel metallization to achieve high performance. However, realizing a multilevel configuration using only the Au system is quite difficult because of Au's poor adhesion to interlayer insulator film. Therefore, the authors developed a Al–Au combined two-level metal system with Cu-doped Al for the first-level metal layer for which the current density is designed to be under 1.5×10^5 A/cm^2.

Murarka *et al.* have reported [6] that the two-level system, Au/Pt/Ti and Al, is stable when both Pt and Ti layers exceed 200 nm preventing Au–Ti and Pt–Al interactions,

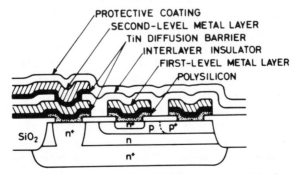

Fig. 16. Schematic transistor cross section. The first- and second-level metal layers are formed by Cu-doped Al and Au/Pt/Ti, respectively.

TABLE II
ALUMINUM CONTACT METAL SYSTEM

Group	First-level metal structure
A	Cu-doped Al(600 nm)/TiN(100 nm)/Ti(50 nm)/poly Si(100 nm)
B	Si-doped Al(600 nm)/poly Si(100 nm)
C	Cu-doped Al(900 nm)/poly Si(100 nm)
D	Pure Al(1.3 μm)/poly Si(100 nm)

In group A, Ti layer is for ohmic contact by titanium silicide formation after in-process annealing.

respectively. A Ti layer with a certain thickness can be a diffusion barrier for Al–Si interaction. However, Ti is essentially a "sacrificial" barrier due to $TiAl_3$ formation which retards failure time depending on the Ti layer thickness [7].

Titanium nitride (TiN) is a fairly stable compound with a high melting point and a high electrical conductivity. It has been reported that TiN films are successful as diffusion barriers and improve the reliability in Au [8] and Al [9] systems. In repeater IC's, the TiN diffusion barrier layers are used under the first- and the second-level metal layers, Cu-doped Al, and Au/Pt/Ti, to prevent Al–Si alloy penetration and Ti–Al reaction, respectively.

Contact Metallurgy Studies

Contact stability in Si shallow junction transistors is required from two viewpoints: thermal stability and electromigration stability at the Si contact. Firstly, high temperature stability of Al contact systems for shallow junctions were investigated. Shallow junction transistors, shown in Fig. 16, with various Al contact systems listed in Table II were prepared. In groups A and B, a multilayer metal system with Au(200 nm)/Pt(100 nm)/Ti(100 nm)/TiN(100 nm)/Ti(50 nm) was used for each second-level metal layer. In groups C and D, a single-layer metal system with Cu-doped Al(1.1 μm) and pure Al(1.3 μm) were used for each second-level metal layer, respectively. The transistor wafer stability at high temperature was investigated by temperature step-stress test in N_2 in a quartz tube furnace, raising the temperature from 300°C up to 600°C in 50°C steps each hour. Junction failures were detected by leakage currents over 100 μA. The Si-doped Al and Cu-doped Al/TiN/Ti systems exhibit the highest thermal stability at 550°C as shown in Fig. 17.

Secondly, electromigration at the anode contact is characterized by Si dissolution and Si transport into the Al layer, followed by shallow junction destruction due to void formation on the Si surface. Electromigration stability at the Si contact has been studied by Mori *et al.* using a high precision diffused resistor with four Al contacts [9]. Resistance increases in diffused resistors due to void formations are shown in Fig. 18. The pure Al- and Si-doped Al systems are poor because of Si transport in the Al layer. Instability in the Al/TiN/Ti system was due to Al disappearance in the contacts as a result of Al electromigration, however, no void formation on the Si surface was observed. Contact microsectioning in the Cu-doped Al system also showed fine void formations. The Cu-doped Al/TiN/Ti system exhibits the highest stability for electromigration at the contacts. This is attributed to Cu-doped Al electromigration-resistance and the TiN diffusion barrier preventing Si dissolution. Thermal and electromigration stabilities in Al contact systems are summarized in Table III.

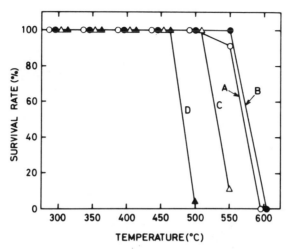

Fig. 17. Survival rate of shallow junction transistors, with various Al contact systems listed in Table II, during temperature step-stress test.

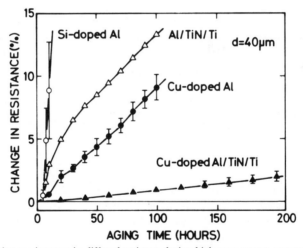

Fig. 18. Resistance increase in diffused resistors during high temperature current-stress test with four Al contact metal systems: Si-doped Al(1.5 μm), Cu-doped Al(1.5 μm), pure Al(1.5 μm)/TiN(50 nm)/Ti(50 nm), and Cu-doped Al(1.5 μm)/TiN(50 nm)/Ti(50 nm). The average contact current density was 4×10^4 A/cm^2 in contacts with 40-μm length(d) and 14-μm width. Resistor temperature was 300°C. Resistance increase for Cu-doped Al/TiN/Ti system was remarkably reduced [9].

TABLE III
THERMAL AND ELECTROMIGRATION STABILITIES IN ALUMINUM
CONTACT METAL SYSTEMS

First-level metal structure	Thermal stability	Electromigration stability	
		Al layer	Si contact
Cu-doped Al/TiN/Ti	excellent	excellent	excellent
Si-doped Al	excellent	poor	poor
Cu-doped Al	good	excellent	good
Pure Al	poor	poor	poor

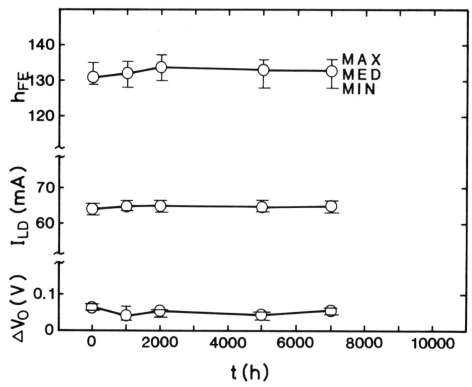

Fig. 19. Main parameter changes of test element circuits during 7,000-h operation tests at 125°C. Parameter h_{FE}, I_{LD}, and ΔV_0 are current gain for unit transistor, laser diode driving current for LD driver, and output voltage difference for differential amplifier, respectively.

Reliability

IC reliability requirements are on the order of a single FIT or less. As a preliminary experiment, test element circuits with a unit transistor, differential amplifier, and LD driver were fabricated and tested. These circuits were highly stable over a 7,000-h test period at 125°C, as shown in Fig. 19.

Reliability evaluation life-test programs are now in progress.

CONCLUSION

Repeater IC's for high-speed optical submarine transmission systems were fabricated using a Si shallow-junction bipolar process with 7 GHz f_T. Repeater circuits are divided into four kinds of IC chips with circuit stability and power dissipation limitation per chip considered. To achieve high reliability, a two-level metallization system using TiN as a diffusion barrier was developed.

Principal results are summarized as follows:

1) The gain (s_{21}), variable gain range, and bandwidth for the equalizing amplifier IC were 65 dB, 30 dB, and 300 MHz, respectively.

2) The timing amplifier IC's gain (s_{21}) was 60 dB and phase deviation was kept within 10°.

3) The rise and fall times, of 300 ps for the decision IC, and 500 ps for the LD driver IC were realized. The decision IC sensitivity was less than 6 mV.

4) It was confirmed that a Cu-doped Al/TiN/Ti contact system is applicable for high current density and exhibited the highest thermal stability at 550°C.

5) Preliminary test element circuits aged at 125°C for up to 7000 h exhibited high stabilities.

From the above results, highly reliable monolithic integration of repeater circuits with high performance appears quite promising.

ACKNOWLEDGMENT

The authors are greatly indebted to H. Mukai, A. Iwata, T. Matsumoto, and M. Aiki. They thank them for their useful suggestions and comments.

REFERENCES

[1] H. Fukinuki, T. Ito, M. Aiki, and Y. Hayashi, "The FS-400M submarine system," this book, ch. 5, p. 69.
[2] R. G. Meyer and R. A. Blauschild, "A 4-terminal wide-band monolithic amplifier," *IEEE J. Solid-State Circuits*, vol. SC-16, no. 6, pp. 634–638, Dec. 1981.
[3] M. Ohara, T. Kamoto, and T. Sakai, "Very wide-band silicon bipolar monolithic amplifiers," in *Dig. Tech. Papers, The 14th Conf. on Solid State Devices* (Tokyo, Japan), 1982.
[4] T. Kamoto, M. Ohara, Y. Kobayashi, and M. Aiki, "Bipolar monolithic very wide-band amplifier," *Trans. IECE Japan*, vol. J66-C, no. 12, pp. 967–973, 1983.
[5] H. Satoh and H. Wano, "Reliability assurance of semiconductor devices for CS-36M submarine cable repeaters," *Rev. ECL*, vol. 22, no. 5–6, pp. 456–463, 1974.
[6] S. P. Murarka, H. J. Levinstein, I. Blech, T. T. Sheng, and M. H. Read, "Investigation of the Ti-Pt diffusion barrier for gold beam leads on aluminum," *J. Electrochem. Soc.*, vol. 125, no. 1, pp. 156–162, 1978.
[7] R. W. Bower, "Characteristics of aluminum-titanium electrical contacts on silicon," *Appl. Phys. Lett.*, vol. 23, no. 2, pp. 99–101, 1973.
[8] S. Kanamori and T. Matsumoto, "Suppression of platinum penetration failure in Ti/Pt/Au beam lead metal systems using a TiN diffusion barrier," *Thin Solid Films*, vol. 110, pp. 205–213, 1983.
[9] M. Mori, S. Kanamori, and T. Ueki, "Degradation mechanism in Si-doped Al/Si contacts and an extremely stable metallization system," *IEEE Trans. Components, Hybrids, Manuf. Technol.*, vol. CHMT-6, no. 2, pp. 159–162, 1983.

38
Reliability of Semiconductor Lasers and Detectors for Undersea Transmission Systems

YOSHINORI NAKANO, HIROMI SUDO, GENZO IWANE,
TADASHI MATSUMOTO, AND TETSUHIKO IKEGAMI, MEMBER, IEEE

INTRODUCTION

Undersea optical cable transmission systems are going to be operating between distant islands and across oceans within the next several years [1], [2]. High reliability is required for the component devices used in the undersea repeaters because of the extraordinary expense of a deep-sea repair operation in comparison to that of a land system. For undersea transmission systems, it has been common to aim at a system life of at least 25 years and a submerged plant MTBF (mean time between failures) of over 10 years. The reliability targets of LD's and APD's available for the first undersea transmission systems are several hundred FIT's and several FIT's, respectively, as reported elsewhere in this book. The most urgent and serious problem is, therefore, to establish a reliability assurance system for optical devices such as semiconductor lasers (InGaAsP-LD's) and detectors (Ge-APD's).

Lifetests of LD's and APD's have been conducted for 3 years in order to realize a large capacity (400 Mbit/s), long-haul (~ 20-km span) optical transmission land system using single-mode fiber cables (F-400M) [3]. At the beginning of the study on the reliability of long wavelength LD's, nonburied heterostructure (non-BH) LD's such as PCW LD's [4] or SAS LD's [5] had been adopted. Soon thereafter, however, buried heterostructure LD's such as DC-PBH LD's [6] or VSB LD's [7] took the place of non-BH LD's because they exhibit excellent performance such as low threshold current, high external differential quantum efficiency, and well stabilized transverse mode oscillation. Therefore, the light sources described here are limited to BH LD's. On the other hand, reliability studies on optical detectors for communication systems have hardly been reported on, although such devices are required to be very highly reliable in undersea transmission systems.

This chapter seeks to establish such a high reliability assurance system for the devices available for the undersea transmission system in the near future. At first, the reliability is considered on the basis of statistical treatment where the failure modes are divided into wear-out and random failures. Then, lifetest results of LD's and APD's for land optical transmission systems are arranged, and some problems in offering highly reliable devices are made clear. Some failure causes which were found in the various aging tests have been studied, and thus a new design for the devices available for undersea transmission systems, including countermeasures against such failures, is presented with its typical characteristics. Finally, the lifetest plans of LD's and APD's to assure high reliability for undersea optical transmission systems are presented, and the preliminary results of such lifetests are reported.

STATISTICAL ASSESSMENTS FOR DEVICE RELIABILITY

For a statistical study of semiconductor device reliability, it is useful to divide the failure modes into wear-out and random failures [8]. Wear-out failures mainly depend on gradual degradation of electrical and optical characteristics and obey lognormal distributions as failure density functions [9]. On the other hand, random failures are due to sudden failures and usually obey exponential distributions.

The lognormal distribution function can be written as follows:

$$\lambda(t) = \frac{\sqrt{2}\,\exp\left(-1/2\sigma^2(\ln t/t_m)^2\right)}{\sqrt{\pi}\,t\sigma\,\mathrm{erfc}\left(1/\sqrt{2}\,\sigma(\ln t/t_m)\right)}. \tag{1}$$

Here, t is the service time, t_m is the median lifetime in hours, and σ is the standard

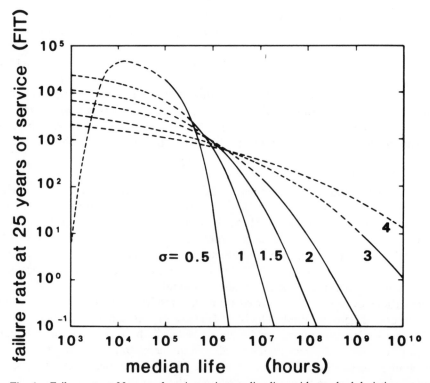

Fig. 1. Failure rate at 25 years of service against median lives with standard deviations as a parameter. The solid part of the λ curves denotes the portion of increase in λ with service time. In contrast, the broken parts indicate the domain of decrease in λ with service time, and the failure rate at 25 years of service drawn by a broken line is not true in our discussion.

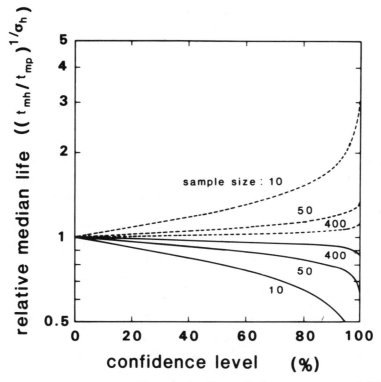

Fig. 2. Median lives versus confidence levels with sample sizes as a parameter. Solid and broken lines indicate the upper and lower limit values, respectively.

deviation in a logarithmic time scale for a lognormal failure distribution. Relations among failure rate at 25 years of service λ, and t_m can be easily calculated by using (1), and typical curves using σ as a parameter are illustrated in Fig. 1. It is apparently shown that the longer t_m is, the lower λ is, and also, the smaller σ is, the lower λ is. The solid part of the λ curves denotes the portion of increase in λ with service time, and this behavior is due to wear-out failure. In contrast, the broken parts indicate the domain of decrease in λ with service time, and the failure rate at 25 years of service drawn by a broken line is not true in our discussion. The relation between t_m and σ implies that the suppression of σ is useful in reducing the value of t_m when the same reliability is required.

Next, let us study the certainty of estimated value against sample number. Sample sizes of 10, 50, and 400, calculated results of upper and lower limit values of selective median value $(t_{mp}/t_{mh})^{1/\sigma h}$, by changing confidence levels, are illustrated in Fig. 2. Here, t_{mp} is the median value of infinite population, and t_{mh} and σ_h are the values estimated from experimental results, respectively. With increasing sample size, the difference between upper and lower limit values becomes small where the experimental result is close to a real value. When the sample size is 400, the difference in t_m can be estimated to be within 8 percent, under a 90-percent confidence level, by assuming $\sigma_h = 1.0$.

The random failure rate is constant throughout the device operating times as the failure density function is the exponential distribution $f(t) = \lambda \exp(-\lambda t)$. A point estimated value of failure rate at a defined time is given as

$$\lambda = r/t \cdot N. \tag{2}$$

Here, r, N, and t denote the number of failed devices, sample size, and testing time, respectively. The upper and lower limit values of random failure rate can be estimated by multiplying some factor determined by the confidence level and the number of failed devices [10].

LIFETEST OF LD'S AND APD'S FOR LAND TRANSMISSION SYSTEMS

Lifetest Results of DC-PBH LD's

LD's used in the lifetests were DC-PBH LD's, of which design and performance have been reported elsewhere [6]. They were selected by a two-step screening procedure prior to the aging test. In the first step, the lasers were operated at 70°C for 100 h under a constant current of 150 mA, and the lasers with an increase in 50°C threshold current, less than 10 percent, were selected. In the second step, a constant power aging at 8 mW/facet was carried out at 70°C for 100 h, and the lasers with more than a 5-percent increase in 50°C threshold current were screened out [11].

The lasers examined here were mounted on a diamond heatsink with the junction side down and packaged in a dry nitrogen atmosphere prior to the screening procedure.

In the lifetest of LD's, the automatic power control driving (APC) method was used where the drive current of each laser is automatically adjusted to maintain a constant optical output power and is monitored and recorded continuously by a computer controlled data acquisition system [3]. Lifetests with a constant power of 5 mW/facet were carried out at temperatures of 10, 50, and 70°C. The number of samples, the aging times, and component hours at each level are summarized in Table I.

The median values of the increase rate in the driving current during 13 000 h at the aging temperatures of 50 and 70°C were estimated to be 2.9×10^{-6} and 4.7×10^{-6} H^{-1}, respectively, and at 10°C showed no appreciable change.

In our liefetest, the end of life, or the time of failure, was defined as a time when the drive current reached 1.2 times the initial value. By now, 1 device out of 90 has failed, which was tested under the condition of 50°C and 5 mW. The device showed a rapid increase of the drive current started at 3000 h and reached the failure level at 6000 h. By assuming a random failure mode for this rapid degradation, the minimum value of the lifetime (MTTF) was about 3×10^5 h with a confidence level of 90 percent by one-side estimation, counting the component hours of the devices and the number of failed devices under test.

In these lifetests, some other lasers showed a gradual drive current increase. Such gradual degradation can be regarded as a wear-out failure mode. Almost linear increases from the beginning dominate in the present time scale, but a tendency toward gradual saturation in the drive current variation was sometimes observed. This saturation behav-

TABLE I
AGING TEST CONDITIONS OF LAND TRANSMISSION SYSTEM LD'S

Ambient Temp. (°C)	Power (mW)	Number of Devices	Aging Time (h) (the longest at present)
10	5	15	15 000
50	5	55	15 000
70	5	20	14 000

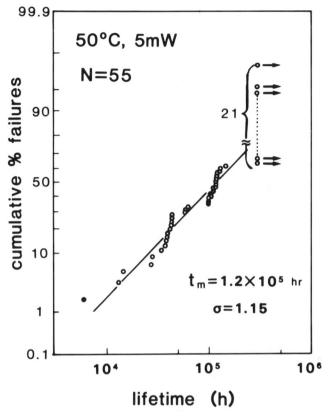

Fig. 3. Lognormal failure distribution of 55 LD's tested under the condition of 50°C and 5 mW. Here, the lifetime is defined as the time when I_d becomes 1.2 times the initial value. $t_m = 1.2 \times 10^5$ h and $\sigma = 1.15$ can be estimated. A closed circle denotes the failed device.

ior leads to extrapolation of a longer median lifetime [12], [13]. Although the saturation behavior is desirable, such classification of degradation behavior has to wait for an analysis based on the longer aging data because the behavior likely depends on the structure of the devices. Since stringent assurance for the reliability of LD's should be required for undersea transmission systems, let us assume a linear extrapolation such that the drive current increases as a linear function of time even after 10^4 h. Using the failure criterion mentioned previously, the lognormal failure distribution for the devices tested under the conditions of 50°C and 5 mW is shown in Fig. 3, including the failed device indicated by the closed circle. The median lifetime t_m and its standard deviation σ can be estimated to be 1.2×10^4 h and 1.15, respectively. Assuming an activation energy of 0.4 eV [11], [14] and a σ of 1.15, the median lifetime of such devices could be 8.9×10^5 h, and the device reliability for 25 years of service would be about 4700 FIT's under the condition of 10°C and 5 mW.

As for the coupling efficiency and pulse response, it has been reported that the BH type laser seems to have high reliability in coupling efficiency stability with single-mode fibers and high stability in pulse response characteristics during long-term constant power aging [3]. Therefore, when the DC-PBH LD's with stable transverse mode oscillation and adequate pulse response are selected prior to implementation, the lasers with the gradual degradation mentioned above will sustain efficient coupling with single-mode fibers and high bit rate modulation throughout the service time.

Lifetest Results of Ge-APD's

The lifetests of germanium APD's for monitoring LD's were conducted under high temperature reverse biased conditions [3]. Measurements were periodically made on dark current, breakdown voltage, quantum efficiency, and the excess noise factor for APD's, and on dark current and quantum efficiency for PD's. 54 APD's under the condition of a reverse bias of 25-V temperatures of 50, 80, and 125°C, and 68 APD's under the reverse biased condition of 100 μA at temperatures of 50 and 100°C were aged during more than 10^4 h. None of them failed where the failure criterion was defined as a time when the dark current exceeds 2 μA. Therefore, fairly long lifetimes and small failure rates can be expected during the practical use of Ge detectors.

ANALYSIS OF THE LIFETEST RESULTS OF LD'S AND AN ADDITIONAL SCREENING METHOD FOR HIGHER RELIABILITY LD'S

The causes of gradual degradation are not fully understood at the present time; however, it is noteworthy that there have been many lasers with very small changes in drive current even after aging over 10^4 h as shown in Fig. 3. Under the limited time of the plan, how to pick up such good LD's with a low degradation had been our issue, and the third screening procedure could be proposed by the following study [15]. The hard

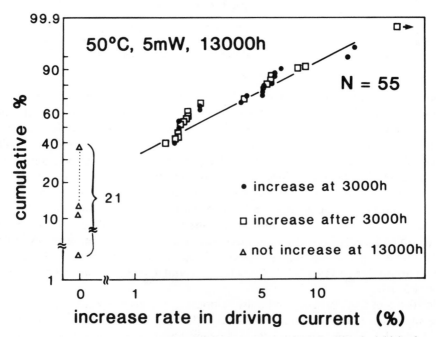

Fig. 4. Weibull plot of the increase rate of driving current to keep 5 mW to its initial value during the aging of 13 000 h at 50°C ambient.

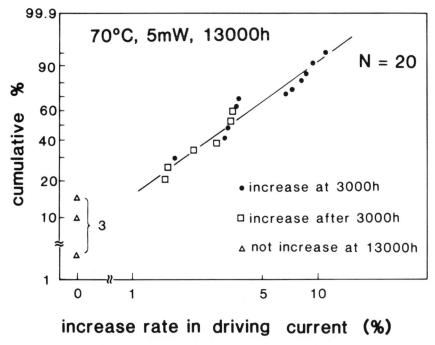

Fig. 5. Weibull plot of the increase rate of driving current to keep 5 mW to its initial value during the aging of 13 000 h at 70°C ambient.

screening method of LD's, mentioned in the previous section was a very useful way to reject bad devices, but it is not sufficient for the undersea transmission systems. Weibull plots to drive current increase rate I_d from its initial value I_{do} under the aging conditions of 50 and 70°C during 13 000 h are shown in Figs. 4 and 5, respectively. Almost the same distribution is obtained in both cases.

In Figs. 4 and 5, triangles and squares indicate devices in which appreciable changes in I_d could be observed after the aging time of 13 000 h and could not be observed at the aging time of 3000 h. These are 17 and 6 devices at 50 and 70°C, respectively.

Figure 4 indicates that after the aging time of 13 000 h at 50°C and 5 mW, 6 devices out of 38 which showed no appreciable change in I_d at 3000 h showed an increase rate of more than 5 percent. This means that: 1) the initial two-step hard screening is not sufficient to pick up long-term stable devices; and 2) even screening with 3000 h aging at 50°C may miss an unacceptable number of undesirable devices. On the other hand, the aging result at 70°C in Fig. 5 shows that six devices out of nine without appreciable change at 3000-h increase their drive current after 13 000 h; however, the change is gradual, less than 5 percent, and almost linear from the beginning. So, bad devices can be picked out more completely if the screening is performed after 3000 h aging at 70°C. Fig. 6 is the lognormal failure distribution for the devices corresponding to squares and triangles in Fig. 5, which showed no appreciable change after the 3000-h aging at 70°C. The median lifetime t_m and its standard deviation σ can be estimated to be 1.2×10^5 h and 0.7, respectively. Assuming an activation energy of 0.4 eV and a σ of 0.7, the median lifetime could be 2.1×10^6 h at 10°C and the reliability of the device around 70 FIT's for 25 years of service could be obtained. The combination of the initial hard screening

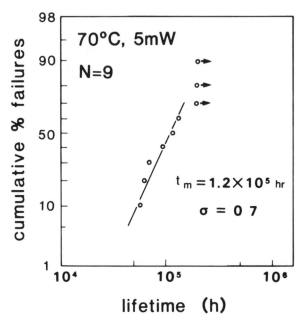

Fig. 6. Lognormal failure distribution of 9 LD's which indicate no appreciable increase of I_d at 3,000 h in Fig. 5. The lifetime t_m using the same definition as that of Fig. 3, and standard deviation σ can be estimated to be 1.2×10^5 h and 0.7, respectively.

method and the selection after additional aging for 3000 h at 70°C could be adequate for selecting LD's suitable for the undersea transmission system from among the devices made for land transmission systems.

MODIFICATION OF LD'S AND APD'S FOR HIGHER RELIABILITY

There have been cases where the long-lived lasers for land systems suddenly failed in the laboratory. Almost all of them occurred for devices being operated at high temperature and during current stressed conditions. They resulted from the growth of tin whiskers, void formations between the heatsink and package stem, separation of metallization metal from the diamond heatsink, and reaction between the solder material and the laser chip [16], [17]. From the viewpoint of practical use of LD's, the removal of these sudden failures is of great importance in offering highly reliable devices.

Three countermeasures have been adopted for high reliability LD's used for undersea transmission system applications. The first one is junction-side-up chip mounting, which is able to avoid short-circuit formation or solder penetration due to solder excursion and also to eliminate stress-induced damage. The second one is the application of Si heat sink material in place of the diamond heatsink. This reduces the risk of incomplete metallization between the heatsink and the laser chip. The last one is the application of an Au rich Au–Sn alloy as an adhesive solder which is more reliable than an Sn rich Au–Sn alloy or pure Sn. There have been no intentional changes concerning the laser chip structure and fabrication processes in comparison to those for land transmission systems.

Several Ge-APD's have failed under the step stress test with increasing temperature and current as parameters. The failure modes were disconnection and/or soft breakdown of

TABLE II
InGaAsP-LD Electrooptical Characteristics ($T = 25°C$)

Characteristic	Test Condition	MIN.	TYP.	MAX.	unit
voltage V	$I = 30$ mA		1.1	1.3	V
threshold current I_{th}			20	30	mA
efficiency ΔI	$P = 5$ mW		18	25	mA
drive current I_{op}	$T = 70°C$ $P = 8$ mW		90	130	mA
wavelength λ	$P = 5$ mW	1300	1315	1330	nm
spectral half width $\Delta\lambda$	$P = 5$ mW			4	nm

$I-V$ characteristics. The former failure is due to the generation of an Au–Al mixture at the contact between an Au wirelead and an Al electrode pad. The latter failure mode is due to Al metal penetration into the Ge depletion layer [18], but as the activation energy of this failure mode is more than 1 eV, the failure rate would be negligibly small when the operating temperature is around 10°C. When higher reliability is required, however, the Au/Pt/Ti ohmic metal system for the Ge-APD's in place of Al seems to have high stability and is being studied [19].

With these modifications, the structures of our devices for the undersea transmission system are fixed. Typical characteristics of the lasers are summarized in Table II. The samples used in the assurance test mentioned in the next section are selected by the initial two-step hard screening method. Typical characteristics of improved Ge-APD's are summarized in Table III.

Lifetest for High Reliability Assurance

LD's Lifetest

It was indicated previously that even among LD's for land systems made more than two years ago, the three-step screening method can pick up good devices for undersea systems. The purpose of following new LD's lifetests are to know the standard deviation of the degradation rate depending on the manufacturing procedure for the modified devices and to assure reliability against random failure by counting the component hours of the devices. Moreover, how much the third screening condition can be relaxed is also practically interesting.

When the required reliability for the LD's with a t_m of 5×10^5 h is 300 FIT's, considering that there is no derating factor of the worst case, a σ of less than 0.35 is needed. However, it would be extremely difficult to attain such a low standard deviation in reference to that of Section III. If the magnitude of σ is 1.0, the median lifetime sufficient to assure 300 FIT's is more than 10^6 h, and a reliable derating factor is necessary. In order to determine a derating factor for LD's, the other aging tests where the LD's are overstressed by temperature and/or current are under consideration.

Against random failures, on the other hand, several hundreds of FIT's reliability can be confirmed by 10^6–10^7 component hours. An example of such a lifetest is one with several

TABLE III
Ge-APD Electrooptical Characteristics ($T = 25°C$)

Characteristic	Test Condition	MIN.	TYP.	MAX.	unit
break down voltage V_B	$I_R = 100\ \mu A$	33	38	43	V
dark current I_R	$V = 0.9$ V		0.3	0.35	μA
capacitance C	$V = 20$ V $f = 1$ MHz		2	3	pF
efficiency η	$\lambda = 1.3\ \mu m$	70	75		%
noise factor X	$M = 10$		0.95	1.0	
multiplication factor M	$M = 10$	50	70		
cut-off frequency f_c	$\lambda = 1.3\ \mu m$ $I_R = 1\ \mu A$	1000			MHz

TABLE IV
Aging Test Plan for Highly Reliable LD's

Ambient Temp. (°C)	Power (mW)	Number of Devices
10	5	50
50	5	400
70	5	50

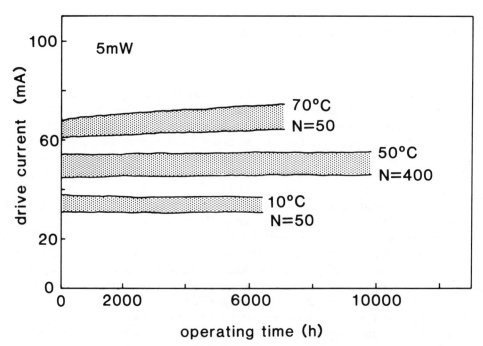

Fig. 7. Aging characteristics for improved LD's for undersea transmission systems under the condition of 5 mW. All the aging data are within the dotted area.

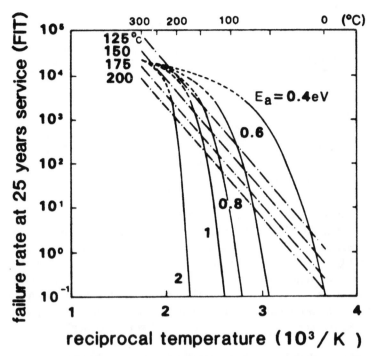

Fig. 8. Relation between failure rate at 25 years of service and reciprocal tempertaure as for Ge-APD's. Dot-dashed lines indicate the estimated values for random failures, assuming 0.5 eV as the activation energy and 1 failure out of 100 devices in the test of 10^4 h. Solid and broken line curves indicate the estimated values for wear-out failures with the activation energy as a parameter, assuming $t_m = 4 \times 10^3$ h and $\sigma = 1.0$ at 260°C.

hundred devices and an aging time of more than 10 000 h. A lifetest plan for LD's aimed to assure reliability as high as 300 FIT's during aging of about 10^4 h is shown in Table IV.

At present, 500 improved LD's for undersea transmission systems have been tested under the condition of 5 mW, and the longest aging time so far is around 10 000 h. All the lasers are operating very stably, as shown in Fig. 7. Even under the condition of 70°C the largest increase rate in drive current at 6000 h is less than 10 percent.

Ge-APD's Lifetest

The reliability required for Ge-APD's is several FIT's. In order to assure such high reliability, the assurance against random failures is more serious than that of the wear-out failures because the activation energy of wear-out failures seems to be relatively large and their failure rate could be estimated to be less than 1 FIT during practical use at around 10°C.

Now, in order to assure several FIT's as reliability against random failures, the sample size becomes enormously large supposing a lifetest aging time of 10^4 h. Therefore, a derating factor for random failures would be necessary, and it is very important to establish this as soon as possible. Here, it is significant to select a test condition such that random failures are dominant rather than wearout failures. Derating curves of random and wear-out failures are illustrated in Fig. 8 as functions of ambient temperatures. The straight lines for random failures at 125, 150, 175, and 200°C denote the estimated values,

TABLE V
AGING TEST PLAN FOR HIGHLY RELIABLE GE-APD

Failure Mode	Ambient Temp. (°C)	Current (μA)	Number of Devices
wear-out	227	no bias	20
	260		20
	295		20
random	150	200	100
	175	200	50
	200	200	30

assuming 0.5 eV as the activation energy and 1 failure out of 100 devices in the censored test of 10^4 h. On the other hand, as far as the wear-out failure rate is concerned, a similar calculation is carried out by assuming the median life t_m of 4×10^3 h and a standard deviation σ of 1.0 at 260°C. It is apparent that if the activation energy of the wear-out failure mode is more than 1 eV, the random failure is dominant rather than wear-out failure in the temperature range below 200°C. As for the lifetest of Ge-APD's to assure reliability against random failure, the sample size is assigned by supposing that the number of failed devices is the same in each level, and a typical example is shown in Table V.

In order to assure reliability against wear-out failures, accelerating the test by temperature overstressing is necessary, using at least three temperature levels with a maximum temperature of 295°C. This is the highest temperature among those previously reported. The sample size in each level should be no less than 20, as shown in Fig. 2. The other temperature levels are chosen such that the reciprocal of each temperature is arranged in the same interval that is shown in Table V.

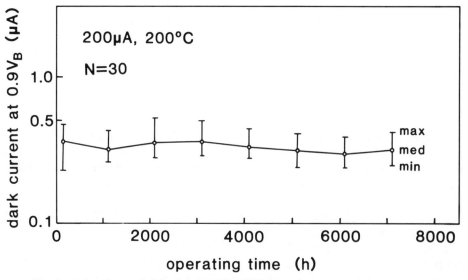

Fig. 9.   Aging characteristic for improved Ge-APD's for undersea transmission systems under the condition of 200°C and 200 μA.

At present, 240 improved Ge-APD's have been tested under the same conditions as those tabulated in Table V. Fig. 9 shows a variation in dark current under the condition of 200°C and 200 μA. No random failure has been observed up to the operation of 7,000 h. As far as wear-out failures are concerned, the detailed analysis on aging data will be reported in near future.

CONCLUSION

The strategy for establishing the reliability assurance system for LD's and Ge-APD's available for undersea transmission systems has been described on the basis of aging data during more than 10^4 h where lifetests are conducted for the devices available for land transmission systems. Through a statistical analysis on the reliability of semiconductor devices, which is conducted by separating the failure modes into wear-out and random failures, it is found that strict reliability assurance at present is more difficult for wear-out failure for LD's and for random failure for Ge-APD's.

However, the three-step screening method could pick up highly reliable devices even among the devices for land systems (LD's with around 70 FIT's at 10°C and 5 mW). The new highly reliable devices, improved for the purpose of reducing random failure modes, are presented, and their initial behaviors are shown to be very promising.

Determination of derating factors for their practical use at around 10°C ambient in undersea repeaters remains as future work. Providing that such a problem is solved, the required reliability of a few tens of FIT's for LD's and a few FIT's for Ge-APD's throughout 25 years of service could be assured by the proposed three-step screening of a population of the devices manufactured at present.

ACKNOWLEDGMENT

The authors would like to thank H. Mukai and M. Fujimoto for their encouragement throughout this work and K. Takahei and K. Kuroiwa for their preparation of aging test equipment and valuable comments. They would also like to thank M. Fukuda and O. Fujita for their measurements during lifetesting.

REFERENCES

[1] Y. Fukinuki, T. Ito, M. Aiki, and Y. Hayashi, "The FS-400M submarine system," pp. 794–800, this issue.
[2] Y. Niiro, "Optical fiber submarine cable system development at KDD," *IEEE J. Select. Areas Commun.*, vol. SAC-1, no. 3, pp. 467–478, Apr. 1983.
[3] K. Takahei, K. Kuroiwa, and T. Ikegami, "Reliability of 1.3 μm semiconductor lasers and Ge-detectors," *Rev. Elec. Commun. Lab.*, vol. 31, no. 3, pp. 321–330, May 1983.
[4] M. Ueno, I. Sakuma, T. Furuse, Y. Matsumoto, H. Kawano, Y. Ide, and S. Matsumoto, "Transverse mode stabilized InGaAsP/InP ($\lambda = 1.3$ μm) plano-convex waveguide lasers," *IEEE J. Quantum Electron.*, vol. QE-17, pp. 1930–1940, Sept. 1981.
[5] M. Yano, H. Nishi, and M. Takusagawa, "Oscillation characteristics in InGaAsP/InP DH lasers with self-aligned structure," *IEEE J. Quantum Electron.*, vol. QE-15, p. 1388–1395, Dec. 1979.
[6] I. Mito, M. Kitamura, K. Kobayashi, S. Murata, M. Seki, Y. Odagiri, H. Nishimoto, M. Yamaguchi, and K. Kobayashi, "InGaAsP double channel planar buried heterostructure laser diode (DC-PBH LD) with effective current confinement," *IEEE J. Lightwave Technol.*, vol. LT-1, no. 1, pp. 195–202, Mar. 1983.
[7] H. Ishikawa, H. Imai, T. Tanahashi, K. Hori, and K. Takahei, "V-grooved substrate buried heterostructure InGaAsP/InP laser emitting at 1.3 μm wavelength," *IEEE J. Quantum Electron.*, vol. QE-18, pp. 1704–1711, Oct. 1982.

[8]   H. Fukui, S. H. Wemple, J. C. Irvin, and W. C. Niehaus, J. C. M. Hwang, H. M. Cox, W. O. Schlosser, and J. V. Dilorenzo, "Reliability of power GaAs field effect transistors," *IEEE Trans. Electron Devices*, vol. ED-29, Mar. 1982.

[9]   A. S. Jordan, "A comprehensive review of the lognormal failure distribution with application to LED reliability," *Microelectron Rel.*, vol. 18, no. 3, pp. 267–279, May 1978.

[10] M. Fujiki and H. Shiomi, "Reliability in electronics," *Trans. IECE Japan* (in Japanese), p. 37, 1978.

[11] T. Ikegami, K. Takahei, M. Fukuda, and K. Kuroiwa, "Stress test on 1.3 μm buried-heterostructure laser diode," *Electron. Lett.*, vol. 19, no. 8, pp. 282–283, 1983.

[12] K. Mizuishi, M. Sawai, S. Todoroki, S. Tsuji, M. Hirao, and M. Nakamura, "Reliability of InGaAsP/InP buried heterostructure 1.3 μm lasers," *IEEE J. Quantum Electron.*, vol. QE-19, pp. 1294–1301, Aug. 1983.

[13] A Rosiewicz and P. A. Kirkby, "The reliability of IRW lasers in 1.3 μm monosystems," in *Proc. 9th Euro. Conf. Opt. Commun.* (Geneva, Switzerland), Oct. 1983.

[14] Y. Nakano, M. Fukuda, H. Sudo, O. Fujita, and G. Iwane, "Thermally accelerated degradation of 1.3 μm BH lasers," *Electron. Lett.*, vol. 19, no. 15, pp. 567–568, July 1983.

[15] Y. Nakano, G. Iwane, and K. Ikegami, "Screening method for laser diodes with high reliability," *Electron. Lett.*, to be published.

[16] M. Fukuda, O. Fujita, and G. Iwane, "Failure modes of InGaAsP/InP lasers due to adhesives," *IEEE Trans. Components, Hybrids, Manuf., Technol.* vol. CHMT-7, no. 2, pp. 202–206, June 1984.

[17] K. Mizuishi, "Some aspects of bonding-solder deterioration observed in long-lived semiconductor lasers: Solder migration and whisker growth," *J. Appl. Phys.*, vol. 55, no. 2, pp. 289–295, Jan. 1984.

[18] K. Chino and K. Fukuda, "Degradation mechanism in breakdown of Ge-APD," in *Tech. Dig., Fall Meeting IECE Japan*, 1983, p. 96.

[19] Y. Tashior, A. Murakami, and H. Iwasaki, "Ti/Pt/(Au) electrode Ge-APD," in *Tech. Dig., Spring Meeting IECE Japan*, 1984, pp. 4–17.

Part IX
Supervisory, Control, and Terminal Equipment

Control of the undersea system from the land terminal is accomplished in a variety of novel techniques. In this part we examine some of these supervisory techniques proposed for use in lightwave systems.

39

Dialogue Channels Used for Remote Supervision and Remote Control of the Underwater Plant in Submarine Digital Telephone Links Using Optical Fibers

JEAN-CLAUDE LACROIX

INTRODUCTION

The planned minimum service life of the underwater plant and the regenerators in particular is 25 years. Consequently, it is essential to be able to monitor changes in the transmission quality of each regeneration section in order to predict any possibility of failure, hence the requirement for remote supervision. Moreover, and in view of the fact that at present the laser reliability is such that four to one redundancy may be required, remote supervision may be complemented by greater or lesser redundancy in the associated electronic circuits and requires, in all cases, remote control facilities to carry out the switching operations needed to keep the link functioning correctly. Identifying the regeneration section which is faulty or which is likely to cause a fault to occur is naturally essential.

The remote supervision criteria and the dialogue channels usable for remote supervision are discussed below. The final section of the description includes a table summarizing the resources used by various manufacturers.

REMOTE SUPERVISION CRITERIA

The criteria selected make it possible to evaluate the transmission quality for each regeneration section. They are generally selected from the following criteria, classified according to three categories:

1) Reception criteria:

- Detection threshold of the receiving diode (optical/electrical conversion);
- Receive amplifier AGC voltage; and
- Absence of receive signal.

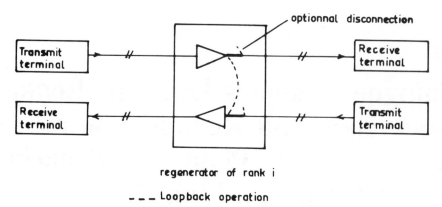

regenerator of rank i

– – – Loopback operation

Fig. 1.   Simplified loopback test block schematic.

2) Transmission criteria:

- Average laser diode current; and
- Average laser diode voltage (optional: complements previous criterion).

3) Transmission quality criteria:

- Error rate based on the redundancy of the line code employed, and
- Signal-to-noise ratio based on a pseudo error count effected by comparing a decision circuit in the regeneration path and a degraded decision circuit.

As a general rule, one criterion is selected from each category, and these make it possible to attribute degradation of transmission quality to the transmission, reception, or regeneration subsystems, and possibly to decide to change over to a backup subsystem.

Loopback Testing

In certain circumstances, such as the sudden failure of a laser or cutting of the cable, for example, the criteria listed above may prove inadequate. It is then necessary to carry out successive loopback tests in order to determine the faulty regeneration section or to resolve any doubt as to the origin of the fault that has occurred.

When these conditions apply, a loopback command is sent from one end of the link to instigate internal switching in the regenerator of rank *i* in order to retransmit the information transmitted in the return direction (see Fig. 1).

Loopback testing may be applied either to the electrical signal or to the optical signal.

Dialogue Channels Usable

The transmission quality must be monitored continuously and, in particular, it is necessary to keep track of changes in the criteria selected so as to be able to predict faults and where necessary to send a command to switch over to backup subsystems.

This requires means for communicating with the regenerators which will not degrade the digital signal transmitted in any significant way.

Changes in the criteria may be monitored by cyclic interrogation (interrogation/response mode), but this approach is not systematically applicable to all channels. It may be affected by the spontaneous transmission of a message or alarm indication in the event that a threshold is exceeded. This does not involve any additional time for acquisition of criteria and may be applied to all channels.

The channels are classified in terms of three categories: 1) ancillary channel; 2) channel formed by modulation of the line signal; 3) channel using the digital line signal.

Ancillary Channel

An ancillary conductor is used and may be an electrical or optical conductor.

1) Electrical Conductor: This may be a coaxial cable, the power feed conductors, etc. The information may be analog or digital, using a bandwidth limited to low frequencies such that the attenuation of the transmission medium makes it possible to use amplification or regeneration sections compatible in terms of length with the main system.

2) Optical Transmission Medium: An additional fiber is used, and the corresponding regeneration system may be simplified through the use of a reduced bandwidth.

In both cases the ancillary channel requires its own amplification and possibly regeneration, thus increasing the complexity of the underwater plant; on the other hand, the ancillary channel is usable for all the fiber pairs of the cable.

Channel Formed by Modulation of the Line Signal

The line signal may be amplitude, phase, and frequency modulated. Pulsewidth modulation is excluded as it significantly degrades the signal-to-noise ratio, which significantly penalizes the system.

1) Amplitude Modulation (Overmodulation): The thresholds corresponding to the high and low levels are assigned a certain tolerance in the regenerator decision circuits. Part of this tolerance is used to modulate the high and low levels by means of a low bit rate digital signal. This penalizes the system by reducing the regeneration section. Further, the low bit rate signal must be extracted, regenerated, and used to overmodulate the reconstituted high bit rate signal (see Fig. 2).

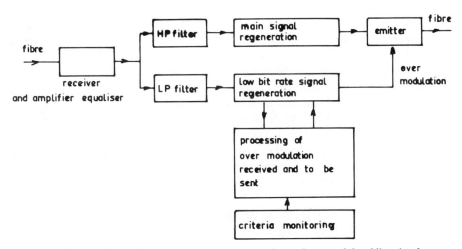

Fig. 2.   Block schematic of regenerator processing and overmodulated line signal.

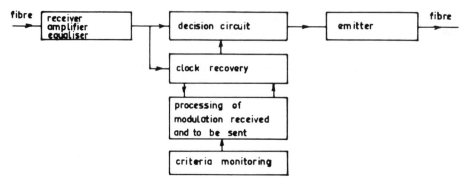

Fig. 3. Block schematic of regenerator processing a phase or frequency modulated signal.

2) Line Frequency Phase or Frequency Modulation: If the line bit rate is phase or frequency modulated with a restricted excursion, the resulting jitter remains moderate (for example, for a peak deviation of 10 kHz the resulting peak jitter is 0.01 percent for a line bit rate of 100 Mbit/s [1].

Also, if the modulation rate is sufficiently low to be contained with the bandwidth of the timing recovery device of the regenerators, the modulation is also regenerated and no further hardware is required.

Only a unit making it possible to modulate the regenerated main signal is required since what is needed is to transmit information concerning a remote supervision criterion of the regenerator. In this case, modulation may be effected by operating on the clock recovery device. This action is embodied in a variable time delay associated with the bandpass filter of the recovery system (this filter may be a SAW filter, a phase-locked loop, etc.) (see Fig. 3).

Channel Using Digital Line Signal

Information may be carried by dedicated bits or in place of the line signal.

1) Dedicated Bits: There are two cases to be distinguished, according to whether the redundancy of the code is high or low.

a) High redundancy code: At present these are generally of the nB/mB block type (in which to n incoming bits there are made to correspond m bits (where $m > n$) to form the line code). Under these conditions certain configurations are illegal, all "1" and all "0," for example, in the case of around m bits and above. As such configurations are simple to detect they may be used when they are available thus recognized for transmission of the rank i of the regenerator and the threshold exceeded, without any interrogation procedure. At the receiving end, the demultiplexer makes it possible to locate the special configurations and thus to exploit their content.

When it is necessary to operate in the interrogation/response mode, the regenerators must incorporate a device for synchronizing on the rate at which special configurations appear. In this case, only the information relating to the remote supervision criteria is sent in response.

Generally speaking, the synchronization device functions at the same timing rate as the line signal. It may therefore be complex, entailing a nonnegligible continuous energy consumption.

b) Low redundancy code: This may be the scrambled binary signal with, for simple redundancy, a parity and/or complementarity indication or simply systematic insertion of bits as in the case of the frame alignment word.

The line signal thus consists of the highest bit rate frame in which the dedicated bits are reserved to transmit remote supervision information; they may be few in number, one per frame, for example. It is essential to be able to recognize the location of the dedicated bits at each regenerator, and a high bit rate interlocking system is also necessary.

An interrogation/response dialogue method may be adopted.

2) Transmission of Remote Supervision Information Instead of the Line Signal: The principle is based on the fact that if the quantity of information to be transmitted is low, it may be substituted for the line signal without significantly degrading the error rate.

A continuous interrogation/response remote supervision method is not permissible. Put simply, when a supervision criterion reaches and exceeds a set threshold, the rank i of the regenerator and the corresponding fault are transmitted in place of and at the same timing rate as the line signal, using a preestablished structure and code. The resulting remote alarm message is made redundant so that the probability of random simulation of this message is very low, even if the line error rate is high (1×10^{-2}, for example). Naturally this nonsimulation ability is easier to achieve using a high redundancy code such as a block code, for example.

TELEMETRY

If the criteria selected do not yield the operation margin(s) with sufficient precision, it may be necessary to deliberately degrade the characteristics of one or more regenerators in order to increase the sensitivity of the criteria employed.

To achieve this the terminal sends a degradation command; scrutinizing the evolution of the remote supervision criteria then makes it possible to evaluate the operating margins.

The degradation command may be a command acting on a unit of the regenerator, the command being handled by the remote supervision channel selected. It may also vary the characteristics of the signal sent. For example, increasing or decreasing the timing rate by varying the justification rate of the incoming bit streams provides for artificially decentering the bandwidth of the timing recovery filters and progressively increasing the error rate.

MEASURE ADOPTED BY THE PRINCIPAL EQUIPMENT MANUFACTURERS

These measures result from a compromise among minimizing the complexity of the underwater plant (and thus its reliability), the system penalty resulting from the remote supervision channel, and the philosophy adopted: interrogation/response method or otherwise.

For these reasons the channel selected for interrogation is not necessarily the same as that selected for response.

Table I summarizes the provisions adopted by three manufacturers (Bell, STC, and Submarcom) in connection with the call for tenders for the TAT 8 transatlantic link.

TABLE I

Manufacturer	In/out of service	Redundancy	Interrogation	Response	Telemetry	Remote control	Criteria selected	Channel
Bell (USA)	In	Complete system (fiber and regenerator) switchable in sections	Essential	On interrogation	No (subject to confirmation)	Section switching	• AGC voltage • Error rate calculated on parity bits (24B/1P) • Laser currents	Ancillary coaxial pair, low-speed digital signal
	Out	As above	Essential	On interrogation	No (subject to confirmation)	Loopback		ditto
ST (UK)	In	Two emitters (subjet to confirmation)	Essential	On interrogation	No (subject to confirmation)	Emitter switching	• Error rate on 7B/8B code parity • Laser current	Phase and frequency modulation
	Out	As above	Essential	On interrogation	No (subject to confirmation)	Loopback	Receiver detection threshold	Transmission of a special frequency
Submarcom (France)	In	Four emitters	Optional	Instantaneous transmission on crossing threshold	By variation of transmit timing rate	Emitter switching	• Signal failure • Pseudo-error rate by comparison with degraded decision circuits • Laser current	• Substitution for transmit digital signal, without loss of alignment at receiving end • Line code 5B/6B • Transmission of specific frequency for switching
	Out	As above	Optional	As above	As above	Loopback		Transmission of a special frequency

After agreement between the three manufacturers Bell, STC, and Submarcom who participate in TAT8 achievement, the following choices have been effected for this link:

Line code: 24B 1P
Dialogue channels in service:

- Phase and frequency modulation Bell and STC; and
- Substitution for transmit digital signal Submarcom.

Dialogue channels out of service:

- Transmission of a special frequency STC and Submarcom; and
- Phase and frequency modulation Bell (subject to confirmation).

CONCLUSION

The remote supervision channel and the criteria adopted, essential to maintaining satisfactory transmission quality on a submarine link for 25 years, are selected so that the complexity of the regenerators and the system penalty are not significantly affected by the remote supervision function.

A compromise is sought allowing for other choices, in particular the line code. On this depends the ease with which errors may be detected and the possibility of substituting messages for the transmitted signal with a very low if not zero probability of simulation.

REFERENCE

[1] J. A. Kitchen and P. Cochrane, "A novel service/supervisory channel for digital trunk transmission systems," in *Proc. BTRS, Inst. Elec. Eng. Conf.*, Mar. 1981.

40
The SL Supervisory System

CLEO D. ANDERSON, MEMBER, IEEE, AND DAVID LOUIS KELLER

INTRODUCTION

The supervisory system we describe provides the means of remotely controlling and monitoring repeaters in a digital transmission system. It is applicable mainly to long-haul facilities such as undersea systems where the use of separate fibers, channels, or conductors is impractical. Because it enables the localization, correction, and possibly prediction of failures, it is an extremely critical part of the repeater. Our approach to achieving the necessary supervisory system reliability combines the use of established technology with redundancy. It features computer-controlled in-service supervision of repeaters by information carried over the optical fibers by the digital data streams.

FUNCTIONS

The supervisory functions may be divided into two broad categories, control and monitor. Examples of control functions include the following:

replacement switching of standby circuit modules such as transmitters or receivers;

replacement switching of a complete standby regenerator span section including a fiber and a regenerator;

switching of common monitor and control circuits onto any one of a multiplicity of regenerators within a given repeater; and

insertion of "stressing networks" which can give an indication of margin in the associated span.

Examples of Monitor Functions are as follows:

measurement of performance-related parameters such as laser bias current or receiver automatic gain control (AGC) voltage;

counting of events such as block parity errors over a given time interval; and

indicating logic states of any of the above-mentioned switching functions.

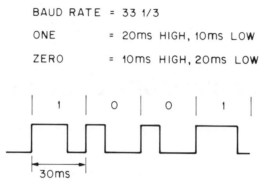

Fig. 1.   Supervisory baseband signal format.

GENERAL DESCRIPTION

The Baseband Supervisory Signal

Supervisory information to and from repeaters is carried digitally by groups of 40 baseband pulses which modulate the data streams as described in the sections "Processing the Parity Bits" and "The Response Channel". The baseband signal, Fig. 1, is a string of long and short return-to-zero (RZ) 33 1/3 baud pulses which represent binary ones and zeros, respectively. This asynchronous format, wherein each pulse carries its own timing information, simplifies burst mode operation. Groups of bits define parts of an instruction as follows:

bit	1	even parity over the 20 bits
bits	2–10	repeater address
bits	11–16	operation code
bits	17–19	regenerator address
bit	20	end bit, always a one.

The 20-bit word is sent twice, separated by one time slot, and the corresponding bits are compared. Instructions are executed (in the addressed repeater) only if the two words are identical. Every executed command results in a 40-bit response to the originating terminal. The first 20 bits are the same as the command, the second 20 bits convey information

bits	21–29	block error count or laser bias current or AGC voltage
bits	30–31	laser power state (indicates which laser transmitter is being biased)
bits	32–39	state of transposition relays or error monitor
bit	40	always a one

Commands may be sent at a rate of about one every 4 s.

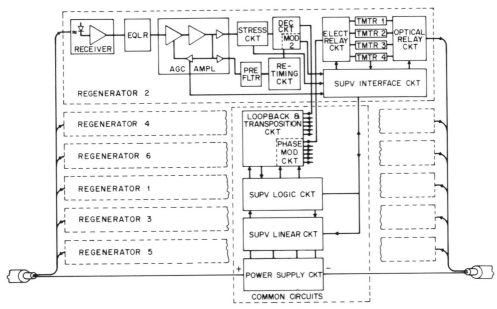

Fig. 2. SL repeater block diagram.

Control Channel

Each of the working fibers in the SL System carries two 139.264-Mbit/s CEPT-4 channels and operates at a line rate of 295.6 Mbit/s. It uses a block line code called 24blp wherein each block of 24 information bits is followed by one even parity bit. In addition to preventing strings of more than 24 ones the parity bits provide a simple low-overhead (5 percent) means of controlling the supervisory circuits and monitoring bit-error ratios in the repeaters. The control (or parity) channel is discussed in the following two sections. Control signals can be sent over any working line.

Repeater Supervisory Circuits

1) General: Fig. 2 is a block diagram of the SL repeater emphasizing the supervisory circuits. The low-frequency supervisory control signal is recovered from the parity bits as discussed in the section "Processing the Parity Bits". The baseband signal, the envelope of 26.393-kHz tone bursts, is recovered in the decision circuit, amplified, bandpass-filtered, and detected in the linear circuit and decoded in the logic circuit. The logic circuit also executes those relay operate commands associated with the loopback and transposition circuit. Loopback is used for localizing faults which interrupt power. Transposition refers to the substitution of the standby line section in place of a failed line section. After each supervisory command is executed a response is returned to the originating terminal. The response contains the requested information for monitor type instructions, or verifies the execution of a switching type instruction. The response channel described in the section "The Response Channel" is also called the jitter channel because information is contained in low-frequency phase jitter sidebands of the line baud rate. Baseband supervisory response information gates a 26.393-kHz subcarrier which phase modulates the data streams in the direction of the originating terminal. The phase modulators are located between the loopback and transposition circuit and the transmitters.

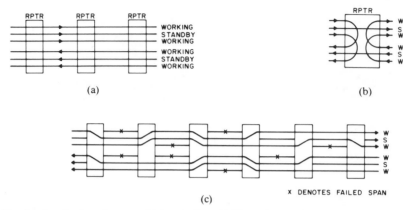

Fig. 3.   Loopback and transposition switching. (a) Normal state. (b) One possible loopback combination. (c) Standby restoration of failed working spans.

2) Supervisory Linear Circuit: This circuit is a 48-pin complementary bipolar integrated circuit (CBIC), which contains the following:

two 26.4-kHz bandpass amplifiers and detectors, one for each direction of transmission;
one A/D converter for measuring AGC voltage and laser bias current; and
one 3200-Hz local clock which drives the logic IC.

3) Supervisory Logic Circuit: This 84-pin Integrated Injection-Logic IC, together with the Linear IC and associated chip capacitors is mounted on an 84-pin film integrated circuit (FIC) forming the common supervisory circuit module. The logic IC is discussed in the section "Decoding and Executing the Received Command".

4) Loopback and Transposition Circuit (L&T): The L&T circuit contains 12 highly reliable miniature magnetic latching relays which provide the line switching functions of Fig. 3. Magnetic latching relays were selected for this function because of their low-loss, good isolation, and nonvolatility. The latter feature avoids the problem of resetting the network following power turndown. This particular relay has a reliability record established over two generations of analog undersea systems.

5) Phase Modulator Circuit: Six phase modulators are mounted on the L&T circuit printed wiring board Fig. 4. Each phase modulator is a varactor-controlled phase shifting network which can impress up to 20 deg of peak phase deviation upon the data stream. The varactor control voltage consists of 26.4-kHz tone bursts generated by the linear-logic hybrid integrated circuit (HIC). The modulation is applied simultaneously to all three lines transmitting towards the originating terminal.

6) Supervisory Interface Circuit: Each regenerator contains a Supervisory Interface Circuit which executes the supervisory instruction addressed to that regenerator. Under the control of data buses emanating from the logic IC this circuit:

1) operates the electrical and optical relays used for transmitter sparing;
2) controls the insertion of the stressing network for determining margin;
3) samples the AGC voltage and laser bias current which is further processed in the linear-logic HIC;
4) detects block parity violations which are counted in the supervisory logic circuit; and
5) encodes the logic state of the transmitter sparing relays.

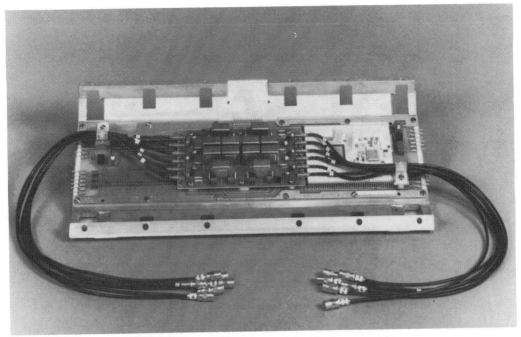

Fig. 4.   Photograph—Repeater common supervisory circuits.

CONTROL OF THE INSERTED PARITY BITS

A simplified logic diagram of the parity bit control and insertion circuit is given in Fig. 5. The inputs are data, data clock, line clock (f_b) and the baseband supervisory signal (X) previously described. The output is serial blocks of M-1 (24) data bits each followed by a parity bit. When X is low all parity bits are even. When X is high, every Nth (224th) block parity bit is odd. Thus when the supervisory baseband signal is high, the parity bit

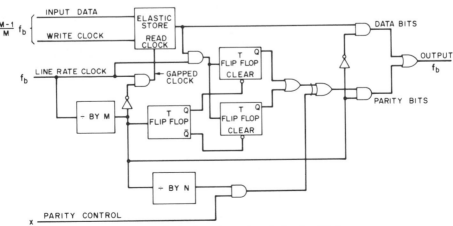

Fig. 5.   Parity insertion and control—simplified.

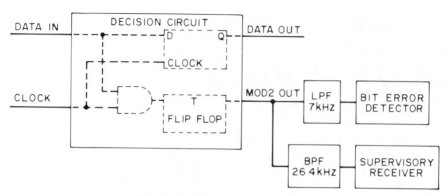

Fig. 6.   Parity bit processing in repeaters.

is synchronously complemented at a rate given by

$$\text{Parity Complement Rate (PCR)} = \frac{f_b}{MN} = 52785.7/\text{s}.$$

The following discussion concerns the operation of the circuit in Fig. 5. Input data are read into an elastic store at a rate of $(M-1)/Mf_b$. The average output rate of the store equals the input rate. However, the read clock is gapped so that one out of every M pulses at the line rate f_b is missing. The parity bit is inserted during the gap interval. Parity is computed by two T-type (toggle) flip–flops which alternate allowing one to be counting while the other is cleared. When the lower input to the exclusive OR gate is high, the output is complemented. This occurs only when the output of the two dividers M and N and the input X are simultaneously high. Each SL multiplex frame contains 56 parity blocks. One parity bit in each frame is dedicated to supervisory signaling and the terminal error detectors ignore parity violations occurring in that slot.

Processing the Parity Bits

Figure 6 shows how the parity bit information is processed in the repeaters. Each decision circuit contains, in addition to the normal regeneration circuits, a modulo-2 divider consisting of an "and" gate and a T flip–flop. The T flip–flop toggles (changes state) on every input "one." The signal at the output of the modulo-2 divider has the following properties, assuming unit amplitude rectangular NRZ pulses:

Let i refer to the time slot. Parity slots occur when i is a multiple of M, where M is the number of bits in a parity block.

First consider the bits in the data slots

$$E\{X_iX_j\} = \tfrac{1}{4} \qquad i \neq M \quad \text{for all } j$$

Next, consider the bits in the parity slots $i = nM$.

a) If the recovered parity bits are "ones"

$$E\{X_{nM}X_j\} = \tfrac{1}{2}, \qquad \text{for all } j \neq kM; \quad k = 1,2,3$$
$$= 1, \qquad \text{for all } j = kM.$$

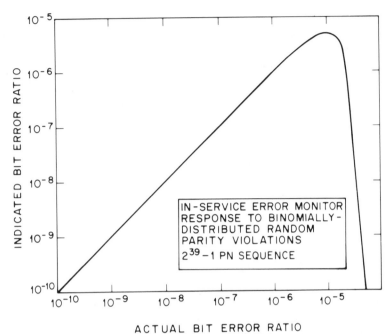

ACTUAL BIT ERROR RATIO

Fig. 7.   Measured response of BER detector.

b) If the recovered parity bits are "zeros"

$$E\{X_{nM}X_j\} = 0, \qquad \text{for all } j.$$

This indicates that the average (dc) value of the signal at the output of the modulo-2 divider changes whenever the bit in the parity slot (at the output) changes. In summary the output signal from the modulo-2 divider has the following properties:

1) Recovered data bits are completely random.
2) Recovered parity bits appear as strings of ones or zeros.
3) The recovered parity bit changes only if:

    i) the parity block contains an odd number of errors, or
    ii) the parity bit was inserted with odd parity.

By properly filtering this signal, changes in the recovered parity bit can be detected. The low-frequency components, such as those caused by infrequent block errors introduced by the line, are passed by the low pass filter, quantized, and counted as block errors. Figure 7 shows the measured response of the bit-error detector for binomially distributed random errors. Higher frequency components, namely those inserted at the terminal by the circuit of Fig. 5, are passed by the bandpass filter. The center frequency of the bandpass filter is one-half the parity bit complement rate

$$f_o = \frac{1}{2} \frac{f_b}{NM}$$

for SL

$$f_o = \frac{1}{2} \frac{295.6 \times 10^6}{224 \times 25} = 26.393 \text{ kHz.}$$

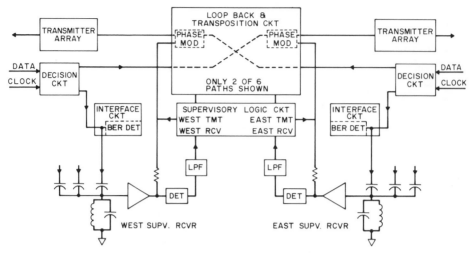

Fig. 8. Parity and response channel—simplified.

A tone of frequency f_o appears at the output of the bandpass filter whenever the X input of Fig. 5 is high. The envelope of the tone, which is the supervisory baseband signal, is recovered by amplifying, rectifying, and low-pass filtering the output of the bandpass filter as shown in Fig. 8.

In order for the parity channel to be useful as a reliable signaling medium, the peak signal-to-rms noise ratio at the LPF outputs of Fig. 8 should be at least 20 dB. Example

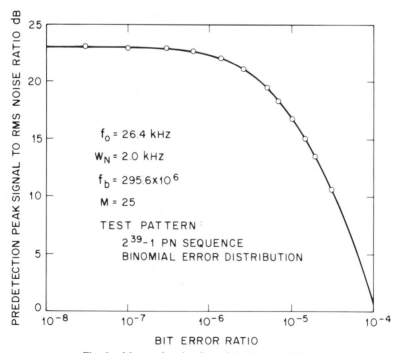

Fig. 9. Measured parity channel S/N versus BER.

calculations are given in Appendix I. As indicated, two or more lines can share common receiver, however, because of possible destructive interference, signaling is applied to only one line at a time. The effect of signaling line bit-error-ratio on the parity channel signal-to-noise ratio is shown in Fig. 9. Errors on the nonsignaling lines have negligible effect on the parity channel S/N.

Supervisory signaling by parity bit complementation has no effect on the normal operation of the regenerator because, prior to the modulo-2 divider, the parity bits are indistinguishable from the data bits. A major advantage of this signaling method is the simplicity of the associated circuits. Only the modulo-2 divider requires high-speed circuits and they are not in the main data signal paths. No separate regeneration of the supervisory signal is involved. Reliability is enhanced by having as many supervisory control channels to a repeater as there are working lines to that repeater.

Decoding and Executing the Received Command

The recovered baseband supervisory signal at the detector output of Fig. 8 is processed by the logic circuit. Besides a local clock signal, which is asynchronous with both the line and supervisory data, there are two control signal inputs labeled West Rcv and East Rcv. The first input circuit to receive a command, which is a packet of pulsewidth modulated pulses representing binary ones and zeros, inhibits the other until the instruction has been processed. This avoids interruptions which might otherwise occur if two or more terminals signaled simultaneously. The pulse string is serially loaded into a 20-cell shift register as ones and zeros. Parallel decoders determine if the address portion of the instruction corresponds to that assigned to the repeater, and that the parity and end bits are correct. If the instruction is valid a flag is set and as the second (and identical) 20-bit word is read in, the incoming and outgoing bits in the shift register are compared. If an error is detected the flag is reset, if not, the instruction is processed under control of sequencing and timing circuits. Magnetic latching relays of the loopback and transposition network are operated directly by the logic circuit relay drivers. Parallel data buses carry control signals to the interface circuits of the individual regenerators. After executing the instruction the original command is transmitted back to the originating terminal. Then status information is loaded and shifted out. The type of status information which is loaded depends upon the instruction. An output steering circuit pulsewidth modulates the shift register output, adds the 26.4-kHz subcarrier and applies the resulting signal to the phase modulators on those lines transmitting towards the originating terminal.

The Response Channel

Because the parity bits are not recovered and reinserted at the regenerators they cannot be used for signaling from repeater to terminal. The method we have developed for conveying these response signals uses a two stage modulation process. First the supervisory baseband signal is converted to a corresponding series of 26.393-kHz subcarrier tone bursts. The tone bursts in turn phase modulate the data stream(s) transmitted by the responding repeater. In SL the subcarrier is actually provided by the signaling terminal by leaving the last pulse of the command high during the interval when a repeater is responding. The peak phase deviation is low, typically 20 deg, so that the line bit-error-ratio is not measurably degraded. The subcarrier frequency is chosen to minimize the effects of interfering noise caused by the phase jitter of all the regenerators on the particular line.

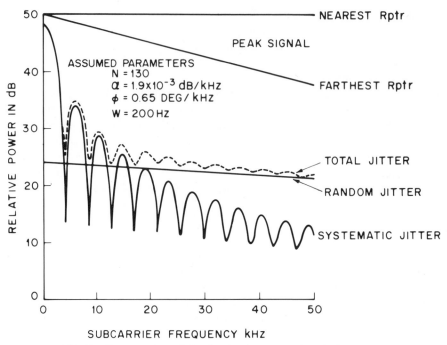

Fig. 10.   Response channel signal and noise versus subcarrier frequency.

The phase modulator consists of a first order all-pass network located between the decision circuit and the transmitter. The control element is a reverse-biased varactor diode which is part of the shunt capacitor branch. The keyed subcarrier modulates the capacitance of the diode which in turn varies the phase shift of the network. This method of phase modulating the data stream causes minimal degradation in the line bit error ratio because the phase modulator follows the decision circuit. Calculations show that the effect of the resulting alignment jitter on downstream regenerators is negligible.

The use of a subcarrier, instead of direct baseband, to phase modulate the transmitted data stream is the means of obtaining a suitable signal-to-noise ratio on the response channel without using relatively large index phase modulation. Appendix II gives expressions for the response channel signal-to-noise ratio in terms of the relevant system parameters. Fig. 10 indicates how the response channel signal and noise varies with subcarrier frequency for a given set of parameters.

RECOVERING THE RESPONSE CHANNEL SIGNAL

Recovery of the response channel signal, Fig. 11, involves a two-step demodulation process: The received clock signal from the receiving terminal regenerator, after division to a convenient frequency, is demodulated in a phase detector whose reference clock is provided by a narrow band (5–10-Hz) phase lock loop. The output of the phase detector contains the same subcarrier tone bursts which were applied at the responding repeater's phase modulators. However, the signal at this point is embedded in a high level of

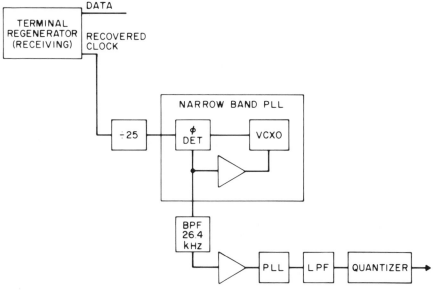

Fig. 11.   Response channel receiver.

background noise. The baseband signal is recovered from the pulsed carrier signal by a high Q bandpass amplifier, a tone decoder (PLL), a low-pass filter, and a quantizer. After conversion from pulsewidth modulated pulses to binary zeros and ones, the response signal is processed by the same computer which originated the command.

APPENDIX I: PARITY CHANNEL PREDETECTION SIGNAL-TO-NOISE RATIO

Supervisory information carried by the periodically complemented parity bits is recovered as shown in Fig. 12: Assume 1-V NRZ pulses at output of mod-2 divider. At this point, the low-frequency component due to the parity bits is a square wave of peak value $1/M$ and frequency f_o. The fundamental component passed by the bandpass filter whose transfer function is $G(jw)$ is

$$\text{Signal} = \frac{2}{M\pi} |G(j2\pi f_o)| \cos(2\pi f_o t).$$

The spectral density of the noise due to the random binary NRZ data bits at the divider

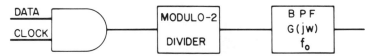

Note: Parity bit complement rate $= 2f_o \text{s}^{-1}$, line data rate $= f_b$ b/s, bits per parity block $= M$

Fig. 12.   Recovery of supervisory signal in regenerators.

output is

$$\Phi(w) = \left(\frac{M-1}{M}\right)^2 \frac{1}{4f_b} \frac{\sin^2\left(\frac{w\tau}{2}\right)}{\left(\frac{w\tau}{2}\right)^2}$$

where

$$\tau = \frac{1}{f_b}.$$

At the filter output the mean-square noise is given by

$$\overline{N^2} = \int_{-\infty}^{\infty} \Phi(w)|G(jw)|^2 \, df.$$

The peak signal to rms noise is given by

$$\text{SNR} = \frac{\left(\frac{2}{M\pi}\right)^2 |G(j2\pi f_o)|^2}{\overline{N^2}}.$$

If $f_o \ll f_b$ then the integral representing the mean-square noise is

$$\overline{N^2} \simeq \left(\frac{M-1}{M}\right)^2 \frac{1}{4f_b} |G(j2\pi f_o)|^2 2W_N$$

where W_N is the filter noise bandwidth

$$\text{SNR} \simeq \frac{\left(\frac{2}{M\pi}\right)^2}{\left(\frac{M-1}{M}\right)^2 \frac{1}{4f_b} \cdot 2W_N}$$

$$\text{SNR} \simeq \frac{8}{\pi^2} \frac{1}{(M-1)^2} \frac{f_b}{W_N}$$

for the case

$$f_b = 295.6 \times 10^6$$
$$W_N = 2000$$
$$M = 25$$
$$\text{SNR} = 208 \text{ or } 23.2 \text{ dB.}$$

APPENDIX II: RESPONSE CHANNEL SIGNAL-TO-NOISE RATIO

Let

$H(f)$ = jitter transfer function of regenerator

$\log_e (H(f)) = \alpha(f) + j\phi(f)$

θ_s = peak signal phase deviation in degrees

θ_u^2 = uncorrelated (random) phase jitter density deg^2/Hz

θ_c^2 = correlated (systematic) phase jitter density deg^2/Hz

W = jitter channel receiver noise bandwidth in Hz

N = total number of regenerators

n = number of regenerators from signal source to receiver

Received Signal $= |\theta_s e^{-na}|^2 = \theta_s^2 e^{-2na}$

Received Random Jitter $= W \sum_{n=1}^{N} |\theta_u H(f)|^{2n}$

$$= \theta_u^2 W e^{-(N-1)a} \frac{\sinh(N\alpha)}{\sinh(\alpha)}$$

Received Systematic Jitter [1]

$$= W \left| \sum_{n=1}^{N} \theta_c H(f)^n \right|^2$$

$$= \theta_c^2 W e^{-(N-1)\alpha} \left[\frac{\cosh(N\alpha) - \cos(N\phi)}{\cosh(\alpha) - \cos(\phi)} \right]$$

$$\text{SNR} = \frac{\theta_s^2 e^{-(N-1-2n)\alpha}}{W \left[\theta_u^2 \frac{\sinh(N\alpha)}{\sinh(\alpha)} + \theta_c^2 \left(\frac{\cosh(N\alpha) - \cos(N\phi)}{\cosh(\alpha) - \cos(\phi)} \right) \right]}$$

REFERENCES

[1] C. J. Byrne, B. J. Karafin, and D. B. Robinson, "Systematic jitter in a chain of digital regenerators," *Bell Syst. Tech. J.*, vol. 42, pp. 2679–2714, Nov. 1963.

41
Line Signals in Submarine Digital Telephone Links Using Optical Fibers

JEAN-CLAUDE LACROIX

INTRODUCTION

The digital transmission of information over long-haul links, especially using optical fibers, requires certain qualities of the signal transmitted over the line.

The signal must be of limited bandwidth at both ends to facilitate the design of the amplifier. It must also be of a random nature to permit the recovery of timing signals in the underwater regeneration devices and must feature redundancy to facilitate error detection irrespective of the information transmitted.

Proper selection of the line code and of the scrambling sequence enables the required qualities of the transmitted signal to be achieved, as discussed below. These are complemented by the characteristics required of the frame, for the jitter to which the signal is subjected due to waiting time to be limited and so that the transmitted signal is relatively insensitive to error packets.

LINE CODE

Codes may be defined and compared in terms of a number of characteristics and the quality to be achieved.

The characteristics required are primarily associated with the sensitivity of the receiver, the required bandwidth, the spectrum of frequencies needed to recreate a local clock, and the redundancy necessary to facilitate error detection.

The main characteristics comprise the following.

Number of Levels: Binary, ternary, and quaternary codes with two, three, and four levels provide for defining two, three, and four different states, that is, a greater or lesser number of units of information per unit of time.

Redundancy: This is the percentage of supplementary information when the word on the line is compared to the uncoded word. It may be used to reduce and periodically eliminate the running digital sum. The low-frequency components of the spectrum are thus reduced, and the f_{max}/f_{min} ratio, through being reduced, facilitates the design of the receiver. Also, it provides for error detection through verification of the transcoding rules.

TABLE I

Code*	Modulation rate	Relative bandwidth	Redundancy	Running digital sum	Error detection	Clock recovery	dB penalty
NRZ	1	1		No limit	No	No	0
RZ (1B/2B)	1	2		No limit	No	Yes	0
CMI (1B/2B)	1	2		No limit	Yes	Yes	−3
Biphase (1B/2B)	1	2		No limit	Yes	Yes	−3
3B/4B block code	1.33	1.33	2	Limited	Yes	Yes	−1.2
5B/6B block code	1.2	1.2	2	Limited	Yes	Yes	−0.8
7B/8B block code	1.14	1.14	2	Limited	Yes	Yes	−0.56
Scrambled NRZ + 1 parity bit (24B 1P)	1.04	1.04		No limit	Yes	Yes	−0.17
Scrambled NRZ + 1 parity bit + 1 complementarity bit (12B 1P 1C)	1.16	1.16		Limited	Yes	Yes	−0.6
1B/1T bipolar	1	1		Limited	Yes	Yes	−6
3B/2T	0.66	0.66	1.12	Limited	Yes	Yes	−4.2
4B/3T	0.75	0.75	1.68	Limited	Yes	Yes	−4.7
6B/4T	0.66	0.66	1.26	Limited	Yes	Yes	−4.2
3B/2Q	0.66	0.66	2	Limited	Yes	Yes	−10.2
7B/4Q	0.5	0.57	2	Limited	Yes	Yes	−9.6

*NRZ   No Return to Zero
RZ   Return to Zero
CMI   Coded Mark Inversion
B   Binary
T   Ternary
Q   Quaternary

Bit Rate Changing: It is the transcoding function which makes *m* binary, ternary, or quaternary elements, etc., correspond to *n* binary elements (bits). According to the number of levels, the line bit rate may be lower or higher than that of the uncoded information.

Transparency: This is the capability of the transmission system to accept all messages without error.

Major Code Types

There are three major code types in use at present: 1) block codes; 2) scrambled NRZ codes including a parity and/or complementarity indication; and 3) multilevel codes.

Table I compares a number of codes to the simple binary (NRZ) code.

Codes for High Bit Rate Optical Transmission

Multilevel codes are appreciated for the necessary bandwidth reduction, but for high bit rate transmission, especially over optical fibers, they are not currently used for the following reasons: 1) changes in laser diodes with aging not yet sufficiently controllable; 2) complexity of regenerators and, in particular, the decision circuits; and 3) system penalty may be higher than with other codes. Consequently, block codes or scrambled NRZ codes with simple parity are preferred at present.

The former are appreciated for the bandwidth limitation in the low-frequency domain and for their redundancy, which facilitates error detection and permits the use of remote

TABLE II

Manufacturer	Code	Line bit rate
Bell (USA)	Scrambled NRZ with parity: 24B 1P	295 600 kbit/s*
STC (UK)	7B/8B block code with parity	324 315 kbit/s
Submarcom (France)	5B/6B block code	339 225 kbit/s

*Choice effected by the three manufacturers who participate at the TAT 8 achievement.

supervision messages which are not likely to be simulated by the line code, even in the presence of a high error rate ($<1 \times 10^{-3}$, for example).

The latter are appreciated for their bandwidth limitation in the high frequency domain, for the ease with which transcoding can be effected on transmission and reception at high bit rates, and for their low system penalty, in spite of the nonlimited character of this type of code.

The selection of a code represents a compromise which can be readily reexamined given technological advances and changes in the criteria selected. For these reasons the manufacturers who answered the call for tenders for the TAT 8 transatlantic system all opted for different choices (see Table II).

SCRAMBLING SEQUENCES

The clock recovery essential to the regeneration of digital signals requires that the line signal comprise a large number of transitions. Certain codes, especially block codes, produce a limited running digital sum and thus a sufficient quantity of transitions. However, if the uncoded signal consists of a long sequence of identical or recurrent bits or groups of bits, the same code word or the same group of code words will be repeated several times consecutively after transcoding. Energy may then become concentrated at specific frequencies in the spectrum.

Scrambling is required in this case and is of key importance in the case of codes with a nonlimited running digital sum.

The scrambling operation consists of adding to the incident binary signal a pseudorandom sequence which has properties very similar to those of a random sequence.

Choice of Scrambling Sequence

The first problem is to select the generator polynomial of the scrambling sequence.

This polynomial $f(x)$ must be irreducible in the Galois body $CG(p)$ and must have p^{n-1} different roots on the extension body $CG(p^n)$; it is then primitive, and the p^{n-1} elements of the extension body form a cyclic group [1].

It is theoretically possible to find at least one irreducible and primitive polynomial per degree n; it remains to select the degree.

Generally speaking, and in order to limit long sequences of identical bits, the degree selected is low, but is higher than a few units to make the sequence long enough to be considered random. A ninth degree polynomial is recommended by the CCITT for 140-Mbit/s links [2].

However, simulation of the scrambling sequence by the incident signal, although extremely rare, remains possible.

In particular, experience has shown that such simulation is possible in practice, in particular when the length of the sequence is not equal to a prime number.

Under these conditions, and given that only degrees 3, 5, 7, 13, etc., generate sequence lengths equal to a prime number, it is the seventh degree which is almost always adopted; the length of the scrambling sequence is then 127 bits. Note that 127 is greater than 50, and that this is a sufficient condition for the binomial probability law to be approximated by the normal or Laplace–Gauss law. Theoretical studies of the line spectrum are generally based on the simplifying assumption that the random variable is governed by the normal law. This hypothesis is therefore legitimate when the length of the scrambling sequence is greater than 50.

Use of Scrambling Sequence

There is little to be said here on synchronous and asynchronous scrambling techniques, which are well known (see Appendix). The following remark is called for, however.

Where a line signal is formed from a number of high bit rate incident signals (at 140 Mbits/s, for example), it is generally desirable to scramble each incident signal before multiplexing, to facilitate the design of the corresponding registers. Also, in this case a single synchronous pseudorandom sequence generator is all that is required. It is then necessary to adhere to a minimum offset between the various scrambling sequences to avoid significant energy losses in the band resulting from multiplexing.

Frame Structure

Generally speaking, the frame structure represents a compromise which is necessary for obtaining a demultiplexer synchronization time which is lower than that for the next level down (specifically, 64 μs for frames at n times 140 Mbits/s).

This compromise has to take account of 1) the length and the structure of the frame alignment word and the demultiplexer synchronization and desynchronization criteria; 2) the length of the frame; and 3) the time for resynchronization of the demultiplexer in 99 percent of cases.

This compromise is sought at each level in the hierarchy and affects the characteristics of the signal sent over the line only through the jitter due to the waiting time and the immunity to error packets, both these factors being further discussed below.

Jitter Due to Waiting Time

Multiplexing systems are designed to accept, within certain limits, jitter present in the component trains in the case of the multiplexer and in the resultant train received in the case of the demultiplexer. The latter acts as a jitter reducer, according to its bandwidth and the capacity of its buffer memory.

However, plesiochronous multiplexing creates jitter with components at very low frequencies, due to the waiting-time phenomenon. The amplitude of these components is directly linked to the justification rate adopted. As a general rule, this jitter is not affected by the reduction effect of the demultiplexer.

Its spectrum results from the convolution product of two spectra, one of which consists of the frame frequency and its harmonics, and the other of which consists of the average justification frequency and its harmonics. Beat frequencies thus appear which, for justification rates adjacent to certain specific values, yield components at very low frequencies that the frequency-selective circuit in the receive channel cannot eliminate. These specific values are of the m/n type where m and n are integers, the peak-to-peak amplitude of the jitter reaching $1/n$ bits [3].

The justification rate must therefore be selected so that the jitter due to the waiting time remains below 0.3 peak-to-peak unit intervals, given the component signal frequency tolerance (the 0.3 peak-to-peak unit interval is the limit adopted by the CCITT in its Recommendation G.703).

Immunity to Error Packets

Error packets on digital links are generally due to the presence of industrial interference in the vicinity of the line and regenerators or in the vicinity of the terminal equipment.

In the case of an optical fiber link, the line signal is only very slightly affected by industrial interference, if at all, and the same applies to the regeneration equipment when it is underwater. This leaves the terminal equipment and regenerators in shallow water or forming part of the terminal equipment.

Given these conditions, there is still a risk of error packets, although this is significantly lower than in the case of terrestrial digital links using coaxial cable.

When an error packet reaches a duration immediately higher than that of $n-1$ sectors of the highest bit rate frame (the shortest sector in the hierarchy), n justification indications may be erroneous, depending on the position of the error packet relative to the frame, and this for a null error rate, distributed randomly. In this case the risk of loss of justification indication is high (if the justification recognition criterion is n of m) and results in the event of loss in the desynchronization of the entire digital hierarchy, in view of the consequent shifting of all the frame alignment words.

It is therefore necessary to propose sectors of the greatest possible duration and, consequently, a maximum frame length.

As a general rule, for multiplexing two component signals at 140 Mbits/s, five justification indications are employed with a three out of five majority detection principle. The frame length before transcoding is then 1344 bits* (Bell and STC proposals in response to TAT 8 call for tenders) or 1960 bits (Submarcom proposal).

In the former case the frame can accept error packets with a duration of 1.35 μs, and in the second it can accept error packets with a duration of 1.98 μs.

Conclusion

The characteristics of the line signals which may be used on submarine digital links using optical fibers are principally concerned with the code, the scrambling sequence, and the frame structure adopted.

While the selection criteria relating to the scrambling sequence and the frame structure are comparable to those for terrestrial digital systems using coaxial cable, those influencing the choice of line code result from a compromise which may be called into question by future developments in technology.

---

*Choice effected by the three manufacturers who participate at the TAT 8 achievement.

information signal   information signal scrambled

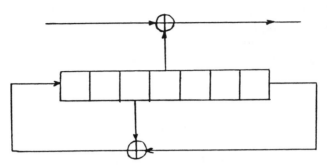

Fig. 1.   Synchronous scrambling register.

APPENDIX

Synchronous Scrambling

The scrambling is obtained by an exclusive OR operation between the information signal and any output of the scrambling register.

This register is initiated, without information signal, by a short sequence which is, in general, the frame alignment word or a part, or an image of it. In this case, the loading of a mark at $t = 0$ is not necessary.

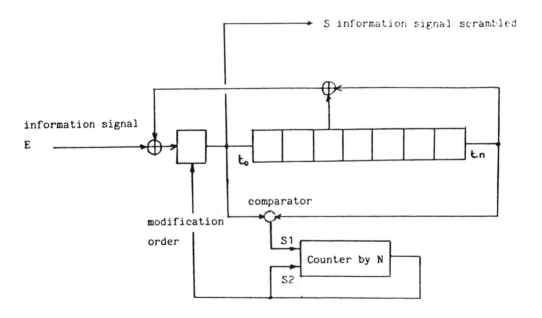

if bit to = bit tn   S1 increase the counter
if bit to ≠ bit tn   S1 reset the counter.

Fig. 2.   Asynchronous scrambling register.

The receiver realignment mode requires the frame alignment word recognition to initiate the receive register.

Asynchronous Scrambling

The register content depends on information signal, there is not register initiation and the receive realignment mode does not require the frame alignment. It is used when multiplexing is not necessary to form a line signal.

However, if information signal is full mark, of full space the register is blocked and recopies the information signal. The resolution difficulty consists of modifying the enter signal when the signals at t_o and t_n are the same N times consecutives.

In reception the same principle is used, the access E and S are reversed.

REFERENCES

[1] J. Clavier, G. Coffinet, M. Niquil, and F. Behr, *Théorie et Technique de la Transmission de Donnees, Tome 1.* Paris: Masson, 1972.

[2] "Equipements de ligne pour système de transmission à 140 Mbit/s sur câbles à paires coaxiales," *Câbles Transmiss.*, no. 2, pp. 278–338, Apr. 1978.

[3] Y. Matsura, S. Kazuka, and K. Yuki, "Jitter characteristics of pulse stuffing synchronisation," in *IEEE Int. Conf. Commun.*, June 1968, pp. 259–264.

42

Terminal Transmission Equipment (TTE) for the SL Undersea Lightwave System

J. LANCE FROMME AND M. D. TREMBLAY

INTRODUCTION

The SL Transmission Terminal Equipment (TTE) provides the interface between the inland toll transmission network and the SL undersea cable. The TTE's central transmission function is multiplexing two asynchronous 139.264 Mbit/s (CCITT 4th level) signals, plus housekeeping and maintenance digits, into each 295.6-Mbit/s fiber-optic SL line signal (and the inverse). The TTE operates with two working fiber-pairs and one standby fiber-pair.

Fig. 1 is a general block diagram of the TTE. The multiplexing, formating, overhead bit addition, generation of line frequency, and several related functions, are performed in the *muldex* (multiplex-demultiplex) subsystem. The muldex protection arrangement is two working units and one protection unit in each direction of transmission.

The line rate electrical-to-optical and optical-to-electrical signal conversion is performed in a *terminal regenerator* subsystem. The transmit side consists of retiming, laser-transmitter, and optical switching circuits; the receive side uses p-i-n receivers and retiming circuits. A supervising circuit common to all Terminal Regenerators places them under supervisory control of the SCOUT[1] system, in common with the SL undersea regenerator equipment.

The SL TTE includes comprehensive maintenance facilities; not only local, but also end-to-end system maintenance functions are performed. The TTE controls and monitors the parity of the line signals, first as a measure of line error performance and second as a means of supervisory signaling. The TTE provides key stimuli for the SL Supervisory System and performs specific actions under its command. The TTE accumulates data suitable for the long-term performance evaluation of the system.

The TTE provides six 64-kbit/s voice or digital service channels on each fiber pair, via dedicated T-bits, for orderwire and telemetry.

---

[1]SCOUT = *Surveillance and Control of Undersea Transmission.* See [1] and [3].

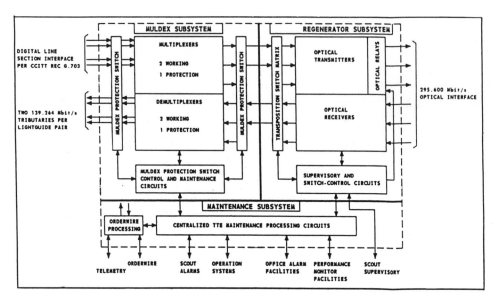

Fig. 1.

The TTE occupies four 7-ft high, 19-in wide bay frames (see Fig. 2) and is constructed according to the AT&T-Technologies UNIPACK[2] standard. It has relatively shallow depth (about 10 in), and it does not require access from the rear during installation or service. This enables equipment lineups to be placed against a wall or back-to-back for better use of office floor space. Shelf assemblies, as well as circuit packs, are equipped with connectors for ease of installation or removal.

The SL TTE also forms the basis for the System Assembly and Laying Test Equipment. This is used in the cable factory during system integration and used at the terminal sites and aboard the cable ship during installation. The bay frames and most equipment shelves are identical in either application, so that the testing system used for installation and laying of a cable system can be easily and quickly converted to a TTE for the commercial operation of that system.

MULTIPLEXING FRAME FORMAT

The multiplexing frame format used in the SL system closely follows the example given in CCITT Recommendation G.922, adapted for the different system capacity and modified to provide parity control.

Table I shows the construction of the frame format; Table II gives the performance characteristics of this frame format. The performance in the way of frame alignment time, burst error resistance, etc., is virtually identical to that of CCITT Recommendation G.922.

FUNCTIONAL DESCRIPTION

The following is a general description of the key subsystems of the TTE.

[2] UNIPACK is a trademark of AT&T-Technologies, Inc.

Fig. 2.   SL TTE block diagram.

TABLE I
SL MULTIPLEXING FRAME FORMAT

Line Rate	295.6 Mbit/s ±0.0003%
Number of Tributaries	2
Tributary Rate	139.264 Mbit/s ±0.0015%
Frame Length	1400 Bits
Frame Rate	211.142 kHz

Frame Contents:

Set 1:   Bits 1-12   Frame Alignment Word, 111110100000
            Bits 24, 50, 75, ··· 200 Parity Bits
            Other Bits   Information from Tributaries

Set 2   Bit 1   Justification Control Bit, Tributary 1
through   Bit 2   Justification Control Bit, Tributary 2
Set 6:   Bits 25, 50, 75, ··· 200 Parity Bits
            Other Bits   Information from Tributaries

Set 7:   Bit 1   First T-Bit
            Bit 2   Second T-bit
            Bit 3   Bit Position Available for
                     Justification of Tributary 1
            Bit 4   Bit Position Available for
                     Justification of Tributary 2
            Bits 25, 50, 75, ··· 200 Parity Bits
            Other Bits   Information from Tributaries

TABLE II
SL MULTIPLEXING PERFORMANCE

Maximum justification rate	211.1 kbit/s
Nominal justification rate	90.8 kbit/s
Nominal justification ratio	0.43
Capacity available for service channel, per fiber-pair	383.9 kbit/s (net)
Minimum error burst length to cause loss of frame alignment	14.2 microseconds
Mean time between loss of frame alignment at BER $= 10^{-4}$	26.5 days
Minimum error burst length to cause incorrect dejustification	1.35 microseconds
Mean time between incorrect dejustification at BER $= 10^{-4}$	5.5 days
Time to detect loss of frame alignment	<18.9 microseconds
Time to re-acquire frame alignment and verify, > 99% probability	< 30 microseconds

Multiplexer

The transmission interface on the inland side is at 139.264 Mbit/s according to CCITT Recommendation G.703. The 139.264-Mbit/s CMI input circuit contains an automatic equalizer that can compensate for up to 12-dB loss at the Nyquist frequency, between this equipment and the connecting equipment. (This amounts to about 450 ft of standard AT&T-Technologies interoffice coaxial cable.) A jitter-reducing circuit is incorporated to enable the system to accept input jitter of at least 5 Unit Intervals peak-to-peak (up to 4 kHz), while keeping output jitter (at the remote end of the system) within the CCITT recommendation of <1.0 Unit Interval peak-to-peak (up to 10 kHz).

Conventional pulse-stuffing (positive justification) is used to synchronize the tributary information with the system clock. Information bits are scrambled with a $(2^7 - 1)$-bit maximal-length pseudorandom sequence that is reset every frame period.

The internally generated system clock has ± 3 ppm short- and long-term stability; in addition, the system can be operated with either 2.048- or 295.6-MHz external clock sources. (Automatic fallback to the internal clock is provided if an external clock is not available).

Terminal Regenerators

The terminal regenerators use the same highly integrated silicon circuits that are used in SL undersea regenerators. Ross *et al.* [2] describe these components.

Independent transmitting and receiving subsystems are provided for up to three pairs of fibers. Each outbound fiber is equipped with a cold-standby laser transmitter and optical relay.

A supervisory transceiving and logic circuit communicates with the SCOUT system, measures key regenerator parameters, and executes switch commands exactly the same as if this were an undersea SL repeater.

Demultiplexer

The demultiplexer performs the inverse function of the multiplexer. In addition to the customary frame-acquisition, de-interleaving, overhead-removal, and frequency-smoothing functions, the demultiplexer detects parity errors and processes them. Parity errors that were intentionally generated by the transmitting TTE, i.e., supervisory signaling, are not logged as "system" errors, but are instead reconverted to the original supervisory signal and routed to the local SCOUT terminal. Other errors are automatically logged.

Protection Systems

The SL availability objective is approximately the same, on a mileage basis, as the current AT&T-Communications long-haul terrestrial facility objective of 99.98 percent per 4,000 miles. For a transatlantic system, for example, this amounts to less than 2 h per year. The SL TTE is designed to account for a small fraction of this (a few minutes per year). To achieve this, two important factors are designed-in: first, protection or standby systems are provided with automatic switchover to minimize service interruption in case of failure; second, faults can be quickly identified to the circuit pack level, and repaired without interruption of service, so the probability of simultaneous failures is minimized.

The protection switching architecture of the SL TTE is divided into two parts; the muldex subsystems, and the terminal regenerator subsystems.

1) Muldex Protection: The TTE contains three multiplexers and three demultiplexers, each trio making up an arrangement of two working and one protection unit. The complete transmission function of each multiplexer and demultiplexer is periodically verified by a time-shared bit-by-bit monitor subsystem; this is designed to detect serious faults (BER worse than 10^{-3}) and keep the service outage to less than 50 ms. Less serious faults (BER of 10^{-6} or above) will take proportionally longer to detect and switch.

The monitor subsystem sequentially examines the operation of all multiplexers and demultiplexers (working and protection) under microprocessor control.

In multiplexer protection, a duplicate of a multiplexer output is processed in a "test demultiplexer," that is part of the monitor subsystem. An output of the "test demultiplexer" is compared (after suitable delay adjustment) with the appropriate multiplexer input, bit-by-bit; normally, the signals will be identical. If so, that signal path is judged to be working correctly and a different tributary is next selected for comparison. Failure to achieve correlation leads to a sequence of cross-checking, validation of the protection demultiplexer, and ultimately a protection switch, if needed.

In demultiplexer protection, a duplicate of the incoming line signal is processed in the same "test demultiplexer" and one of its outputs is compared, after suitable delay adjustment, to the corresponding output of the demultiplexer-under-test. Again, a test for correlation at the tributary level is used to determine the need for a protection switch.

2) Terminal Regenerator Protection: The terminal regenerators reside in the TTE frame but are maintained by SCOUT as part of the underwater line shore-regenerator section. SCOUT supervises and controls the terminal regenerator subsystem the same way that it handles an undersea repeater. Identical supervisory signaling means are used, and identical circuit parameters are monitored (these include laser bias, receiver AGC, parity violation counts, and switch status). The supervising of all terminal regenerator circuits is centralized in one common supervising circuit using the same set of highly integrated supervisory chips found in an SL repeater.

A "transposition switch" network separates the terminal regenerators from the muldexes. With this switching network, the third, standby regenerator span in each direction can replace either of the two working spans. Loopback switching (toward the undersea link) is provided, as well.

Orderwire System

A total of 422.284 kbit/s is available in the frame format for T-bit usage. The T-bits carry up to six, nominally 64-kbit/s, channels plus T-bit framing information. The orderwire subsystem consists of two T-bit muldexing subsystems (one per working fiber), several (plug-compatible) types of channel units, and orderwire terminating units that provide telephone sets, signaling, and network connections (if required).

1) T-Bit Muldex Subsystem: Each T-bit muldex subsystem synchronously multiplexes and demultiplexes six, nominally 64 kbit/s, channels into the T-bit slots of the line signal.

The T-bit stream is automatically routed to the traffic-bearing high-speed multiplexers and demultiplexers, whenever a muldex protection switch is made; this ensures that high-speed muldex protection switching does not seriously impair the orderwire function.

2) Orderwire Channel Units: The Orderwire Channel Units each provide a conversion between the 64-kbit/s digital interface, on the muldex side, and a standard 4-wire, 600-Ω data or voice interface on the network side. Channel units can be selected to provide a digital four-wire duplex interface at 64 kbit/s, two digital four-wire duplex interfaces at 32-kbit/s, or a four-wire duplex analog (300–3000-Hz) interface; the analog signal is encoded into 64-kbit/s PCM, or Mu-law.

These channel units are plug-compatible and may be selected to fit the Customer's needs.

3) Orderwire Terminating Equipment: The Orderwire Terminating Shelf provides a TouchTone handset, a speaker, extension jacks, and space for up to six orderwire terminating units. Each terminating unit provides TouchTone signaling to up to 55 similar remote units, access to a DDD port (out-dial and in-dial), and several types of local extension ports.

Parity Control and Supervisory Signaling

The primary function of the 25-bit parity blocks on the SL line signal is to provide a means of error monitoring in the undersea repeaters, (see [3]), and for in-service, end-to-end performance monitoring. Most of the time, all parity blocks are made to have even parity.

However, two other modes of operation call for specific kinds of intentional parity violations. First, intentional parity violations are used for signaling from the SCOUT terminal to the repeater supervisory circuits; second, it is occasionally necessary to transmit a line signal with a known, constant parity violation rate in order to calibrate the parity detection and alarm facilities.

1) Supervisory Command Transmission: When the SCOUT terminal must send a command to an undersea repeater, a low-frequency pulsewidth-modulated baseband signal is generated in the TTE. During the pulse-on times of this signal, the multiplexer complements the parity bit in one out of every 224 parity blocks. The repeater supervisory circuitry detects and decodes this pattern of parity violations, thus receiving the command.

In order that these intentional violations do *not* cause unwanted alarms at the receiving TTE, the position of the parity bits that are used for signaling is fixed with respect to the frame marker in the line signal. In the receiving (demultiplexer) parity circuit, violations occurring in this position are not logged in the end-to-end error counter. Instead, parity violations that occur in this position are demodulated to the original command. This is relayed to the local SCOUT terminal.

Violations detected in other parity blocks are steered to the demultiplexer's shelf maintenance processor, where they are logged. Thus, the intrinsic error-monitoring function of the parity blocks is preserved, yet supervisory signaling is allowed at any time. (The exclusion of the supervisory signaling parity block from end-to-end monitoring results is less than 2-percent error; this will be approximately constant, so it can be corrected with a simple scaling factor.)

2) Parity Violation Testing: Occasionally it will be necessary to cause "test" parity violations in the line signal, to verify the operation of the repeater monitors and the remote TTE alarms. In this instance, upon command of SCOUT, the multiplexer causes parity violations, at an appropriate rate, in the parity blocks that are not reserved for signaling. Thus, both the undersea repeaters and the remote TTE will interpret these violations as transmission errors, and their response to such errors can be verified.

3) Supervisory Response Reception: The return supervisory signaling (from repeaters back to SCOUT) is somewhat different than command transmission. In the repeater, the response message phase-modulates the high-speed line signal. The office regenerator passes the recovered clock to a phase-modulation receiver circuit that recovers the original PWM pulse train, which is relayed to the local SCOUT terminal.

Maintenance and Operations Functions

Maintenance functions performed by the SL TTE include TTE equipment frame maintenance, monitoring the end-to-end performance of the SL system, relaying alarm and status indications to SCOUT to help in the maintenance of the undersea cable, and communicating with external Remote Operations Systems or centralized maintenance centers.

1) End-to-End Maintenance and Performance Monitoring: The SL TTE detects fault conditions and performs consequent actions in accord with CCITT Recommendation G.922, Paragraph 5 and Table 2/G.922. In addition, other status indications are relayed from the receiving station to the transmitting station, including line power feed equipment status, received parity violation rate, loss of incoming tributary signal, and application of AIS.

The information available locally at each TTE is quite extensive. The TTE keeps a continuous record of the received parity violations, and processes this (as well as its alarm and status indications) to provide measures of overall system performance as required by the customer. For example, counts of errored seconds or minutes, seconds or minutes having more than a threshold number of errors, and similar types of system performance measures can be generated from TTE provided data.

2) TTE Frame Maintenance: The TTE provides extensive fault detection and fault-locating capabilities. Each of the multiplex and demultiplex shelves includes a shelf maintenance processor that processes alarm indications from the circuit packs of the shelf, performs local fault isolation, and also communicates with the central maintenance and protection switch-control processors.

A central maintenance processor serves as the overall coordinator for TTE maintenance and is responsible for notifying the office alarm system of major and minor alarm conditions. The central maintenance processor also drives the Control and Display panel; this panel provides the operator with full visibility of all equipment status, as well as manual control of protection switching actions.

3) SCOUT Assistance: The TTE provides the transmission access for supervisory command and response, and can generate parity errors for calibration, all controlled by SCOUT.

In addition, the TTE forwards information regarding the performance of the system so that SCOUT can react accordingly. This information includes prompt notification of overall system status (i.e., line failed, TTE failed), notification of (real time) line errors, and messages received from a remote SCOUT terminal (through an orderwire channel).

4) Operation with Remote Operations Support Systems: Each TTE can provide the above maintenance data, over a serial data link, to a remote terminal or computer so that the operators can obtain a complete picture of the system's status and performance from a single location.

SUMMARY

The Terminal Transmission Equipment designed for the SL Undersea Lightwave System provides line terminating and multiplexing/demultiplexing functions in accord with all applicable CCITT Recommendations. Microprocessor-controlled protection switching and maintenance subsystems ensure high continuity of service. Parity control of the line signal is used not only for error detection at the repeaters and end-to-end performance monitoring, but also for supervisory signaling. In addition, the system is designed to accommodate the Customers' growing need for remote (centralized) monitoring and maintenance.

REFERENCES

[1] P. K. Runge and P. R. Trischitta, "The SL undersea lightwave system," this book, ch. 4, p. 51.
[2] D. G. Ross, R. M. Paski, D. G. Ehrenberg, G. M. Homsey, "A highly integrated regenerator for 295.6 Mb/s undersea optical transmission," this book, ch. 26, p. 377.
[3] C. D. Anderson and D. L. Keller, "The SL supervisory system," this book, ch. 40, p. 567.

Part X
Future Undersea Lightwave Systems

The future of undersea lightwave communication systems is a bright one. We conclude this book by examining the possibilities.

43

Future 1.55-μm Undersea Lightwave Systems

RICHARD E. WAGNER

INTRODUCTION

Over the past thirty years, the demand for transoceanic telecommunications traffic has been increasing at a steady pace. In the North Atlantic, which has shown the largest and earliest traffic demands, undersea systems have been installed on a cycle of every four to five years to keep pace with the demands. This growth, as measured by the number of voice trunks provided, has continued at an exponential rate that ranges somewhere between 20 and 27 percent per year. Figure 1 shows the transatlantic growth of real voice trunk capability over the last 30 years, not including circuit multiplication from TASI or digital compression techniques. It illustrates a total capacity of about 20,000 voice trunks after the installation in 1988 of the first transatlantic lightwave system, referred to as TAT-8. Other areas, such as the Pacific, Mediterranean, and Carribean have shown steady growth also, but the required capacity is generally smaller and the demand lags the Atlantic demand by a few years. Consequently, it is appropriate to use the North Atlantic to gauge growth in demand for lightwave undersea systems in the next ten years.

If these growth trends remain into the 1990's, then it is possible to estimate an upper and lower limit on the size of system that will be needed. If the cycle is four years at a growth rate of 20 percent per year, then the total capacity should double requiring installation of a new system comparable in capacity to TAT-8 in the year 1992. If the cycle is five years at a growth rate of 25 percent per year, then the total capacity should triple requiring a new transatlantic system of twice the capacity of TAT-8 to be installed in 1993. Such projections, of strong growth in undersea systems, indicate the need to study carefully the technologies available for providing the next generation of lightwave undersea systems.

Two of the strongest motivating factors in the design of the next generation system are to continue to provide high reliability and to drive the costs down as much as possible. Since the length of cable is fixed by choice of landing sites, the latter objective must be met primarily by reducing the number of repeaters needed. This means, in turn, that one of the strongest motivating factors in the design of the next generation of undersea lightwave systems is to design for maximum repeater spacing. On the other hand the effort involved in developing new technology for such systems is expensive, so that it is

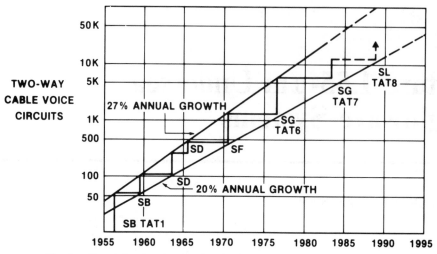

Fig. 1.   Growth in transatlantic voice circuit capacity in the past thirty years.

important to design with as little difference as possible from existing systems. The final choice of system design will be set by considering the tradeoff between these two conflicting motivations. As expected, a large number of technical design options exist for future systems, and these options are discussed in the paragraphs that follow.

TECHNICAL TRENDS

In order to understand what technology options might be available for 1992–1993, it is informative to examine the development of events leading up to the provision of service for TAT-8 in 1988. As shown in Table I, this development spans a time period of almost ten years, beginning in 1979 with the demonstration of the key technology—a single-mode

TABLE I
TECHNOLOGY DEVELOPMENT

System	SL	Options for Future 1.55 μm Undersea Systems	
Laser	1.3 μm Multi-longitudinal mode	1.55 μm Multi-longitudinal mode	1.55 μm Single-longitudinal mode
Fiber	Conventional	Dispersion-shifted	Lowest loss
Key Technology	BH laser	Triangular core fiber	SF laser
Technology Demonstration	1.3 μm mesa structure 1979	9 km × 4 fiber cable 1983	C^3, EC, or DFB structure 1983
System Demonstration	274 Mb/s 35 km 1981	296 Mb/s 150 km 1984	420 Mb/s 203 km 1984
Final Design	Deep sea trial 1982	——	——
System Ready for service	TAT-8 1988	——	——

BH laser operating at room temperature at 1.3 μm [1]. This was followed in 1980 by a proposal for an undersea system [2], and by laboratory system demonstrations in 1981 at a bit rate of 274 Mbit/s [3]. Very soon after that, in late 1982, a deep-sea trial [4], [5] was performed that established the feasibility of the technology for undersea applications. This allowed final design and reliability programs to begin in confidence. By 1985, AT&T will be installing a prototype system in the Canary Islands [6], and by 1988 the TAT-8 system will be available for service using 1.3-μm single-mode technology [7].

While it seems obvious that the design goal of longer repeater spacing can be met best with 1.55-μm technology, it is clear from such a timetable that laboratory demonstration of such systems should be occurring now, and reliability programs should be beginning soon in order to deploy systems for 1992–1993 service. In addition, this means that key technology should already have been demonstrated in order to meet such a schedule. Fortunately this is so, in that appropriate laser, fiber, and receiver technology has already been reported for 1.55-μm systems.

Other, newer technologies are beginning to emerge and these may compete for such applications. One of these is coherent detection techniques [8], [9] which promise the possibility of line amplifiers without baseband electronic regenerators. The key technology requires a phase-stable laser and line amplifier, but these have not yet been adequately demonstrated. Another technology of interest is that of systems in the Mid-IR, at a wavelength of 1.5–4 μm, where new fiber materials theoretically promise extremely low loss [10]. In this case the key technology milestone is demonstration of a fiber with loss less than that now available at 1.55 μm. Again, this is not yet demonstrated even in short lengths of fiber. Consequently, while these new technologies bear consideration for the second half of the next decade, they are extremely unlikely to be deployed for transoceanic service in the 1992–1993 period, and thus will not be mentioned further.

SYSTEM RELIABILITY

It is well known that the overall reliability of the optical transmission line is of paramount importance in undersea systems. In the 1.3-μm technology this requirement has been addressed in three ways: by qualifying all components to be of the highest possible reliability, by providing laser transmitter sparing, and by providing span redundancy. This has been a departure from previous analog systems where technology was established and reliability was assured solely by qualification and certification of components, and not by redundancy. In 1.55-μm single-mode fiber technology it may still be necessary to provide laser sparing and span redundancy, because the technology is again new, although the level of confidence may be higher than for the first 1.3-μm system deployed. However, the total number of components in a system will be fewer since fewer repeaters will be needed, and so the reliability requirements are slightly less stringent for 1.55-μm systems.

While the TAT-8 system will have about 130 repeaters, a similar system at 1.55 μm will likely require only about 65 repeaters. With this few repeaters, adequate reliability can be achieved with 2×1 laser sparing and a span redundancy of two working lines and one protection line, as shown in the reliability assessment of Table II. With this arrangement the required transmitter reliability is about 2400 FIT's (corresponding to a median life of about 1/3 million h), the required receiver reliability is about 10 FIT's and the IC reliability is about 1 FIT per IC.

TABLE II
RELIABILITY ASSESSMENT

	1.55 μm Future System	1.3 μm TAT-8 System
Number of Repairs	3	3
Working Lines	2	2
Protection Lines	1	1
Number of Repeaters	65	130
FITS*/Repeater	200	100
FITS/1-Way Line Section**	520	350
Receiver FITS	10	10
Regenerator Electronics FITS	70	70
Spared Transmitter FITS	440	270
Required Transmitter Reliability		
2 Transmitters/Regenerator	2400 FITS (3×10^5 hours)	1800 FITS (4×10^5 hours)
4 Transmitters/Regenerator	——	6600 FITS (1×10^5 hours)

*One FIT represents one failure in 10^9 device hours.
**Allows 6 FITS for repeater circuits common to all lines.

For transmitters and receivers, these requirements are reasonable to prove within the required time provided that aging mechanisms are accelerated by elevated temperatures. At 60°C with an activation energy of 1 eV, corresponding to an acceleration factor of about 450, the required laser reliability can be demonstrated in 10 000 h with less than 50 devices. This device-hour estimate applies only to qualification of a population of lasers representative of the final design that have been properly screened, and it assumes a constant device failure rate. It does not include the iterative effort needed to identify aging mechanisms and adjust device design, to establish proper purge and screening programs, to determine activation energy of devices, and to define proper certification procedures, which is a lengthy process.

For receivers, which can be operated at temperatures up to about 200°C, the acceleration factor is well over 1000 so that the required reliability can be demonstrated in 10 000 h with about 50 devices. Again, this estimate applies only to the qualification of properly screened devices.

For IC's the reliability requirements suggest that silicon technology will be favored because a large body of reliability data is already available, while newer technologies such as GaAs must wait until significant reliability data becomes available.

LASER OPTIONS

Four basic laser types potentially suitable for undersea applications have been demonstrated at 1.55 μm: Fabry-Perot (FP) lasers, Cleaved Coupled Cavity lasers (C³) lasers, External Cavity (EC) lasers, and Distributed feedback (DFB) lasers. These laser types can each be made from a variety of laser device structures, including ridge guide and buried heterostructures such as the mesa, channeled substrate, and dual-channel planar structures. The number of possible combinations is immense, but large development efforts and reliability programs are needed to qualify a specific device for undersea systems, so it is essential to make a wise choice early. The functional characteristics of each of the laser

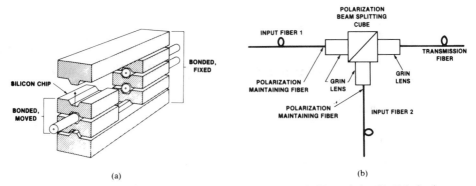

Fig. 3. Sketch of two sparing component options. (a) Si-chip switch. (b) Polarization combiner.

Theoretically DFB lasers operate with two longitudinal Bragg modes, but one of the modes can be suppressed by spoiling the symmetry of the cavity. This is normally done by altering the back facet [21] so that it has a different reflection than the front facet. The result is that it may be necessary to couple light from the front facet in order to provide accurate power control circuitry for laser transmitters.

SPARING COMPONENT OPTIONS

Reliability considerations indicate that 2×1 sparing is adequate, and this gives two major options for sparing of transmitters. The first option is an optical relay similar to that developed for the TAT-8 system [22]. It consists of a pair of stationary fibers sandwiched between three precision-etched silicon V-groove chips and a moveable fiber sandwiched between two V-groove chips [23]. This second sandwich is moved from one position to another by self-latching electrical coils, aligning the output fiber with either of the pair of input fibers. A sketch of such a component is shown in Fig. 3(a).

The insertion loss of such optical relays is less than 1 dB, the switching speed about 5–10 ms, and the drive voltage about 20 V. These characteristics are adequate for undersea applications, but proving the reliability for a mechanical moving switch required a major development. A comprehensive development program for TAT-8 components has established the reliability of such switches.

Another alternative, which avoids the reliability concerns posed by moving parts, relies on the polarization characteristics of semiconductor lasers. With this alternative, two orthogonally polarized laser outputs are combined in a polarization multiplexer [24], as shown in Fig. 3(b). The two input fibers of the polarization combiner are polarization-maintaining fibers, and the output fiber is identical to the transmission fiber. A disadvantage of this configuration is that the laser package must have a polarization-maintaining output fiber [24], thus requiring another fiber alignment in the laser coupling stage of manufacture. In addition, the polarization-maintaining characteristics of the laser, fiber, and combiner must be established to provide 25-year reliability. However, polarization-combining components can be made with about 1 dB of insertion loss and obviously require no switching power.

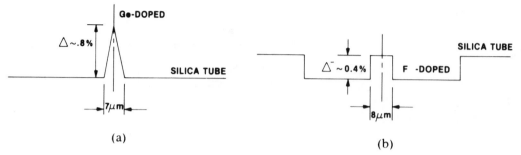

(a)   (b)

Fig. 4.   Sketch of fiber index profiles for (a) triangular-profile fiber and (b) silica-core fiber.

FIBER OPTIONS

A dispersion-shifted fiber is needed for use with FP lasers. So far the most practical design is one with a triangular index profile [25], as shown in Fig. 4(a). This fiber has a zero dispersion wavelength that can be tailored to be at any desired wavelength in the 1.5–1.6-μm region. It has good bending properties and can be cabled with negligible added loss [26]. The lowest loss reported for such fiber is 0.21 dB/km in a short length. The measured spectral loss curve of a 30-km spliced section is shown in Fig. 5, indicating a minimum loss of 0.25 dB/km at a wavelength of 1.565 μm.

For single-longitudinal-mode lasers, a dispersion-shifted fiber is not needed. In this case it is possible to improve the bending and loss characteristics with a design that is nearly pure silica in the core [27]. A fiber of this design, as illustrated in Fig. 4(b), has achieved the lowest loss reported to date—0.155 dB/km in a short unspliced length [27]. The measured spectral loss of a 30-km spliced section is shown in Fig. 5, indicating a minimum loss of 0.19 dB/km at a wavelength of 1.565 μm.

Flame fusion splicing studies of these two fiber types indicate that very low loss splices can be made. For triangular index profile fibers the mean and standard deviation of the

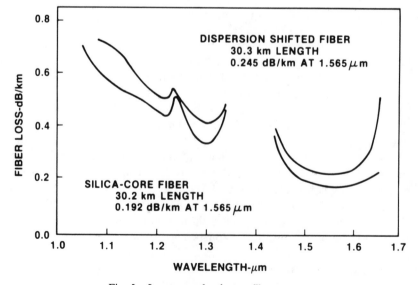

Fig. 5.   Loss curves for the two fiber types.

loss values of seven splices was 0.08 and 0.07 dB, respectively. For the silica-core fibers, eight splices produced mean and standard deviations of 0.06 and 0.04 dB, respectively. The slight difference between the two occurs because the spot size for the triangular profile fiber is smaller.

Both fiber designs have low phosphorus doping, which is particularly important for 1.55-µm systems. At 1.55 µm the radiation darkening [28] and hydrogen [29] effects are two to three times as large as at 1.3 µm and the fiber span distances are twice as long. Since these effects are less pronounced when the phosphorus level in the fiber is low, the two design are advantageous from this perspective.

RECEIVER OPTIONS

Receivers built using InGaAs p-i-n diodes [30] followed by a silicon bipolar transimpedance amplifier IC [31] have shown superior performance at 1.3 µm. These same receivers perform equally well at 1.55 µm, and have measured sensitivities of about −36-dBm incident average power for 300 Mbit/s with an NRZ data format.

Recently APD receivers are also showing promise. Although considerable effort has been expended on Ge APD's [32], these suffer from reduced efficiency and response speed at 1.55 µm. In addition, the use of Ge requires that a third material system, besides the InP and silicon that are used in lasers and electronics, be qualified for undersea reliability.

The most recent APD results in InGaAs are very encouraging. These devices have separate absorbing and multiplication regions [33]–[35], resulting in high speed, very low

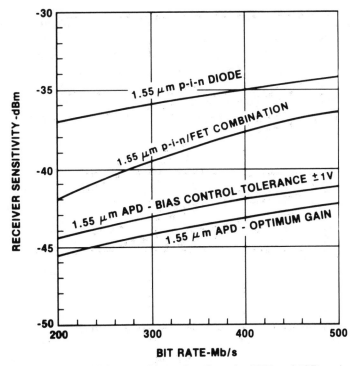

Fig. 6.   Receiver sensitivity versus bit rate for p-i-n, p-i-n/FET, and APD receivers.

capacitance, and low dark currents. Devices have been reported that allow receiver sensitivities of −44 dBm at 300 Mbit/s. This represents a potential improvement over a p-i-n diode of about 8 dB at optimum gain, but a bias voltage of about 100 V is required. In addition, the bias voltage must be controlled electronically to maintain optimum gain, which means extra regenerator control electronics.

Several p-i-n/FET combinations provide alternate and promising receiver designs. In this case the receiver sensitivities are expected to be better than for p-i-n diodes with silicon transimpedance amplifier IC's, but poorer than for APD receivers. Monolithic InP devices with the photodetector and first amplifier stage integrated on the same chip [36], [37] look attractive. However, devices suitable for high-performance receivers have not yet been reported. Another alternative uses a separate photodetector with an integrated GaAs FET amplifiers. Such receivers have been built that match the predicted sensitivity improvements relative to silicon IC's, but a strong reliability program will be necessary before such a new IC technology can be introduced to undersea systems.

Receiver sensitivities, as calculated using measured device parameters, are shown in Fig. 6 for InGaAs photodiodes using p-i-n, p-i-n/FET, and APD combinations within a speed range of 200–500 Mbit/s. The APD calculations assume that the bias voltage is controlled to within 1 V of the optimum gain. The improvement in sensitivity for APD receivers is about 7 dB in this speed range, and for the p-i-n/FET combination about 2–5 dB.

System Options

Two major options emerge for future 1.55-μm systems. The one option uses dispersion-shifted fibers with FP lasers [38], and the other uses single-longitudinal-mode lasers with a silica-core fiber [39]. In both cases the repeater spacing depends on which receiver technology is used. The range is set by APD and p-i-n technologies which correspond to longest and shortest spacings, respectively, with the p-i-n/FET combinations in between. Consequently four system configurations are of interest, consisting of the two major system options, each combined with the APD and p-i-n receiver options.

The 1.3-μm system loss budgets provide a convenient reference for estimating span distances at 1.55 μm. Table III shows a loss budget for a deep-water span that is

TABLE III
SYSTEM LOSS BUDGETS
(300 Mbit/s)

System Type Receiver Type		1.3 μm BH Laser p-i-n	1.55 μm FP Laser p-i-n	1.55 μm FP Laser APD	1.55 μm SF Laser p-i-n	1.55 μm SF Laser APD
Transmitter Output	(dBm)	0.0	0.0	0.0	−3.0	−3.0
Sparing Component and Splices	(dB)	2.5	2.0	2.0	2.0	2.0
Receiver Sensitivity	(dBm)	−35.0	−36.0	−43.0	−36.0	−43.0
Total Available Loss	(dB)	32.5	34.0	41.0	31.0	38.0
Aging and Penalties	(dB)	6.0	6.0	6.4	6.0	6.4
Margins	(dB)	4.0	4.0	4.0	4.0	4.0
Total Available to Cable	(dB)	22.5	24.0	30.6	21.0	27.6
Cables Fiber Loss	(dB/km)	0.45	0.25	0.25	0.20	0.20
Span Length	(km)	50	96	122	105	138

representative of current 1.3-μm technology operating at about 300 Mbit/s, along with corresponding data for the four system options at 1.55 μm. The performance characteristics discussed in the sections on laser, sparing component, fiber, and receiver are used to construct the table. In all cases the span distances are about twice what could be achieved with 1.3-μm technology.

The span distances are 26–33 km longer when APD receivers are used instead of p-i-n receivers, corresponding to a 30-percent reduction in the total number of repeaters needed for a system. This would represent a considerable system cost reduction. The span distances are 9–16 km longer when single-longitudinal-mode laser systems are used instead of FP laser systems, which corresponds to an additional 10-percent reduction in the number of repeaters needed. The final choice of which technology is best depends on the complex economic tradeoff between repeater costs, span distances, and technology development costs.

System Penalties

There are a large variety of system penalties that need to be accounted for in the system loss budgets. These include extinction ratio penalties, optical feedback effects, and chromatic dispersion effects. Normally, extinction ratio effects are quite small, since the lasers are modulated from below threshold. Exceptions to this occur for the C^3 and EC lasers, but this has been accounted for by the lowered effective launched power. In other cases, the extinction ratio may degrade with aging of the laser if the transmitter control circuit can not perfectly adjust the drive conditions to match laser characteristic changes. For an extinction ratio of 20:1, these effects are less than 0.5 dB for p-i-n receivers and less than 0.9 dB for APD receivers when operated at or below optimum gain [40]. These factors are included in the penalty allocations of the loss budget, and account for the differences in aging values shown in Table III.

The effects of optical feedback are also small [41], if the system is designed to minimize optical reflections. The main contributors to reflections are the sparing component and fiber splices near the transmitter. These can typically be held to less than a -15-dB return loss, in which case the system penalty for FP lasers is less than 0.5 dB. For single-longitudinal-mode lasers, optical reflection penalties have not yet been adequately characterized.

Chromatic dispersion in single-mode fibers introduces three different effects in these systems. These are called misequalization [42], mode partition noise [43], and chirp [44] penalties. The first two are predominant in FP laser systems, and all three can be apparent in single-longitudinal-mode laser systems.

Misequalization produces intersymbol interference at the receiver. This effect can arise even when the receiver band shape is perfectly equalized to accommodate the waveform from a single line of the laser, because many laser lines may contribute to the actual received waveform. Each of the lines contributes an identical waveform advanced or delayed according to the amount of dispersion it experiences, and weighted according to the relative power in the line. The combined waveform is different than any of the individual waveforms, but it is of a fixed shape as long as the weights of the lines are not fluctuating from bit to bit. The combined waveform is misequalized, producing a system penalty P_m, given by [42], [12]

$$P_m = 30(m\sigma BZ)^2 \tag{1}$$

where σ is the rms spectral width of the laser, m is the fiber chromatic dispersion, B is the bit rate, and Z is the span distance.

When the weights of the lines are fluctuating from bit to bit, then the received waveform for each bit is different and the penalty is increased over that given above. This extra penalty is referred to as mode partition noise penalty P_p. It is given by [43]

$$P_p = -5\log\left[1 - 11k^2\pi^4(m\sigma BZ)^4\right] \tag{2}$$

where k^2 corresponds to the fraction of laser power that is fluctuating from bit to bit.

For a perfect single-longitudinal-mode laser, these two effects would not occur. However, they can be apparent in real single-longitudinal-mode lasers because the Fabry–Perot or Bragg modes adjacent to the main mode may not be perfectly suppressed. For example, if there is a single unsuppressed longitudinal mode spaced a distance of $\Delta\lambda$ from the main mode with suppression ratio of S, then the rms spectral width is $\sigma = \Delta\lambda\sqrt{1/S}$. For DFB lasers, typical suppression ratios are 1000 or greater, and mode spacings are about 20–40 Å, while for C^3 and EC lasers mode spacings are about 10–15 Å and suppression ratios are greater than 200. This leads to rms spectral widths for a single unsuppressed mode of from 0.6 to 1.2 Å, depending on the laser. In fact, real single-longitudinal-mode lasers may have more than one side mode, each potentially contributing to misequalization and mode partition noise penalties.

Even if single-longitudinal-mode lasers had infinite mode suppression, laser dynamics would broaden the laser line during modulation. The laser wavelength decreases temporarily when it is turned on and then increases temporarily again when it is turned off, a transient effect called chirp. The duration of these decreases and increases in wavelength is comparable to the relaxation oscillation time of the laser, which is typically 150–250 ps. During this time the wavelength shift can be large enough that dispersion delays or advances the received power in the pulse edges more than a bit interval. This occurs particularly at high bit rates and long distances. When it does occur, the power lost from the bit is not effective in producing a signal at the receiver. The system penalty P_c is directly related to the fraction of power shifted out of the bit, and is given by [44]

$$P_c = 10\log\left[\frac{1}{1 - 4t_c B}\right], \qquad \text{for } Z > t_c/m\delta\lambda \tag{3}$$

where t_c, the duration of the wavelength shifts, is assumed to be half the relaxation oscillation time, and $\delta\lambda$ is the amount of wavelength shift.

To give a feel for the size of these penalties, Table IV lists penalties calculated using (1)–(3) for a bit rate of 300 Mbit/s and a span distance of 125 km. The FP results assume

TABLE IV
ESTIMATED SYSTEM PENALTIES
(300 Mbit/s, 125 km)

Laser	Penalty (dB)		
	Misequalization	Partition Noise	Chirp
FP	0.38	0.09	0.00
C^3 or EC	0.16	0.02	0.41
DFB	0.05	0.00	0.41

a dispersion of 1.0 ps/km·nm corresponding to a difference between laser and zero dispersion wavelengths of 20 nm, a laser rms spectral width of 3 nm, and a k-factor of 0.5. The single-longitudinal-mode laser results assume that the equations for misequalization and mode partition noise are applicable, although these have not yet been verified extensively for such lasers. For C^3 and EC lasers the mode spacings have been taken to be 15 Å with a side-mode suppression of 200, and for DFB lasers 20 Å with side-mode suppression of 1,000. For all three single-longitudinal-mode lasers, the relaxation oscillation time has been taken to be 150 ps and the chromatic dispersion 18 ps/km·nm. Penalties comparable to these have been included in the loss budgets used to determine span distances for the four system alternatives in Table III.

For dispersion-shifted fiber systems, it is necessary to set limits on laser wavelength, spectral width, and fiber chromatic dispersion in order to maintain acceptable penalties. For single-longitudinal-mode laser systems, it is necessary to set limits on mode suppression, and chirp characteristics to maintain acceptable penalties.

COMMENTS

There are two options for future 1.55-µm undersea lightwave systems that result in repeater spacing greater than 100 km—dispersion-shifted fiber systems and single-longitudinal-mode laser systems. In addition, there are many options regarding the best choice of devices to use in these systems. The resulting repeater spacing can range from 96 to 138 km depending on these choices. The business decisions regarding which systems and which devices to develop depend on the complex tradeoff between development costs and installed system costs, and on the manufacturability and early availability of suitable devices for use in reliability qualification programs. While all of the key technologies upon which such decisions must be based have been demonstrated, the required reliability for the various technological options remains to be proven.

ACKNOWLEDGMENT

The author would like to thank N. S. Bergano, R. L. Easton, V. J. Mazurczyk, and A. L. Simons for valuable contributions, including reliability information, receiver calculations, and dispersion penalty calculations.

REFERENCES

[1] M. Nakamura, "Semiconductor injection lasers for long wavelength optical communications," in *Proc. Mtg. Integrated and Guided-Wave Optics*, paper MD1 (Incline Village), Jan 28–30, 1980.

[2] C. D. Anderson, R. F. Gleason, P. T. Hutchison, and P. K. Runge, "An undersea communication system using fiberguide cables," *Proc. IEEE*, vol. 68, p. 1299, 1980.

[3] R. E. Wagner, S. M. Abbott, R. F. Gleason, R. M. Paski, A. G. Richardson, D. G. Ross, and R. D. Tuminaro, "Lightwave undersea cable system," in *Proc. IEEE Int. Conf. Commun.*, paper 7D.6, June 13–17, 1982.

[4] P. K. Runge, "Deep-sea trial of an undersea lightwave system," in *Tech. Dig. OFC '83* (New Orleans, LA), Feb. 28–Mar. 2, 1983, p. 8.

[5] H. J. Schulte, "Transmission tests during the SL lightwave submarine cable system sea trial," in *Proc. SPIE*, vol. 425 (San Diego, CA), Aug. 23–24, 1983.

[6] S. W. Dawson, J. Riera, and E. K. Stafford, "CTNE undersea lightwave inter-island system," this book, ch. 8, p. 109.

[7] P. K. Runge and P. R. Trischitta, "The SL undersea lightwave system," this book, ch. 4, p. 51.

[8]   T. G. Hodgkinson, R. Wyatt, D. W. Smith, D. J. Malyon, and R. A. Harmon, "Studies of 1.5 μm coherent transmission systems operating over installed cable links," in *Conf. Rec. IEEE Global Telecommun. Conf.*, paper 21.3 (San Diego, CA), Nov. 28–Dec. 1, 1983.

[9]   K. Emura, M. Shikada, S. Fujita, I. Mito, H. Honmou, and K. Minemura, "Optical FSK heterodyne-single filter detection experiments using a directly modulated DFB-laser diode," in *Conf. Proc. 10th ECOC* (Stuttgart, Germany), Sept. 3–6, 1984, p. 228.

[10] S. Yoshida, "Review of new materials for infrared fibers," in *Tech. Dig. IOOC '83* (Tokyo, Japan), June 27–30, 1983, p. 98.

[11] P. Besomi, R. B. Wilson, R. L. Brown, N. K. Dutta, P. D. Wright, and R. J. Nelson, "High-temperature operation of 1.55 μm InGaAsP double-channel buried-heterostructure lasers grown by LPE," *Electron Lett.*, vol. 20, p. 417, 1984.

[12] V. J. Mazurczyk, "202 km transmission spans at 1550 nm with multilongitudinal mode lasers," in *Tech. Dig. OFC '84*, Postdeadline paper WJ8 (New Orleans, LA), Jan. 23–25, 1984.

[13] W. T. Tsang, N. A. Olsson, and R. A. Logan, "High-speed direct single-frequency modulation with large tuning rate and frequency excursion in cleaved-coupled-cavity semiconductor lasers," *Appl. Phys. Lett.*, vol. 42, p. 650, 1983.

[14] L. A. Coldren, T. L. Koch, C. A. Burrus, and R. G. Swartz, "Intercavity coupling gap width dependence in coupled-cavity lasers," *Electron Lett.*, vol. 20, p. 350, 1984.

[15] L. A. Coldren, K. J. Ebeling, R. G. Swartz, and C. A. Burrus, "Stabilization and optimum biasing of dynamic-single-mode coupled-cavity lasers," *Appl. Phys. Lett.*, vol. 44, p. 169, 1984.

[16] W. T. Tsang, R. A. Logan, N. A. Olsson, H. Temkin, J. P. van der Ziel, I. P. Kaminow, B. L. Kasper, R. A. Linke, V. J. Mazurczyk, B. I. Miller, and R. E. Wagner, "119-km, 420 Mb/s transmission with a 1.55 μm single-frequency laser," in *Tech. Dig. OFC '83*, postdeadline paper PD9 (New Orleans, LA), Feb. 28–Mar. 2, 1983.

[17] B. L. Kasper, R. A. Linke, J. C. Campbell, A. G. Dentai, R. S. Vodhandel, P. S. Henry, I. P. Kaminow, and J-S. Ko, "A 161.5 km transmission experiment at 420 Mb/s," in *Conf. Proc. 9th ECOC*, postdeadline paper (Geneva, Switzerland), Oct. 23–26, 1983.

[18] K-Y. Liou, C. A. Burrus, R. A. Linke, I. P. Kaminow, S. W. Granlund, C. B. Swan, and P. Besomi, "Single-longitudinal-mode stabilized graded-index-rod external coupled-cavity laser," *Appl. Phys. Lett.*, vol. 45, p. 729, 1984.

[19] M. Kitamura, M. Yamaguchi, S. Murata, I. Mito, and K. Kobayashi, "Low-threshold and high-temperature single-longitudinal-mode operation of 1.55 μm-band DFB-DC-PBH LDs," *Electron Lett.*, vol. 20, p. 595, 1984.

[20] Y. Ichihashi, H. Nagai, T. Miya, and Y. Miyajima, "Transmission experiment over 134 km of single mode fiber at 445.8 Mb/s," in *Proc. IOOC '83*, postdeadline paper 29C5-2 (Tokyo, Japan), June 27–30, 1983.

[21] K. Utaka, S. Akiba, K. Sakai, and Y. Matsushima, "Effect of mirror facets on lasing characteristics of distributed feedback InGaAsP/InP laser diodes at 1.5 μm range," *IEEE J. Quantum Electron.*, vol. QE-20, p. 236, 1984.

[22] S. Kaufman, R. L. Reynolds, and G. C. Loeffler, "Optical switch for the undersea lightwave system," ch. 34, p. 487.

[23] W. C. Young and L. Curtis, "Cascaded multipole switch for single mode and multimode optical fibers," *Electron. Lett.*, vol. 17, p. 571, 1981.

[24] S. Tsutsumi, Y. Ichihashi, M. Sumida, and H. Kano, "LD redundant system using polarization components for a submarine optical transmission system," ch. 35, p.497.

[25] M. A. Saifi, S. J. Jang, L. G. Cohen, and J. Stone, "Triangular-profile single-mode fiber," *Opt. Lett.*, vol. 7, p. 43, 1982.

[26] A. D. Pearson, L. G. Cohen, W. A. Reed, J. T. Krause, E. A. Sigety, F. V. DiMarcello, and A. G. Richardson, "Transmission, splicing, and cabling performance of dispersion-shifted single-mode fiber," in *Tech. Dig. OFC '84* (New Orleans, LA), Jan. 23–25, 1984, p. 56.

[27] R. Csencsits, P. J. Lemaire, W. A. Reed, D. S. Shenk, and K. L. Walker, "Fabrication of low-loss single-mode fibers," in *Tech. Dig. OFC '84* (New Orleans, LA), Jan. 23–25, 1984, p. 54.

[28] E. J. Friebele, K. J. Long, and M. E. Gingerich, "Radiation damage in single-mode optical-fiber waveguides," *Appl. Opt.*, vol. 22, p. 1754, 1983.

[29] E. W. Mies, D. L. Philen, W. D. Reents, Jr., and D. A. Meade, "Hydrogen susceptibility studies pertaining to optical fiber cables," in *Tech. Dig. OFC '84*, postdeadline paper WI3 (New Orleans, LA), Jan. 23–25, 1984.

[30] T. P. Lee, C. A. Burrus, and A. G. Dentai, "InGaAs/InP p-i-n photodiodes for lightwave communications at the 0.95–1.65 μm wavelength," *IEEE J. Quantum Elect.*, vol. QE-17, p. 232, 1981.

[31] M. L. Snodgrass and R. Klinman, "A high reliability high sensitivity lightwave receiver for the SL undersea lightwave system," ch. 33, p.475.

[32] T. Mikawa, S. Kagawa, T. Kaneda, T. Sakurai, H. Ando, and O. Mikami, "A low-noise n + np germanium avalanche photodiode," *IEEE J. Quantum Electron.*, vol. QE-17, p. 210, 1981.

[33] Y. Matsushima, S. Akiba, K. Sakai, Y. Kushiro, Y. Koda, and K. Utaka, "High-speed-response InGaAs/InP heterostructure avalanche photodiode with InGaAsP buffer layers," *Electron. Lett.* vol. 18, p. 945, 1982.

[34] J. C. Campbell, A. G. Dentai, W. S. Holden, and B. L. Kasper, "High performance avalanche photo diode with separate absorption, grading, and multiplication regions," *Electron. Lett.*, vol. 19, p. 818, 1983.

[35] T. Torikai, Y. Sugimoto, K. Taguchi, K. Makita, H. Ishihara, K. Minemura, T. Iwakami, and K. Kobayashi, "Low noise and high speed InP/InGaAsP/InGaAs avalanche photodiodes with planar structure grown by vapor phase epitaxy," in *Conf. Proc. 10th ECOC* (Stuttgart, Germany), Sept. 3–6, 1984, p. 220.

[36] R. F. Leheny, R. E. Nahory, M. A. Pollack, A. A. Ballman, E. D. Beebe, J. C. DeWinter, and R. J. Martin, "Integrated $In_{0.53}Ga_{0.47}As$ p-i-n FET photoreceiver," *Electron. Lett.*, vol. 16, p. 353, 1980.

[37] K. Kasahara, J. Hayashi, K. Makita, K. Taguchi, A. Suzuki, H. Nomura, and S. Matushita, "Monolithically integrated $In_{0.53}Ga_{0.47}As$ PIN/InP MISFET photoreceiver," *Electron. Lett.*, vol. 20, p. 314, 1984.

[38] N. S. Bergano, R. E. Wagner, H-T. Shang, and P. F. Glodis, "150 km, 296 Mb/s dispersion shifted fiber system experiment," in *Tech. Dig. OFC '85* (San Diego, CA), Feb. 11–13, 1985.

[39] V. J. Mazurczyk, N. S. Bergano, R. E. Wagner, K. L. Walker, N. A. Olsson, L. G. Cohen, R. A. Logan, and J. C. Campbell, "420 Mb/s transmission through 203 km using silica-core fiber and a DFB laser," in *Conf. Proc. 10th ECOC*, postdeadline paper PD7 (Stuttgart, Germany), Sept. 3–6, 1984.

[40] R. G. Smith and S. D. Personick, "Receiver design for optical fiber communication systems," in *Semiconductor Devices for Optical Communication*, H. Kressel, Ed. Berlin, Germany: Springer-Verlag, 1980.

[41] V. J. Mazurczyk, "Sensitivity of single mode buried heterostructure lasers to reflected power at 274 Mb/s," *Electron. Lett.*, vol. 17, p. 143, 1981.

[42] L. G. Cohen and S. Lumish, "Effects of water absorption peaks on transmission characteristics of LED-based lightwave systems operating near 1.3 μm wavelength," *IEEE J. Quantum Electron.*, vol. QE-17, p. 1270, 1981.

[43] K. Ogawa, "Analysis of mode partition noise in laser systems," *IEEE J. Quantum Electron*, vol. QE-18, p. 849, 1982.

[44] R. A. Linke, "Transient chirping in single frequency lasers: Lightwave systems consequences," *Electron. Lett.*, vol. 20, p. 472, 1984.

Subject Index

Editors' Biographies

Patrick R. Trischitta (S'77–M'83) was born in Pittston, PA, on May 29, 1958. He received the B.E.E. degree in 1979 and the M.S.E.E. degree in 1980 from the Georgia Institute of Technology, Atlanta, and the Ph.D. degree in electrical engineering from Rutgers University, New Brunswick, NJ, in 1986.

Since 1980, he has been a member of the Technical Staff in the Undersea Systems Laboratory at AT&T Bell Laboratories, Holmdel, NJ, where he has been engaged in the system development and characterization of undersea lightwave communication systems.

Dr. Trischitta is a member of Eta Kappa Nu and is a licensed Professional Engineer. His IEEE Communications Society activities include Communication's Society delegate to the steering committee of the *IEEE/OSA Journal of Lightwave Technology*, past chairman of the New Jersey Coast Communications Society Chapter and past Editor of Fiber Optics of the IEEE TRANSACTIONS ON COMMUNICATIONS.

Peter K. Runge received the M.S. and Ph.D. degrees in electrical engineering from the Technical University of Braunschweig, Germany, in 1963 and 1967, respectively.

He has been with Bell Laboratories since 1967. He has been engaged in research of He–Ne, and organic dye lasers and exploratory development of fiber optic repeaters and single fiber optic connectors. In 1976, he became supervisor, responsible for the development of single fiber optic connectors. He was also engaged in exploratory development of active and passive fiber optic components. In 1979, he led a group responsible for research on undersea lightwave systems. In 1980, he assumed his present position, "Head, Undersea Lightwave Systems Development Department" and has been responsible for the development of the undersea portion of the SL system.

Credits

583